30·00

OXFORD MEDICAL PUBLICATIONS

Advanced Renal Medicine

Advanced Renal Medicine

Edited by

A.E.G. RAINE
Professor of Renal Medicine,
St Bartholomew's Hospital Medical College
London

OXFORD NEW YORK TOKYO
OXFORD UNIVERSITY PRESS
1992

Oxford University Press, Walton Street, Oxford OX2 6DP
Oxford New York Toronto
Delhi Bombay Calcutta Madras Karachi
Kuala Lumpur Singapore Hong Kong Tokyo
Nairobi Dar es Salaam Cape Town
Melbourne Auckland Madrid
and associated companies in
Berlin Ibadan

Oxford is a trade mark of Oxford University Press

Published in the United States
by Oxford University Press Inc., New York

A catalogue record for this book is available from the British Library

Library of Congress Cataloging in Publication Data
(Data available)
ISBN 0 19 262102 5 h/b
0 19 262101 7 p/b

Typeset by Colset Pte Ltd, Singapore
Printed in Great Britain
by Bookcraft (Bath) Ltd,
Midsomer Norton, Avon

Preface

The past decade has seen immense progress in both the scientific basis and clinical practice of renal medicine, ranging from the elucidation of the molecular aetiology of certain primary renal diseases, to the routine use of recombinant human erythropoietin for the treatment of the anaemia of renal failure. Despite such progress, many important questions remain unanswered. Why, for example, should renal dysfunction, once established, progress inexorably? Are recent trends towards high flux, short duration, haemodialysis justifiable by risk–benefit analysis? Is it true that the current provision of renal replacement therapy in the UK is inadequate? These are amongst the wide-ranging issues considered here.

The aim of this book is to review the current science and clinical practice of nephrology, and to highlight the areas in which we still have much to learn. This compilation of chapters has derived from the Advanced Course in Nephrology, held in London under the aegis of the British Postgraduate Medical Federation, University of London, UK. All credit goes to the contributors, not only for stimulating lectures, but for so willingly condensing their thoughts on paper here. The result, I hope, is that there will be something of value both for the renal specialist and the general physician, as well as for those engaged in basic research related to nephrology.

Thanks are due to Wendy Riley of the British Postgraduate Medical Federation, for her invaluable assistance with organization; to Noreen Creedon, for unstinting secretarial help; and to the staff of the Oxford University Press, for constructive advice and encouragement.

London
July 1992

A.E.G.R.

Contents

Part 3 Hypertension

Part 4 Molecular Genetics in Renal Disease

Part 5 Glomerulonephritis

Part 6 Vasculitis

Contributors

D. Adu *Consultant Physician and Nephrologist, Queen Elizabeth Hospital, Birmingham, UK.*

L.R.I. Baker *Consultant Physician and Nephrologist, Department of Nephrology, St Bartholomew's Hospital, London, UK.*

F.W. Ballardie *Consultant Physician and Nephrologist, Manchester Royal Infirmary, Manchester, UK.*

L.A. Brawn *Senior Registrar in Medicine for the Elderly, Leicester General Hospital, Gwendolen Road, Leicester, UK.*

A. Burns *Research Fellow, Renal Unit, Department of Medicine, Royal Postgraduate Medical School, London, UK.*

H.S. Cairns *Senior Registrar, Institute of Urology, St Philip's Hospital, London, UK.*

J.S. Cameron *Professor of Renal Medicine, Guy's Hospital, London, UK.*

V. Cattell *Department of Pathology, St Mary's Hospital Medical School, London, UK.*

W.R. Cattell *Consultant Physician and Nephrologist, St Bartholomew's Hospital, London, UK.*

G.R.D. Catto *Professor, Department of Medicine and Therapeutics, University of Aberdeen, UK.*

J. Cunningham *Consultant Nephrologist, Royal London Hospital, London, UK.*

C.J. Danpure *Head, Biochemical Genetics Research Group, Clinical Research Centre, Harrow, Middlesex, UK.*

M. Davies *Senior Lecturer, Institute of Nephrology, University of Wales, College of Medicine, Cardiff, UK.*

J. Feehally *Consultant Nephrologist, Department of Nephrology, Leicester General Hospital, Leicester, UK.*

J.D. Firth *Clinical Lecturer, Nuffield Department of Clinical Medicine. John Radcliffe Hospital, Oxford, UK.*

G. Gaskin *MRC Training Fellow, Renal Unit, Department of Medicine, Royal Postgraduate Medical School, London, UK.*

D.W.R. Gray *Reader in Transplantation Surgery, Oxford Transplant Centre, The Churchill Hospital, Oxford, UK.*

R.N. Greenwood *Consultant Nephrologist, Lister Hospital, Stevenage, UK.*

B.M. Hendry *Professor of Renal Medicine, King's College School of Medicine and Dentistry, London, UK.*

D.R.W. Jayne *Clinical Research Fellow, Gonville and Caius College, Department of Medicine, University of Cambridge, UK.*

M.C. Jones *Lecturer, Department of Medicine and Therapeutics, University of Aberdeen, UK.*

J.G.G. Ledingham *Professor and May Reader in Medicine, Nuffield Department of Clinical Medicine, John Radcliffe Hospital, Oxford, UK.*

C.M. Lockwood *Senior Lecturer, Department of Medicine, School of Clinical Medicine, University of Cambridge, UK.*

I.C. Macdougall *Senior Registrar, Department of Nephrology, St Bartholomew's Hospital, London, UK.*

G.A. MacGregor *Professor of Cardiovascular Medicine, Blood Pressure Unit, Department of Medicine, St George's Hospital, London, UK.*

M.N. Maisey *Professor of Radiological Sciences, Guy's Hospital, London, UK.*

J. Michael *Consultant Physician, Queen Elizabeth Hospital, Birmingham, UK.*

S.H. Morgan *Senior Registrar in Renal Medicine, St George's Hospital, London, UK.*

G.H. Neild *Professor of Nephrology, Institute of Urology and Nephrology, Middlesex Hospital, London, UK.*

A.O. Ogunlesi *Research Fellow, Molecular Immunology Group, Institute of Molecular Medicine, John Radcliffe Hospital, Oxford, UK.*

V. Parsons *Consultant Physician, King's College Hospital Renal Unit, Dulwich Hospital, London, UK.*

M.B. Pepys *Professor of Immunological Medicine, Immunological Medicine Unit, Department of Medicine, Royal Postgraduate Medical School, London, UK.*

R.E. Phillips *Senior Research Fellow, Molecular Immunology Group, Institute of Molecular Medicine, John Radcliffe Hospital, Oxford, UK.*

L. Poston *Senior Lecturer, Sherrington School of Physiology, St Thomas' Hospital, London, UK.*

C.D. Pusey *Senior Lecturer, Renal Unit, Department of Medicine, Royal Postgraduate Medical School, London, UK.*

A.E.G. Raine *Professor of Renal Medicine, St Bartholomew's Hospital Medical College, London, UK.*

L.E. Ramsay *Professor in Clinical Pharmacology and Therapeutics, Consultant Physician, University Department of Medicine and Pharmacology, Royal Hallamshire Hospital, Sheffield, UK.*

P.J. Ratcliffe *Wellcome Senior Fellow, Institute of Molecular Medicine, John Radcliffe Hospital, Oxford, UK.*

A.J. Rees *Professor of Nephrology, Department of Medicine, Royal Postgraduate Medical School, Hammersmith Hospital, London, UK.*

C.O.S. Savage *MRC Clinical Scientist and Consultant Physician, Vascular Biology Team, Clinical Research Centre, Harrow, Middlesex, UK.*

D.R.J. Singer *BHF Intermediate Research Fellow, Blood Pressure Unit, Department of Medicine, St George's Hospital, London, UK.*

J.H. Turney *Consultant Nephrologist, Leeds General Infirmary, Leeds, UK.*

G.C. Viberti *Professor of Diabetic Medicine, Guy's Hospital, London, UK.*

D.M. Vigushin *Research Fellow, Immunological Medicine Unit, Department of Medicine, Royal Postgraduate Medical School, Hammersmith Hospital, London, UK.*

J.D. Walker *Senior Registrar, Department of Endocrinology, St Bartholomew's Hospital, London, UK.*

P.C. Waller *Senior Medical Adviser, Medicines Control Agency, Nine Elms Lane, London, UK.*

J. Walls *Professor of Nephrology, Leicester General Hospital, Leicester, UK.*

D.C. Wheeler *Senior Registrar, Institute of Nephrology, University of Wales, College of Medicine, Cardiff, UK.*

H.N. Whitfield *Consultant Urologist, Department of Urology, St Bartholomew's Hospital, London, UK.*

R. Wilkinson *Professor, Department of Nephrology, Freeman Hospital, Newcastle-upon-Tyne, UK.*

J.D. Williams *Reader in Nephrology, Institute of Nephrology, University of Wales College of Medicine, Cardiff Royal Infirmary, Cardiff, UK.*

A.J. Wing *Consultant Physician, St Thomas' Hospital, London, UK.*

W.W. Yeo *Lecturer in Clinical Pharmacology and Therapeutics, Department of Medicine and Pharmacology, Royal Hallamshire Hospital, Sheffield, UK.*

Part 1 Physiology and Pathophysiology

1 Regulation of water metabolism, thirst, and plasma sodium concentration

J.G.G. Ledingham

Disorders of water homoeostasis are common in clinical practice. These range from minor disturbances, in which recovery occurs whether or not management is appropriate, to rarer, potentially life-threatening, conditions, such as the neurological syndromes associated with acute hyponatraemia. The factors commonly underlying disturbed water balance include abnormalities of thirst, of fluid intake unrelated to thirst (often iatrogenic), of arginine vasopressin (AVP) metabolism, and of renal function. Of these, thirst is the factor least recognized and least commonly investigated.

Thirst

In experimental animals thirst and water intake may be equated,[1] but in humans, palatability, drinking with food, drinking on social occasions, and the enjoyment of such substances as alcohol, tea, and coffee, are likely to override primary thirst, such that the kidney is more commonly concerned with the excretion of excess free water than, as in animals, with forming a concentrated urine. Satiety and the physiological stimuli which tend to stop drinking in the presence of overhydration are weak and easily overridden.[2] Far stronger are the mechanisms invoking thirst in the presence of an absolute or relative deficit of water.

Osmoregulation

The osmoregulation of thirst is closely analogous to the mechanisms that are more widely recognized to control the secretion of AVP. Osmoreceptors for thirst are exquisitely sensitive to any increase in extracellular fluid osmolality, unaccompanied by similar changes in the cells of the osmoreceptor and thus leading to cellular dehydration. Infusions of sodium chloride or sucrose solutions were shown by Gilman in 1937 to induce drinking in dogs,[3] whilst a similar osmotic load in the form of glucose or urea did not. In humans, infusions of hypertonic mannitol or saline have been shown to increase both thirst and AVP secretion, with a remarkable correlation with plasma osmolality in each case, whereas hypertonic urea and glucose did not give such effects[4].

Osmoreceptors subserving thirst are now known to be separate from those subserving AVP secretion.[5] They are sensitive to changes in osmolality of the blood, rather than of the cerebrospinal fluid. In humans osmoreceptors are situated in the region of the anterior hypothalamus, and in animals in the organum vasculosum of the lamina terminalis and the subfornical organ of the circumventricular region. These structures are perfused by fenestrated capillaries, providing a local break in the blood–brain barrier. In humans, absence of thirst, with preservation of osmoregulation of AVP secretion, has been described in relation to lesions of the anterior hypothalamus that spare the supraoptic nucleus but include the anterior wall of the third ventricle. How osmotic stimuli are translated into thirst and whether there is localization of function in higher centres has yet to be investigated.

Measurement of thirst in humans depends on the use of visual analogue scales, such as those devised by Rolls *et al.*[6] These can be used to assess thirst after water deprivation or infusion of hypertonic saline. (Fig. 1.1). The set point at which plasma osmolality stimulates thirst can be detected by slow intravenous infusion of hypertonic saline.[7] This threshold tends to be higher than that for AVP secretion; it varies between individuals, and within individuals in the presence of large changes in plasma volume.[8]

Volume regulation

As in the case of the control of AVP secretion, hypovolaemia or hypotension may induce thirst, although the mechanism is relatively insensitive, requiring a fall of at least 10 per cent in either blood pressure or plasma volume. Detection of such changes are probably mediated through low pressure volume receptors in capacitance vessels and in the cardiopulmonary vasculature. Persistent large changes in blood volume tend to reset the osmostat downward in the case of volume contraction, and upward in chronic expansion.

Angiotensin II

Infusion of angiotensin, either systemically or directly into the brain, induces drinking in animals,[9] but an essential role for the renin–angiotensin system in the thirst provoked by volume depletion in humans is unproven. Brown *et al.* have described patients with hypertension and renal failure in whom thirst was insatiable and renin activity extraordinarily high, and in whom binephrectomy immediately relieved the thirst.[10] However, intravenous Asp^1-Val^5 angiotensin II amide (Hypertensin, Ciba), given to water-replete men, in doses up to 16 ng/kg/min, had no effect on drinking or thirst in six of ten male subjects and had much less effect than did hypertonic saline in the other four who did experience some thirst. These observations do not exclude a synergistic effect of angiotensin II with volume depletion, or in association with osmotic stimuli, but any such effect of angiotensin II must be small.[7]

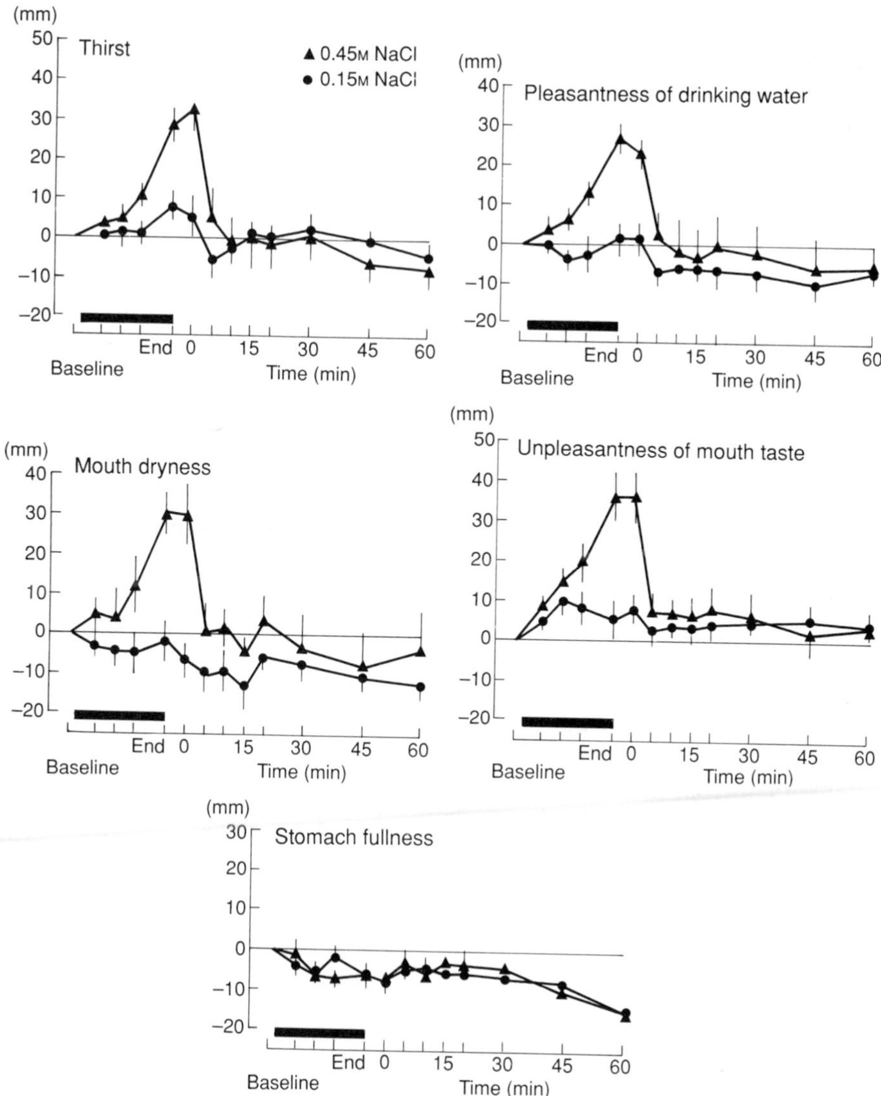

Fig. 1.1 Effect of double-blind infusions of hypertonic saline (0.45 M NaCl) or isotonic saline (0.15 M NaCl) on changes in subjective ratings of thirst from the baseline. Values are means ± SE. Bar = infusions. (From Phillips *et al.*[7], with permission.)

Arginine vasopressin

Osmoregulation

Since the pioneering work of Verney (1947),[11] involving the use of carotid loops on conscious dogs, the prime importance of osmoregulation of AVP secretion has been established. As in the case of thirst, the cells detecting change in hydration in response to osmotic stimuli are situated in the anterior hypothalamus in the organum vasculosum of the lamina terminalis and the subfornical organ. These cells, which are served by many afferent nerve fibres from the brain stem, are also influenced by a number of substances, including opioids, angiotensin II, atrial natriuretic peptide, glucocorticoids, catecholamines, and gamma aminobutyric acid.[12] The osmoreceptor cells integrate their various stimulant or inhibitory influences, and thus regulate the formation of AVP by the supraoptic and paraventricular nuclei, and its secretion by the posterior pituitary gland. Demonstration of the precision of control, the detection of the set point, and the close mathematical relationship between plasma concentrations of AVP and plasma osmolality has been possible since the development of highly sensitive immunoassays, which are capable of detecting the very small amounts of vasopressin that circulate in human plasma.

The set point for AVP secretion lies near 285 mosmol/kg, and above that set point a concentration of the peptide between 0.3 pmol/l and 4 pmol/l spans the range of urinary concentration from isosomotic to maximal concentration with respect to plasma. In contrast to the precise linear relationship between plasma osmolality, thirst, and plasma AVP concentration, demonstrated by infusion of hypertonic saline (Fig. 1.2), restoration of a fluid deficit by drinking leads to a rapid fall in plasma AVP concentration and in perception of thirst, unrelated to the change in plasma osmolality. The mechanism of this phenomenon is not yet known, but it may involve sensory stimuli from the oropharynx, rather than from the stomach.[13]

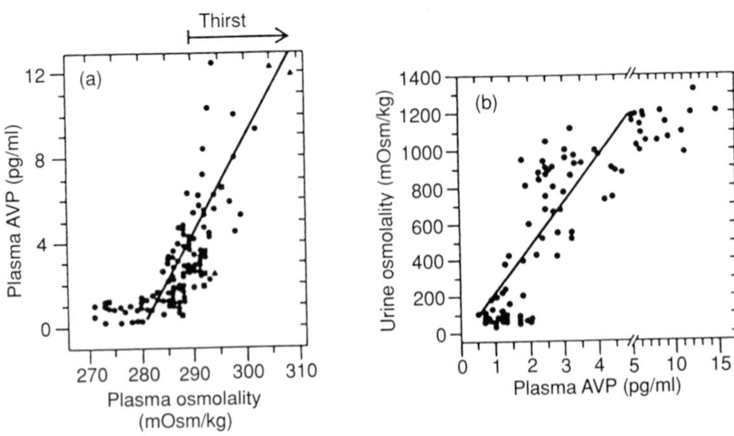

Fig. 1.2 Relationship of AVP to ***a*** plasma osmolality and ***b*** urine osmolality in healthy adults in varying states of hydration. The level of plasma osmolality at which thirst begins is shown in ***a***. (From Robertson *et al.*[8], with permission.)

Volume regulation

As in the case of thirst, hypovolaemia with a fall in plasma volume or arterial pressure of around 10 per cent or more lowers the threshold for osmotic stimulation of AVP, and a fall of volume of 25 per cent or more induces AVP secretion independent of osmolality.

A number of other factors may result in the secretion of AVP, unrelated either to volume or osmotic stimuli, and these include nausea, hypoglycaemia, and probably increased concentrations in the brain or systemic circulation of angiotensin II.

Renal water excretion and conservation

The excretion or conservation of water by the kidney, in relation to plasma osmolality, is best assessed by measurement of free water clearance (C_{H_2O}) in states of overhydration and of tubular reabsorption of solute-free water (T_{CH_2O}) in dehydration. These values are determined by considering the relationships between osmolar clearance (C_{osm}) and urine volume as follows:

$$\mathbf{C}_{osm} = \frac{\text{urine Osm} \times \text{urine ml/min}}{\text{plasma Osm}}$$

$$\mathbf{C}_{H_2O} = \text{urine volume/min} - \mathbf{C}_{osm}$$

$$\mathbf{T}_{CH_2O} = \text{the negative value of } \mathbf{C}_{H_2O}.$$

After maximal water loading in an healthy adult (conventionally 20 ml of water per kg body weight) urine flow rates reach around 20 ml/min and free water clearance 15–18 ml/min. In hydropenia, in contrast, urine flow declines to around 0.5 ml/min, and maximal tubular reabsorption of solute-free water rarely exceeds 1.0–1.5 ml/min. The kidneys in health, therefore, are more effective in protecting against overhydration than dehydration, so that thirst and access to water are of particular importance in states of water deficit.

A number of factors may inhibit the ability of the kidney to excrete or conserve free water. Any cause of a reduction in glomerular filtration rate, whether renal or extrarenal, may have a profound effect. In chronic renal failure the capacity of the kidney to conserve or excrete water normally is compromised at glomerular filtration rates as high as 60 ml/min.[14] The potential importance of the loss of glomerular filtration rate, which occurs with age in this context (Table 1.1), is illustraed in a report by Lindeman *et al.*[15] Comparable attenuation of urinary concentration also occurs with age, Rowe *et al.* reporting minimal urine flow rates of 1 ml/min in the elderly (compared with 0.49 ml/min in the young) and urine osmolalities of 882 mosmol/kg (compared with 1109 mosmol/kg in the young).[16]

Diuretics also drastically reduce the ability of the kidney to excrete or conserve free water. Agents inhibiting Na-K-Cl cotransport within the ascending

Table 1.1 Maximal diluting capacity in men according to age (from Lindeman *et al.*)[15]

Mean age (years)	31	60	84
Mean GFR (ml/min)	149 ±9	92 ± 8	65 ± 4
Maximal C_{H_2O}(ml/min)	16.2 ± 1.4	8.4 ± 1.3	5.9 ± 1
C_{H_2O}/100 ml GFR (ml/min)	10.2	9.1	9.1

limb of the loop of Henle impair both urinary concentration and dilution. Thiazides, which inhibit transport at the cortical diluting site, much reduce maximal free water clearance.

Failure of delivery of sodium to the diluting sites of the loop of Henle and the distal convoluted tubule, as probably occurs in heart failure, hepatic cirrhosis, and nephrotic syndrome, will inhibit the excretion of free water. The same occurs when AVP, or AVP-like substances, are present in plasma, and when secretion of AVP is inappropriate to plasma volume or osmolality (SIADH) or is appropriate to volume but not necessarily to osmolality, as in severe sodium depletion.

States of absolute or relative overhydration

Cardiac failure, hepatic cirrhosis, and nephrotic syndrome

Although these oedematous conditions are primarily states of sodium retention, hyponatraemia may complicate those cases particularly resistant to diuretic therapy. Relative water retention in this situation appears to be due partly to secretion of AVP, inappropriate to osmolality (but perhaps not to 'effective' plasma volume), and partly to the effects of diuretics inhibiting the excretion of free water. Enhanced sodium reabsorption may also occur at proximal sites in the nephron, with lesser distal delivery. Drinking in this situation must be inappropriate both to osmotic and volume status, but it is not clear how often there is inappropriate thirst or inappropriate social drinking. Ramsay *et al.*[17] reported increased drinking by dogs in models of heart failure, and there are also some reports of increased thirst in congestive cardiac failure in humans.[18] On rare occasions hyponatraemia in these conditions may be due to an absolute deficiency of sodium, as well as to relative overhydration. In these circumstances, it may occasionally be justified to contemplate administration of sodium to oedematous patients, but much more commonly water restriction is the proper approach. In most patients, hyponatraemia is primarily an index of the severity of the underlying disorder, and it may need to be managed conservatively. Symptoms from this chronic form of hyponatraemia are very rare.

True sodium depletion

Hyponatraemia may accompany sodium depletion resulting from misuse of diuretics, gastrointestinal or renal sodium loss, or Addison's disease, when relative water retention occurs due to continued or enhanced water intake during impairment of the excretion of free water by the kidney. Secretion of AVP appropriate to volume status, but not to osmotic status enhances water retention. This situation is commonly mistaken for true SIADH, because the need to demonstrate a high rate of renal sodium excretion, one of the original criteria of Bartter and Schwartz for SIADH, is ignored.[19] Restoration of plasma volume and of renal perfusion will allow the increased formation of free water and ultimate resolution of the hyponatraemia.

True syndrome of inappropriate ADH secretion (SIADH)

The multiple causes of this syndrome have been recently reviewed.[20] It is particularly important to demonstrate normal renal and adrenal function before making this diagnosis. In this context, the recent demonstration that adrenal insufficiency secondary to hypopituitarism may result in inappropriate AVP secretion, corrected by glucocorticoid replacement[21] is particularly pertinent.

Emphasis has been placed on abnormalities of AVP secretion, although there must be coincident abnormalities of thirst. Zerbe *et al.* described four separate patterns of abnormal response to infusion of hypertonic saline in this state:[22]

(1) Type A – in which AVP secretion is erratic and unrelated to plasma osmolality;

(2) Type B – when the osmostat is reset at a lower level;

(3) Type C – where there is a constant 'leak' of AVP at low plasma osmolalities, increasing above values of 285 mosmol/kg;

(4) Type D – when the AVP response appears normal and renal water retention occurs despite the absence of measurable AVP in plasma.

Explanations advanced for the paradox of type D include the possibility of upregulation of renal vasopressin receptors, deficiency of local prostaglandins, or perhaps most likely, the presence of a vasopressin-like substance undetected by immunoassay.

In contrast to the volume of work performed on AVP metabolism in SIADH, little has been done to evaluate thirst in this syndrome. It must be abnormal in every case and in some with compulsive polydipsia or an inability to reduce fluid intake appropriately, it may be of prime importance in aetiology and in management.

It is uncommon for patients with mild hyponatraemia (plasma sodium concentration more than 120 mmol/l) to develop symptoms of water intoxication or to need particular treatment for it. In these patients, the important step is to find a cause and to attend to that, rather than to the electrolyte disturbance *per se*. Some patients may complain of lethargy or may become confused, ataxic, tremulous or drowsy. In these patients fluid restriction may be effective and well tolerated; if not, the problem is best managed (pending the availability of specific antagonists) by the use of demethylchlortetracycline (demeclocycline) which, in a dose of 600–1200 mg per day, induces a degree of nephrogenic diabetes insipidus, allowing the excretion of free water despite continued high plasma levels of AVP. The management of more severe hyponatraemia by the use of hypertonic saline infusion is rarely necessary or safe in the chronic syndrome, as discussed below.

Water intoxication or acute hyponatraemia

On rare occasions, hyponatraemia may be sufficiently severe and acute to present a medical emergency and a difficult therapeutic dilemma. Most cases have arisen in the USA in young women given inappropriate amounts of intravenous fluids after elective, and often minor, surgery, in psychotic patients with compulsive polydipsia or, on occasion, in elderly women treated with diuretics, particularly Moduretic.[23]

Postoperative hyponatraemia

Although hyponatraemia is the commonest disorder of water and electrolyte balance in hospital patients, and is particularly common postoperatively, it is rarely severe or symptomatic. However, there have now been 51 reports of symptomatic hyponatraemia after elective surgery in previously healthy young women, all of whom died or were disabled with subsequent neurological problems.[24] In 11 of these these fall in plasma sodium was from 140 mmol/l to 116 mmol/l and death was caused by cerebral oedema, with herniation through the tentorial hiatus, together with pituitary, hypothalamic, and medullary infarction.[25] This resulted in terminal central diabetes mellitus and diabetes insipidus. Initial symptoms included severe headache, nausea, and vomiting. Respiratory arrest occurred within 36 h of recovery from anaesthesia, at which time net fluid intake (mainly 5 per cent dextrose) was calculated to be 6.3 l. In the previous report, the neurological symptoms of water intoxication developed some 48 h postoperatively when the average intake of 5 per cent dextrose had been 8 l.

Clearly, this appalling problem has a major iatrogenic component and it is preventable by awareness of the well-known fluid and electrolyte metabolic response to injury, combined with proper attention to perioperative and postoperative administration of fluid. Nearly 40 years ago, Le Quesne drew attention to the inevitability of water retention, with concentrated urine independent of sodium retention, in the first 24–36 h postoperatively. The recommendation was that in the first 24 h postoperatively not more than 1.0–1.5 l of 5 per cent dextrose should be given, other than to replace unusual losses.[26]

Other causes of acute symptomatic hyponatraemia

Perhaps the most important of these is in psychotic patients with severe compulsive polydipsia. This can be associated with almost any psychiatric condition, ranging from personality disorders to the psychoses, particularly schizophrenia. Reports from the USA have suggested that 7–18 per cent of patients in mental hospitals there drink excessively, with around half of these drinking enough to cause symptoms of water intoxication, sometimes with fatal outcome.[27,28] Patients may drink as much as 30–40 l/d. In these patients there is no need to invoke SIADH or impairment of renal function to explain profound hyponatraemia. In some, however, symptoms arise after much lower intakes, for example 7–10 l/d. There have been reports of associated SIADH, perhaps related to the psychosis itself, perhaps to the psychotropic drugs used to treat it, or to both.

There is some evidence of resetting of the osmostat in occasional patients. Goldman *et al.* have reported reduced free water clearance after water loading in some psychotic subjects, despite normal suppression of plasma AVP and normal renal function and osmolar clearance.[29] This shift in relationship between urinary osmolality and plasma AVP suggests altered receptor or postreceptor sensitivity, but the defect is so far unexplained. The symptoms of severe acute water intoxication are heralded by headache, nausea and vomiting, and, if unchecked, progression to drowsiness, fits, and ultimately respiratory arrest. In the untreated cases reported by Fraser and Arieff,[25] respiratory arrest was followed by some combination of unilateral or bilateral dilatation of the pupils, apnoea, and upgoing plantar responses prior to the development of diabetes insipidus and mellitus with hypernatraemia.

Diuretics and hyponatraemia

Hyponatraemia may be observed quite commonly some 3–14 d after beginning treatment with a thiazide diuretic, for example Moduretic (hydrochlorothiazide 50 mg and amiloride 5 mg), in elderly subjects, particularly in women over the age of 75 years. Although the change is usually mild and asymptomatic this is not always the case, and therefore persistent neurological damage may occur. Since the change in plasma sodium occurs so quickly and without evidence of volume depletion, this is usually a dilutional hyponatraemia. It is likely to be due to a combination of reduced maximal free water excretion by the kidneys and, at least in some cases, secretion of AVP inappropriate to osmolality. Excessive thirst and consequent water intoxication may be an important and hitherto largely unrecognized cause; a study of thirst using visual analogue scales in subjects who have suffered previous episodes of diuretic-associated hyponatraemia needs to be made. In the 11 cases studied by Friedman *et al.*[23], a single dose of Moduretic reduced plasma osmolality by 15 mmol/l with a coincident gain in weight of 0.3 kg.

Management of hyponatraemia

The physician encountering a patient in whom hyponatraemia is associated with neurological symptoms is faced with a dilemma. Untreated acute hyponatraemia may, at its worst, result in cerebral oedema sufficiently severe and acute to cause herniation of the brain through the tentorial hiatus, with consequent infarction, not only of the anterior and posterior pituitary, but also of the medulla and hypothalamus.[25] On the other hand, treatment by infusion of hypertonic saline has been associated with the condition of central pontine myelinolysis, alternatively described as the osmotic demyelination syndrome.[30] This condition, first described in malnourished alcoholic patients, complicates correction of the hyponatraemic state, symptoms developing characteristically 24–48 h later. Its features include quadriparesis, respiratory arrest, pseudobulbar palsy, mutism, and, uncommonly, fits. Post-mortem, areas of demyelination are found not only in the central pons but also, on occasion, in the basal ganglia, internal capsule, lateral geniculate body, or even the cerebral cortex. Magnetic resonance imaging techniques may be used to detect lesions in non-fatal cases, these becoming discernible some seven to ten days after the correction of the electrolyte disturbance.

Evidence suggests that this syndrome is the consequence of treatment of hyponatraemia, rather than of hyponatraemia itself, but how it occurs is uncertain. It is not an hypoxic lesion, nor has every case been preceded by hyponatraemia. Most commonly, however, it does occur after correction of severe hyponatraemia and then only in those cases (both in animal experiments and in clinical medicine) in which correction has taken place more than 24 h after the development of the state. An explanation for the differences in outcome between subjects treated within 24 h or later, lies in the work of Arieff *et al.*, who were the first to describe the mechanisms whereby the brain may adapt to an hypotonic environment. Animal studies have shown that, after initial buffering of the effects of hyponatraemia by an increase of flow of water from the interstitial tissues of the brain to the cerebrospinal fluid, there is a critical change in cerebral intracellular osmolality. A fall in intracellular sodium chloride is measurable within 30 min of the start of the electrolyte disturbance and of potassium within 3 h. In addition to the 10–15 per cent fall in intracellular electrolyte content, there are losses of other osmotically active substances, predominantly amino acids, so that brain water content is ultimately restored to near normal.

These adaptive mechanisms are probably the reason for the greater susceptibility to the demyelinating syndrome of patients in whom treatment is undertaken more than 24 h after the onset of water intoxication and for the relative lack of neurological symptoms associated with chronic (more than 48 h duration) hyponatraemia. In the chronic state there may be no symptoms, but common manifestations include confusion, ataxia, muscle cramps or, in more severe cases, tremor and coma.

In considering how to treat an individual case, two important questions arise. Given that some correction of hyponatraemia must be made in those with an acute symptomatic clinical course, does it matter how fast the disturbance is corrected or to what degree? This remains an area of controversy. There are still those who consider rapid correction hazardous,[31] while others suggest that it is not the initial rate of correction which determines the risk of the demyelinating syndrome; rather that the risk is increased by four factors, as follows:

(1) an increase in plasma sodium concentration to normal or supranormal levels in less than 48 h;

(2) an increase in plasma sodium concentration by more than 25 mmol/l in the first 48 h;

(3) any episode of hypoxia;

(4) an increase in plasma sodium concentration to high levels particularly in patients with hepatic encephalopathy.[32]

There is probably less disagreement than at first appears, since these latter comments were made without distinguishing whether patients were treated for acute (less than 24 h duration) or chronic hyponatraemia. All would surely agree that hyponatraemia, of whatever numerical severity, in the absence of symptoms should not be treated, or should only be treated with water restriction or demethylchlortetracycline. Most would now accept that in acute symptomatic hyponatraemia the risk of demyelination from rapid partial correction is small and that the problem really lies in the management of chronic hyponatraemia accompanied by alarming neurological symptoms, when the chance of causing demyelination by rapid correction is probably high. An approach suggested as prudent by Berl is to follow the old adage that disturbances in milieu interieur developing quickly need to be treated quickly, while those developing slowly should be equally slowly corrected. In acute and life-threatening situations the risk of cerebral oedema proceeding to respiratory arrest must outweigh the degree of risk associated with rapid correction. There are many suggested protocols, though no generally agreed formula has been devised; for instance, 100–300 ml of hypertonic saline, adjusted to increase plasma sodium by some 12 mmol/l at a rate not exceeding 1–2 mmol/l/h. Most would suggest that the plasma concentration of sodium should not be increased above 120 mmol/l in the first 24 h, or that the increase should exceed 10 per cent of the initial value in the first 24 h.

In the chronic hyponatraemia state, when adaptation is fully developed, it is sensible to proceed more slowly, restoring concentrations at a rate of some 0.5–0.75 mmol/l/h over the first 24–48 h.

Hypernatraemic states

Neither neurological nor nephrogenic diabetes insipidus leads to any significant hypernatraemia, provided thirst mechanisms are intact and there is access to fluid, despite urine volumes which may be as great as 20 l/24 h. When hypernatraemia does occur, it is nearly always the consequence of a grossly inadequate intake of water. On rare occasions it may be due to infusion of hypertonic saline solutions or other unintended overdoses of sodium, but in most cases defects of thirst or access to water, accompanied also by defects of AVP secretion, are the consequence of lesions in the region of the hypothalamus. Baylis and Thompson have reviewed the four recognized patterns of disturbance in hypernatraemic patients revealed by infusion of hypertonic saline.[13] These include:

(1) resetting of the osmostat for both thirst and vasopressin secretion, so that the threshold for both is set at a higher plasma osmolality than normal;

(2) a decrease in the degree of thirst and of AVP secretion at high plasma osmolalities but with a normal threshold (decreased slope);

(3) patients with total absence of thirst but with a persistent leak of vasopressin, a particularly serious, if rare condition, which results in neurological disturbances unless water is regularly prescribed or taken according to urinary losses and changes in body weight;

(4) a child in whom persistent hypernatraemia was caused by absence of osmoregulation of thirst, with complete preservation of osmoregulation of vasopressin release.[5]

The management of these rare cases has been well reviewed.[33]

References

1. Fitzsimons, J.T. (1976). The physiological basis of thirst. *Kidney Int.* **10**, 3–11.
2. Rolls, B.J. and Rolls, E.T. (1982). *Thirst.* Cambridge University Press.
3. Gilman, A. (1937). The relation between blood osmotic pressure, fluid distribution and voluntary water intake. *Amer. J. Physiol.* **120**, 323–8.
4. Zeibe, R.L. and Robertson, G.L. (1983). Osmoregulation of thirst and vasopressin secretion in human subjects: effect of various solutes. *Amer. J. Physiol.* **244**, E607–14.
5. Hammond, D.N., Moll, G.W., Robertson, G.L., and Chelmicka-Schorr, E. (1986). Hypodipsic hypernatraemia with normal osmoregulation of vasopressin. *N. Engl. J. Med.* **315**, 433–6.
6. Rolls, B.J., Wood, R.J., Rolls, E.T., Lind, H., Lind, W., and Ledingham, J.G.G. (1980). Thirst following water deprivation in humans. *Amer. J. Physiol.* **239**, R476–82.
7. Phillips, P.A., Rolls, B.J., Ledingham, J.G.G., Forsling, M.L., and Morton, J.J. (1985). Osmotic thirst and vasopressin release in humans: a double-blind crossover study. *Amer. J. Physiol.* **248**, R645–50.
8. Robertson, G.L. (1983). Thirst and vasopressin function in normal and disordered states of water balance. *J. Lab. Clin. Med.* **101**, 351–71.
9. Fitzsimons, J.T. (1979). *The physiology of thirst and sodium appetite.* Cambridge University Press.
10. Brown, J.J., Curtis, J.R., Lever, A.F., Robertson, J.I.S., de Wardener, H.E., and Wing, A.J. (1966). Plasma renin concentration and the control of blood pressure in patients on maintenance haemodialysis. *Nephron*, **6**, 329–49.
11. Verney, E.B. (1947). The antidiuretic hormone and the factors which determine its release. *Proc. Roy. Soc. London*, Series B, **135**, 25–106.
12. Ramsay, D.J. (1989). The importance of thirst in maintenance of fluid balance. *Baillieres Clin. Endocrinol. Metab.* **3**, 371–91.
13. Baylis, P.H. and Thompson, C.J. (1988). Osmoregulation of vasopressin secretion and thirst in health and disease. *Clin. Endocrinol.* **29**, 549–76.
14. Baldwin, D.S., Berman, H.J., Heinemann, H.O., and Smith, H.W. (1955). The

elaboration of osmotically concentrated urine in renal disease. *J. Clin. Invest.* **34**, 800–7.
15. Lindeman, R.D., Lee, D.T., Yiengst, M.J., and Shock, N.W. (1966). Influence of age, renal disease, hypertension, diuretics and calcium on antidiuretic responses to suboptional infusions of vasopressin. *J. Lab. Clin. Med.* **68**, 202–23.
16. Rowe, J.W., Shock, N.W., and de Fronzo, R.A. (1976). The influence of age on the renal response to water deprivation in man. *Nephron*, **17**, 270–78.
17. Ramsay, D.J., Rolls, B.J., and Wood, R.J. (1975). The relationship between elevated water intake and oedema associated with congestive cardiac failure in the dog. *J. Physiol. (London)*, **244**, 303–15.
18. Holmes, J.H. (1960). The thirst mechanism and its relation to edema. In Meyer J.H. & Fuchs M (eds) *Edema: mechanisms and management* (ed. J.H. Meyer and M. Fuchs), pp. 95–102. WB Saunders, Philadelphia.
19. Bartter, F.C., and Schwartz, W.B. (1967). The syndrome of inappropriate secretion of antidiuretic hormone. *Amer. J. Med.* **42**, 790–806.
20. Robertson, G.L. (1989). Syndrome of inappropriate antidiuresis. *New Engl. J. Med.* **321**, 538–9.
21. Oelkers, W. (1989). Hyponatraemia and inappropriate secretion of vasopressin (antidiuretic hormone) in patients with hypopituitarism. *New Engl. J. Med.* **321**, 492–6.
22. Zerbe, R., Stopes, L., and Robertson, G.L. (1980). Vasopressin function in the syndrome of inappropriate antidiuresis. *Ann. Rev. Med.* **31**, 315–27.
23. Friedman, E., Schadel, M. Hackin, H., and Fartel, Z. (1989). Thiazide-induced hyponatraemia. *Ann. Int. Med.* **110**, 24–31.
24. Arieff, A.I. (1986). Hyponatraemia, convulsions, respiratory arrest and permanent brain damage after elective surgery in healthy women. *New Engl. J. Med.* **314**, 1529–35.
25. Fraser, C.L., and Arieff, A.I. (1990). Fatal central diabetes mellitus and insipidus resulting from untreated hyponatraemia: a new syndrome. *Ann. Int. Med.* **112**, 113–19.
26. Le Quesne, L.P. (1957). Fluid balance. In *Surgical practice* (2nd edn) Lloyd-Luke, London.
27. Ferrier, I.N. (1985). Water intoxication in patients with psychotic illness. *Brit. Med. J.* **2**, 1594–5.
28. Berl, T. (1988). Psychosis and water balance. *New. Engl. J. Med.* **318**, 441–2.
29. Goldman, M.B., Luchins, D.L., and Robertson, G.L. Mechanisms of altered water metabolism in psychotic patients with polydipsia and hyponatraemia. *New Engl. J. Med.* **318**, 397–403.
30. Berl, T. (1990). Treating hyponatraemia: damned if we do and damned if we don't. *Kidney Int.* **37**, 1006–18.
31. Sterns, R.H. (1990). The treatment of hyponatraemia: first do no harm. *Amer. J. Med.* **88**, 557–60.
32. Ayns, J.C., Krothapalli, R.K., and Arieff, A.I. (1987). The treatment of symptomatic hyponatraemia and its relation to brain damage. *New Engl. J. Med.* **317**, 1190–95.
33. Robertson, G.L. (1984). Abnormalities of thirst regulation. *Kidney Int.* **25**, 460–69.

2 Membrane transport in uraemia

B.M. Hendry

Certain aspects of the clinical syndrome of uraemia are partially explicable on a cellular, or molecular, basis. For example, the anaemia of chronic renal failure is in some part due to a failure of renal erythropoietin secretion. Nevertheless, most manifestations of uraemia remain unexplained. One possible explanation for certain features of uraemia is that they arise from alterations in cell membrane transport function. This might occur because of the direct action of plasma toxins on cell membranes, or by other mechanisms. Cell function is intimately dependent on cell membrane function and there is evidence that in renal failure cell membranes are abnormal in both composition and properties. This chapter provides an outline of the alterations in membrane transport reported in renal failure, and discusses their possible significance.

The original observations

There have been numerous attempts to study membrane transport in renal failure. This interest arose from the observations made by Welt *et al.* in the 1960s,[1] who demonstrated that erythrocyte Na^+ K^+ ATPase activity was reduced in patients with end stage renal failure. The patients had not received dialysis and many were terminally ill. The observation suggested a mechanism by which renal failure might injure cells throughout the body: if a plasma factor which inhibits the sodium pump builds up in uraemia, then, because of the central and ubiquitous role of the pump in cell metabolism and function, widespread cell damage, and eventually cell death, might result. This led to the concept of a plasma toxin or toxins in uraemia, which acted by inhibition of membrane transport processes, and more particularly by inhibition of the sodium pump.

In parallel with this work on uraemia, there were also reports suggesting a role for a circulating inhibitor of the sodium pump (endogenous digoxin) in volume-expanded states and perhaps in hypertension.[2] It was proposed that a circulating endogenous digoxin might be natriuretic by direct renal actions (reducing sodium chloride reabsorption), and might play a role in the physiological response to volume-overload and in the pathophysiology of essential hypertension. Various mechanisms for the extra-renal biological activity of endogenous digoxin were proposed. One idea was that it might cause a rise in intracellular free calcium due to inhibition of Na^+/Ca^{2+} exchange secondary to increased intracellular sodium. Raised intracellular calcium might in turn be responsible for vasoconstriction and increased systemic peripheral vascular resistance.

These hypotheses concerning the role of circulating inhibitors of membrane transport in uraemia, volume-overload and hypertension stimulated a great deal of research. None of them has been proved entirely correct, and much of the literature remains controversial. This chapter will not consider the topics of hypertension and volume-overload further, but a number of issues in this field have recently been reviewed by Swales.[3] In uraemia, the original deductions made from the results of Welt can now be seen to be oversimplified. The alterations in membrane transport do not appear to be confined to the sodium pump. The clinical importance in uraemia of circulating factors and of the membrane transport changes themselves remain ill-defined.

The range of membrane transport abnormalities

The erythrocyte is the most accessible model cell for the study of membrane transport in human renal failure. Accordingly, the majority of studies concern this cell type. The relevance of the findings made in erythrocyte experiments to membrane function in other tissues remains the subject of debate. In the erythrocyte model a wide variety of systems have been studied; in some cases the results are supported by data from other cells, particularly leucocytes. The range of abnormalities found is illustrated in Table 2.1.

The Na^+ K^+ATPase (sodium pump)

Recent work has confirmed the original observation that membrane Na^+ K^+ATPase activity is reduced in renal failure. There are defects in both

Table 2.1 Membrane transport systems in uraemia. Data from human erythrocyte transport, except where indicated

Systems exhibiting abnormal properties
Na^+ K^+ATPase (activity reduced)
y^+ system amino acid transport (V_{max} increased)
ASC system amino acid transport (V_{max} decreased)
gly system amino acid transport (K_m decreased)
A system amino acid transport (activity reduced) (data from rat skeletal muscle, following acute uraemia)
Choline transport (V_{max} increased)
Systems with normal (or near normal) properties
Na^+/Li^+ countertransport
Na^+/H^+ countertransport (data from human leucocytes)
Nucleoside transport
L system amino acid transport
T system amino acid transport

ATPase enzyme activity and in the ion-pumping process.[4] The pump inhibition is present in erythrocytes, both from patients with chronic renal failure (CRF) who have not undergone dialysis and from patients receiving renal replacement therapy by haemodialysis or by continuous ambulatory peritoneal dialysis (CAPD).[5] The abnormalities appear to reverse completely following successful renal transplantation, probably within 3 weeks of the transplant.[5,6] Whether the pump defect in renal failure is due to a reduced number of pump molecules, or a reduced turnover per pump molecule is not clear, and this may vary between patients who have undergone dialysis and those who have not. Fervenza *et al.*[5] have reported that erythrocyte sodium pump number (measured as specific [^{3}H]-ouabain binding) is reduced by 35 per cent in patients with CRF who have not undergone dialysis, but is normal in patients receiving maintenance haemodialysis three times a week. Conversely, patients on haemodialysis exhibit a 30–40 per cent reduction in turnover per pump, and therefore continue to have a defect in overall pump activity per cell, albeit with a different origin. The abnormalities of sodium pump function in patients on CAPD appear to be quantitatively smaller with 10–20 per cent reductions in activity.

Although there is broad agreement about the presence of a defect in erythrocyte sodium pump activity in uraemia,[4] some authors have reported normal pump function.[7] The origins of these discrepancies are unclear. Differences in cell washing and preparation may play a role as, for example, a plasma toxin or inhibitor could be removed by these procedures. Plasma from patients with CRF is capable of inhibiting both the isolated ATPase enzyme and ion translocation by the sodium pump *in situ*.[5,8] Similarly, erythrocytes from patients on haemodialysis demonstrate improved sodium pump function after two hours of incubation in normal plasma.[5] These results support the idea of an inhibitor of the sodium pump in uraemic plasma, but there are many conflicting reports. The uraemic defect in erythrocyte sodium pump function appears to be representative of changes in a number of tissues and cell types. In humans, leucocyte sodium pump activity is depressed in renal failure. In animal models of uraemia, there is evidence of sodium pump dysfunction in skeletal muscle, neurones and gastrointestinal tract cells.[4]

Amino acid transport

Human erythrocytes possess a range of amino acid transport systems which model the behaviour of amino acid transporters in other tissues. Three of these systems have been demonstrated to be abnormal in renal failure. The y^+ amino acid transporter carries basic amino acids (L-lysine, L-arginine) and exhibits an increased V_{max} in uraemia.[9] The ASC amino acid transporter (preferred and eponymous substrates L-alanine, L-serine and L-cysteine) has a reduced V_{max}, while the gly transport system (specific for glycine and sarcosine) demonstrates an increased affinity for extracellular substrate.[10] Two other amino acid transport systems have been examined in human renal failure.

The L system, transporting L-leucine and other neutral amino acids, and the T system, transporting aromatic amino acids including L-tryptophan, exhibit normal activities.[10]

In experimental uraemia in animals, it is possible to examine amino acid transport in cells other than the erythrocyte. In acute uraemia in rats, a defect in skeletal muscle amino acid uptake by system A develops. System A is an insulin-stimulated transporter of neutral amino acids which is not present in erythrocytes. The effects of chronic uraemia on amino acid transport in animal models have not been extensively studied. It seems probable that extensive abnormalities of the membrane transport of amino acids occur in renal failure, although in humans direct evidence is available only for the erythrocyte.

Choline transport

The membrane transport of choline has recently been shown to be abnormal in erythrocytes from patients with renal failure.[11] Choline transport is vital for the synthesis of phosopholipids and, in neuronal cells, for the synthesis of the neurotransmitter, acetylcholine. The relationship between erythrocyte choline transport and membrane transport in other cell types is not established. Patients with end stage chronic renal failure exhibit a markedly increased V_{max} for erythrocyte choline influx, with values two or three times higher than those in controls. The abnormality is not diminished by maintenance dialysis, either by CAPD or haemodialysis. The V_{max} for erythrocyte choline transport shows an inverse correlation with creatinine clearance, both in patients with chronic renal failure and in renal transplant recipients. These changes in choline transport are very marked, and this is the only example where a membrane transport abnormality can be successfully correlated with the degree of renal dysfunction. Nevertheless the origins of the altered choline transport and the clinical significance of the findings remain unknown.

Nucleoside transport

The human erythrocyte nucleoside transporter has been studied in patients on haemodialysis and appears to behave normally. Both V_{max} and substrate affinity are unaltered in renal failure.

Erythrocyte Na^+/Li^+ countertransport and leucocyte Na^+/H^+ countertransport

The erythrocyte Na^+/Li^+ countertransport system has no obvious physiological function, although it has been likened to the Na^+/H^+ countertransporter present in a number of tissues. In diabetic patients with early nephropathy (microalbuminuria) both the erythrocyte Na^+/Li^+ countertransporter and the leucocyte and fibroblast Na^+/H^+ countertransporters exhibit increased V_{max} values[12,13] (see Chapter 9). The behaviour of these systems in non-diabetics with end stage renal failure appears close to normal. Figueiredo *et al.* have demonstrated that the leucocyte Na^+/H^+

countertransporter exhibits a normal V_{max} in end-stage renal failure in diabetic and non-diabetic patients alike.[14,15] Leucocytes from non-diabetic patients on haemodialysis do show an intracellular acidosis, but the scatter between subjects is large and the origins of the acidosis are not clear. It is of great interest that the raised leucocyte Na^+/H^+ countertransport V_{max} found in diabetics with early nephropathy appears to be lost in end stage diabetic nephropathy. The V_{max} for erythrocyte Na^+/Li^+ countertransport has been reported to be normal or slightly increased in non-diabetic uraemic patients and any abnormality, if present, appears to be small.

Membrane transport *in vivo*

All the findings described above were obtained from cells studied *in vitro*, and alterations in behaviour due to the preparation, washing and handling of the cells are extremely difficult to exclude. The function of the human sodium pump has been studied *in vivo* by Boon *et al.*, using sequential measurements of the disposition of an oral load of rubidium chloride.[16] The translocation of an oral load of rubidium into cells is essentially dependent upon sodium pump activity. This work demonstrated that *in vivo* erythrocyte sodium pump function is clearly inhibited in renal failure and suggested that there were widespread reductions in sodium pump activity in other tissues. The function of other membrane transport systems *in vivo* has proved difficult to measure directly. It has become possible to employ phosphorous nuclear magnetic resonance in humans to obtain evidence about the intracellular pH and indirectly about Na^+/H^+ countertransport activity in skeletal muscle *in vivo*. In this regard Syme *et al.*[17] have employed a combination of ^1H and ^{87}Rb nuclear magnetic resonance to measure the concentration of rubidium in rat skeletal muscle. This method opens the possibility of monitoring sodium pump function in a variety of tissues *in vivo* in humans. These techniques and other similar ones may provide more detailed information about membrane transport *in vivo* in the future.

The origins of altered membrane transport in uraemia

The origins of altered membrane transport in uraemia remains a complicated and controversial area. The bulk of evidence and opinion favours the view that plasma factors or toxins in uraemia are responsible for altering membrane properties. There is a considerable literature supporting the existence of an 'endogenous digoxin' both in uraemic plasma and in plasma obtained in other settings, such as volume expansion or pregnancy.[18] Certainly, there are substances in plasma which cross-react with antibodies directed against digoxin. Similarly, there are factors which can be obtained following separation and extraction from plasma which will inhibit Na^+ K^+ATPase activity *in vitro*. Nevertheless, there remain doubts about whether the substances assayed in plasma by antibody binding as 'endogenous digoxin' are inhibitors of the sodium pump *in vivo*. Numerous substances have been suggested as possible circulating inhibitors of the sodium pump, but in no case is the evidence compelling. Recent attention has focused on the possibility that plasma non-esterified fatty acids and lysophospholipids may be endogenous sodium pump inhibitors[19], and this possibility is currently being explored in detail. In addition, a compound with properties and structure indistinguishable from the cardiac glycoside ouabain has been identified in human plasma (see Chapter 13).

The existence of a wide range of membrane transport changes in uraemia indicates that the alterations in membrane function cannot be explained solely by the presence of a digoxin-like substance. Either there are intrinsic changes in membrane properties due to altered cell development, or extrinsic (circulating) factors are affecting a number of different membrane transport proteins. If the latter explanation is correct, then the hypothesis that these factors are lipids or related substances is attractive, as such molecules will adsorb readily to the cell membrane and are likely to interact with numerous transport proteins following free and rapid diffusion in the plane of the membrane. Ideas as to how these interactions alter the activity of some but not all membrane transport systems remain speculative.

Clinical importance of altered membrane transport

In all tissues, cell function is intimately dependent upon normal membrane transport function, and thus the possible significance of altered membrane transport in generating some of the clinical and pathological features of uraemia is very great. The sodium pump in particular is crucial to numerous metabolic, transport and electrical events vital to organ function. Nevertheless, the clinical consequences of the observed 30–40 per cent inhibition of sodium pump activity in uraemia are hard to predict. There is great inter-individual variability and measured values of sodium pump activity in patients with uraemia overlap with values obtained in normal subjects. There is no demonstrable correlation between sodium pump dysfunction and the clinical or biochemical severity of renal failure. It has been reported that oral supplementation with carnitine reverses the abnormalities of sodium pump function in patients on haemodialysis.[20] No dramatic improvement in uraemic symptoms was reported. It seems likely that some manifestations of uraemia are worsened by sodium pump dysfunction. On physiological grounds it is likely that skeletal muscle weakness and impaired central nervous system function might be particularly exacerbated by this mechanism.

The inter-organ transport of amino acids is altered in renal failure both in the post-prandial and post-absorptive states.[21] For example, hepatic uptake of L-serine after a protein meal is impaired. There are also altered plasma and intracellular concentrations of amino acids in uraemia. Normal amino acid translocation and metabolism clearly require adequate function of membrane amino acid transporters and the altered behaviour of systems y^+, ASC and gly could contribute to the abnormal organ handling of amino acids and to the increased protein catabolism found in uraemia.

Membrane choline transport is clearly vital in certain cell types (such as cholinergic neurones), but the altered erythrocyte membrane handling of choline may not be reproduced in other tissues. Choline transport changes might play a role in the pathogenesis of some of the neurological and muscular manifestations of uraemia, but a great deal of new information is required about neuronal choline transport in renal failure before such a hypothesis can be established.

Conclusions

The abnormalities of membrane transport in uraemia are not confined to the sodium pump but extend at least to amino acid and choline transport. Considerable further work is required to define the nature of the defects in humans in cells other than the erythrocyte. Cell membrane changes may contribute to the pathogenesis of the clinical syndrome of uraemia, but the evidence for this remains indirect.

References

1. Welt, L.G., Sachs, J.R., and McManus, T.J. (1964). An ion transport defect in erythrocytes from uremic patients. *Trans. Assoc. Am. Physicians*, **77**, 169–81.
2. De Wardener, H.E. and MacGregor, G.A. (1980). Dahl's hypothesis that a saluretic substance may be responsible for a sustained rise in arterial pressure: its possible role in essential hypertension. *Kidney Int.* **18**, 1–9.
3. Swales, J.D. (1988). Blood pressure: from cells to populations. *J. Royal Coll. Physicians*, **22**, 11–15.
4. Kaji, D. and Kahn, T. (1987). Na-K pump in chronic renal failure. *Am. J. Physiol.* **252**, F785–93.
5. Fervenza, F.C., Hendry, B.M., and Ellory, J.C. (1989). Effects of dialysis and transplantation on red cell Na pump function in renal failure. *Nephron*, **53**, 121–8.
6. Zannad, F., Royer, R.J., Kessler, M., *et al.* (1982). Cation transport in erythrocytes of patients with chronic renal failure. *Nephron*, **32**, 347–50.
7. Corry, D.B., Tuck, M.L., Brickman, A.S. *et al.* (1986). Sodium transport in red blood cells from dialysed uremic patients. *Kidney Int.* **29**, 1197–202.
8. Cole, C.H., Balfe, J.W., and Welt, L.G. (1968). Induction of a ouabain-sensitive ATPase defect by uremic plasma. *Trans. Assoc. Am. Physicians*, **81**, 213–20.
9. Fervenza, F.C., Harvey, C.M., Hendry, B.M., and Ellory, J.C. (1989). Increased lysine transport capacity in erythrocytes from patients with chronic renal failure. *Clinical Science*, **76**, 419–22.
10. Fervenza, F.C., Meredith, D., Ellory, J.C., and Hendry, B.M. (1990). A study of the membrane transport of aminoacids in erythrocytes from patients on haemodialysis. *Nephrol. Dial. Transplant.* **5**, 594–99.
11. Fervenza, F.C., Merdith, D., Ellory, J.C., and Hendry, B.M. (1991). Abnormal

erythrocyte choline transport in patients with chronic renal failure. *Clinical Science*, **80**, 137–41.

12. Mangili, R., Bending, J.J., Scott, G., Li, L.K., Gupta, A., and Viberti G.-C. (1988). Increased sodium-lithium countertransport activity in red cells of patients with insulin-dependent diabetes and nephropathy. *N. Engl. J. Med.* **318**, 140–5.
13. Ng, J., Simmons, D., Frighi, V., Garrido, M.C., and Bomford, J. (1990). Effect of protein kinase C modulators on the leucocyte Na/H antiport in type I diabetic subjects with albuminuria. *Diabetologia*, **33**, 278–84.
14. Ng, L.L., Figueiredo, C.E.P., Garrido, M.C., Bomford, J., Ellory, J.C. and Hendry, B.M. (1990). Leucocyte intracellular pH and Na/H antiporter activity in patients with chronic renal failure. *Clinical Science*, **78**, (22), 9P.
15. Figueiredo, C.E.P., Ng, L.L., Garrido, M.C., Davies, J.E., Ellory, J.C., and Hendry, B.M. (1990). Leucocyte intracellular pH and Na/H antiporter activity in end-stage diabetic nephropathy. *Nephrol. Dial. Transplant.* **5**, 667.
16. Boon, N.A., Aronson, J.K., Hallis, K.F., White, N.J., Raine, A.E.G., and Grahame-Smith, D.G. (1984). A method for the study of cation transport *in vivo*: effects of digoxin administration and of chronic renal failure on the disposition of an oral load of rubidium chloride. *Clinical Science*, **66**, 569–74.
17. Syme, P.D., Dixon, R.M., Allis, J.L., Aronson, J.K., Grahame-Smith, D.G., and Radda, G.K. (1990). A non-invasive method of measuring concentrations of rubidium in rat skeletal muscle *in vivo* by ^{87}Rb nuclear magnetic resonance spectroscopy: implications for the measurement of cation transport activity *in vivo*. *Clinical Science*, **78**, 303–9.
18. Graves, S.W., and Woodward, G.H. (1987). Endogenous digitalis-like natriuretic factors. *Ann. Rev. Med.* **38**, 433–44.
19. Kelly, R.A., Canessa, M.L., Steinman, T.I. *et al.* (1989). Hemodialysis and red cell cation transport in uremia: role of membrane free fatty acids. *Kidney Int.* **35**, 595–603.
20. Labonia, W.D., Morelli, O.H., Giminez, M.I. *et al.* (1987). Effects of L-carnitine on sodium transport in erythrocytes from dialysed uremic patients. *Kidney Int.* **32**, 754–9.
21. Tizianello, A., Deferrari, G., Garibotto, G. *et al.* (1987). Abnormal amino acid metabolism after amino acid ingestion in chronic renal failure. *Kidney Int.* **32**, (22), S181–5.

3 Atrial natriuretic peptide and the pathophysiology of chronic renal disease

D.R.J. Singer and G.A. MacGregor

The kidney is of fundamental importance in the control of sodium and water balance and of blood pressure. More than 90 years ago, Tigerstedt and Bergman (1898) found that extracts of renal cortex have potent vasoconstrictor effects. This led to the discovery of the renin-angiotensin-aldosterone system, which is now well-established as an anti-natriuretic mechanism involved in the renal homoeostasis of sodium excretion. During the last decade, it has been shown that the heart has an endocrine function in the control of sodium balance, with secretion of the biologically active atrial natriuretic peptide (ANP), alpha-human ANP 99–126, and brain natriuretic peptide (Saito *et al.* 1988). This chapter will review the role of ANP and the renin-angiotensin-aldosterone system in the physiology of the control of sodium balance and in the pathophysiology and treatment of renal disease.

Physiology of the control of sodium excretion

Atrial natriuretic peptide

The 28 amino acid *C*-terminal alpha-human ANP and 98 amino acid *N*-terminal pro-ANP, peptide fragments of ANP pro-hormone, are co-secreted from dense granules in atrial myocytes in the normal heart and are also secreted by ventricular myocytes in cardiac failure. There is now clear evidence in support of a role for ANP in the physiology of the control of sodium balance. Intravenous infusion of ANP in physiological doses in humans causes a natriuresis. Potentiation of endogenous ANP secretion by administration of inhibitors of neutral endopeptidase 24.11 (NEP-I) also results in natriuresis (Jardine *et al.* 1989). ANP levels alter with changes in dietary sodium intake, plasma ANP increasing when dietary sodium is increased from low or normal to high intake, and these increased levels are sustained when the high sodium diet is continued (Fig. 3.1). Furthermore, antagonism of ANP by monoclonal antibodies in animals inhibits the natriuretic response to intravenous saline (see Singer *et al.* 1991).

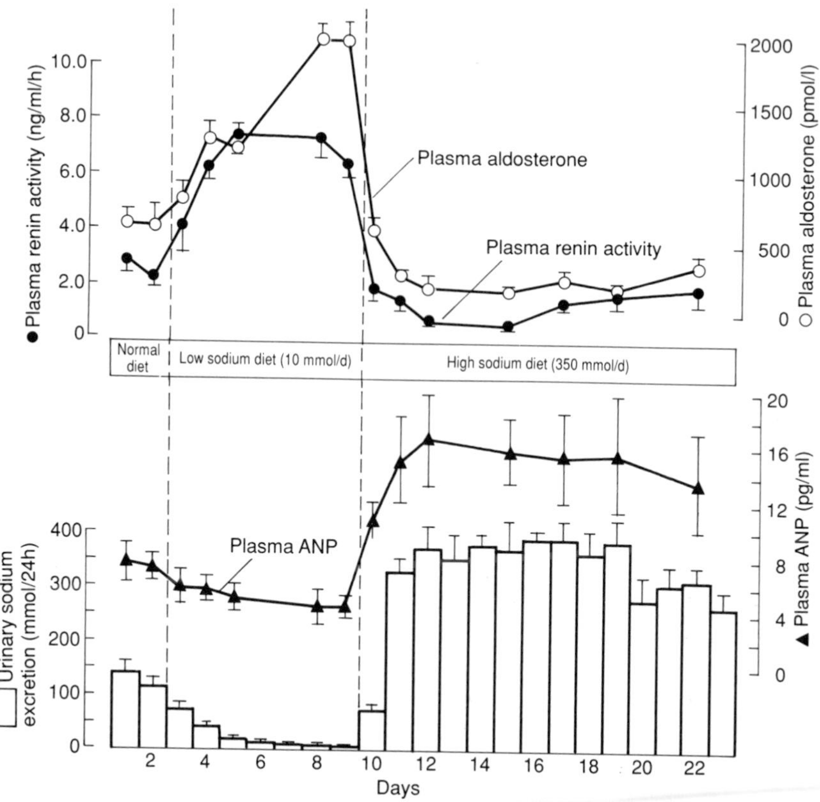

Fig. 3.1 Plasma ANP, renin activity, aldosterone and urinary sodium excretion on normal diet, 7 days of low sodium diet (10 mmol/d) and 14 days of high sodium diet (350 mol/d); normal subjects, n = 8; means ± SEM. (Adapted from Sagnella *et al.* 1987.)

The renin-angiotensin-aldosterone system

Renin is an enzyme secreted by the juxta-glomerular apparatus in the kidney under the control of hormonal and neural factors, as well as in response to changes in the contents of the renal tubular fluid. In addition, local tissue renin systems are widely distributed at extra-renal sites, including vessels, the heart and the central nervous system (Campbell 1987). Renin acts on angiotensinogen to generate angiotensin I, which is cleaved by converting enzyme to the major active peptide of the renin-angiotensin-aldosterone system, the octapeptide angiotensin II (AII). AII is a vasoconstrictor, and also enhances proximal tubular reabsorption of sodium and stimulates adrenal cortical secretion of the

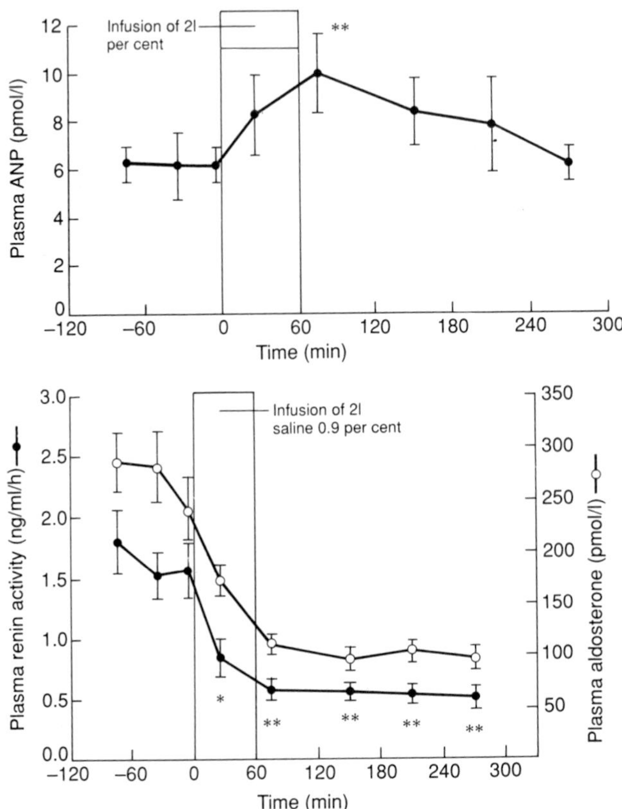

Fig. 3.2 Plasma ANP, renin activity and aldosterone in response to a 2 l infusion of saline 0.9 per cent (308 mol sodium); normal subjects n = 8. *$p < 0.05$, **$p < 0.001$ compared with baseline; means ± SEM. (Adapted from Singer *et al.* 1991.)

mineralocorticoid aldosterone, which acts principally by increasing fractional distal tubular rejection of sodium.

An intravenous sodium load causes rapid, sustained suppression of plasma renin activity and aldosterone (Fig. 3.2; Singer *et al.* 1991). Dietary sodium restriction stimulates, and high sodium intake suppresses the renin-angiotensin-aldosterone system (Brown *et al.* 1963; Sagnella *et al.* 1987; Fig. 3.1), which is sensitive to small changes in dietary sodium intake with a significant decrease in plasma aldosterone within 24 h of a 50 mmol increase in dietary sodium intake (Sagnella *et al.* 1989).

Mechanisms of action of ANP

Renal mechanisms for the natriuresis caused by an increase in plasma ANP include an increase in glomerular filtration rate, tubular effects, and altered tubulo-glomerular balance. The net effect is an increase in both fractional proximal and distal tubular rejection of sodium. These mechanisms are mediated directly by binding to guanylate cyclase-linked high affinity receptors for ANP in the glomerulus, the epithelial cells of the distal nephron, including the distal convoluted tubules and collecting ducts, and in the vasae rectae, with generation of the second messenger cGMP (Raine *et al.* 1989*a*).

ANP also promotes natriuresis by suppressing renin secretion, inhibiting vasopressor effects of angiotensin II, and reducing stimulated secretion of aldosterone by angiotensin II and other secretagogues (Raine *et al.* 1989*a*). In the central nervous system, ANP inhibits the pressor and anti-natriuretic response to angiotensin II (Al-Barazanji and Balment 1990) and ANP has been shown in humans to inhibit angiotensin II-stimulated proximal tubular reabsorption of sodium (McMurray and Struthers 1988).

Aldosterone does not appear to modulate stimulated ANP secretion directly (Singer *et al.* 1991). However, chronic mineralocorticoid supplements result in increased plasma ANP, in association with sodium and water retention (Cappuccio *et al.* 1987).

Factors influencing the response to an increase in plasma ANP

Dietary sodium intake

In the 24 h after treatment with a neutral endopeptidase inhibitor, there was six-fold greater increase in urinary sodium excretion on high compared with low sodium intake in humans (Singer *et al.* 1990). This may in part reflect the greater increase in levels of ANP achieved on a high compared with a low sodium diet with this inhibitor of ANP catabolism. However, the natriuresis in response to ANP infusion (Fig. 3.3) is also much greater on a high, compared with a low, sodium intake (Singer *et al.* 1989).

Activity of the renin-angiotensin-aldosterone system

One mechanism for the blunted response to ANP infusion on a low sodium diet is activation of the renin-angiotensin-aldosterone system. When the renin system is blocked by an angiotensin converting enzyme inhibitor, the response to exogenous ANP in heart failure may be enhanced (Abassi *et al.* 1990). Similarly enhanced effects on natriuresis in heart failure have been reported in response to the increase in plasma ANP when a neutral endopeptidase inhibitor is combined with angiotensin converting enzyme inhibition, compared with neutral endopeptidase inhibitor alone (Margulies *et al.* 1990).

Renal perfusion pressure

The renal response to administered ANP is critically sensitive to renal perfusion pressure, with marked reduction of natriuresis when renal perfusion pressure is reduced (Raine *et al.* 1989*a*).

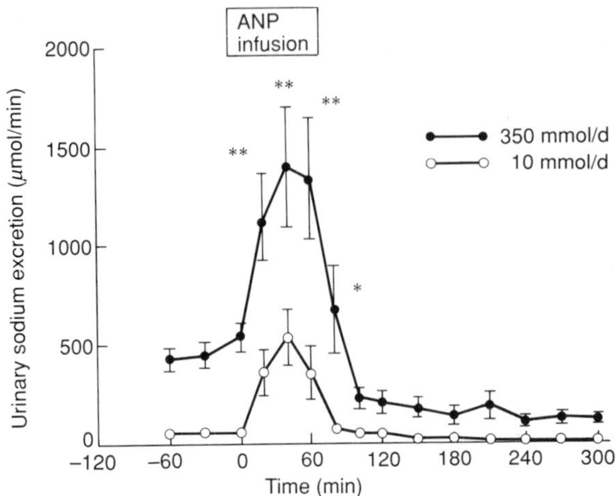

Fig. 3.3 Urinary sodium excretion in response to a 60 min infusion of alpha-human ANP 102–126 (37 pmol/kg/min) on the 5th day of low sodium diet (10 mmol/day), and on the 5th day of high sodium diet (350 mol/d); essential hypertension on no other treatment, n = 8. *$p < 0.05$, **$p < 0.001$ for ANP compared with placebo on high sodium diet; **$p < 0.001$ for ANP compared with placebo on low sodium diet; means ± SEM (Adapted from Singer *et al.* 1989.)

Kinetics of the control of sodium excretion

There are alterations in the relative importance of different mechanisms with time after a sodium load. Plasma ANP increases transiently after a sodium load, however, suppression of the renin system is more sustained (Fig. 3.2; Singer *et al.* 1991). This is consistent with a role for ANP in the early excretion of an intravenous sodium load. Even after plasma ANP has returned to basal values, there may be sustained biological effects mediated in part by second messenger activation including stimulation of guanylate cyclase (Hamet *et al.* 1986) and in part by recycling by cells of unaltered, and thus still biologically active, ANP (Panday and Inagami 1990).

The importance of the renin-angiotensin-aldosterone system in the natriuretic response can be assessed by clamping plasma levels of components of the system. When a continuous low dose infusion of aldosterone was given to healthy subjects to prevent the levels of aldosterone from falling, the natriuresis from 60–240 min after the start of saline infusion was significantly less than when placebo was infused instead of aldosterone (Fig. 3.4). The results of this aldosterone clamp study suggest that inhibition of the renin-angiotensin system and in particular suppression of aldosterone secretion may explain much of the later increase in sodium excretion after saline loading. As actions of ANP include inhibition of the renin system, suppression of this system with saline may be mediated in part by the initial increase in plasma ANP.

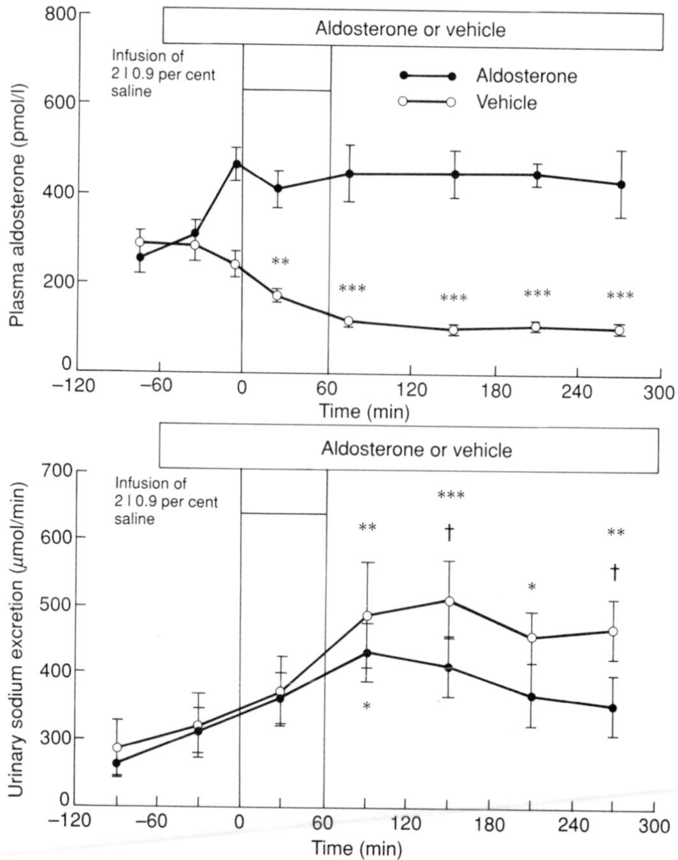

Fig. 3.4 Effects of intravenous aldosterone infusion on the response of urinary sodium excretion to a 2 l infusion of saline 0.9 per cent (308 mol sodium); normal subjects n = 8. ANOVA: treatment effect: F = 5.6, $p < 0.05$; for paired means $^*p < 0.05$, $^{**}p < 0.01$, $^{***}p < 0.001$ compared with baseline; $^{\dagger}p < 0.05$ compared with placebo; means ± SEM. (Adapted from Singer *et al.* 1991.)

ANP and the renin-angiotensin-aldosterone system in renal disease

Plasma ANP is elevated in both acute and chronic renal failure. These raised ANP levels may give insight into the pathophysiology of renal disease, and may be useful both as an index of disease severity and of response to treatment (Table 3.1).

Table 3.1 Plasma ANP levels by methods using an extraction step and glomerular filtration rate (GFR – mean ± SEM or mean (range)) in patients with chronic renal failure not requiring dialysis compared with healthy control subjects. $^*P < 0.01$ compared with controls.

Chronic renal failure				Normal controls				
Number	Age (years)	GFR (ml/min/1.73 m^2)	Plasma ANP (pmol/l)	Number	Age (years)	Plasma creatinine (μmol/l)	Plasma ANP (pmol/l)	Reference
25	48 ± 9	4.4 ± 3.0**	31.4 ± 3.2	40	21–70	normal	*13.6 ± 0.7	Ref. 4 in Woolf *et al.* 1989*a*
13	46 ± 4	39 ± 5	35.0 ± 5.7	18	40 ± 5	normal	*15.0 ± 1.4	Suda *et al.* 1988
36	43 ± 3	21(4–67)	14 ± 2	24	42 ± 3	<120	*6.9 ± 0.8	Woolf *et al.* 1989*a*
15	54 ± 3	22(6–54)†	36.8 ± 3.5	47	42 ± 2	83 ± 2	3.1 ± 0.2	Spencer *et al.* 1989

** Plasma creatinine – mg/dl
† GFR – ml/min

Mechanisms for elevated plasma ANP in renal disease

Reduced catabolism of ANP

Loss or dysfunction of renal tissue results in reduced renal clearance of ANP by renal clearance receptors (Porter *et al.* 1990) and by renal neutral endopeptidase 24.11. Thus fractional renal excretion of ANP is significantly correlated with creatinine clearance (Marumo *et al.* 1988), and in patients with unilateral renal ischaemia, renal clearance of ANP is reduced in the affected kidney (Więçek *et al.* 1990). The increased ratio of *N*- to *C*-terminal ANP from 20-fold in normal subjects to 36-fold in dialysis-independent renal failure (Buckley *et al.* 1989) allows insight into the relative importance of reduced clearance for the increase in plasma ANP in renal disease.

Altered pre-load and cardiac function

A major stimulus for ANP secretion is atrial distension. This increases with atrial stretch because of volume overload in renal disease. In addition, cardiac disease is commoner in renal failure and cardiac failure is associated with increased atrial pressure and stretch.

Possible influence of elevated AII

Volpe *et al.* (1990), found that elevated plasma levels of AII in dogs achieved by low dose AII infusion were associated with increased plasma ANP levels. Animals with the greatest increase in ANP had greatest blunting of AII-induced sodium retention and aldosterone stimulation. A role for AII in direct stimulation of ANP secretion in renal disease in humans remains to be confirmed.

Pathophysiology

Sodium and water overload

Mechanisms by which ANP could be implicated in the sodium and water overload of renal disease include impaired secretion of ANP by the heart, reduced sensitivity of the kidney to ANP because of acquired abnormalities in receptor numbers and function (for example via ANP-guanylate cyclase uncoupling), and indirect mechanisms such as renal ischaemia-induced activation of anti-natriuretic mechanisms including the renin-angiotensin-aldosterone system and renal sympathetic outflow.

Spencer *et al.* (1989) reported raised levels of ANP in patients with dialysis-independent chronic renal failure on their usual diet and similar proportional increase in plasma ANP from a low to a high sodium diet compared with normal subjects (Fig. 3.5). Studies with water immersion-induced increases in central blood volume in acute and chronic renal failure showed no blunting in the expected increase in plasma ANP in humans (Kokot *et al.* 1989). Thus, cardiac ANP secretion does not appear to be abnormal in renal disease, at least in response to these stimuli.

Loss of nephrons results in reduced functional capacity of tubules and altered tubulo-glomerular feedback, as well as impaired responses to hormonal regulators of tubular and glomerular function. Bacay *et al.* (1989) found no differences in glomerular ANP receptors in biopsies from patients with minimal change disease, membranous nephropathy or focal glomerulo-sclerosis compared with normal human glomeruli. Benigni *et al.* (1990) found decreased renal cortical ANP binding in experimental diabetes. Thus in some conditions, abnormal ANP binding may contribute to impaired renal handling of sodium and water in renal disease.

Relative changes in the renin system in response to altered dietary sodium intake are similar to normal, with suppression of plasma renin activity and aldosterone by high, and stimulation by low, sodium diet (Spencer *et al.* 1989; Fig. 3.5).

Hypertension

As in essential hypertension, one correlate of plasma ANP in renal disease is the level of blood pressure (Woolf *et al.* 1989*a*; Tonolo *et al.* 1989). Furthermore, the decrease in ANP levels with dialysis is correlated with the degree of postural hypotension which develops (Raine *et al.* 1989*b*). Thus, ANP may be an index of volume-dependent hypertension and of the adequacy of fluid removal with dialysis. Exogenous ANP has potent blood pressure-lowering effects, when infused in large doses to increase plasma ANP levels to the range seen in patients with severe or dialysis-dependent renal disease (Singer *et al.* 1989). It remains unresolved whether the elevated ANP could be a direct compensatory response to the level of blood pessure, preventing otherwise more severe hypertension.

Proteinuria

High dose ANP infusion (Woolf *et al.* 1989*b*) or bolus ANP injection (Fukui *et al.* 1990) causes a significant increase in albuminuria. Fukui *et al.* (1990) reported an increase of 48 per cent in the $U_{Pr}vU_{Cr}V$ ratio, an index of glomerular protein permeability. Thus, high levels of endogenous ANP could exacerbate albuminuria in renal parenchymal disease, including diabetes mellitus.

Progression of renal disease

High ANP levels, by causing hyperfiltration across the glomerulus, could play a role in the progression and scarring which occurs in many forms of parenchymal renal disease. ANP does not appear to mediate dietary protein-induced hyperfiltration in normal subjects, as assessed from the response to short-term dietary protein loading (Doorenbos *et al.* 1990). In patients with a single kidney, focal sclerosis has been described. However, adaptation over 1–11 years in the remaining kidney after unilateral nephrectomy does not appear to be associated with major changes in plasma ANP, AII, or aldosterone (Sorenson *et al.* 1990). Nonetheless, the administration of ANP to achieve an increase in plasma

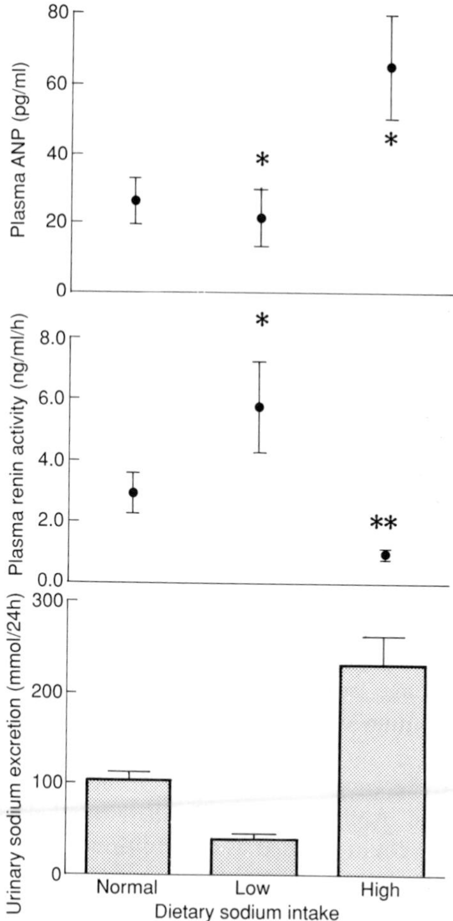

Fig. 3.5 Plasma ANP, renin activity, and urinary sodium excretion on normal diet, 5th day of low sodium diet (10 mmol/d) and 14 days of high sodium diet (350 mol/d); chronic renal failure, n = 9. *$p < 0.05$, **$p < 0.005$ compared with normal diet; means ±SEM. (Adapted from Spencer *et al.* 1989.)

ANP to levels found in chronic renal disease is associated with significant glomerular hyperfiltration (Fukui *et al.* 1990).

ANP as a diagnostic index

Acute renal failure

Plasma ANP (Kanfer *et al.* 1989; Cho *et al.* 1989) and cyclic GMP (Kanfer *et al.* 1989) are elevated during the early phase of acute renal failure, returning to normal with improvement in renal function. Plasma ANP and fractional excretion of sodium were maximal at the onset of the diuretic phase in patients not given diuretics or dialysis (Kanfer *et al.* 1989).

Chronic renal failure

Plasma ANP levels are only weakly related to the severity of renal impairment, in part because of the heterogenous aetiology of renal disease. In particular, in interstitial renal disease there is often a salt-losing state which would be expected to result in relatively lower ANP levels. Other factors which influence plasma ANP need to be considered, including age, exercise, posture, sodium intake, level of blood pressure, and treatment (Woolf *et al.* 1989*a*; Spencer *et al.* 1989).

In the sodium and water-loaded state before haemodialysis, plasma ANP levels are particularly high (Wilkins *et al.* 1986). Volume removal plays an important role in the large fall in plasma ANP with dialysis. Plasma ANP decreases with salt and water removal by ultrafiltration alone (Wilkins *et al.* 1986; Shiota *et al.* 1990) but is not significantly altered by dialysis when salt and water balance is unchanged (Shiota *et al.* 1990). Some authors report a small contribution of loss of ANP across the dialysis membrane (Deray *et al.* 1990). However where there is sustained weight decrease at 24 h post dialysis, ANP levels remain lower compared with pre-dialysis (Buckley *et al.* 1989).

However post-dialysis, plasma ANP is still greatly elevated compared with controls matched for age, sex and blood pressure (Fig. 3.6). Possible reasons include persisting volume overload, cardiac dysfunction and reduced renal ANP clearance.

There are large differences in basal ANP levels between individuals, thus predicting an 'optimal' ANP level is difficult. Nonetheless, within individual subjects plasma ANP could be used as an additional index of the adequacy of volume removal with dialysis.

ANP or potentiation of ANP as treatment

Potential applications of the manipulation of the ANP system in patients with renal failure include treatment of sodium and water overload and hypertension, and use of the ANP system to prevent osmotic damage during angiography, to protect the transplanted kidney from ischaemia and to reduce morbidity in acute or in acute on chronic renal failure.

Exogenous ANP

Studies in models of renal disease In the rat, exogenous ANP blunts renin secretion in response to hypotension (Scheuer *et al.* 1987). Intra-renal artery ANP infusion in the anaesthetized rabbit protects the kidney from the usual reductions in natriuresis and diuresis, glomerular filtration rate, and in renal blood flow with hypoxia (Wiesel *et al.* 1990). Treatment with atrial natriuretic peptides has been shown to improve renal function in experimental acute renal failure. Atriopeptin III improves ischaemic acute renal failure *in vitro* and

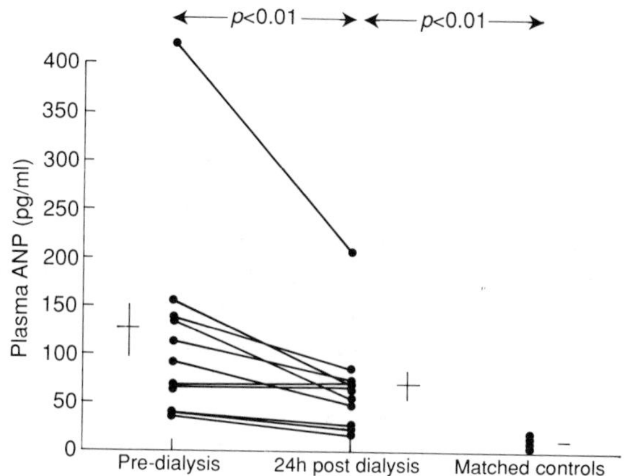

Fig. 3.6 Plasma ANP before and 24 h after haemodialysis in 11 patients with end-stage renal failure, and in 11 healthy controls matched for age, sex, ethnic group and post-dialysis blood pressure. Wilcoxon; means ± SEM.

in vivo in the rat (Nakamoto *et al.* 1987). ANP increases glomerular filtration rate in cis-platinum-induced experimental acute renal failure (Capasso *et al.* 1987). ANP in a rat cyclosporin A model reversed nephrotoxicity, with less reduction in glomerular filtration rate, in urine flow, natriuresis and kaliuresis compared with control animals (Capasso *et al.* 1990). Repeated bolus applications of ANP improved post-ischaemic acute renal failure in dogs (Neumayer *et al.* 1990).

In rat models of experimental nephrosclerosis and anti-glomerular basement membrane disease, natriuresis, diuresis and hypotensive effects following bolus atriopeptin III were similar to the response in normal animals. Responses were similarly exaggerated in normal and chronic renal failure animals on high, compared with normal, sodium intake (Sterzel *et al.* 1987).

Studies in humans Elevation of plasma ANP levels by infusion of low dose ANP in patients with chronic renal failure results in significant natriuresis and diuresis, associated with sustained elevation of plasma cyclic GMP after plasma ANP has returned to basal values (Woolf *et al.* 1989*a*), and, at high dose, reduction in blood pressure (Suda *et al.* 1988).

The effects of ANP and mannitol on renal blood flow and radio-contrast-induced nephropathy were compared in patients with chronic renal failure (Kurnik *et al.* 1990). ANP increased renal blood flow in contrast to mannitol. However, the occurrence of acute on chronic renal impairment with contrast medium was similar with ANP and mannitol. Thus, ANP appears no different from conventional prophylaxis in patients undergoing angiography.

There is no clear evidence of improvement in renal transplant outcome by ANP administration in protocols so far studied (Smits *et al.* 1989).

Neutral endopeptidase inhibitors (NEP-I)

Treatment with exogenous ANP is limited, because the peptide is denatured in the gut and is therefore not adequately bioavailable orally. In normal subjects (Jardine *et al.* 1989) and patients with essential hypertension (O'Connell *et al.* 1989; Singer *et al.* 1990), NEP-I increase plasma ANP 2–3 fold, with a greater absolute increase in plasma ANP levels at high compared with low basal ANP levels (Singer *et al.* 1990). NEP-I also inhibit catabolism of other biologically active peptides, including brain natriuretic peptide (Saito *et al.* 1988), levels of which are in the same range as ANP in chronic renal failure. Therefore NEP-I could be particularly effective in renal failure, in which basal plasma ANP levels become elevated with increasing severity of the renal disease.

Thus, there may be a place for the use of ANP as acute parenteral treatment and for neutral endopeptidase 24.11 inhibitors as chronic oral treatment either alone, or in combination with drugs such as angiotensin converting enzyme inhibitors, which block mechanisms that limit the response to elevation of plasma ANP levels.

Conclusions

There is now clear evidence of a role for the cardiac hormone ANP in the physiology of the control of sodium secretion and in the pathophysiology of chronic renal disease. An understanding of ANP, including the interactions of ANP with the renin-angiotensin-aldosterone system, may be applied to assessment of the severity of renal disease and of the response to treatment. Experimental studies with exogenous ANP suggest potential applications in the treatment of sodium and water overload, and hypertension and possibly also in preserving renal function in patients with renal disease. Further work is required to establish whether the orally active neutral endopeptidase 24.11 inhibitors will be of clinical value in the management of chronic renal disease.

References

Abassi, Z., Haramati, A., Hoffman, A., Burnett, J.C. Jr, and Winaver, J. (1990). Effect of converting-enzyme inhibition on renal response to ANF in rats with experimental heart failure. *American Journal of Physiology*, **259**, R84–9.

Al-Barazanji, K.A., and Balment, R.J. (1990). The renal and vascular effects of central angiotensin 11 and atrial natriuretic factor in the anaesthetized rat. *Journal of Physiology*, **423**, 485–93.

Bacay, A.C., Mantyh, C.R., Cohen, A.H., Mantyh, P.W., and Fine, L.G. (1989) Glomerular atrial natriuretic factor receptors in primary glomerulopathies: studies on human renal biopsies. *American Journal of Kidney Diseases*, **14**, 386–95.

Benigni, A., Perico, N., Dadan, J. *et al.* (1990). Functional implications of decreased renal cortical atrial natriuretic peptide binding in experimental diabetes. *Circulation Research*, **66**, 1453–60.

Brown, J.J., Davies, D.L., Lever, A.F., and Robertson, J.I.S. (1963). Influence of sodium loading and sodium depletion on plasma-renin in man. *Lancet*, **ii**, 278–9.

Buckley, M.G., Sagnella, G.A., Markandu, N.D., Singer, D.R.J., and MacGregor, G.A. (1989). Immunoreactive N-terminal pro-atrial natriuretic peptide in human plasma: plasma levels and comparisons with alpha-human atrial natriuretic peptide in normal subjects, patients with essential hypertension, cardiac transplant and chronic renal failure. *Clinical Science*, **77**, 573–9.

Campbell, D.J. (1987). Circulating and tissue angiotensin systems. *J. Clin. Invest.*, **79**, 1–6.

Capasso, G., Anastasio, P., Giordano, D., Albarano, L., and De Santo, N.G. (1987). Beneficial effects of atrial natriuretic factor on cisplatin-induced acute renal failure in the rat. *American Journal of Nephrology*, **7**, 228–34.

Cappuccio, F.P., Markandu, N.D., Buckley, M.G., Sagnella, G.A., Shore, A.C., and MacGregor, G.A. (1987). Changes in the plasma levels of atrial natriuretic peptides during mineralocorticoid escape in man. *Clin. Sci.* **72**, 531–9.

Cho, K.W., Kim, S.H., Koh, G.Y. *et al.* (1989). Plasma concentration of atrial natriuretic peptide in different phases of Korean hemorrhagic fever. *Nephron*, **51**, 215–9.

Deray, G., Maistre, G., Cacoub, P. *et al.* (1990). Renal and hemodialysis clearances of endogenous natriuretic peptide. A clinical and experimental study. *Nephron*, **54**, 148–53.

Doorenbos, C.J., Iestra, J.A., Papapoulos, S.E., Odink, J., and Van Brummelen, P. (1990). Atrial natriuretic peptide and chronic renal effects of changes in dietary protein and sodium intake in man. *Clinical Science*, **78**, 565–72.

Fukui, K., Hirata, Y., Kimura, K. *et al.* (1990). The effects of atrial natriuretic peptide on urinary protein excretion in patients with renal parenchymal disease and those with diabetic mellitus. *Nippon Jinzo Gakkai Shi.* **32**, 199–205.

Hamet, P., Tremblay, J., Pang, S.C. *et al.* (1986). Cyclic GMP as mediator and biological marker of atrial natriuretic factor. *J Hypertension*, **4**, (2), S49–56.

Jardine, A., Connell, J.M.C., Dilly, S.G. *et al.* (1989). Inhibition of ANP degradation by the atriopeptidase inhibitor UK 69,578 in normal man. *Lancet*, **ii**, 591–3.

Kanfer, A., Dussaule, J.C., Czekalski, S., Rondeau, E., Sraer, J.D., and Ardaillou, R. (1989). Physiological significance of increased levels of endogenous atrial natriuretic factor in human acute renal failure. *Clinical Nephrology*, **32**, 51–6.

Kokot, F., Grzeszczak, W., Zukowska-Szczechowska, E., and Wiecek, A. (1989). Water immersion-induced alterations of plasma atrial natriuretic peptide level and its relationship to the renin-angiotensin-aldosterone system and vasopressin secretion in acute and chronic renal failure. *Clinical Nephrology*, **31**, 247–52.

Kurnik, B.R., Weisberg, L.S., Cuttler, I.M., and Kurnik, P.B. (1990). Effects of atrial natriuretic peptide versus mannitol on renal blood flow during radiocontrast infusion in chronic renal failure. *Journal of Laboratory and Clinical Medicine*, **116**, 27–36.

McMurray, J. and Struthers, A.D. (1988). Effects of angiotensin II and atrial natriuretic peptide alone and in combination on urinary water and electrolyte excretion in man. *Clin. Sci.* **74**, 419–25.

Margulies, K.B., Perrella, M.A., McKinley, L.J., and Burnett, J.C. (1990). Converting enzyme inhibition enhances the renal responses to neutral endopeptidase inhibition (NEP-I) in heart failure. *J. Am. Soc. Nephrol.* **1**, 420.

Marumo, F., Sakamoto, H., Umetani, N., and Okubo, M. (1988). Atrial natriuretic peptide in kidney of renal disease patients and healthy persons. *Endocrinologia Japonica*, **35**, 523–9.

Nakamoto, M., Shapiro, J.I., Shanley, P.F., Chan, L., and Schrier, R.W. (1987). *In vitro* and *in vivo* protective effect of atriopeptin III on ischemic acute renal failure. *J. Clin. Invest.* **80**, 698–705.

Neumayer, H.H., Blossei, N., Seherr-Thohs, U., and Wagner, K. (1990). Amelioration of post-ischaemic acute renal failure in conscious dogs by human atrial natriuretic peptide. *Nephrology, Dialysis, Transplantation*, **5**, 32–8.

O'Connell, J.E., Jardine, A., Davidson, G., Doyle, J., Lever, A.F., and Connell, J.M.C. (1989). UK79,300, an orally active atriopeptidase inhibitor, raises plasma atrial natriuretic peptide and is natriuretic in essential hypertension. *J. Hypertension*, **7**, 923.

Pandey, K.N. and Inagami, T. Receptor-mediated endocytosis and intracellular trafficking of atrial natriuretic factor: evidence for retroendocytosis of intact ligand. Proceedings of Ottowa symposium on atrial natriuretic factor, *Satellite Symposium of the 13th Meeting of the International Hypertension Society*, Canada, June 1990, 17A.

Porter, J.G., Arfsten, A., Fuller, F., Miller, J.A., Gregory, L.C., and Lewicki, J.A. (1990). Isolation and functional expression of the human atrial natriuretic peptide clearance receptor cDNA. *Biochemical and Biophysical Research Communications*, **171**, 796–803.

Raine, A.E.G., Firth, J.G., and Ledingham, J.G.G. (1989*a*). Renal actions of atrial natriuretic factor. *Clin. Sci.* **76**, 1–8.

Raine, A.E., Bock, A., Muller, F.B. *et al.* (1989*b*) Comparative effects of haemodialysis and haemofiltration on plasma atrial natriuretic peptide. *Nephrology, Dialysis, Transplantation*, **4**, 222–7.

Sagnella, G.A., Markandu, N.D., Shore, A.C., and MacGregor, G.A. (1987). Plasma immunoreactive atrial natriuretic peptide and changes in dietary sodium intake in man. *Life Sci.* **40**, 139–43.

Sagnella, G.A., Markandu, N.D., Buckley, M.G., Miller, M.A., Singer, D.R.J., and MacGregor, G.A. (1989). Hormonal responses to gradual changes in dietary sodium intake in man. *Am. J. Physiol.* **256**, R1171–5.

Saito, Y., Nakao, K., Itoh, H. *et al.* (1988). Brain natriuretic peptide is a novel cardiac hormone. *Biochem. Biophys. Res. Commun.* **155**, 740–6.

Shiota, J., Kubota, M., Hamada, C., and Koide, H. (1990). Plasma atrial natriuretic peptide during hemodialysis with or without fluid removal. *Nephron*, **55**, 283–6.

Singer, D.R.J., Markandu, N.D., Sagnella, G.A., and MacGregor, G.A. (1989). Prolonged decrease in blood pressure after atrial natriuretic peptide infusion in essential hypertension: a new anti-pressor mechanism? *Clin. Sci.* **77**, 253–8.

Singer, D.R.J., Markandu, N.D., Buckley, M.G., Miller, M.A., Sagnella, G.A., and MacGregor, G.A. (1991). Dietary sodium and inhibition of neutral endopeptidase 24.11 in essential hypertension. *Hypertension*, **18**, 798–804.

Singer, D.R.J., Shirley, D.G., Markandu, N.D., *et al.* (1991). How important are suppression of aldosterone secretion and stimulation of atrial natriuretic peptide in the natriuretic response to an acute sodium load in man? *Clin. Sci.* **80**, 293–9.

Smits, P., Huysmans, F., Hoitsma, A., Tan, A., and Koene, R. (1989). The effect of alpha-human atrial natriuretic peptide on the incidence of acute renal failure in cadaveric kidney transplantation. *Transplant International*, **2**, 73–7.

Sorensen, S.S., Jespersen, B., and Pedersen, E.B. (1990). Atrial natriuretic factor, angiotensin II, aldosterone, arginine vasopressin and urinary prostaglandin E2 excretion in hyperfiltrating unilaterally nephrectomized humans. *Scandinavian Journal of Clinical and Laboratory Investigation*, **50**, 371–8.

Sosa, R.E., Volpe, M., Marion, D.N. *et al.* (1986). Relationship between renal haemodynamic and natriuretic effects of atrial natriuretic factor. *Am. J. Physiol.* **250**, F520–4.

Spencer, S., Sagnella, G.A., Markandu, N.D. *et al.* (1989). Plasma atrial natriuretic peptide in chronic renal failure and effects of changes in dietary sodium intake. In *Progress in atrial peptide research*, Vol. III (American Society of Hypertension Symposium Series, ed. B.M. Brenner and J.H. Laragh) Raven Press, New York.

Sterzel, R.B., Luft, F.C., Lang, R.E., and Ganten, D. (1987). Effects of atrial natriuretic factor in rats with renal insufficiency. *Journal of Laboratory and Clinical Medicine*, **110**, 63–9.

Suda, S., Weidmann, P., Saxenhofer, H., Cottier, C., Shaw, S.G., and Ferrier, C. (1988). Atrial natriuretic factor in mild to moderate chronic renal failure. *Hypertension*, **11**, 483–90.

Tigerstedt, R. and Bergman, P.G. (1898). Niere und Kreislauf. *Skand. Arch. Physiol.* **8**, 223–71.

Tonolo, G., Soro, A., Scardaccio, V. *et al.* (1989). Correlates of atrial natriuretic factor in chronic renal failure. *Journal of Hypertension*, **7**, S238–9.

Volpe, M., Atlas, S.A., Sosa, R.E. *et al.* (1990). Angiotensin II-induced atrial natriuretic factor release in dogs is not related to hemodynamic responses. *Circulation Research*, **67**, 774–9.

Wieçek, A., Kokot, F., Kuczera, M., Klin, M., and Kiersztejn, M. (1990). Plasma level of atrial natriuretic peptide in renal venous blood—marker of kidney ischemia? *Clinical Nephrology*, **34**, 26–9.

Wilkins, M.R., Wood, J.A., Adu, D., Lote, C.J., Kendall, M.J., and Michael, J. (1986). Change in plasma immunoreactive atrial natriuretic peptide during sequential ultrafiltration and haemodialysis. *Clin. Sci.* **71**, 157–160.

Woolf, A.S., Kingswood, J.C., Mansell, M.A., and Moult, P.J.A. (1989*a*). Plasma levels of atrial natriuretic factor in chronic renal disease. *Eur. J. Int. Med.* **1**, 121–6.

Woolf, A.S., Mansell, M.A., Hoffbrand, B.I., Cohen, S.L., and Moult, P.J. (1989*b*). The effects of low dose intravenous 99–126 atrial natriuretic factor infusion in patients with chronic renal failure. *Postgraduate Medical Journal*, **65**, 362–6.

4 Endothelium-derived relaxing factor and endothelin: their actions and possible roles in renal disease

H.S. Cairns

The endothelium is now known to play a major role in the regulation of both vascular tone and platelet reactivity. The demonstration in the 1970s that the endothelium produces the vasodilator prostaglandin prostacyclin (PGI_2) has led to the discovery of a number of other endothelium-derived vasoactive factors. In this chapter two of these factors will be discussed, the vasodilating endothelium-derived relaxing factor (EDRF) and the vasoconstrictor peptide endothelin. These two factors have potent effects on vascular tone in the kidney and are under active investigation as mediators of the haemodynamic changes that occur in a variety of renal conditions.

Endothelium-derived relaxing factor

General actions and identity

The effect of acetylcholine (ACh) on vascular tone in various *in vitro* models has long been known to be variable, sometimes constricting and sometimes dilating. The reason for this was unclear until Furchgott and Zawadski demonstrated in 1980 that the endothelium, when stimulated by ACh, releases a soluble vasodilator which is not a cyclo-oxygenase product.[1] Furthermore, they showed that an artery with damaged or absent endothelium constricts in response to ACh, explaining the previously noted variable effect of ACh on vascular tone. They called this factor EDRF. Many vasodilators including ACh, bradykinin, substance P and adenosine diphosphate (ADP) have now been shown to act by stimulating the endothelium to produce EDRF.

Studies of EDRF, both in isolated vessels and organs and *in vivo*, have depended on the generation of EDRF from the endothelium of the tissue being studied, using promoters and inhibitors of EDRF synthesis or, *in vitro*, the removal of the endothelium to confirm that the effect is the result of EDRF generation.

EDRF is a vasodilator in all vessel beds tested, has a short half life *in vitro* of approximately 6 seconds and is inactivated by a variety of factors, including superoxide radicals and haemoglobin.[2,3] The short half life of EDRF suggests that it acts primarily in a paracrine fashion, although this is not completely clear.[4] EDRF exerts its effect on vascular smooth muscle by stimulating the enzyme soluble guanylate cyclase which increases intracellular cyclic guanosine monophosphate (cGMP).[5] This contrasts with the vasodilator prostaglandins which activate adenyl cyclase, increasing cyclic adenosine monophosphate (cAMP).

Soluble guanylate cyclase and cGMP production in vascular smooth muscle is also activated by the nitrovasodilators, such as nitroprusside and glyceryl trinitrate. These agents are thought to act by generating nitric oxide (NO) which then stimulates guanylate cyclase.[5] This observation lead Furchgott and Ignarro independently to suggest that EDRF might be NO. Moncada *et al.* demonstrated shortly afterwards that EDRF and exogenous NO are very similar; they have similar half lives *in vitro*, produce similar patterns of vasodilatation and are inactivated by the same agents.[6,7] Furthermore, NO generation can be detected *in vitro* where EDRF is released.[7,8] The nitrogen comes from one of the terminal guanidino-nitrogen atoms of the amino acid L-arginine[8,9] and a novel citrulline-forming enzyme has been identified as catalysing this reaction.[10]

It has been suggested recently that EDRF is not NO but rather an NO-containing compound. The behaviour of EDRF and exogenous NO in non-vascular tissues, such as tracheal smooth muscle, is not identical and now two reports indicate that exogenous NO is less potent than EDRF. NO may be measured either by chemiluminescence or by electron paramagnetic resonance of NO-Hb, and EDRF produces less NO than equally vasodilating doses of exogenous NO.[11,12] Other NO-containing compounds have therefore been considered; S-nitrosothiols such as S-nitroso-L-cysteine are more potent vasodilators than NO and behave in a very similar fashion to EDRF.[12] EDRF may be several, rather than a single, S-nitrosothiol, which could account for the variety of effects of EDRF on non-vascular tissues.

The exact identity of EDRF, therefore, is still unclear although this should be resolved in the near future.

Platelets

EDRF is also a potent inhibitor of platelet aggregation, an effect mediated by stimulation of platelet soluble guanylate cyclase and elevation of cGMP;[13] S-nitrosothiols and NO produce a similar effect.[14] This effect was first shown *in vitro* using both intact vessels[15] and cultured endothelial cells.[14] More recently *ex vivo* inhibition of platelet aggregation has been demonstrated following the *in vivo* stimulation of EDRF release.[16] It is unclear whether the EDRF-platelet effect is predominantly local, at the endothelial cell luminal surface, or whether EDRF affects the systemic aggregability of platelets. Pharmacological doses of endothelium-dependent vasodilators administered *in vivo* do decrease *ex vivo* platelet aggregation[16] although EDRF inhibitors do not increase *ex vivo* aggregability.[17] This suggests that normally EDRF does not significantly alter systemic platelet aggregability. The effect of EDRF on platelet–endothelial interaction and its disturbance in atherosclerotic plaques may be important in the pathogenesis of coronary artery disease.

The kidney

The early studies of EDRF utilized large vessels or cultured endothelial cells, both of which are suitable for superfusion cascade experiments. More recent studies have indicated the importance of EDRF in the control of vascular tone in the microcirculation and particularly the kidney. EDRF reduces renal vascular resistance both *in vitro*[18,19] and *in vivo*.[4] The effect on the glomerular filtration rate (GFR) is still unclear, although endothelium-dependent vasodilators increase the GFR within the rabbit kidney *in vivo*; in the same model vasodilators acting directly on vascular smooth muscle reduce the glomerular filtration rate.[20] This suggests that endothelium-dependent vasodilators, and therefore EDRF, have a heterogeneous effect on the renal microvasculature, preferentially dilating the afferent glomerular arteriole.

A number of recent studies have addressed a possible pathogenic role for EDRF in acute renal failure, cyclosporin nephrotoxicity and essential hypertension.

Acute renal failure and cyclosporin nephrotoxicity are both characterized, at least in part, by an increase in renal vascular resistance and a fall in glomerular filtration rate. Studies in animal models of ischaemic acute renal failure demonstrate a reduction of EDRF-dependent vasodilatation in the affected kidney whereas the effect of endothelium-independent dilators, which act directly on vascular smooth muscle, is not affected.[21,22] This suggests that acute renal failure damages the endothelium and its ability to produce EDRF and that this contributes to the fall in glomerular filtration rate which occurs.

Cyclosporin nephrotoxicity, on the other hand, appears to reduce to a similar extent the effect of EDRF-dependent and EDRF-independent vasodilation,[18,23] indicating that the vascular resistance changes are unlikely to be a direct consequence of reduced EDRF generation.

Two studies on the forearm vasculature of patients with essential hypertension have shown that EDRF-dependent vasodilatation is reduced, whereas endothelium-independent dilatation is not affected.[24,25] Although the importance of increased peripheral vascular resistance in the pathogenesis of essential hypertension is unclear, reduced EDRF generation may have an important role.

These attempts to find a pathogenic role for EDRF do not indicate whether any impairment of EDRF vasodilatation is a primary event or is merely secondary to other processes. Furthermore, if the endothelium and its production of EDRF and other vasodilators is damaged, this may not be readily amenable to treatment with exogenous agents. The effect of EDRF in the microcirculation may depend upon its local generation in particular parts of the microcirculation, and the administration of exogenous agents which mimic the effect of EDRF on vascular smooth muscle may not have the same effect.

Endothelin

Identity

The identification of EDRF led to an expansion of interest in the effect of the endothelium on vascular tone. There followed several demonstrations that, in addition to its release of vasodilators, such as EDRF and prostacyclin, the endothelium produces a vasoconstrictive factor which is not a cyclo-oxygenase product. Yanagisawa *et al.* then published a report detailing the sequence, origin and actions of an endothelium-derived vasoconstrictor which they called endothelin.[26]

Endothelin is a 21 amino acid peptide which is cleaved from a 203 residue precursor, preproendothelin. A 39 residue intermediate, called big endothelin, is generated from preproendothelin by the action of dibasic pair-specific endopeptidases. Endothelin itself is then produced by an unusual proteolytic cleavage involving a specific endopeptidase; the identity and localization of this putative endothelin converting enzyme is still unclear.[26,27] Endothelin has 4 cysteine residues with 2 disulphide bonds, giving it a structure previously unknown in mammalian physiology, although similar to a group of cardiac toxins of non-mammalian origin.

It is now apparent that human and other mammalian cells contain gene sequences for three distinct endothelin peptides. These have been termed endothelin 1, 2 and 3 (ET-1, ET-2, and ET-3).[28] ET-1 is the peptide described by Yanagisawa, which is produced by endothelial cells;[26,29] it has also been detected in other sites, including cerebrospinal fluid[30], where its function is unknown. ET-2 differs from ET-1 by two residues and has similar vascular smooth muscle effects. Despite this similarity ET-2 is apparently not produced by endothelial cells.[28] ET-3 differs from ET-1 by 6 amino acids and is a much less potent vasoconstrictor.[28] It is probably produced mainly in the brain[30] although detectable levels are present in plasma; its actions are as yet unknown.

Actions

The effect of endothelin on vascular smooth muscle was initially thought to result from the stimulation of dihydropyridine-sensitive Ca^{2+} channels.[26] It is now clear that, although endothelin-dependent vasoconstriction does depend on an increase in intracellular Ca^{2+}, this is due primarily to a release of Ca^{2+} from intracellular stores.[31] This effect is phosphoinositol-dependent,[32] and there are high affinity receptors for endothelin (apparent K_D 2–4 × 10^{-10}M) on vascular smooth muscle which mediate the effect.[33,34] There is a secondary influx of Ca^{2+} via both non-specific[35] and dihydropyridine-sensitive channels.[33,36] Endothelin receptors are also present in extravascular sites in the renal interstitium, intestine, brain and lung,[37,38] although the significance of these is unknown.

The *in vivo* stimulus for the release of endothelin is unclear, although hypoxia, other vasoconstrictors, thrombin, and changes in transmural pressure increase *in vitro* endothelium-dependent vasoconstriction[39,40] and endothelin release as detected by radioimmunoassay.[29,41]

Studies of the effect of endothelin, in contrast to those of EDRF activity, have used commercially-available exogenous ET-1. As with all studies of this type, the effects of exogenous ET-1 may not accurately reflect those of the endogenous peptide. When ET-1 is administered *in vitro* or *in vivo* it has a very prolonged effect even though the plasma half life is short[42]; presumably this reflects binding to receptors which are widely distributed. ET-1 can be detected in low concentrations in the plasma of normal individuals and its plasma concentration increases in a number of conditions where there may be endothelial damage; for example acute myocardial infarction and endotoxaemia.[43,44] Nevertheless the increase observed is less than that produced by a vasoactive dose of exogenous ET-1. These facts suggest that endogenous ET-1 may produce its effect mainly at a local level in different microvascular beds.

The kidney

ET-1 is a very potent vasoconstrictor, being approximately 100 times more potent in molar terms than angiotensin II. ET-1 in animal experiments produces a potent vasoconstriction in the kidney both *in vitro*[45,46] and *in vivo*.[47,48] This increase in renal vascular resistance is associated with substantial falls in the glomerular filtration rate and the fractional excretion of sodium. Micropuncture studies in rats have shown that ET-1 reduces intraglomerular pressure by constricting both afferent and efferent glomerular arterioles,[47] although the major effect may be on the afferent arteriole.[49] The effect on the filtration coefficient K_f is as yet uncertain, although ET-1 contracts mesangial cells in culture.[47] In various models the vasoconstrictive effect of ET-1 is partially offset by the intrarenal generation of vasodilators such as PGI_2 and EDRF.[42,46] As with EDRF, a possible pathogenic role for ET-1 has been investigated in acute renal failure and cyclosporin nephrotoxicity.

In a rat model of ischaemic acute renal failure the selective infusion, into part of an affected kidney, of a blocking antibody to ET-1 resulted in a reduction in vascular resistance and an increase in single nephron glomerular filtration rate in that part of the kidney.[50] The single nephron glomerular filtration rate in other parts of the kidney was not affected and infusion of the antibody into a normal kidney did not change the baseline parameters. In another rat model of acute renal failure ET-1 concentrations in both plasma and renal tissue rose significantly and prolonged infusion of anti-ET-1 during the induction of acute renal failure reduced the elevation in plasma urea.[51] Thus ET-1 appears to be at least partially responsible for the increased renal vascular resistance and fall in glomerular filtration rate of the early phase of acute renal failure.

A similar study in a rat model of acute cyclosporin nephrotoxicity has shown that anti-ET-1 infusion reduces renal vascular resistance and increases glomerular filtration rate.[52] Furthermore, plasma ET-1 concentrations increase significantly following the induction of acute cyclosporin nephrotoxicity. Therefore, a causal role for ET-1 in acute cyclosporin nephrotoxicity appears likely although other vasoactive mediators may also play a role.

Summary

EDRF and endothelin are potent vasoactive factors with major effects on renal vascular resistance and renal function. A number of preliminary reports indicate that they may be involved in the development of acute renal failure, cyclosporin nephrotoxicity and essential hypertension. Further knowledge about EDRF, endothelin and other endothelium-derived factors may be useful in the prevention or treatment of these and other conditions.

References

1. Furchgott, R.F., and Zawadski, J.V. (1980). The obligatory role of endothelial cells in the relaxation of arterial smooth muscle by acetylcholine. *Nature*, **288**, 373–6.
2. Griffith, T.M., Edwards, D.H., Lewis, M.J., Newby, A.C., and Henderson, A.H. (1984). The nature of endothelium-derived vascular relaxant factor. *Nature*, **308**, 645–7.
3. Gryglewski, R.J., Moncada, S., and Palmer, R.M.J. (1986). Bioassay of prostacyclin and endothelium-derived relaxing factor (EDRF) from porcine aortic endothelial cells. *Br. J. Pharmac.* **87**, 685–94.
4. Kon, V., Harris, R.C., and Ichikawa, I. (1990). A regulatory role for large vessels in organ circulation: endothelial cells of the main renal artery modulate intrarenal haemodynamics in the rat. *J. Clin. Invest.* **85**, 1728–33.
5. Ignarro, L.J., Harbison, R.G., Wood, K.S., and Kadowitz, P.J. (1986). Activation of purified soluble guanylate cyclase by endothelium-derived relaxing factor from intrapulmonary artery and vein: stimulation by acetylcholine, bradykinin and arachidonic acid. *J. Pharmac. Exp. Ther.* **237**, 893–900.
6. Palmer, R.M.J., Ferrige, A.G., and Moncada, S. (1987). Nitric oxide release accounts for the biological activity of endothelium-derived relaxing factor. *Nature*, **327**, 524–6.
7. Ignarro, L.J., Buga, G.M., Wood, K.S., Byrns, R.E., and Chaudhuri, G. (1987). Endothelium-derived relaxing factor produced and released from artery and vein is nitric oxide. *Proc. Natl. Acad. Sci. USA*, **84**, 9265–9.
8. Palmer, R.M.J., Ashton, D.S., and Moncada, S. (1988). Vascular endothelial cells synthesize nitric oxide from L-arginine. *Nature*, **333**, 664–6.
9. Schmidt, H.H.H.W., Klein, M.M., Niroomand, F., and Bohme, E. (1988). Is arginine a physiological precursor of endothelium-derived nitric oxide. *Eur. J. Pharmacol.* **148**, 293–5.
10. Palmer, R.M.J., and Moncada, S. (1989). A novel citrulline-forming enzyme implicated in the formation of nitric oxide by vascular endothelial cells. *Biochem. Biophys. Res. Comm.* **158**, 348–52.
11. Rubanyi, G., Johns, A., Harrison, D.G., and Wilcox, D. (1989). Evidence that EDRF may be identical with an S-nitrosothiol and not with free nitric oxide (abstract). *Circulation*, **80**, II281.
12. Myers P.R., Minor, R.L.Jr, Guerra, R.Jr, Bates, J.N., and Harrison, D.G. (1990).

Vasorelaxant properties of the endothelium-derived relaxing factor more closely resemble S-nitrosocysteine than nitric oxide. *Nature*, **345**, 161–3.

13. Hawkins, D.J., Meyrick, B.O., and Murray, J.J. (1988). Activation of guanylate cyclase and inhibition of platelet aggregation by endothelium-derived relaxing factor released from cultured cells. *Biochim. Biophys. Acta*, **969**, 289–96.
14. Radomski, M.W., Palmer, R.M.J., and Moncada, S. (1987). The anti-aggregating properties of vascular endothelium: interactions between prostacyclin and nitric oxide. *Br. J. Pharmac.* **92**, 639–46.
15. Furlong, B., Henderson, A.H., Lewis, M.J., and Smith, J.A. (1987). Endothelium-derived relaxing factor inhibits in-vitro platelet aggregation. *Br. J. Pharmac.* **90**, 687–92.
16. Hogan, J.C., Lewis, M.J., and Henderson, A.H. (1988). In vivo EDRF activity influences platelet function. *Br. J. Pharmacol.* **94**, 1020–22.
17. Rees, D.D., Palmer, R.M.J., and Moncada, S. (1989). Role of endothelium-derived nitric oxide in the regulation of blood pressure. *Proc. Natl. Acad. Sci. USA*, **86**, 3375–8.
18. Cairns, H.S., Fairbanks, L.D., Westwick, J., and Neild, G.H. (1989). Cyclosporin therapy in vivo attenuates the response to vasodilators in the isolated perfused rabbit kidney. *Br. J. Pharmacol.* **98**, 463–8.
19. Bhardwaj, R. and Moore, P.K. (1988). Endothelium-derived relaxing factor and the effects of acetylcholine and histamine on resistance blood vessels. *Br. J. Pharmacol.* **95**, 835–43.
20. Cairns, H.S., Rogerson, M.E., Westwick, J., and Neild, G.H. (1990). Possible regional heterogeneity of endothelium-dependent vasodilatation in the rabbit kidney. *J. Physiol.* (In press).
21. Conger, J.D., Robinette, J.B., and Schrier, R.W. (1988). Smooth muscle calcium and endothelium-derived relaxing factor in the abnormal vascular responses of acute renal failure. *J. Clin. Invest.* **82**, 532–7.
22. Lieberthal, W., Wolf, E.F., Rennke, H.G., Valeri, C.R., and Levinsky, N.G. (1990). Renal ischaemia and reperfusion impair endothelium-dependent vascular relaxation. *Am. J. Physiol.* **256**, F894–900.
23. Gerkens, J.F. (1989). Cyclosporine treatment of normal rats produces a rise in blood pressure and decreased renal vascular responses to nerve stimulation, vasoconstrictors and endothelium-dependent dilators. *J. Pharmacol. Exp. Ther.* **250**, 1105–12.
24. Linder, L., Kiowski, W., Buhler, F.R., and Luscher, T.F. (1990). Indirect evidence for release of endothelium-derived relaxing factor in human forearm in vivo: blunted response in essential hypertension. *Circulation*, **81**, 1762–7.
25. Panza, J.A., Quyyumi, A.A., Brush, J.E.Jr, and Epstein, S.E. (1990). Abnormal endothelium-dependent vascular relaxation in patients with essential hypertension. *N. Engl. J. Med.* **323**, 22–7.
26. Yanagisawa, M., Kurihara, H., Kimura, S. *et al.* (1988). A novel potent vasoconstrictor peptide produced by vascular endothelial cells. *Nature*, **332**, 411–5.
27. Emori, T., Hirata, Y., Ohta, K., Shichiri, M., Shimokado, K., and Marumo, F. (1989). Concomitant secretion of big endothelin and its C-terminal fragment from human and bovine endothelial cells. *Biochem. Biophys. Res. Comm.* **162**, 217–23.
28. Inoue, A., Yanagisawa, M., Kimura, S. *et al.* (1989). The human endothelin family: three structurally and pharmacologically distinct isopeptides predicted by three separate genes. *Proc. Natl. Acad. Sci. USA*, **86**, 2863–7.
29. Emori, T., Hirata, Y., Ohta, K., Schichiri, M., and Marumo, F. (1989). Secretory

mechanisms of immunoreactive endothelin in cultured bovine endothelial cells. *Biochem. Biophys. Res. Comm.* **160**, 93–100.

30. Shinmi, O., Kimura, S., Yoshizawa, T. *et al.* (1989). Presence of endothelin-1 in porcine spinal cord: isolation and sequence determination. *Biochem. Biophys. Res. Comm.* **162**, 340–46.
31. Miasiro, N., Yamamoto, H., Kanaide, H., and Nakamura, M. (1988). Does endothelin mobilize calcium from intracellular store sites in rat aortic vascular smooth muscle cells in primary culture?. *Biochem. Biophys. Res. Comm.* **156**, 312–7.
32. Kasuga, Y., Takuwa, Y., Yanagisawa, M., Kimura, S., Goto, K., and Masaki, T. (1989). Endothelin-1 induces vasoconstriction through two functionally distinct pathways in porcine coronary artery: contribution of phosphoinositide turnover. *Biochem. Biophys. Res. Comm.* **161**, 1049–55.
33. Hirata, Y., Yoshimi, H., Takata, S. *et al.* (1988). Cellular mechanism of action by a novel vasoconstrictor endothelin in cultured rat vascular smooth muscle cells. *Biochem. Biophys. Res. Comm.* **154**, 868–75.
34. Hirata, Y., Yoshimi, H., Takaichi, S., Yanagisawa, M., and Masaki, T. (1989). Binding and receptor down-regulation of a novel vasoconstrictor endothelin in cultured rat smooth muscle cells. *FEBS Letters*, **239**, 13–17.
35. Van Renterghem, C., Vigne, P., Barhanin, J., Schmid-Alliana, A., Frelin, C., and Lazdunski, M. (1988). Molecular mechanism of action of the vasoconstrictor peptide endothelin. *Biochem. Biophys. Res. Comm.* **157**, 977–85.
36. Goto, K., Kasuya, Y., Matsuki, N. *et al.* (1989). Endothelin activates the dihydropyridine-sensitive, voltage-dependent Ca^{2+} channel in vascular smooth muscle. *Proc. Natl. Acad. Sci. USA*, **86**, 3915–8.
37. Koseki, C., Imai, M., Hirata, Y., Yanagisawa, M., and Masaki, T. (1989). Autoradiographic distribution in rat tissues of binding sites for endothelin: a neuropeptide?. *Am. J. Physiol.* **256**, R858–66.
38. Koseki, C., Imai, M., Hirata, Y., Yanagisawa, M., and Masaki, T. (1989). Binding studies for ET-1 in rat tissues: an autoradiographic study. *J. Cardiovasc. Pharmacol.* **13**(5), S153–4.
39. Hickey, K.A., Rubanyi, G., Paul, R.J., and Highsmith, R.F. (1985). Characterization of a coronary vasoconstrictor produced by cultured endothelial cells. *Am. J. Physiol.* **248**, C550–56.
40. Harder, D.R. (1987). Pressure-induced myogenic activation of cat cerebral arteries is dependent on intact endothelium. *Circ. Res.* **60**, 102–7.
41. Schini, V.B., Hendrickson, H., Heublein, D.M., Burnett, J.C.Jr, and Vanhoutte, P.M. (1989). Thrombin enhances the release of endothelin from cultured porcine aortic endothelial cells. *Eur. J. Pharmacol.* **165**, 333–4.
42. De Nucci, G., Thomas, R., D'Orleans-Juste, P., *et al.* (1988). Pressor effects of circulating endothelin are limited by its removal in the pulmonary circulation and by release of prostacyclin and endothelium-derived relaxing factor. *Proc. Natl. Acad. Sci. USA*, **85**, 9797–800.
43. Salminen, K., Tikkanen, I., Saijonmaa, O., Nieminen, M., Fyhrquist, F., and Frick, M.H. (1989). Modulation of coronary tone in acute myocardial infarction by endothelin. *Lancet*, **2**, 747.
44. Sugiura, M., Inagami, T., and Kon, V. (1989). Endotoxin stimulates endothelin-release in vivo and in vitro as determined by radioimmunoassay. *Biochem. Biophys. Res. Comm.* **161**, 1220–27.
45. Firth, J.D., Ratcliffe, P.J., Raine, A.E.G., and Ledingham, J.G.G. (1988). Endothelin: an important factor in acute renal failure? *Lancet*, **2**, 1179–82.

46. Cairns, H.S., Rogerson, M.E., Fairbanks, L.D., Neild, G.H., and Westwick, J. (1989). Endothelin induces an increase in renal vascular resistance and a fall in glomerular filtration rate in the rabbit isolated perfused kidney. *Br. J. Pharmacol.* **98**, 155–60.
47. Badr, K.F., Murray, J.L., Breyer, M.D., Takahashi, K., Inagami, T., and Harris, R.C. (1989). Mesangial cells, glomerular and renal vascular responses to endothelin in the rat kidney. *J. Clin. Invest.* **83**, 336–42.
48. Miller, W.L., Redfield, M.M., and Burnett, J.C.Jr. (1989). Integrated cardiac, renal, and endocrine actions of endothelin. *J. Clin. Invest.* **83**, 317–20.
49. Loutzenhiser, R., Epstein, M., Hayashi, K., and Horton, C. (1990). Direct visualisation of effects of endothelin on the renal microvasculature. *Am. J. Physiol.* **258**, F61–8.
50. Kon, V., Toshioka, T., Fogo, A., and Ichikawa, I. (1989). Glomerular actions of endothelin in vivo. *J. Clin. Invest.* **83**, 1762–6.
51. Shibouta, Y., Suzuki, N., Shino, A. *et al.* (1990). Pathophysiological role of endothelin in acute renal failure. *Life Sci.* **46**, 1611–18.
52. Kon, V., Sugiura, M., Inagami, T., Harvie, B.R., Ichikawa, I., and Hoover, R.L. (1990). Role of endothelin in cyclosporine-induced glomerular dysfunction. *Kidney Int.* **37**, 1487–91.

5 Acidosis in uraemia; metabolic consequences

J. Walls

The kidney plays a vital role in the maintenance of the acid–base balance, both by its ability to excrete an acid load and by its preservation or excretion of buffer substances, such as bicarbonate or phosphate. In normal human subjects approximately 60 milliequivalents of hydrogen are excreted per day, two thirds in the form of ammonia and one third as titratable acids. The ammonia is derived predominantly from the amino acids glutamine and asparginine, as demonstrated by the elegant extraction experiments of Pitts many years ago.[1] Titratable acids are derived predominantly from phosphate excretion, the control of which is influenced by parathyroid hormone. During metabolic acidosis there is increased renal extraction and catabolism of glutamine,[2] which is deaminated and deamidated to give two ammonium ions. The continued extraction and metabolism of glutamine is due to increased levels of mitochondrial glutaminase and cytoplasmic phosphoenolpyruvate carboxykinase, occurring exclusively in the proximal tubule. The source of the increased glutamine required for this process is derived from increased amino acid release from muscle[3] and increased hepatic synthesis.[4]

Development of uraemic acidosis

In renal failure there is a decrease in hydrogen ion excretion with the development of a metabolic acidosis. Hydrogen ion excretion decreases to around 20 meq/24 h. The decrease in ammonia excretion is greater than the decrease in titratable acid and has been demonstrated in experimental animals to show a linear decline with inulin clearance.[5] In addition there is a decrease in titratable acid excretion, although of a lower magnitude than that seen for ammonia. This decrease is presumably a result of the increased parathyroid hormone levels necessary for the maintenance of phosphate excretion. A further mechanism which may play a role in uraemic acidosis is a urinary bicarbonate leak due to a decrease in proximal tubular bicarbonate reabsorption. It has been demonstrated that maximal tubular reabsorption of bicarbonate decreases as parathyroid hormone levels increase, a defect which can be reversed by suppression of parathyroid hormone secretion or by parathyroidectomy.[6] However, experimental data have recently shown that acute infusions of parathyroid hormone increase net acid excretion by increasing both ammonia and titratable acid excretion.[7] Therefore, the role of hyperparathyroidism in uraemic metabolic acidosis remains to be fully determined.

Uraemic acidosis, when severe, typically causes hyperventilation, and it is also well known to increase serum potassium levels, presumably as a consequence of intracellular buffering of hydrogen ions in exchange for potassium. The administration of intravenous bicarbonate is long-established as an emergency treatment regimen for hyperkalaemia. There is also a shift in the oxygen dissociation curve to the right with acidosis, and it may be possible that acidosis plays a role in insulin resistance which is seen in uraemia, similar to that demonstrated in diabetic keto-acidosis. The exact mechanism for this abnormality has yet to be determined. In addition, it is now clear that acidosis exerts major effects on bone and protein metabolism.

Effect of metabolic acidosis on bone metabolism

Renal tubular acidosis in the absence of renal impairment is a well recognized, although uncommon, cause of short stature in children and of osteomalacia in adults. In 1978 McSherry and Morris clearly demonstrated a 2–3 fold increase in growth velocity by correcting the systemic acidosis,[8] with some children moving from below the 5th to above the 35th percentile on growth charts. More recently, the natural history of five such children has been published in detail, confirming this beneficial effect.[9] In adults it is possible to treat the osteomalacia, with a resolution of bone pain and a return to normal of the elevated alkaline phosphatase levels, with oral bicarbonate supplementation alone, i.e. without the use of vitamin D or vitamin D analogues.

There was considerable interest in the role of acidosis in renal osteodystrophy in the 1960s before the details of vitamin D metabolism and the mechanism of parathyroid hormone secretion had been established. Using calcium balance studies, it was demonstrated that acidosis increased urinary and faecal calcium losses and that bicarbonate therapy reduced these losses and returned the patients to normal calcium balance. However, it was not possible to induce net calcium retention.[10] The mechanism of these changes in urinary calcium excretion has not been fully elucidated and may not relate solely to systemic pH.[11] Metabolic acidosis has also been demonstrated to suppress the activity of 1-hydroxylase activity in normal rats, but to have no effect on 24-hydroxylase activity.[12] In thyro-parathyroidectomized non-uraemic rats, metabolic acidosis produced an increase in bone resorption and a decrease in bone formation, although there was no conclusive evidence of osteomalacia at the end of the study.[13]

Recently, Barsotti *et al.*[14] demonstrated a significant fall in serum phosphate when bicarbonate was infused into patients undergoing haemofiltration. The same group also demonstrated that in non-dialysed uraemic patients the administration of an oral alkali increased plasma pH and reduced serum phosphate with no change in phosphate excretion. The mechanism of these changes is as yet undetermined.

Effect of metabolic acidosis on protein metabolism

More than 60 years ago it was suggested that 'alkali therapy' in chronic nephritis was beneficial in lowering blood urea levels, although the mechanism responsible was uncertain.[15] Some 40 years later it was reported that correcting metabolic acidosis in children with renal tubular acidosis stimulates growth, both with increased muscle mass and height.[16] Only in the past five years, however, have there been significant observations to explain such findings. May *et al.*[17] demonstrated that metabolic acidosis, induced by feeding non-uraemic rats diets containing ammonium chloride or hydrochloric acid, stimulated muscle proteolysis, thereby impairing nitrogen retention and stunting growth. These findings have recently been confirmed, and the time course of this effect has been described using the urinary excretion of 3-methyl histidine as a marker of skeletal muscle metabolism.[18]

Following the observations of Lyon *et al.*[15] others have shown that correcting metabolic acidosis in uraemic patients lowers blood urea levels. Initially this was attributed to an improvement in renal function induced by volume expansion brought about by sodium loading consequent on sodium bicarbonate administration. However, Papadoyannakis *et al.*[19] concluded from a short term study that the lowering of blood urea was a result of altered muscle metabolism. These clinical observations were extended by Jenkins *et al.*[20] in a longer term study, which also demonstrated significant decreases in urinary urea excretion and decreased urea production. A decrease in muscle protein degradation was found in rats with chronic renal failure and metabolic acidosis when the acidosis was corrected by sodium bicarbonate, in keeping with the clinical observations.[21]

The necessity to correct uraemic metabolic acidosis has recently been emphasized in a study of uraemic patients with metabolic acidosis who were subjected to protein restriction. When the dietary protein intake was decreased from 80 g/d to 40 g/d (0.6 g/kg body weight/d) there was an expected decrease in blood urea and urinary urea and total nitrogen excretion. However, there was a significant increase in urinary 3-methyl histidine excretion, indicating an increase in muscle protein degradation. When the blood pH was increased to 7.35 by the administration of sodium bicarbonate (1 meq/kg/d) there was a further decrease in blood urea and urinary urea and nitrogen excretion. On this regime the excretion of 3-methyl histidine decreased to below the initial values, indicating a decrease in muscle protein degradation. No changes in renal function, blood pressure or body weight occurred, confirming that the mechanism could be ascribed to an improvement in metabolic parameters rather than changes in excretory function.[22]

The proposed mechanism for these observations is that during metabolic acidosis the liver has an increased need for nitrogen substrates to form more glutamine, which is required for increased ammoniagenesis in the kidney. These nitrogen substrates come from skeletal muscle, the major protein source within the body and, during metabolic acidosis, there is increased amino acid release from skeletal muscle. Correction of metabolic acidosis decreases the need for increased ammonia excretion, and hence reduces the rate of muscle protein degradation.

The above studies, both experimental and clinical, indicate the necessity of correcting metabolic acidosis in uraemic patients, especially if they are to be subjected to protein restriction diets.

Treatment

Previously there has been little attention to the treatment of metabolic acidosis in uraemia unless patients appeared to be symptomatic, by which time the serum bicarbonate level had usually fallen below 15 mmol/l. It is now proposed that positive attempts should be made to keep the serum bicarbonate above 24 mmol/l. This can be achieved with all sodium bicarbonate therapy in increasing doses from 600 mg bd upwards. There is obvious concern with regard to the increased sodium load and careful assessment of fluid status and blood pressure is required. However, it has been demonstrated that patients with chronic renal failure can handle the sodium loading from sodium bicarbonate administration better than a comparable sodium load derived from sodium chloride, with no increase in blood pressure or body weight.[23]

In recent years there has been increasing interest in the use of calcium carbonate as an oral phosphate binder to avoid aluminium ingestion and intoxication. The added benefit of calcium carbonate administration is an increase in serum bicarbonate levels, as demonstrated by Williams *et al.*[24] The major complication of calcium carbonate therapy is hypercalcaemia and this may limit its use as the sole agent for the correction of metabolic acidosis.

References

1. Pitts, R.F. (1988). Renal regulation of acid–base balance. In *Physiology of the kidney and body fluids.* (2nd edn). Year Book Medical Publishers, Chicago.
2. Brosnan, J.T., Vinay, P., Gougoux, A., and Halperin. (1988). Renal ammonium production and its implication for acid–base balance. In *pH homeostasis: mechanisms and control.* (ed. D Haussinger), pp. 281–304. Academic Press, New York.
3. Schrock, H., Chu, C.J., and Goldstein, L. (1980). Glutamine release from midlimb and uptake by the kidney in the acutely acidotic rat. *Biochem. J.* **188**, 557–60.
4. Melbourne, T.C., Phromphetcharat, V., Givens, G., and Joshi, S. (1986). Regulation of interorganal glutamine flow in metabolic acidosis. *Am. J. Physiol.* **250**, E457–63.
5. Walls, J., Lubowitz, H., and Bricker, N.S. (1972). Ammonia excretion in experimental renal disease. *J. Clin. Invest.* **51**, (6), 100a.
6. Muldowney, F.P., Donohoe, J.F., Carroll, D.V., Powell, D., and Freaney, R., (1972). Parathyroid acidosis in uraemia. *Quart. J. Med., New Series XLI*, **163**, 321–42.
7. Bichara, M., Mercier, O., Borensztein, P., and Paillard, M. (1990). Acute meta-

bolic acidosis enhances circulating parathyroid hormone, which contributes to the renal response against acidosis in the rat. *J. Clin. Invest.* **86**, 430–43.
8. McSherry, E. and Morris, R.C. (1978). Attainment and maintenance of normal stature with alkaline therapy in infants and children with classic renal tubular acidosis. *J. Clin. Invest.* **61**, 509–27.
9. Rodriguez-Soriano, J., Vallo, A., Castillo, G., and Oliveros, R. (1982). Natural history of primary distal renal tubular acidosis treated since infancy. *J. Paed.* **101**, 669–76.
10. Litzow, J.R., Lemann, J., and Lennon, E.J. (1967). The effect of treatment of acidosis on calcium balance in patients with chronic azotaemic renal failure. *J. Clin. Invest.*, **46**, 280–86.
11. Lau, K., Nickols, R., and Tannen, R.L. (1987). Renal excretion of divalent ions in response to chronic acidosis: Evidence that systemic pH is not the controlling variable. *J. Lab. Clin. Med.* **109**, 27–33.
12. Reddy, G.S., Jones, G., Koon, S.W., and Fraser, D. (1982). Inhibition of 25 Hydroxyvitamin D_3-1-hydroxylase by chronic metabolic acidosis. *Am. J. Physiol.* **243**, (*Endocr. Metab.*), E265–71.
13. Kraut, J.A., Mishler, D.R., Singer, F.R., and Goodman, W.G. (1986). The effects of metabolic acidosis on bone formation and bone reabsorption in the rat. *Kidney Int.* **30**, 694–701.
14. Barsotti, G., Lazzeri, M., Cristofano, C., Cerri, M., Lupetti, S., and Giovanetti, S. (1968). The role of metabolic acidosis in causing uraemic hyperphosphataemia. *Mineral Electrolyte Metab.* **12**, 103.
15. Lyon, P.M., Dunlop, D.M., and Stewart, C.P. (1931). The alkaline treatment of chronic nephritis. *Lancet,* **ii**, 1009–13.
16. Nash, M.A., Tourada, A.D., Griefer, I., Spitzer, A., and Edelmann, C.M. (1972). Renal tubular acidosis in infants and children. *J. Paed.* **80**, 738–48.
17. May, R.C., Kelly, R.A., and Mitch, W.E. (1986). Metabolic acidosis stimulates protein degradation in rat muscle by a glucocorticoid dependent mechanism. *J. Clin. Invest.* **77**, 614–21.
18. Williams, B., Layward, E., and Walls, J. (1990). Skeletal muscle degradation and nitrogen wasting in chronic metabolic acidosis. *Clin. Sci.* (In press).
19. Papadoyannakis, N.J., Stefanidis, C.J., and McGeown, M.G. (1984). The effect of correction of metabolic acidosis on nitrogen and potassium balance in patients with chronic renal failure. *Am. J. Clin. Nutr.* **40**, 623–7.
20. Jenkins, D., Burton, P.R., Bennett, S.E., Baker, F., and Walls, J. (1988). The metabolic consequences of the correction of acidosis in uraemia. *Nephrol. Dial. Transpl.* **4**, 92–5.
21. May, R.C., Kelly, R.A., and Mitch, W.E. (1987). Mechanism for defects in muscle protein metabolism in rats with chronic renal failure. *J. Clin. Invest.* **79**, 1099–103.
22. Walls, J., and Williams, B. (1990). Metabolic role of acidosis in uraemia. In *Contributions to nephrology*. **81**, pp. 136–141. (ed. A. Albertazzi *et al.*), Karger, Basel.
23. Husted, F.C., Nolph, K.D., and Maher, J.F. (1975). $NaHCO_3$ and NaCl tolerance in chronic renal failure. *J. Clin. Invest.* **56**, 414–9.
24. Williams, B., Vennegoor, M., O'Nunan, T., and Walls, J. (1989). The use of calcium carbonate to treat the hyperphosphataemia of chronic renal failure. *Nephrol. Dial. Transplant.* **8**, 725–9.

6 Lipid abnormalities in renal disease

D.C. Wheeler

The association between abnormal lipid metabolism and renal disease was first recognized more than 160 years ago by Richard Bright, who described lactescent serum in patients with the nephrotic syndrome. His observations were confirmed 90 years later when plasma lipid assays became generally available. With recent advances in the treatment of renal disease it has been possible to study this metabolic complication of uraemia and heavy proteinuria in more detail. Interest has been stimulated by growing recognition that dyslipidaemia is a major risk factor for cardiovascular disease, and by the results of animal studies suggesting that hyperlipidaemia *per se* may contribute directly to renal injury (Grundy 1986; Keane *et al.* 1988). With the availability of new lipid-lowering agents it is now possible to treat lipid abnormalities more effectively. Whether the introduction of such therapeutic strategies will have any impact on the high cardiovascular mortality among renal patients, or will influence the progression of kidney disease, remains to be determined.

Lipids, lipoproteins and apoproteins

Cholesterol and triglyceride are virtually insoluble in water and are transported in the circulation in macromolecular complexes called lipoproteins (Deckelbaum 1987). These comprise an inner core of non-polar lipids (triglyceride and esterified cholesterol) surrounded by an outer coating of polar molecules (phospholipids, free cholesterol and apoproteins). Differences in physical properties and chemical composition distinguish various classes of lipoprotein particle, although there is considerable heterogeneity within each group (Table 6.1). The metabolism of chylomicrons, very low density lipoprotein (VLDL), intermediate density lipoprotein (IDL), low density lipoprotein (LDL) and high density lipoprotein (HDL) particles is a complex dynamic process that can be divided into an exogenous pathway, responsible for transport and distribution of dietary lipid and an endogenous pathway which shuttles cholesterol and triglyceride between the liver and tissues (Fig. 6.1). The genetically determined apoprotein components serve to maintain the structural integrity of these particles, influence the activity of enzymes responsible for their metabolism and act as ligands for cell surface lipoprotein receptors (Table 6.2) (Mahley *et al.* 1984).

Table 6.1 Physicochemical properties of human lipoproteins (Adapted from Deckelbaum 1987.)

Class	Density (g/ml)	Electrophoretic mobility (agarose gel)	Triglyceride (% weight)	Esterified Cholesterol (% weight)
Chylomicrons	<0.95	origin	90–96	1–3
VLDL	0.95–1.006	pre-β	50–65	8–14
IDL	1.006–1.019	β	25–40	20–35
LDL	1.019–1.063	β	6–12	35–45
HDL_2	1.063–1.125	α	3–8	15–20
HDL_3	1.125–1.210	α	3–6	10–18

Lipid and lipoprotein abnormalities in uraemia

The predominant plasma lipid disturbance associated with chronic renal failure is hypertriglyceridaemia, which is reported to occur in between 20 and 75 per cent of patients (Chan *et al.* 1981b). Although a raised plasma triglyceride concentration has not emerged as a clear risk factor for cardiovascular disease in the general population, in uraemic individuals this abnormality reflects an underlying disturbance of lipoprotein metabolism (or dyslipoproteinaemia) with many features associated with accelerated atherosclerotic disease. Abnormalities of the lipoprotein profile include triglyceride enrichment of VLDL and to a lesser extent LDL and HDL particles. In addition, there is an accumulation of IDL and of abnormal particles that result from the incomplete metabolism of VLDL (β-VLDL) and chylomicrons. These 'remnant' lipoproteins are chemically similar to particles detected in the plasma of patients with certain primary hyperlipidaemias and are known to be atherogenic (Nestel *et al.* 1982). Decreased HDL cholesterol concentrations are also found in uraemic plasma and represent another recognized risk factor for cardiovascular disease (Savdie *et al.* 1979).

Apoprotein abnormalities may be present even in patients with normal plasma lipids and include reduced concentrations of the HDL apoproteins AI and AII, normal or slightly increased levels of apo B and CI, increased apo CII, a marked elevation of apo CIII and reduced apo E. The characteristic finding is a reduced apo AI to apo CIII ratio. In addition to these quantitative apoprotein abnormalities, the presence of apo E and C in LDL particles is a further reflection of deranged lipoprotein metabolism (Attman *et al.* 1987; Nestel *et al.* 1982).

The pathogenesis of these lipoprotein abnormalities is not fully understood (Chan *et al.* 1981b) (Fig. 6.2). There is some evidence to suggest that hepatic VLDL production is enhanced, possibly stimulated by increased provision of fatty acid precursors or by impaired carbohydrate tolerance. However, it is now

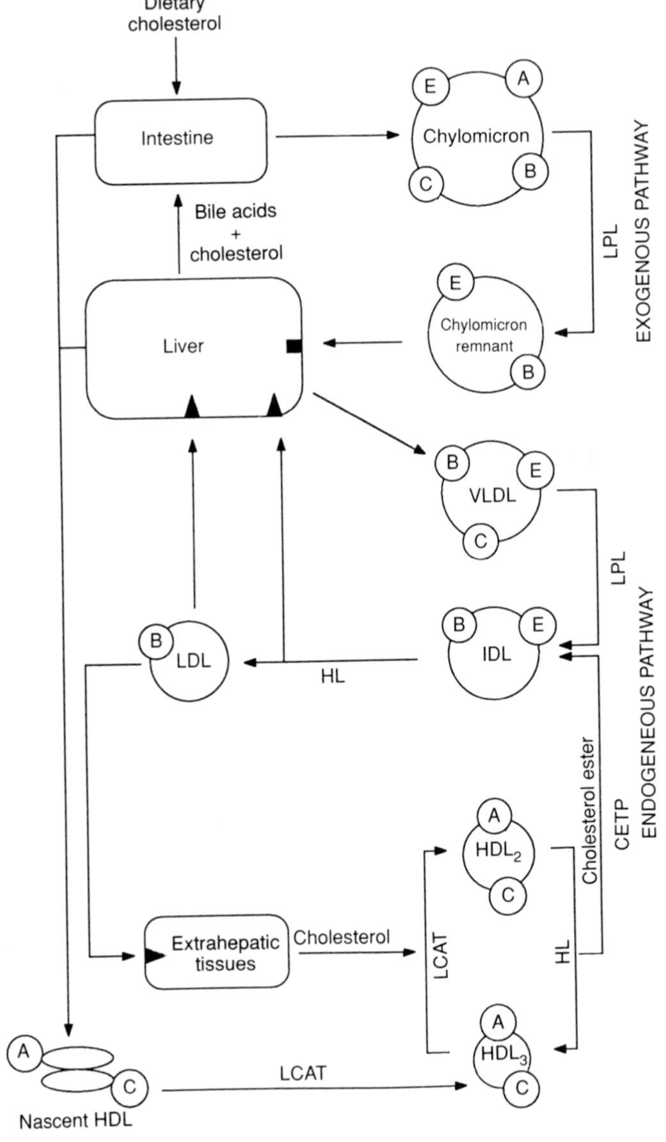

Fig. 6.1 Normal lipoprotein metabolism. Chylomicrons are synthesized in the intestinal wall from dietary lipids and are hydrolysed in the capillary circulation by the endothelial-bound enzyme lipoprotein lipase (LPL). The resultant remnant particles are

Table 6.2 Major human apoproteins and their functions

Class	Distribution	Molecular weight (da)	Functions
AI	CM, HDL	28 000	LCAT activator
AII	HDL	17 000	
B_{48}	CM	264 000	
B_{100}	VLDL, IDL, LDL	594 000	Ligand for LDL receptor
CI	CM	6500	
CII	CM, VLDL, HDL	8800	Activator of lipoprotein lipase
CIII	VLDL, HDL	8750	Inhibitor of lipoprotein lipase
E	CM, VLDL, IDL, HDL	~36 000	Ligand for LDL and chylomicron remnant receptors

CM = chylomicron.

clear that defective catabolism of triglyceride-rich particles is the predominant abnormality in such patients. A number of studies have reported reduced activity of both lipoprotein and hepatic lipase enzymes, which may result from an imbalance of apoprotein cofactors or from the presence of circulating enzyme inhibitors such as parathyroid hormone in uraemic plasma. The activity of lecithin-cholesterol acyltransferase (LCAT) is also impaired in uraemia and may contribute to deranged HDL metabolism.

Finally, diuretic and antihypertensive drugs have been associated with elevated plasma triglyceride and cholesterol concentrations and may contribute to lipid abnormalities in uraemic and non-uraemic patients with renal disease.

Lipid and lipoprotein abnormalities in the nephrotic syndrome

Hypercholesterolaemia is found in the majority of patients with heavy proteinuria and is often regarded as a feature of the nephrotic syndrome (Wheeler

---

taken up by specific hepatic receptors. Very low density lipoprotein (VLDL) is derived from endogenous lipid in the liver and is metabolized by both lipoprotein and hepatic lipase (HL) to intermediate (IDL) and low density lipoprotein (LDL). Both IDL and LDL are taken up by hepatic apoprotein B, E receptors thereby returning cholesterol to the liver. LDL particles also deliver lipid to extrahepatic tissues by the same receptor mechanism. Nascent high density lipoprotein (HDL) is synthesized by the intestine and liver and matures in the circulation acquiring cholesterol from cells. This process requires cholesterol esterification by lecithin-cholesterol acyltransferase (LCAT) and generates HDL_3. Further accumulation of cholesterol results in the formation of HDL_2. Cholesterol ester is then transported to IDL in exchange for triglyceride by cholesterol ester transfer protein (CETP), thereby regenerating HDL_3.

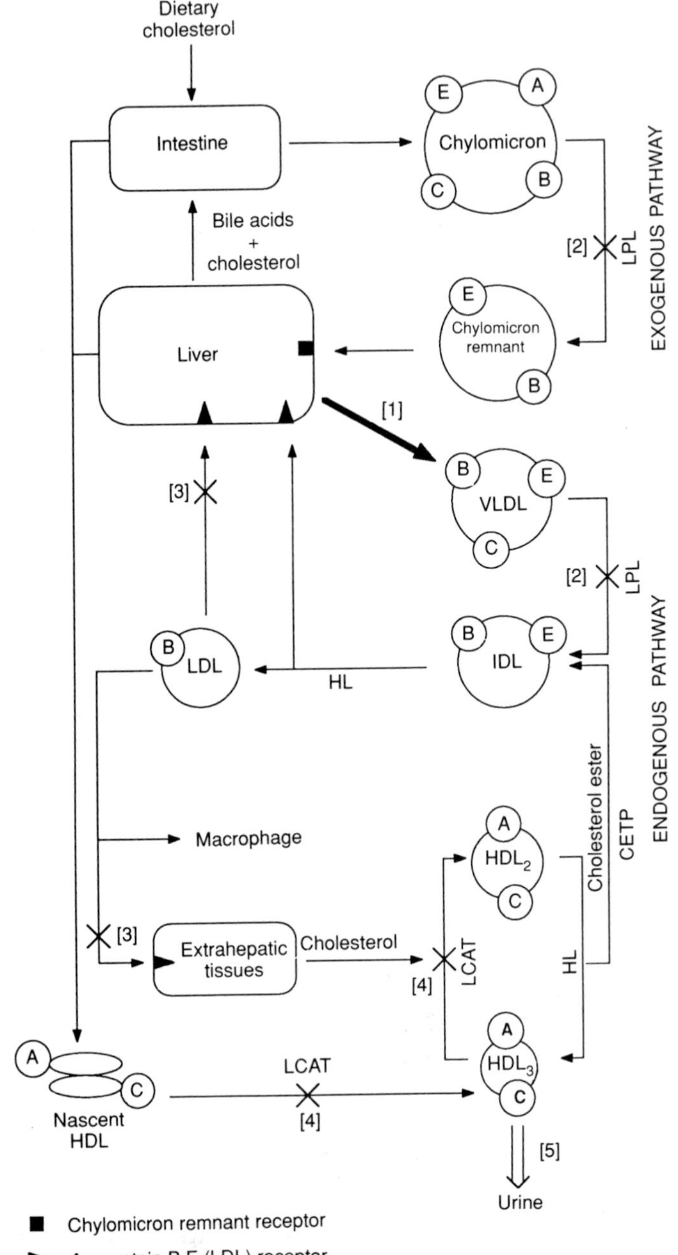

Fig. 6.2 Abnormalities of lipoprotein metabolism in renal disease. Several defects of lipoprotein metabolism have been described in patients with renal disease. Increased hepatic VLDL production [1] may be compounded by a catabolic defect resulting from

et al. 1989; Appel 1991). In the absence of uraemia, hypertriglyceridaemia is less common but may develop later in the course of the disease. Plasma lipid levels generally correlate with urinary protein loss and may fall if patients enter remission. Whilst most investigators agree that the cholesterol content of VLDL and LDL particles is raised, HDL cholesterol has been reported as high, normal or low. Analysis of HDL subfractions in nephrotic plasma has revealed that HDL_2 is reduced whilst HDL_3 remains normal. This is surprising since HDL_3 particles are smaller than HDL_2 (175 000 versus 360 000 da) but may result from urinary loss of immature HDL_3 before conversion to HDL_2 can occur (Fig. 6.2). Apoprotein AI concentrations reflect those of HDL cholesterol whilst apoproteins B and CII are usually elevated.

Nephrotic hyperlipidaemia may result from mechanisms similar to those described in uraemia (Fig. 6.2) (Wheeler *et al.* 1989; Appel 1991). Animal studies conducted more than 30 years ago suggested that hypoalbuminaemia acts as a generalized stimulus for increased protein (and therefore lipoprotein) synthesis by the liver. Because lipoproteins are relatively large, increased production is not matched by urinary loss causing plasma levels to rise. However, more recent studies in humans have shown that dietary-induced changes in the rate of albumin synthesis do not lead to alterations in plasma lipids, suggesting that other factors may be important. One alternative hypothesis is that a fall in plasma oncotic pressure acts as the trigger for hepatic protein production. This might explain why dextran infusions lower cholesterol in both nephrotic rats and humans. Whilst increased synthesis of lipoprotein has been well documented in nephrotic patients, the precise stimulus still remains unclear (Joven *et al.* 1990). In a study of the synthetic rate of albumin in adult nephrotics, plasma cholesterol concentrations were shown to be dependent only on renal albumin clearance and were not influenced by the rate of synthesis. Triglyceride concentrations were very weakly correlated with albumin synthetic rate but also depended primarily on renal clearance of this protein. It was concluded that renal loss of macromolecules cleared in parallel with albumin contributed to a metabolic defect that resulted in raised plasma cholesterol and triglyceride levels (Kaysen *et al.* 1987).

Defects in lipoprotein catabolism are also well recognized in nephrotic patients and the importance of lipoprotein overproduction has been challenged

---

reduced lipoprotein lipase activity [2] to cause the accumulation of partially metabolized triglyceride-rich chylomicron and VLDL remnant particles in the circulation. Plasma cholesterol levels may rise because of impaired receptor-mediated uptake of LDL particles [3], some of which are taken up by macrophages via a scavenger receptor pathway that may contribute to lipid deposition in the arterial wall. Reduced LCAT activity [4] impairs HDL maturation whilst loss of HDL_3 in nephrotic urine [5] may further impair removal of cholesterol from tissues. The contribution made by these various defects to the abnormal lipoprotein profiles associated with uraemia, the nephrotic syndrome and following renal transplantation is discussed in the text.

(Chan *et al.* 1981a). As in uraemia, the metabolism of triglyceride-rich lipoproteins is impaired because of reduced lipoprotein lipase enzyme activity. The latter may result from accumulation of inhibitory fatty acids which are normally bound to albumin, or from urinary loss of a factor important in the synthesis of heparan sulphate, a proteoglycan responsible for anchorage of the enzyme to the capillary endothelium. In addition to impaired catabolism of triglyceride-rich particles, defective clearance of LDL by the normal receptor pathway has been described in these patients and may contribute to hypercholesterolaemia (Warwick *et al.* 1990). Finally, reduced LCAT activity may impair HDL maturation and thereby inhibit cholesterol transport from peripheral tissues to the liver.

Lipid and lipoprotein abnormalities post transplantation

After successful renal transplantation the pattern of lipid abnormality changes rapidly. In the majority of patients, plasma triglyceride levels fall and hypercholesterolaemia develops, reflecting an increase in both HDL and LDL cholesterol concentrations (Appel 1991; Kasiske and Umen 1987). HDL cholesterol is usually restored to normal, but these particles may be enriched with triglyceride or apo AI and analysis of lipoprotein subfractions reveals a persistent depression of HDL_2 levels (Ettinger *et al.* 1987). Such abnormalities suggest that HDL metabolism remains abnormal following renal transplantation.

Post-transplant hyperlipidaemia is thought to be multifactorial in origin and both increased hepatic lipoprotein production and defective VLDL removal have been documented in these patients (Fig. 6.2) (Appel 1991). Steroids have been implicated, particularly in the pathogenesis of hypertriglyceridaemia since plasma triglyceride levels correlate closely with steroid dose (Vathsala *et al.* 1989). Cyclosporin-treated patients may be more at risk of developing hypercholesterolaemia whether or not they are on steroid therapy (Raine *et al.* 1988). Other possible factors include increased food intake, known to stimulate hepatic lipoprotein production and proteinuria which even when averaging less than 1 g/d, has been shown to correlate with plasma cholesterol levels (Kasiske and Umen 1987).

Consequences of hyperlipidaemia in patients with renal disease

Cardiovascular disease

Cardiovascular disease is the most common cause of death in patients with renal failure. Data derived from the European Dialysis and Transplant Association Registry show that myocardial infarction, heart failure and cerebrovascular accident together account for 57 per cent of total deaths on haemodialysis. In the transplant population the death rate from cardiovascular causes now equals that due to infections complicating immunosuppressive therapy (Raine

1988). There is no doubt that when compared with an age-matched population, patients on renal replacement therapy suffer a considerably higher mortality from cardiovascular events (Ritz *et al.* 1985). Nephrotic patients also appear to be at similar risk (Mallick and Short 1981). Whether such individuals experience accelerated atherosclerosis over and above that attributable to their profile of risk factors has been debated for a number of years. It has been argued that the incidence of cardiovascular disease in the renal patient population is no greater than that of non-uraemic individuals with similar risk profiles. Such risk factors have now been defined in the general population and include a number of the abnormal lipid parameters associated with renal disease, notably raised total and LDL cholesterol, reduced HDL cholesterol and the accumulation of circulating remnant particles (Grundy 1986). Hyperlipidaemia has also been correlated with increased cardiovascular disease in small studies of dialysis and transplant patient populations (Kasiske and Umen 1987; Hahn *et al.* 1983). There is now compelling evidence based on the results of recently published trials to suggest that treatment of hyperlipidaemia will reduce the incidence of death resulting from the complications of atherosclerosis (Rifkind 1987). The effect of long-term lipid lowering regimens on mortality in renal patients has not been assessed, and it remains unclear whether the results of trials based on the general population are directly relevant to these individuals.

Accelerated renal disease

The presence of lipid in the diseased kidney was first noted in 1860 by Virchow, who postulated that such changes may complicate an inflammatory process. Since this original description, renal lipid deposition has impressed a number of investigators over the years and is now a recognized feature of chronic renal disease (Keane *et al.* 1988). More recently it has been proposed that lipids may actually play a pathogenic role in progressive glomerular and tubulointerstitial injury (Moorhead *et al.* 1982). Accumulation of lipid material is one of several histological features common to both glomerulosclerosis and atherosclerosis, suggesting that these lesions share a common pathogenesis (Diamond and Karnovsky 1988). This analogy is further strengthened by the anatomical similarities between small arteries and the glomerular 'capillary' tuft (strictly a modified arteriole with the intima stripped away to form the mesangium) and by the fact that vascular smooth muscle cells and mesangial cells are closely related. Recent animal studies have linked hyperlipidaemia with renal injury (Schmitz *et al.* 1989). High cholesterol diets induce spontaneous glomerulosclerosis in a number of small mammals including guinea pigs, rabbits and rats. Such dietary regimens also exacerbate glomerular scarring induced experimentally in rats by administration of puromycin aminonucleoside or surgical removal of kidney tissue. Conversely, pharmacological correction of the lipid abnormalities that accompany renal dysfunction ameliorates glomerulosclerosis in these models. Finally, in the genetically obese Zucker rat, a hyperlipidaemic strain that develops spontaneous glomerulosclerosis, two unrelated lipid lowering agents reduce renal injury without significantly altering glomerular haemodynamic parameters.

Based on our knowledge of atherogenesis it has been possible to postulate likely mechanisms of lipid-induced renal injury (Moorhead *et al.* 1989). LDL particles probably penetrate the mesangium between endothelial cell fenestrations and become bound to mesangial matrix components where they undergo oxidation. The presence of lipoprotein is thought to promote macrophage infiltration and proliferation of intrinsic contractile mesangial cells, thereby leading to mesangial hypercellularity. Cellular uptake of lipoproteins via cell surface receptors may lead to foam cell formation and modify cell secretion of inflammatory mediators, vasoactive agents and matrix components. Study of the effects of LDL particles on cultured mesangial cells may shed further light on how lipids contribute to renal injury. Whether correction of lipid abnormalities will slow progression of human renal disease remains unproven, although trials addressing this issue are currently underway.

Treatment of lipid abnormalities associated with renal disease

Dietary modification is regarded as the first line of therapy for most forms of hyperlipidaemia (Goldberg and Schonfeld 1985). The general principles are to reduce total fat and cholesterol intake, replacing saturated fats with polyunsaturated and monounsaturated varieties or unrefined carbohydrate. In addition, obese patients should also be advised to restrict calorie intake. Such measures lead to modest reductions in both plasma cholesterol and triglyceride concentrations in patients with primary hyperlipidaemias but have not been fully evaluated in the treatment of lipid abnormalities associated with renal disease. If dietary changes prove insufficient, a number of different lipid-lowering drugs are now available (O'Connor *et al.* 1990).

Bile acid sequestrants (cholestyramine and colestipol)

These compounds are non-absorbable anion exchange resins which bind bile acids in the intestinal lumen, interrupting their enterohepatic circulation and thereby enhancing synthesis from cholesterol in the liver. Reduced hepatic cholesterol content increases expression of hepatic LDL receptors causing plasma LDL cholesterol concentrations to fall. These drugs may have unfavourable effects on plasma triglyceride levels since they increase hepatic VLDL synthesis, although addition of a fibric acid derivative usually overcomes this problem. Bile acid sequestrants have the disadvantages of being unpalatable, causing nausea and constipation in many patients. They also interfere with absorption of a number of other drugs notably digoxin, thyroxine and warfarin.

Fibric acid derivatives (clofibrate, bezafibrate, gemfibrozil)

Although clofibrate was developed more than 30 years ago its value as a lipid lowering agent has been limited by side effects which include the formation of cholesterol-rich gallstones and an acute myositic syndrome. Bezafibrate and gemfibrozil, two newer additions to this class, appear to be considerably safer in this respect. These drugs lower both cholesterol and triglyceride by reducing hepatic VLDL synthesis. They also enhance lipoprotein lipase activity and increase receptor-mediated clearance of LDL.

Probucol

This compound is a moderately effective cholesterol-lowering agent which acts by enhancing receptor-mediated uptake of LDL. It has no effect on plasma triglyceride levels and tends to lower HDL cholesterol. Despite this potentially unfavourable effect, probucol has been shown to inhibit atherosclerosis in animal models and to cause regression of xanthomata in humans. This apparent paradox can be explained by the fact that the compound has anti-oxidant properties and may thereby inhibit oxidation of LDL in the arterial intima, a process now thought to play a key role in atherogenesis (Steinberg *et al.* 1989).

Nicotinic acid (and acipimox)

Nicotinic acid (niacin) is an effective cholesterol and triglyceride lowering agent but its use has been limited by a number of unpleasant side effects. The major action of this compound is to inhibit release of non-esterified fatty acids from adipose tissue which in turn decreases hepatic VLDL production and leads to reduced synthesis of LDL. The side effects include flushing, nausea, abdominal discomfort and diarrhoea, and many patients develop rashes and pruritis. Biochemical disturbances include hyperglycaemia, hyperuricaemia, and hepatic dysfunction. Acipimox, a derivative of nicotinic acid, is currently under evaluation in the UK and may prove more potent and better tolerated than the parent compound.

Fish oil supplements (Maxepa)

Marine oils are rich in two long chain polyunsaturated n-3 fatty acids, eicosapentaenoic acid and docosahexaenoic acid. Dietary supplementation with these compounds in concentrated form (10–15 g/d) effectively reduces plasma triglyceride concentrations. The major action of n-3 fatty acids is to reduce hepatic VLDL triglyceride synthesis but they may also increase both LDL and HDL cholesterol levels. Fish oils are unpalatable, but side effects are minimal. These compounds may also have a number of other actions, unrelated to their effects on plasma lipids, that may benefit patients with renal or cardiovascular disease (Von Schacky 1987).

Statins (lovastatin, simvastatin, pravastatin)

These relatively new drugs act by inhibiting the enzyme 3-hydroxy-3-methyl-glutaryl co-enzyme A (HMG-CoA) reductase which is responsible for the rate limiting step of cholesterol biosynthesis. Their major site of action is the liver, where a compensatory increase in LDL receptor expression leads to enhanced hepatic uptake of cholesterol-rich particles. This is accompanied by a reduction in VLDL triglyceride and an increase in HDL cholesterol. The cholesterol lowering effect of these compounds is enhanced by addition of a bile acid sequestrant. Statins appear to be well tolerated although a few patients experience gastrointestinal symptoms. The most serious adverse reaction has been the development of myositis characterized by myalgia, sometimes accompanied by a rise in creatine phosphokinase, that reverses on withdrawal of the drug. The presence of liver disease or concomitant administration of fibric acid derivatives, nicotinic acid or cyclosporin seems to predispose to this syndrome.

Treatment of lipid abnormalities associated with uraemia

A number of different dietary regimens have proved effective in correcting lipid abnormalities associated with uraemia (Ritz *et al.* 1985). Low protein diets with high polyunsaturated to saturated fatty acid ratios reduce plasma triglyceride concentrations and raise HDL cholesterol. Carbohydrate restriction or direct substitution of polyunsaturated for saturated fats may also be beneficial.

Institution of dialysis therapy does not correct lipid abnormalities in uraemic individuals. On the contrary, HDL cholesterol concentrations may fall further on long-term haemodialysis whilst during the first few months of continuous ambulatory peritoneal dialysis (CAPD), triglyceride levels frequently rise along with total, VLDL and LDL cholesterol. A number of other factors associated with dialysis may further contribute to deranged lipid metabolism (Chan *et al.* 1981b). Acetate and glucose loads from dialysis fluids provide additional substrate for hepatic VLDL production. Haemodialysis leads to depletion of carnitine, an amino acid known to play an important role in fatty acid oxidation. Carnitine deficiency may stimulate lipoprotein synthesis by increasing fatty acid delivery to hepatocytes. Heparin releases lipoprotein lipase from the capillary endothelium and repeated administration could deplete enzyme stores. Finally, peritoneal loss of LDL, HDL and of apoproteins, AI, AII, B, CII and CIII may contribute directly to deranged lipid metabolism in individuals on CAPD (Saku *et al.* 1989). Attempts to modify dialysis regimens in the hope of improving lipid abnormalities have met with limited success (Chan *et al.* 1981b). Increasing the frequency of haemodialysis has little impact on hypertriglyceridaemia, although haemofiltration may be beneficial in this respect. The use of non-acetate haemodialysis fluids and avoidance of heparin do not appear to have a major effect on lipid abnormalities in individuals treated by long-term haemodialysis. Similarly, peritoneal solutions based on glucose polymers or amino acids do not dramatically improve either plasma triglyceride or cholesterol concentrations in patients on CAPD.

A number of lipid-lowering drugs have been assessed in uraemic patients, of which fibric acid derivatives are the best studied (Grundy 1990; Appel 1991). Early trials of clofibrate demonstrated that myopathy was common when this drug was administered to haemodialysis patients in regular doses.

However, a reduced dose of 1–2 g/week effectively lowered plasma triglyceride and increased HDL cholesterol levels without significant muscle damage. Gemfibrozil (commenced at 600–900 mg daily in divided doses) has been used successfully in uraemic predialysis patients and in individuals on CAPD with only mild myotoxicity and may prove a safer alternative. However, even in reduced doses, fibric acid derivatives must be used with caution in patients with impaired renal function, and plasma creatine phosphokinase levels should be monitored regularly. Fish oil supplements have also been reported to lower triglyceride levels in patients on haemo- and peritoneal dialysis but effects on HDL cholesterol have not been consistent and altered platelet function may cause unwanted bleeding. Statins have no effect on plasma triglyceride concentration in this group of patients but may raise HDL cholesterol and promote direct removal of remnant particles via LDL receptors.

Treatment of lipid abnormalities associated with the nephrotic syndrome

Dietary therapy has not been properly assessed in patients with hyperlipidaemia complicating the nephrotic syndrome, although diets low in cholesterol and saturated fatty acids are frequently recommended. It seems unlikely that such regimens will have a major impact on the severe hypercholesterolaemia associated with heavy proteinuria, and lipid lowering drugs should be considered in such individuals. A number of agents have proved safe and effective in clinical trials (Grundy 1990). When used in conjunction with an appropriate diet, both cholestyramine (8 g twice daily) and colestipol (15–25 g daily) were shown to reduce LDL cholesterol by 20 per cent but did not significantly elevate HDL cholesterol. Probucol (500 mg twice daily) may be even more effective at lowering LDL cholesterol but also decreased HDL cholesterol by 12–24 per cent. Gemfibrozil (600 mg twice daily) has been used safely in this group of patients and reduced total cholesterol by 26 per cent, LDL cholesterol by 36 per cent, and triglyceride by 17 per cent with no change in HDL cholesterol. In short-term studies, statins have proved to be the most effective drugs in the treatment of nephrotic hyperlipidaemia. Lovastatin and simvastatin (both 20 mg twice daily) lower total cholesterol by 18–36 per cent, LDL cholesterol by 18–47 per cent and slightly increase HDL cholesterol without significant adverse effects. Nicotinic acid and related drugs have not been properly assessed in nephrotic patients.

Treatment of lipid abnormalities following renal transplantation

Many patients gain weight after renal transplantation and calorie restriction in obese individuals may lower both triglyceride and cholesterol concentrations. Low cholesterol, low fat diets with high polyunsaturated to saturated fat ratios have proved effective in a number of studies and should be instituted before lipid-lowering drugs are considered (Shen *et al.* 1983). Changing steroid therapy to an alternate day regimen may be beneficial but does not always result in a significant reduction in plasma lipids, whilst conversion from cyclosporin to azathioprine has produced mixed results (Appel 1991). There have been few studies assessing drug treatment of lipid abnormalities in renal transplant recipients. Statins have been used successfully in azathioprine-treated patients, but reports of rhabdomyolysis and acute renal failure in cardiac transplant recipients given lovastatin with cyclosporin have discouraged the use of this combination. However, recent studies suggest that statins can be used safely in cyclosporin-treated transplant patients with normal liver function although low doses are recommended and plasma creatine phosphokinase levels should be monitored (Kasiske *et al.* 1990). Other lipid-lowering drugs await assessment in this group of patients.

Conclusions

Abnormal lipid metabolism commonly complicates renal disease and often persists after renal transplantation. The resultant lipid and lipoprotein profiles have now been well defined and our knowledge of the underlying mechanisms has improved considerably in recent years. It is also becoming clear that lipid lowering diets and drugs can be used safely and successfully in renal patients. Whether lipid abnormalities contribute to the high incidence of cardiovascular disease or to the progression of kidney damage in such individuals remains to be determined. Whilst such issues are being addressed the potential benefits of treating lipid abnormalities in renal patients must be weighed against the possible adverse effects of therapy.

Acknowledgements

The author would like to thank Cheryl Patterson for secretarial assistance.

References

Appel, G. (1991). Lipid abnormalities in renal disease. *Kidney Int.* **39**, 169–83.

Attman, P.-O., Alaupovic, P., and Gustafson, A. (1987). Serum apolipoprotein profile of patients with chronic renal failure. *Kidney Int.* **32**, 368–75.

Chan, M.K., Persaud, J.W., Ramdial, L., Varghese, Z., Sweny, P., and Moorhead, J.F. (1981a). Hyperlipidaemia in untreated nephrotic syndrome, increased production or decreased removal? *Clin. Chim. Acta*, **117**, 317–23.

Chan, M.K., Varghese, Z., and Moorhead, J.F. (1981b). Lipid abnormalities in uremia, dialysis, and transplantation. *Kidney Int.* **19**, 625–37.

Deckelbaum, R.J. (1987). Structure and composition of human plasma lipoproteins. In: *Atherosclerosis: biology and clinical science.* (ed. A.G. Olsson), pp. 251–5. Churchill Livingstone, Edinburgh.

Diamond, J.R. and Karnovsky, M.J. (1988). Focal and segmental glomerulosclerosis: Analogies to atherosclerosis. *Kidney Int.* **33**, 917–24.

Ettinger, W.H., Bender, W.L., Goldberg, A.P., and Hazzard, W.R. (1987). Lipoprotein lipid abnormalities in healthy renal transplant recipients: Persistence of low HDL_2 cholesterol. *Nephron*, **47**, 17–21.

Goldberg, A.C. and Schonfeld, G. (1985). Effects of diet on lipoprotein metabolism. *Ann. Rev. Nutr.* **5**, 195–212.

Grundy, S.M. (1986). Cholesterol and coronary heart disease: a new era. *J.A.M.A.* **256**, 2849–58.

Grundy, S.M. (1990). Management of hyperlipidemia of kidney disease. *Kidney Int.* **37**, 847–53.

Hahn, R., Oette, K., Mondorf, H., Finke, K., and Sieberth, H.G. (1983). Analysis of cardiovascular risk factors in chronic hemodialysis patients with special attention to the hyperlipoproteinemias. *Atherosclerosis*, **48**, 279–88.

Joven, J., Villabona, C., Vilella, E., Masana, L., Alberti, R., and Valles, M. (1990). Abnormalities of lipoprotein metabolism in patients with the nephrotic syndrome. *N. Engl. J. Med.* **323**, 579–84.

Kasiske, B.L., Tortorice, K.L., Heim-Duthoy, K.L., Goryance, J.M., and Rao, K.V. (1990). Lovastatin treatment of hypercholesterolemia in renal transplant recipients. *Transplantation*, **49**, 95–100.

Kasiske, B.L. and Umen, A.J. (1987). Persistent hyperlipidemia in renal transplant patients. *Medicine*, **66**, 309–16.

Kaysen, G.A., Gambertoglio, J., Felts, J., and Hutchison, F.N. (1987). Albumin synthesis, albuminuria and hyperlipemia in nephrotic patients. *Kidney Int.* **31**, 1368–76.

Keane, W.F., Kasiske, B.L., and O'Donnell, M.P. (1988). Lipids and progressive glomerulosclerosis: a model analogous to atherosclerosis. *Am. J. Nephrol.* **8**, 261–71.

Mahley, R.W., Innerarity, T.L., Rall, S.C., and Weisgraber, K.H. (1984). Plasma lipoproteins: apolipoprotein structure and function. *J. Lipid Res.* **25**, 1277–94.

Mallick, N.P. and Short, C.D. (1981). The nephrotic syndrome and ischaemic heart disease. *Nephron,* **27**, 54–7.

Moorhead, J.F., Chan, M.K., El Nahas, A.M., and Varghese, Z. (1982). Lipid nephrotoxicity in chronic progressive glomerular and tubulo-interstitial disease. *Lancet,* **ii**, 1309–11.

Moorhead, J.F., Wheeler, D.C., and Varghese, Z. (1989). Glomerular structures and lipids in progressive renal disease. *Am. J. Med.* **87**, (5N), 12–20N.

Nestel, P.J., Fidge, N.H., and Tan, M.H. (1982). Increased lipoprotein-remnant formation in chronic renal failure. *N. Engl. J. Med.* **307**, 329–33.

O'Connor, P., Feely, J., and Shepherd, J. (1990). Lipid lowering drugs. *Br. Med. J.* **300**, 667–72.

Raine, A.E.G. (1988). Hypertension, blood viscosity, and cardiovascular morbidity in renal failure: implications of erythropoietin therapy. *Lancet*, **i**, 97–100.

Raine, A.E.G., Carter, R., Mann, J.I., and Morris, P.J. (1988). Adverse effect of cyclosporin on plasma cholesterol in renal transplant recipients. *Nephrol. Dial. Transplant.* **3**, 458–63.

Rifkind, B.M. (1987). Gemfibrozil, lipids, and coronary risk. *N. Engl. J. Med.* **317**, 1279–81.

Ritz, E., Augustin, J., Bommer, J., Gnasso, A., and Haberbosch, W. (1985). Should hyperlipemia of renal failure be treated? *Kidney Int.* **28** (17), S84–7.

Saku, K., Sasaki, J., Naito, S., and Arakawa, K. (1989). Lipoprotein and apolipoprotein losses during continuous ambulatory peritoneal dialysis. *Nephron*, **51**, 220–24.

Savdie, E., Gibson, J.C., Stewart, J.H., and Simons, L.A. (1979). High-density lipoprotein in chronic renal failure and after renal transplantation. *Br. Med. J.* **1**, 928–30.

Schmitz, P.G., Kasiske, B.L., O'Donnell, M.P., and Keane, W.F. (1989). Lipids and progressive renal injury. *Semin. Nephrol.* **9**, 354–69.

Shen, S.Y., Lukens, C.W., Alongi, S.V., Sfeir, R.E., Dagher, F.J., and Sadler, J.H. (1983). Patient profile and effect of dietary therapy on post-transplant hyperlipidemia. *Kidney Int.* **24**, (16), S147–52.

Steinberg, D., Parthasarathy, S., Carew, T.E., Khoo, J.C., and Witztum, J.L. (1989). Beyond cholesterol. Modifications of low-density lipoprotein that increase its atherogenicity. *N. Engl. J. Med.* **320**, 915–24.

Vathsala, A., Weinberg, R.B., Schoenberg, L., Grevel, J., Goldstein, R.A., van Buren, C.T., Lewis, R.M., and Kahan, B.D. (1989). Lipid abnormalities in cyclosporine-prednisone-treated renal transplant recipients. *Transplantation*, **48**, 37–43.

Von Schacky, C. (1987). Prophylaxis of atherosclerosis with marine omega-3 fatty acids. A comprehensive strategy. *Ann. Intern. Med.* **107**, 890–9.

Warwick, G.L., Caslake, M.J., Boulton-Jones, J.M., Dagen, M., Packard, C.J., and Shepherd, J. (1990). Low-density lipoprotein metabolism in the nephrotic syndrome. *Metabolism*, **39**, 187–92.

Wheeler, D.C., Varghese, Z., and Moorhead, J.F. (1989). Hyperlipidemia in nephrotic syndrome. *Am. J. Nephrol.* **9**, (1), 78–84.

7 Hyperparathyroidism and chronic renal failure

J. Cunningham

Almost all patients with significant impairment of renal function have demonstrable abnormalities of bone structure and function, parathyroid overactivity and disturbed vitamin D metabolism. In general, the extent of these abnormalities relates to the severity of the impairment of renal function; in patients with end stage renal disease on dialysis, the derangements to bone and to mineral metabolism are invariably striking.

The pathogenesis is dominated and driven by two abnormalities at the level of the kidney, both exerting downward pressure on extracellular fluid calcium concentration, and thereby promoting secondary hyperparathyroidism. These abnormalities are (1) a real or threatened decrease of phosphaturia as glomerular filtration rate falls, and (2) a relative reduction of renal calcitriol (1,25-dihydroxyvitamin D) synthesis. This latter disturbance is partly the result of a reduced mass of proximal tubular cells which inevitably reduces also the capacity of the kidney to 1-alpha hydroxylate 25-hydroxyvitamin D to 1,25 dihydroxyvitamin D (the active hormonal form of vitamin D) and is further exacerbated by phosphate-related inhibition of the renal 1-alpha hydroxylase enzyme. The whole is further complicated by the likely, though not fully defined, impact of other aspects of the uraemic state[1] and also by the accumulation of aluminium at the mineralization front in some patients (Fig. 7.1).

Phosphate, calcitriol, and parathyroid hormone secretion

Although the early work of Bricker and Slatopolsky demonstrated convincingly the importance of phosphate retention as a trigger for uraemic hyperparathyroidism,[2] evidence has accummulated over the past two decades (slowly in the 1970s and rapidly in the 1980s) that points to a more complex picture. In addition to parathyroid hormone, changes in intracellular phosphate and transepithelial phosphate fluxes in the kidney are also determinants of the rate of 1-hydroxylation of 25-hydroxyvitamin D. Furthermore, the early enthusiasum for 1-alpha hydroxylated analogues of vitamin D as treatment for these patients has subsequently been tempered by disappointing long-term results; despite excellent early responses in most cases, some patients later 'escape' and

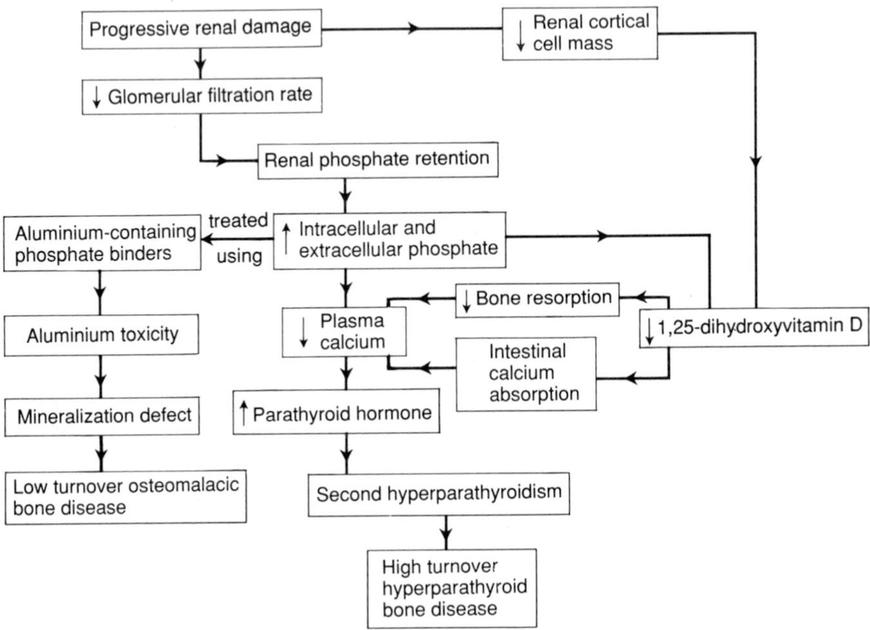

Fig. 7.1 Pathogenesis of uraemic bone disease. High turnover hyperparathyroid bone disease is the lesion found most frequently. In a minority of patients low turnover predominates, usually associated with hyperosteoidosis—low turnover osteomalacic bone disease. In some cases, features of both lesions are present—mixed uraemic osteodystrophy.

develop inexorable hyperparathyroidism with hypercalcaemia, finally requiring parathyroidectomy.[3]

Undoubtedly some of these treatment failures result from inadequate control of hyperphosphataemia, now made even more difficult by increasing concerns with aluminium toxicity.[4–7] The search for safer, better and more palatable phosphate binders continues, thus far with limited success. Many now favour calcium carbonate, which is effective and well tolerated in most patients, though leads to troublesome hypercalcaemia in some. This latter complication usually responds well to a reduction of dialysate calcium concentration and a strategy utilizing high dose calcium carbonate with dialysate calcium reduction in selected patients has been found to control hyperphosphataemia effectively and to lessen hyperparathyroidism.[8]

It is also clear that in chronic uraemia there is a progressive change in the way that the parathyroid glands respond to ambient calcium concentration, and furthermore that lack of calcitriol is likely to be a determinant of this change. It has been known for some time that the parathyroids are likely target organs for calcitriol. This view was based initially on the finding of specific calcitriol receptors in the cytosol of parathyroid cells[9] and later by *in vitro* and *in vivo* data from animal studies showing that calcitriol could reduce parathyroid hormone secretion and pre pro parathyroid hormone mRNA.[10,11] The first really convincing clinical demonstration that calcitriol could modulate parathyroid hormone secretion directly (i.e. independent of calcium) came from Madsen *et al.* in 1982) who showed that parenteral doses of calcitriol strikingly reduced elevated parathyroid hormone concentrations in patients with acute renal failure, in whom blood ionized calcium was held constant and low by continuous peritoneal dialysis.[12] Surprisingly, this work did not initially receive the attention that it deserved and it was only later that the ability of calcitriol (particularly when given parenterally) to reduce parathyroid hormone, independent of calcium elevation, in hyperparathyroid renal patients was widely appreciated.[13] Subsequent studies have confirmed this work[14,15] and it is now fairly clear, although not yet absolutely certain, that intermittent intravenous doses of calcitriol given twice or three times weekly can suppress hyperplastic parathyroid glands better than can conventional regimens using daily orally administered calcitriol. It remains to be seen whether this apparent advantage of intravenous calcitriol is maintained in the long-term.

Mechanisms of hyperparathyroidism in uraemia

In parallel with these clinical developments, much attention has been focused on the basic mechanisms that might underlie progressive hyperparathyroidism in patients with renal failure. Brown *et al.* showed that in parathyroid tissue taken from uraemic patients with secondary hyperparathyroidism, there was an increase in the 'set point' for calcium (the extracellular calcium concentration at which 50 per cent maximal parathyroid hormone release was observed), an abnormality that was also evident in parathyroid adenomas taken from patients with primary hyperparathyroidism.[16] The hyperplastic glands from uraemic patients were still responsive to changes in ambient calcium concentration, but over a range of concentrations slightly higher than that for normal human parathyroid tissue. A number of studies have since indicated clearly that the addition of calcitriol to various *in vitro* models decreases the parathyroid hormone response to lowered ambient calcium concentration in a manner consistent with an action of calcitriol to reduce the parathyroid set point for calcium towards physiological levels (10), and furthermore it now seems likely that intravenous calcitriol given twice or three times weekly may also move the set point towards normal in haemodialysed subjects.[14,15] We have recently found that favourable modification of the abnormal relationship between parathyroid hormone and calcium ion concentration in dialysis patients may be achieved using 5–10 μg of calcitriol given orally once a week as a single dose (Fig. 7.2). This suggests that the apparent advantage of the intravenous route may depend on the intermittent nature of the regimen and that the route of administration could be irrelevant.

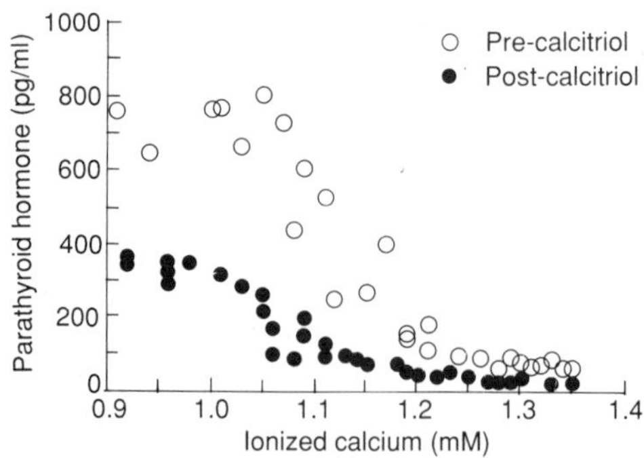

Fig. 7.2 Effect of 'pulse' oral calcitriol (10 μg once weekly) given to a stable haemodialysis patient for one month. There is left shift of the Ca-parathyroid hormone curve, indicating relative suppression of parathyroid hormone during acute sequential hypercalcaemia and hypocalcaemia. Parathyroid hormone 1–84 (whole molecule) specific assay. Normal 10–55 pg/ml.

The above evidence points to lack of calcitriol as being a likely factor favouring 'runaway' hyperparathyroidism in these patients, probably compounded by a reduction in the absolute number of calcitriol receptors in parathyroid tissue in patients with uraemia.[17] Thus the stage is set for progressive hyperparathyroidism which, using currently available therapies, ought to respond well to a combination of rigorous control of hyperphosphataemia together with calcitriol therapy, the latter almost certainly working best when given intermittently rather than daily. However, in practise phosphate control is often suboptimal, is rendered more difficult by concurrent administration of calcitriol which increases the intestinal transport of phosphate as well as of calcium, and calcitriol, given in doses sufficient to control hyperparathyroidism, often leads to unacceptable hypercalcaemia. A treatment is needed that can retard parathyroid hormon secretion selectively, without the accompanying calcaemic effects on intestine and bone seen during calcitriol therapy.

New therapeutic approaches

Although intravenous or oral pulse administration of calcitriol may achieve a degree of target organ specificity the development of new analogues of calcitriol with intrinsically different target organ specificities offers a much better hope for the future. A number of such analogues are currently being evaluated, mainly with a view to application in the field of oncology as promoters of mononuclear cell differentiation.[18] Of these, 22-oxacalcitriol suppresses the parathyroids with a potency comparable with that of calcitriol, while demonstrating only 1–2 per cent of the potency of calcitriol when used to mobilize calcium from bone *in vitro* or to raise blood calcium *in vivo*.[18,19]

Preliminary work in uraemic dogs suggests that remarkable suppression of hyperparathyroidism is achievable in this model also, again without change in blood calcium (Slatopolsky, personal communication).

Precisely how these non-calcaemic analogues achieve target organ specificity is unclear at present. It has been found that their relative potencies on target tissues *in vitro* do not necessarily correlate with receptor binding affinity. Alternative possibilities include: (1) differences in pharmacokinetics affecting presentation to the target cells, possibly the result of the relatively low affinity of 22-oxacalcitriol for plasma vitamin D binding protein;[20] (2) accelerated catabolism in some, but not all, target tissues; (3) altered translocation of the ligand into the target cell, and (4) tissue specific differences of intracellular metabolism of the ligand or of the stability of the receptor–ligand complex.

In conclusion, it seems that the inherent limitations of calcitriol as an agent for the management of uraemic hyperparathyroidism mean that only minor therapeutic refinements can be expected over the next few years. However, the advent of new agents, such as 22-oxacalcitriol, opens up exciting possibilities and it is likely that uraemic hyperparathyroidism will be controlled much better when these compounds reach the clinical arena. A final note of caution is appropriate, however. Very recent reports have drawn attention to forms of uraemic bone disease in which extreme hypocellularity and low bone turnover predominate, in the absence of aluminium or excess osteoid.[21] This disorder has been provisionally termed 'adynamic bone disease', is associated with relatively low parathyroid activity, and its recognition raises the possibility that new therapies capable of eliminating parathyroid hormone excess in uraemic patients may, paradoxically, render these patients more susceptible to adynamic bone disease.[22] Put another way, it is conceivable that parathyroid hormone, with its various osteoblast stimulating actions, may be required in modest excess to optimize bone cell metabolism in uraemia. Careful long-term clinical evaluation will be needed before these new vitamin D analogues, such as 22-oxacalcitriol, can be accepted as the panacea for uraemic hyperparathyroidism.

References

1. Hsu, C.H. and Patel, S. (1990). Effects of uremic toxins on calcitriol metabolism. *J. Am. Soc. Nephrol.* **1**, 586. (Abstr.).
2. Slatopolsky, E., Caglar, S., Pennell, J.P., Taggart, D.D., Canterbury, J., Reiss, E., and Bricker, N.S. On the prevention of secondary hyperparathyroidism in experimental chronic renal disease using proportional reductions of dietary phosphorous intake. *Kidney Int.* **2**, 147–51.
3. Sharman, V.L., Brownjohn, A.M., Goodwin, F.J., Hateley, W., Manning, R.M., O'Riordan, J.L.H., Papapoulos, S.E., and Marsh, F.P. (1982). Longterm experience of alfacalcidol in renal osteodystrophy. *Quart. J. Med.* **203**, 271–8.
4. Alfrey, A.C., Le Gendre, G.R., and Kaehny, W.D. (1976). The dialysis encephalopathy syndrome. Possible aluminium interaction. *N. Engl. J. Med.* **294**, 184–8.

5. Altmann, P., Al-Salihi, F., Butter, K., Cutler, P., Blair, J., Leeming, R., Cunningham, J., and Marsh, F. (1987). Aluminium and erythrocyte dihydropteridine reductase activity in non-encephalopathic hemodialysis patients. *N. Engl. J. Med.* **317**, 80–84.
6. Altmann, P., Plowman, D., Marsh, F., and Cunningham, J. (1988). Aluminium chelation therapy in dialysis patients: evidence for inhibition of haemoglobin synthesis by low levels of aluminium. *Lancet*, **1**, 1012–15.
7. Altmann, P., Dhanesha, U., Hamon, C., Cunningham, J., Blair, J., and Marsh, F. (1989). Disturbance of cerebral function by aluminium in haemodialysis patients without overt aluminium toxicity. *Lancet*, **11**, 7–12.
8. Sawyer, N., Noonan, K., Altmann, P., Marsh, F., and Cunningham, J. (1989). High dose calcium carbonate with stepwise reduction in dialysis calcium concentration: effective phosphate control and aluminium avoidance in haemodialysis patients. *Nephrol. Dial. Transpl.* **4**, 105–9.
9. Brumbaugh, P.F., Hughes, M.R., and Haussler, M.R. (1975). Cytoplasmic and nuclear binding components for 1,25-dihydroxyvitamin D_3 in chick parathyroid glands. *Proc. Nat. Acad. Sci. USA*, **72**, 4871–5.
10. Chan, Y.L., McKay, C., Dye, E., and Slatopolsky, E. (1986). The effect of 1,25-dihydroxycholecalciferol on parathyroid hormone secretion by monolayer cultures of bovine parthyroid cells. *Calcif. Tiss. Int.* **38**, 27–32.
11. Silver, J., Russell, J., and Sherwood, I.M. (1985). Regulation by vitamin D metabolites of messenger ribonucleic acid pre-proparathyroid hormone in isolated bovine parathyroid cells. *Proc. Nat. Acad. Sci. USA*, **82**, 4270–3.
12. Madsen, S., Olgaard, K., and Ladefoged, J. (1981). Suppressive effect of 1,25 $(OH)_2D_3$ on circulating parathyroid hormone in acute renal failure. *J. Clin. Endocrinol Metab.* **53**, 823–7.
13. Slatopolsky, E., Weerts, C., Thielan, J., Horst, R., Harter, H., and Martin, K. (1984). Marked suppression of secondary hyperparathyroidism by intravenous administration of 1,25-dihydroxycholecalciferol in uremic patients. *J. Clin. Invest.* **74**, 2136–43.
14. Delmez, J., Tindira, C., Groomes, P., Dusso, A., Windus, D.W., and Slatopolsky, K. (1989). Parathyroid hormone suppression by intravenous $1,25(OH)_2D$. A role for increased sensitivity to calcium. *J. Clin. Invest.* **83**, 1349–55.
15. Dunlay, R., Rodriguez, M., Felsenfeld, A.J., and Llach. F. (1989). Direct inhibitory effect of calcitriol on parathyroid function (sigmoidal curve) in dialysis. *Kidney Int.* **36**, 1093–8.
16. Brown, E.M., Wilkson, R.E., Eastman, R.C., Pallotta, J., and Marynick, S.P. (1982). Abnormal regulation of parathyroid hormone release by calcium in secondary hyperparathyroidism due to chronic renal failure. *J. Clin. Endocrinol. Metab.* **54**, 172–9.
17. Korkor, A.B. (1987). Reduced binding of $[^3H]$ 1,25-dihydroxyvitamin D_3 in the parathyroid glands of patients with renal failure. *N. Engl. J. Med.* **316**, 1573–7.
18. Abe, J., Takita, Y., Nakano, T., Miyaura, C., Suda, T., and Nishii, Y. (1989). A synthetic analogue of vitamin D_3, 22 oxa 1, 25-dihydroxyvitamin D_3 is a potent modulater of in vivo immunoregulating activity without inducing hypercalcemia in mice. *Endocrinology*, **124**, 2645–7.
19. Brown, A.J., Ritter, C.R., Finch, J.L., Morrissey, J., Martin, K.J., Murayama, E., Nishii, Y., and Slatopolsky, E. (1989). The non-calcemic analogue of vitamin D, 22-oxacalcitriol, suppresses parathyroid hormone synthesis and secretion. *J. Clin. Invest.* **84**, 728–32.

20. Brown, A.J., Dusso, A.S., Gunawardhana, S., Negrea, L., Finch, J., Lopez-Hilker, S., Mori, T., Nishii, Y., and Slats, E. (1990). Biological relevance of the low affinity of non calcaemic vitamin D analogs to serum vitamin D binding protein. *J. Am. Soc. Nephrol.* **1**, 584, (Abstr.).
21. Moriniere, P., Cohen-Solal, M., Belbrik, S., Boudailliez, B., Marie, A., Westeel, P.F., Renaud, H., Fievet, P., Lalan, J.D., Sebert, J.L., and Fournier, A. (1989). Disappearance of aluminic bone disease in a longterm asymptomatic dialysis population restricting $Al(OH)_3$ intake: emergence of an idiopathic adynamic bone disease not related to aluminium. *Nephron*, **53**, 93–101.
22. Hodsman, A.B., Sherrard, D.J., Wong, E.G.C., Brickman, A.S., Lee, D.B.N., Alfrey, A.C., Singer, F.R., Norman, A.W., and Coburn J.W. (1981). Vitamin D resistant osteomalacia in hemodialysis patients lacking secondary hyperparathyroidism. *Annals Int. Med.* **94**, 629–37.

8 Phosphate homoeostasis in uraemia

V. Parsons

Origin of phosphate in plasma and urine

Phosphate is an important intracellular ion which moderates intermediary metabolism and acts as an energy store in association with nucleotides. Phosphoproteins are found in many structural proteins, such as bone and enamel, while nucleoproteins are phosphorylated during replication. For this reason the phosphate intake is an important nutritional requirement and the main sources are milk, milk products, eggs, meat, and fish. Fats and carbohydrates contain less phosphate, and some vegetable proteins such as phylate may contain high-phosphate compounds which chelate calcium and interfere with its absorption. Phosphate compounds, once hydrolysed in the gut, become available for absorption unless they in turn are chelated by the presence of a high concentration of divalent ions, such as calcium and magnesium, or trivalent ions, such as aluminium. There is thus an immediate need when looking at the phosphate intake to take account of the protein intake and the mineral content of the diet; the latter can be modified therapeutically.

Once available for absorption, phosphate is transported both actively and passively across the mucosa of the gut. Those factors known to affect its transport in the renal tubule also affect gut transfer, but not to such a great extent as phosphate secretion. Losses into the gut also counteract a great deal of the phosphate absorbed (Marshall 1976). Binding of the phosphate to protein and lipids in plasma takes place and leads to two causes of pseudohyperphosphataemia, hypergammaglobulin and hyperlipidaemia, which may confuse the picture initially (Weinberg and Adler 1989). Part of the phosphate is ionized as salts of sodium and potassium, a minor buffering function, part is bound to proteins, and a small amount is complexed with calcium and magnesium. It is partly the complexing with calcium and magnesium which interferes with the ionized calcium concentrations sensed by the parathyroid gland, thereby stimulating release of one of the main hormonal controls of phosphate homeostasis, parathyroid hormone (Raman 1971).

Parathyroid hormone suppresses the transfer of phosphate back across the renal tubular cells, leading to a depression in the $TmPO_4$ and hypophosphataemia. Urinary phosphate losses through this mechanism are also seen in myxoedema, hypopituitarism and cortisol excess. Increased phosphate retention is seen in the opposite states of hypoparathyroidism, hyperthyroidism, acromegaly and Addison's disease (Brewer *et al.* 1987). Occasionally following trauma, rhabdomyolysis or diabetic coma, large phosphate movements occur which stabilize quite quickly after a few days of resuscitation.

Side effects of hyperphosphataemia

Although plasma phosphate levels in these latter conditions never reach the concentrations seen in renal failure, it is reassuring that levels of around 1.8 to 2.0 mmol/l may occur without any serious systemic side effects. In renal failure however, phosphate retention exceeds that seen physiologically in the young growing person and gives rise to three main complications: (1) secondary hyperparathyroidism. (2) suppression of 1-alpha hydroxylation of $25(OH)D_3$. (3) deposition of excessive quantities of crystalline phosphate and hydroxyapatite in the skin (itching), eye (red eye) and possibly bones (osteosclerosis, the rugger jersey sign) (Parfitt 1969). It is for these reasons rather than a particular plasma concentration of phosphate that hyperphosphataemia is to be avoided. The plasma phosphate concentration is often measured predialysis (i.e. after 48 hours' negligible renal function) in the case of patients on haemodialysis and often postprandially in continuous ambulatory peritoneal dialysis (CAPD) patients which gives the highest concentrations during the day. Hence, attempts to bring the phosphate concentrations into the 'normal range' probably imply excessive chelation and may even expose the patient to periods of hypophosphataemia sufficient to create a danger of osteomalacia (Bloom and Flinchum 1960).

An equally important aspect of the treatment of hyperphosphataemia is the prevention of any degree of hyperparathyroidism, as this may have deleterious effects on bone, cardiac function, and occasionally skin. High plasma phosphate is known to stimulate parathyroid hormone release independently of hypocalcaemia, and thus prevention of hyperphosphataemia is an essential part of the control of hyperparathyroidism. Hyperphosphataemia also causes poor 1-alpha hydroxylation of vitamin D2 and D3, and this can be corrected by the administration of either 1-alpha $(OH)D_3$ or $1{:}25(OH)2D_3$ in appropriate dosage.

The only factor known to influence the deposition of calcium salts besides the correction of the calcium and phosphate product is the maintenance of a high to normal magnesium concentration. Thus, the control of excessive hyperphosphataemia by the mechanisms outlined in Table 8.1. is important in avoiding these complications.

Options for phosphate control

For growing children with renal failure, the need for adequate quantities of calcium and phosphate make it important to choose an excellent source of protein, an adequate calcium and phosphate intake and sufficient vitamin D3 to ensure adequate calcification of growing bone. Similarly, muscle must have

Table 8.1 Binders used to control hyperphosphataemia

Method/Binder	Efficiency	Side effects
Dietary restriction of protein	Not always easy	Protein restriction may lead to muscle mass reduction
Aluminium $(OH)_3$	Good	Aluminium Toxicity
$CaCO_3 Ca(OH)_2$	Fair	Hypercalcaemia } Reduce Dialysate Ca
Ca acetate	Excellent	Hypercalcaemia } Reduce Dialysate Ca
$Mg(OH)_2 MgCO_3$	Moderate	Hypermagnesaemia; reduce dialysate Mg
$CaCO_3$ and $MgCO_3$	Good	Monitor Ca or Mg
Resins	Fair	Acceptability Occasional hypercalcaemia.

sufficient supply of these elements to ensure vitamin D dependent calcium movement in muscle membranes. Most paediatricians use calcium carbonate as the phosphate binder together with vitamin D metabolites. Regular monitoring of plasma calcium concentration is essential, as are yearly checks of calcification of unfused epiphyses.

For adult patients there may be a long period of chronic renal failure before dialysis commences which can lead to significant renal bone disease if hyperphosphataemia and secondary hyperparathyroidism are not controlled. Phosphate retention, and with it inadequate vitamin D hydroxylation, take place when the glomerular filtration rate falls below 25 ml/min. At this level significant urea and creatinine retention may not have occurred (Kanis *et al.* 1977, Andreoli *et al.* 1981). Although bone biopsy may reveal early osteomalacia, bone scanning and measurement of plasma alkaline phosphatase may detect progressive renal bone disease that requires more active therapy. This is discussed in Chapter 7.

Nutritional advice will probably involve lipid and protein restriction, as this will lower phosphate intake. Certain foods contain high amounts of phosphate, and these include meat extracts, pastes, yeast products, coffee concentrates, sardines where the whole fish is consumed, and wholemeal bread. Protein restriction must not lead to relative malnutrition, with loss of body weight and muscle mass.

If hyperphosphataemia does not respond to dietary manipulation then a phosphate binder may help. Forty years ago calcium carbonate was successfully used in large doses for this purpose. Balance studies showed that the negative calcium balance occurring in renal failure would be reversed over a period of time, and phosphate absorption reduced. When maintenance dialysis became available to prolong the life of the patient, calcium was infused during the dialysis procedure and aluminium hydroxide was substituted for calcium carbonate. It took another 15 years before this presumably innocuous unabsorbed aluminium was implicated in a series of complications. The first to be described was indolent, painful, crippling osteomalacic bone disease, followed by a neurological syndrome with apraxia and later dementia. Anaemia with hypoplasia of the bone marrow and parathyroid unresponsiveness completed the picture.

These complications were seen mainly in patients who were dialysed against fluids containing aluminium. Orally administered aluminium hydroxide was also a factor, while some batches of albumin and some continuous ambulatory peritoneal dialysis fluids have in the past contained toxic concentrations of aluminium. Studies of the half life of aluminium after patients have been converted to deionized water and alternative phosphate binders have shown that it may be as long as a year. This suggests that most of aluminium is bound in bone rather than soft tissues.

There are now a series of alternative binders to enable control of hyperphosphataemia (Table 8.1). Calcium-containing salts include carbonate, acetate and citrate. The latter unfortunately causes more hypercalcaemia than the other two salts and may enhance residual aluminium absorption from food sources (Schaefer 1989). Calcium carbonate has been in use longer than calcium acetate and has been successful in certain patients who can tolerate the rather dry and numerous tablets required. Some centres have employed a mixture of calcium carbonate and aluminium hydroxide either given together or sequentially and have reduced though not eliminated the phosphate burden. A chewable form of calcium carbonate with and without magnesium carbonate has also been used successfully (Coburn and Saluski 1989).

Use of magnesium salts alone creates risks of hypermagnesaemia and diarrhoea. The latter is commoner with magnesium hydroxide, and is eased if magnesium carbonate is used, particularly together with calcium carbonate. Up to 5 g of calcium or magnesium salts may be used a day as phosphate binders, and it is important to mix them with the food as it is eaten to achieve maximum binding (Schiller *et al.* 1989). Compliance is often a problem and usually those patients found to have the highest phosphates are those who eat well and forget their binders at meal times. If magnesium salts are used for control of hyperphosphataemia and compliance is good, it is quite safe to omit the magnesium from the dialysate, whether haemodialysis or CAPD is the mode of treatment (O'Donovan *et al.* 1986; Brewer *et al.* 1987; Fournier and Moriniere 1988).

With increasing use of calcium salts as phosphate binders there has been a general move to reduce the dialysate calcium concentration from 1.65 mmmol/l to as low as 1.2 or 1.0 mmmol/l. Such concentrations avoid hypercalcaemia when large doses of calcium are used, together with vitamin D analogues (Sawyer *et al.* 1989). On the whole, hypercalcaemia can be avoided by giving vitamin D intravenously (1 μg of 1:25 (OH)2D3) or orally (5 μg 1 alpha) once weekly. This approach achieves suppression of parathyroid hormone synthesis without necessarily raising the serum calcium.

Frequency of monitoring

One of the difficulties in manipulating calcium and magnesium salts as binders is the tendency occasionally to swing outside the acceptable concentrations for both ions. Frequent monitoring is therefore probably more necessary than in the days when aluminium salts were used. Plasma calcium and magnesium need to be monitored at monthly intervals at the commencement of therapy. Plasma phosphate and alkaline phosphatase are usually readily available also with routine automated methods (Brewer *et al.* 1987). Measurement of plasma aluminium concentration is useful initially to look at the aluminium load and occasionally a DFO (desferrioxamine) test will reveal more heavy tissue loading than can be suspected from the plasma concentration alone.

Parathyroid hormone estimations, bone X-rays, bone scans and possibly bone biopsies are much less frequently required. The parathyroid hormone assays now available are much more reliable than previously and can be done every four months. The other estimations can be made once a year and bone biopsy should be reserved for patients with symptomatic bone disease or those in whom a differential diagnosis between renal osteodystrophy and possible infiltrative lesions has to be made.

Personal experience has shown that parathyroid hormone concentrations have been remarkably normal in a large proportion of patients on dialysis who use calcium and or magnesium salts as phosphate binders. Of over 100 patients treated with haemodialysis or CAPD over a period of four years, only five have had to be considered for parathyroidectomy and some of these had been on dialysis for a number of years.

Optimum phosphate and parathyroid hormone concentrations

Phosphate concentrations need to be kept below 2.0 mmol/l if possible, though not below 1.0 mmol/l. Parathyroid hormone concentrations (intact assay) are probably raised in a proportion of patients with renal failure without any adverse effects. As Table 8.2 illustrates, twice the normal concentration (160–200 pg/ml) appears to be quite an acceptable level to aim at. Current experience suggests that the relationship between phosphataemia and hyperparathyroidism is not clear cut. With the newer ways of giving vitamin D, including once-weekly and intravenous administration, parathyroidectomy can probably be avoided in the majority of patients in the future.

Table 8.2 Comparison of patients on CAPD receiving three different regimens for phosphate binding, with the mean plasma levels over a year of various elements and parathyroid hormone concentrations (mean ±S.D.).

Regimen	Number of patients	Ca (mmol/l)	PO_4 (mmol/l)	Mg (mmol/l)	Aluminium (μml/l)	Parathyroid hormone (pg/ml)
$CaCO_3$ and $MgCO_3$	32	2.41 ±0.15	1.36 ±0.41	0.97 ±0.21	0.35 ±0.32	121
$CaCO_3$	10	2.43 ±0.19	1.38 ±0.27	0.96 ±0.26	0.20 ±0.1	141
$AL(OH)_3$	8	2.53 ±0.07	1.46 ±0.27	1.12 ±0.27	1.86 ±0.88*	121

* Significant difference $p < 0.001$. All other comparisons not significant (Parsons *et al.* 1992).

References

Andreoli, S.P., Dunson, J.W., and Bergstein, J.M. (1981). Calcium carbonate is an effective phosphorus binder in children with chronic renal failure. *Amer. J. Kid. Disease*, **9**, 206–10.

Bloom, W.L. and Flinchum, D. (1960). Osteomalacia with pseudo fractures caused by ingestion of aluminium hydroxide. *J. Am. Med. Ass.* **174**, 1327–30.

Brewer, J., Monitz, C., Baldwin, D., and Parsons, V. (1987). The effects of zero magnesium dialysate and magnesium supplements on ionised calcium concentrations in patients on regular dialysis treatment. *Nephrology Dialysis and Transplantation*, **2**, 347–50.

Coburn, J.W. and Saluski, I. (1989). Control of serum phosphorus in uraemia. *New England Journal of Medicine*, 1140–41.

Fournier, A. and Moriniere, P. (1988). Magnesium hydroxide is a useful complementary aluminium-free phosphate binder to moderate doses of oral calcium in uraemic patients on chronic haemodialysis. *Clinical Nephrology*, **29**, 319.

Kanis, J.A., Henderson, R.G., Heynen, G., Ledingham, J.G.G., Russell, R.G., Smith, R., and Walton, R.J. (1977). Renal osteodystrophy in non dialysed adolescents; longterm effects of treatment with 1-alpha hydroxycholecalciferol. *Arch. Dis. Child.* **52**, 473–81.

Marshall, D.H. (1976). *Calcium and phosphate kinetics in calcium, phosphate and magnesium metabolism*. (ed. B.E.C. Nordin). pp. 257–97. Churchill Livingstone, Edinburgh.

O'Donovan, R., Baldwin, D., Hammer, M., Monitz, C., and Parsons, V. (1986). Substitution of aluminium salts by magnesium salts in control of dialysis hyperphosphataemia. *Lancet*, **1**, 880–81.

Parsons, V., Baldwin, D., Monitz, C., Marsden, J., Ball, E., and Rifkin, I. (1992). The successful control of hyperparathyroidism in patients on continuous ambulatory peritoneal dialysis using magnesium and calcium carbonate as phosphate binders. *Nephron*, **62**, 489–94.

Parfitt, A.M. (1969). Soft tissue calcification in uraemia. *Arch. Int. Med.*, **124**, 544–56.

Raman, A. (1971). Calcium fractions of normal serum. *Clinical Biochemistry*, **4**, 141–6.

Sawyer, N., Noonan, K., Altmann, P., Marsh, F., and Cunningham, J. (1989). High dose calcium carbonate with stepwise reduction in dialysate calcium concentration. Effective phosphate control and aluminium avoidance in haemodialysis patients. *Nephrology Dialysis Transplantation*, **4**, 105–9.

Schaefer, K. (1989). Uraemic hyperphosphataemia: what is the theraphy of choice? *Nephrology, Dialysis Transplantation*, **4**, 1005–7.

Schaefer, K., Von Herrath, D., and Erley, C. (1988). Treatment of uraemic hyperphosphataemia. Is there still a need for aluminium salts? *Am. J. Nephrol.* **8**, 173–8.

Schiller, L.R., Santakna, C.A., Sheich, M.S., Emmett, M., and Fordtran, J.S. (1989). Effect of the time of administration of calcium acetate in phosphorus binding. *New England. J. Med.* **320**, 1110–3.

Weinberg, J. and Adler, M.J. (1989). Spurious hyperphosphataemia in patients with dysglobulinaemia. *Mineral and Electrolyte Metabolism*, **15**, 185–6.

9 Pathogenesis of diabetic nephropathy

G.C. Viberti and J.D. Walker

Diabetic nephropathy, defined by persistent proteinuria, a progressive loss of the glomerular filtration rate over time, arterial hypertension, and usually accompanying retinopathy, is a serious complication of both insulin-dependent and non-insulin-dependent diabetes mellitus. It affects up to 30 per cent of insulin-dependent patients (IDDs) and 15 per cent or more of non-insulin-dependent patients (NIDDs) depending on their ethnic origins.[1-3] In IDDs this complication often supervenes when the professional, social, domestic and financial responsibilities of the individual are at a peak. Diabetic nephropathy is also associated with most of the excess cardiovascular mortality seen in patients with diabetes: IDDs with persistent proteinuria have a relative mortality from cardiovascular events 37 times that of the non-diabetic population and 9 times that of the non-proteinuric diabetic population.[4]

End stage renal failure is the final result of diabetic nephropathy for 600 diabetic patients in the UK each year[5] and diabetic patients represent approximately 11 per cent of patients treated in most renal units in the UK. In North America this figure may be as high as 35 per cent.

The aim of all physicians treating patients with diabetes is to prevent renal and other complications, and to reduce the rate of decline in renal function in those with established nephropathy. A thorough study of the pathogenesis of diabetic nephropathy is therefore helpful in order to plan rational interventions.

Clinical diabetic nephropathy

The diagnosis of clinical diabetic nephropathy is usually based on clinical symptoms and signs. Renal biopsy is rarely considered necessary to give pathological confirmation, but this is useful if the clinical presentation is atypical, for example the absence of retinopathy, the presence of haematuria or the finding of persistent proteinuria in the IDD of short disease duration. The characteristic finding in diabetic nephropathy is of persistent proteinuria, usually detected by Albustix® and corresponding to a urinary albumin concentration of at least 250 mg/l and a total urinary protein loss of > 0.5 g/24 h. An elevated blood pressure, retinopathy (frequently proliferative) and lipid abnormalities are accompanying features and the condition characteristically leads to a progressive decline in renal function. The morphological correlate of the declining glomerular filtration rate is mesangial expansion and sclerosis of glomeruli. The rate of decline in the glomerular filtration rate varies from 0.5 to 2.4 ml/min/month between individuals, yet appears to be constant in a given individual. Therapeutic interventions known to reduce the rate of decline include dietary protein restriction and blood pressure control; strict glycaemic control seems to exert little influence at this stage of diabetic nephropathy.

Epidemiological considerations

The incidence of diabetic nephropathy peaks after 15–17 years of insulin-dependent diabetes, but declines thereafter. A cumulative risk of approximately 30 per cent has been reported in cohorts studied after 1950, indicating that a subset, albeit a substantial one, of diabetic patients are affected by nephropathy. The factors that predispose this group to renal disease are still not elucidated.[6] Although poor glycaemic control is likely to play some part in the susceptibility to nephropathy it may not be the sole determinant.[7]

Microalbuminuria

Markers of incipient diabetic nephropathy have provided insights into both its pathogenetic mechanisms and susceptibility to this complication. The earliest such marker is a subclinical elevation in the urinary albumin excretion rate which has been termed 'microalbuminuria' or 'incipient nephropathy'.[8,9]

Urinary albumin excretion in healthy individuals ranges between 2.5 and 26 mg/24 h, with a geometric mean around 9.5 mg/24 h and 92 per cent of values falling below 18 mg/24 h.[10] These levels are defined as normoalbuminuria. Diabetic patients with urine positive to commercial albumin-testing sticks (for example Albustixs) generally have a urinary albumin excretion rate in excess of 250 mg/24 h and these levels are classified as macroalbuminuria or clinical proteinuria.

Microalbuminuria covers the range of urinary albumin excretion rates between normo- and macroalbuminuria. The lower boundary of the microalbuminuric range still, however, remains somewhat controversial. This is in part due to the different urinary albumin excretion rates which were found, in prospective studies of both IDDs and NIDDs, to predict later nephropathy and/or mortality. All of these studies suggested a threshold level above which the progression to nephropathy was increased yet in all studies this threshold was slightly different, due possibly to differences in methods of urine collection and length of follow-up. Diabetics with an overnight or 24 h albumin excretion between 50 and 250 mg/24 h have a 20-fold greater risk of developing nephropathy than do patients with rates below this level.

The origins of microalbuminuria

The glomerular blood–urine barrier acts as a functional membrane with pores of an average size of 5 nm and with a negative electrical charge.[11,12] Factors which determine the passage of proteins across the barrier include the size and charge of the molecule and the transglomerular pressure gradient ($\bar{\Delta}P$) facilitating filtration. When microalbuminuria is present the clearance of albumin, a polyanion, and IgG, a larger and electrically neutral molecule are both increased. This is likely to be due to an elevation of $\bar{\Delta}P$ which has been demonstrated using micropuncture techniques in diabetic rats.[13] As albumin excretion increases to levels greater than 90 μg/min (130 mg/24 h) there is a disproportionate increase in the clearance of albumin which is probably due to a loss of the fixed negative glomerular charge.[14] These changing clearances can be studied by means of the selectivity index, i.e. the clearance of IgG/clearance of albumin. This starts to fall as the excretion of albumin increases, reaching a nadir when it is about 90 μg/min. With increasing levels of albuminuria the selectivity index again begins to rise. This signals the excretion of relatively more IgG and the transition to heavier proteinuria.

Associations of microalbuminuria

Albumin excretion is up to 15 per cent higher in the upright ambulatory position than at rest, and acute strenuous exercise causes a further increase. This has led to the widespread use of timed overnight urinary collections for assessment of albumin excretion; by eliminating these factors this approach minimizes the already large biological intra-individual variation in urinary albumin excretion rate, which approaches 40 per cent.

Microalbuminuria is rarely seen in the first five years of diagnosis in IDDs (although may be present at diagnosis in non-insulin dependent patients) suggesting that it represents early disease rather than a marker of susceptibility to the disease.[15]

Higher levels of glycosylated haemoglobin and 24-h plasma glucose have been found to be associated with microalbuminuria in IDD patients.[16] Two independent case-control studies involving IDDs have shown elevations in concentrations of low-density lipoprotein cholesterol, very low-density lipoprotein cholesterol and total triglyceride in microalbuminuric IDDs and a positive correlation between atherogenic lipoprotein fractions and urinary albumin excretion rate has been reported.[17,18]

In addition fibrinogen concentrations are higher in IDD patients with microalbuminuria and a reduced rise in tissue plasminogen activator (tPA) in response to exercise occurs in these patients. These findings have been taken to suggest that patients with microalbuminuria have evidence of generalized endothelial dysfunction or damage.[18,19] IDDs with microalbuminuria have an increased transcapillary escape of albumin, suggesting increased vascular leakiness, compared with matched normoalbuminuric patients.[19] Proliferative retinopathy is also associated with elevated albumin excretion although a raised blood pressure may be a common factor in this association.

These associations are summarized in Table 9.1.

Blood pressure and microalbuminuria

In IDDs microalbuminuria is associated with an elevation in arterial blood pressure (Table 9.1) compared with matched normoalbuminuric patients.[16] This association is of particular interest, as at this stage renal function is not impaired. It has been proposed that microalbuminuria and an increase in blood pressure may be joint manifestations of a common determinant, even though blood pressure often still remains within the so-called normal range. Whether a rise in blood pressure chronologically precedes, follows or accompanies the elevation of urinary albumin excretion is still somewhat controversial. One recent publication reported that albumin excretion rose first[20], but others found that the two occurred together[21,22] or that blood pressure increased first.[23] Recently, 40 per cent of patients with microalbuminuria and modest elevations of blood pressure have been shown to have intraventricular septal hypertrophy.[24]

Familial/genetic factors and diabetic nephropathy

Diabetic nephropathy clusters in families, with diabetic siblings of diabetic probands with nephropathy having a frequency of nephropathy five times higher than diabetic siblings of diabetic probands without nephropathy.[25] Among the Pima Indians with non-insulin dependent diabetes the prevalence

Table 9.1 Associations of microalbuminuria

Variable	Effect
Sex	More frequent in males; 2:1 ratio
Duration of diabetes	Very rare in first 5 years
Arterial pressure	Raised compared with matched normoalbumiuric IDDs, but may be still within 'normal' range
Serum lipoproteins	'Atherogenic' profile
Exercise	May increase albumin excretion rate
Poor glycaemic control	May increase albumin excretion rate
Transcapillary escape rate of albumin	Increased
Red blood cell Na^+/Li^+ countertransport activity	Increased in IDDs with microalbuminuria as a group
Diabetic retinopathy	Associated

of nephropathy in diabetic offspring of parents who both had diabetes and proteinuria is as high as 46 per cent.[26]

The importance of familial factors, and thus possibly heredity, has been further emphasized by the demonstration of a raised arterial pressure in the parents of proteinuric insulin-dependent diabetic patients.[27,28] These observations have been complemented by the finding of raised red blood cell sodium–lithium countertransport activity, a marker for essential hypertension, in insulin-dependent diabetic patients with either microalbuminuria or established nephropathy.[28-30] A higher level of red cell sodium–lithium countertransport activity was found in the parents of insulin-dependent diabetics with nephropathy.[31] In half of the sets of the parents of the proteinuric diabetic patients both parents had a red blood cell sodium–lithium countertransport activity above the median value for the whole parental group, while this occurred in only 10 per cent of sets of parents of normoalbuminuric patients. These recent findings strongly support the view that familial, possibly genetic, factors related to a predisposition to hypertension are important in the susceptibility to diabetic nephropathy.

The pathogenic mechanism by which an elevation of red blood cell sodium–lithium countertransport activity might contribute to renal disease in diabetes remains to be elucidated. It is believed that the red blood cell sodium–lithium countertransport activity is a mode of operation of the physiological sodium–hydrogen antiport. This ubiquitous cell membrane cation transport system is involved in the regulation of renal proximal tubule sodium reabsorption, and smooth muscle and mesangial cell growth and reactivity to a variety of vasoactive mediators and growth factors. *In vitro* studies have suggested there may be a genetically determined hyper-responsiveness of this system to the hormonal and metabolic disturbances of diabetes.[32] If so, this hyper-responsiveness may be implicated in the pathogenesis of the haemodynamic and structural abnormalities associated with arterial hypertension and renal disease in diabetes.

The relationship between the red blood cell sodium–lithium countertransport activity and arterial pressure is complex, and it may be that it functions as a marker of risk for hypertension or its complications, renal and cardiovascular disease.

Glomerular hyperfiltration

The idea that diabetic patients may exhibit glomerular hyperfiltration is over 50 years old. Using accurate techniques for estimation of glomerular filtration rate several more recent studies have found that up to 25 per cent of patients have a glomerular filtration rate exceeding the upper limit of the normal range. The role of glomerular hyperfiltration as a forerunner to diabetic nephropathy is not clear at present. Although an earlier study[8] suggested a relationship between early hyperfiltration and the later development of proteinuria and reduced renal function, a recent report of an 18 year follow-up of 29 insulin-dependent diabetic patients found no association between increased urinary albumin excretion rate and early glomerular hyperfiltration.[33] Additionally, a 5 year prospective study of two matched cohorts of insulin-dependent diabetic patients with and without hyperfiltration found no evidence of progression to persistent proteinuria in those with glomerular hyperfiltration, even though the glomerular filtration rates fell more rapidly in this group.[34] A recent report has described an association between glomerular hyperfiltration and red blood cell sodium–lithium countertransport rates in insulin-dependent diabetics.[35]

Treatment

The methods available for the treatment of diabetic renal disease have improved considerably over the last 15 years. Early identification of those 'at-risk' or with incipient disease (microalbuminuria) has increased, coupled with better insulin delivery systems which can produce improved metabolic control. These developments have enabled the effect of 'tight' metabolic control to be studied at an early and thus perhaps reversible or arrestable stage of the disease process. The results of well-conducted trials involving small numbers of patients have suggested that at the stage of microalbuminuria the elevated urinary albumin loss is at least arrestable by strict control of blood glucose.[36–38] Lowering blood pressure also lowers urinary albumin excretion[39] but again long-term prospective studies looking at the effect of this treatment on the decline of renal function at this stage of diabetic nephropathy are not available.

Lowering of blood pressure in established diabetic nephropathy has been shown to be of benefit in reducing the rate of decline of glomerular filtration rate in one non-controlled study involving a small number of patients.[40] Whether angiotensin converting enzyme (ACE) inhibitors offer particular advantages in the treatment of hypertension in patients with diabetes remains to be proven. We have recently compared hydrochlorothiazide and captopril therapy in IDDs with diabetic nephropathy and have demonstrated that while both agents lower blood pressure and urinary albumin excretion, the ACE inhibitor reduces the filtration of neutral dextrans of radius 50–74A. This suggests that captopril has an additional anti-proteinuric effect by acting on glomerular permselectivity.[41] Other workers have found similar effects.[42]

Long-term reduction in dietary protein intake retards the rate of loss of glomerular filtration rate in IDDs with diabetic nephropathy, independently of changes in blood pressure, but this effect is heterogenous and patients placed on such diets need careful and accurate assessments to ensure a response is achieved.[43]

Summary

The precise pathogenesis of diabetic nephropathy and the factors responsible for the susceptibility to it remain to be established. There is, however, increasing evidence that diabetic renal disease clusters in families and that genetic factors related to the predisposition to arterial hypertension are likely to exert a critical influence on the susceptibility to this important complication. The massive increases in cardiovascular mortality seen in patients with diabetic nephropathy suggests that the pathological process leading to glomerulopathy affects other, more widespread, vascular beds and that perhaps a common determinant is responsible for the aggregation of these different clinical manifestations in nephropathic patients.

References

1. Andersen, A.R., Sandhl-Christiansen, J., Andersen, J.K., Kreiner, S., and Deckert, T. (1983). Diabetic nephropathy in type 1 (insulin-dependent) diabetes: an epidemiological study. *Diabetologia*, **25**, 496–501.
2. Mogensen, C.E. (1984). Microalbuminuria predicts clinical proteinuria and early mortality in maturity-onset diabetes. *N. Engl. J. Med.* **310**, 356–60.
3. Allawi, J., Roa, P.V., Gilbert, R. *et al.* (1988). Microalbuminuria in non-insulin-dependent diabetes: its prevalence in Indian compared to European patients. *Br. Med. J.* **296**, 462–4.
4. Borch-Johnsen, K. and Kreiner, S. (1987). Proteinuria: value as a predictor of cardiovascular in insulin dependent diabetes. *Br. Med. J.* **294**, 1651–4.
5. Renal failure in diabetics in the UK: deficient provision of care in 1985. Joint working party on diabetic renal failure of the British Diabetic Association, the Renal Association and the Research Unit of the Royal College of Physicians. (1988). *Diabetic Medicine*, **5**, 79–84.
6. Viberti, G.C. and Walker, J.D. (1988). Diabetic Nephropathy: etiology and prevention. *Diabetes/Metabolism Reviews*, **4**(2), 147–62.
7. Nyberg, G., Blohme, G., and Norden, G. (1987). Impact of glycaemic control in progression of clinical diabetic nephropathy. *Diabetologia*, **30**, 82–6.
8. Mogensen, C.E. and Christensen, C.K. (1984). Predicting diabetic nephropathy in insulin-dependent diabetic patients. *N. Engl. J. Med.* **311**, 89–93.
9. Viberti, G.C., Jarrett, R.J., Mahmud, U., Hill, R.D., Argyropoulos, A., and Keen, H. (1982). Microalbuminuria as a predictor of clinical nephropathy in insulin-dependent diabetes mellitus. *Lancet*, **1430**–32.
10. Viberti, G.C. and Keen, H. (1984). The patterns of proteinuria in diabetes mellitus. *Diabetes*, **33**, 686–92.
11. Brenner, B.M., Hostetter, T.H., and Humes, H.D. (1978). Molecular basis of proteinuria of glomerular origin. *N. Engl. J. Med.* **298**, 826–33.
12. Myers, B.D., Winetz, J.A., Chui, E., and Michaels, A.S. (1982). Mechanisms of proteinuria in diabetes mellitus: a study of glomerular barrier function. *Kidney Int.* **21**, 633–41.
13. Hostetter, T.H., Troy, J.C., and Brenner, B.M. (1981). Glomerular haemodynamics in experimental diabetes mellitus. *Kidney Int.* **19**, 410–15.

14. Viberti, G.C., Wiseman, M.J., and Redmond, S. (1984). Microalbuminuria: its history and potential for prevention of clinical nephropathy in diabetes mellitus. *Diabetic Nephropathy*, **3**, 70–82.
15. Close, C.F. (1987). Sex, diabetes duration and microalbuminuria in type 1 (insulin-dependent) diabetes mellitus. *Diabetologia*, **30**, 508A.
16. Wiseman, M.J., Viberti, G.C., Mackintosh, D., Jarrett, R.J., and Keen, H. (1984). Glycaemia, arterial pressure and microalbuminuria in Type 1 (insulin-dependent) diabetes mellitus. *Diabetologia*, **26**, 401–5.
17. Jones, S.L., Close, C.F., Mattock, M.B., Jarrett, R.J., Keen, H., and Viberti, G.C. (1989). Plasma lipid and coagulation factor concentrations in insulin dependent diabetics with microalbuminuria. *Br. Med. J.* **298**, 487–90.
18. Jensen, T., Stender, S., and Deckert, T. (1988). Abnormalities in plasma concentrations of lipoproteins and fibrinogen in type 1 (insulin-dependent) diabetic patients with increased urinary albumin excretion. *Diabetologia*, **31**, 142–5.
19. Deckert, T., Feldt-Rasmussen, B., Borch-Johnson, K., Jensen, T., and Kofoed-Envoldsen, A. (1989). Albuminuria reflects widespread vascular damage: the Steno hypothesis. *Diabetologia*, **32**, 219–26.
20. Mathiesen, E.R., Ronn, B., Jensen, T., Storm, B., and Deckert, T. (1990). Relationship between blood pressure and urinary albumin excretion in development of microalbuminuria. *Diabetes*, **39**, 245–9.
21. Messant, J. (1990). Progression to microalbuminuria in normoalbuminuric Type 1 (insulin-dependent) diabetic patients. *Diabetologia*, **33**, A30.
22. Berglund, J. and Newberg, K. (1990). Incipient diabetic nephropathy in a Swedish province—impact of past glycaemic control, blood pressure and smoking habits. *Diabetologia*, **33**, A140.
23. Knowler, W.C., Bennett, P.H., Nelson, R.G., Saad, M.F., and Pettitt, D.J. (1988). Blood pressure before the onset of diabetes predicts albuminuria in type 2 (non-insulin-dependent) diabetes. *Diabetologia*, **31**, A509.
24. Sampson, M.J., Chambers, J., Sprigings, D., and Drury, P.L. (1990). Intraventricular septal hypertrophy in type 1 diabetic patients with microalbuminuria or early proteinuria. *Diabetic Medicine*, **7**, 126–31.
25. Seaquist, E., Goetz, E.C., Rich, S., and Barbosa, J. (1989). Familial clustering of diabetic kidney diseases. *N. Engl. J. Med.* **320**, 1161–5.
26. Pettitt, D.J., Saad, M.F., Bennett, P.H., Nelson, R.G., and Knowler, W.C. (1990). Familial predisposition to renal disease in two generations of Pima Indians with type 2 (non-insulin-dependent) diabetes mellitus. *Diabetologia*, **33**, 438–43.
27. Viberti, G.C., Keen, H., and Wiseman, M.J. Raised arterial pressure in parents of proteinuric-insulin-dependent diabetics. (1987). *Br. Med. J.* **295**, 515–17.
28. Krolewski, A.S., Canessa, M., Warram, J.H., Laffel, L.M.B., Christlieb, A.R., Kowler, W.C., and Rand, L.I. (1988). Predisposition to hypertension and susceptibility to renal disease in insulin-dependent diabetes mellitus. *N. Engl. J. Med.* **318**, 140–5.
29. Mangili, R., Bending, J.J., Scott, G., Li, L.K., Gupta, A., and Viberti, G.C. (1988). Increased sodium-lithium countertransport activity in red cells of patients with insulin-dependent diabetes and nephropathy. *N. Engl. J. Med.* **318**, 146–50.
30. Jones, S.L., Trevisan, R., Tariq, T. *et al.* (1990). Sodium lithium countertransport activity is increased in microalbuminuric diabetics. *Hypertension*, **19**(6), 570–5.
31. Walker, J.D., Tariq, T., and Viberti, G.C. (1990). Sodium-lithium countertransport activity in red cells of patients with insulin-dependent diabetes and nephropathy and their parents. *Br. Med. J.* **301**, 635–8.

32. Li, L.K., Trevisan, R., Walker, J.D., and Viberti, G.C. (1990). Overactivity of sodium-hydrogen antiport and enhanced cell growth in fibroblasts of Type 1 (insulin-dependent) diabetics with nephropathy. *Kidney Int.* **37**, 199A.
33. Lervang, H.H., Jensen, S., Brochner-Mortensen, Ditzel, J. (1988). Early glomerular hyperfiltration and the development of late nephropathy in Type 1 (insulin-dependent) diabetes mellitus. *Diabetologia*, **31**, 723–9.
34. Jones, S.L., Wiseman, M.J., Viberti, G.C., and Keen, H. (1987). Glomerular hyperfiltration and albuminuria-a 5 year prospective study in Type 1 (insulin-dependent) diabetes mellitus. *Diabetologia*, **30**, 536A.
35. Carr, S., Mbanya, J.-C., Thomas, T. *et al.* (1990). Increase in glomerular filtration rate in patients with insulin-dependent diabetes and elevated erythrocyte sodium–lithium countertransport. *N. Engl. J. Med.* **322**, 500–5.
36. Kroc collaberative study group. (1984). Blood glucose control and the evolution of diabetic retinopathy and albuminuria. *N. Engl. J. Med.* **311**, 365–72.
37. Feldt-Rasmussen, B., Mathiesen, E.R., and Deckert, T. (1986). Effect of two years strict metabolic control on progression of incipient nephropathy in insulin-dependent diabetics. *Lancet*, **ii**, 1300–4.
38. Dahl-Jorgensen, K., Hanssen, K., Kierulf, P., Bjoro, T., Sandvik, L., and Aageneas, O. (1988). Reduction of urinary albumin excretion after 4 years of continuous subcutaneous insulin infusion in insulin-dependent diabetes mellitus. The Oslo study. *Acta Endocrinologica (Copenh.)*, **117**, 19–25.
39. Marre, M., Chatellier, G., Leblanc, H., Guyene, T.T., Menard, J., and Passa, P. Prevention of diabetic nephropathy with enalapril in normotensive diabetics with microalbuminuria. (1988). *Br. Med. J.* **297**, 1092–5.
40. Parving, H.H., Andersen, A.R., Schmidt, U.M., Hommel, E., Mathiesen, E.R., and Svendsen, P.A. (1987). Effect of antihypertensive treatment on kidney function in diabetic nephropathy. *Br. Med. J.* **294**, 1443–7.
41. Pinto, J.R., Walker, J.D., Turner, C., Beesley, M., and Viberti, G.C. (1990). Renal response to lowering arterial pressure by angiotensin-converting enzyme inhibitor or diuretic therapy in insulin-dependent diabetic patients with nephropathy. *Kidney Int.* **37**(1), 516A.
42. Morelli, E., Loon, N., Meyer, T., Peters, W., and Myers, B. (1990). Effects of converting-enzyme inhibition on barrier function in diabetic glomerulopathy. *Diabetes*, **39**, 76–82.
43. Walker, J.D., Bending, J.J., Dodds, R.A., Mattock, M.B., Murrells, T.J., Keen, H., and Viberti, G.C. (1989). Restriction of dietary protein and progression of renal failure in diabetic nephropathy. *Lancet*, **ii**, 1411–14.

10 Idiopathic retroperitoneal fibrosis

L.R.I. Baker

The term retroperitoneal fibrosis is an unfortunate one, both because there are so many causes of fibrosis in the retroperitoneal area (Table 10.1) and because it is anatomically misleading and says nothing about pathogenesis. Evidence has now accumulated to suggest that the condition is an auto-allergic periaortitis and the advent of computerized tomography scanning has emphasized the periaortic nature of the condition (Fig. 10.1). The term periaortitis is, therefore, in many ways preferable to retroperitoneal fibrosis. The condition was first described in 1905 by Albarran, and the classic description came from John K. Ormond, who was a urologist at the Henry Ford Hospital in Detroit, Michigan, USA, in 1948.

Pathogenesis

The periaortic nature of retroperitoneal fibrosis has long been known to surgeons and pathologists. Histological findings include atheroma, medial thinning, splits in the media and an increase in the adventitia, which contains an inflammatory infiltrate. These findings are present to some extent within the aorta of some patients with advanced atherosclerosis who have not suffered a clinical illness and who may reasonably be classified as having 'subclinical periaortitis'. The fibrous tissue itself contains macrophages and plasma cells but not polymorphonuclear leucocytes.

It now seems likely that periaortic fibrosis is an auto-allergic response to leakage of material derived from atheromatous plaques in the diseased aorta. The substance ceroid may be involved in the reaction. Ceroid is an insoluble

Table 10.1 Causes of fibrosis in the retroperitoneum

Diverticulitis	Endometriosis	Infection
Post-irradiation	Crohn's disease	Sarcoma
Lymphoma	Metastatic cancer	
Fibromatoses of infancy and childhood		
Mesenteric fibromatosis		
Idiopathic retractile (sclerosing) mesenteritis		
Inflammatory pseudotumour		
Drugs (methysergide, ergotamine)		

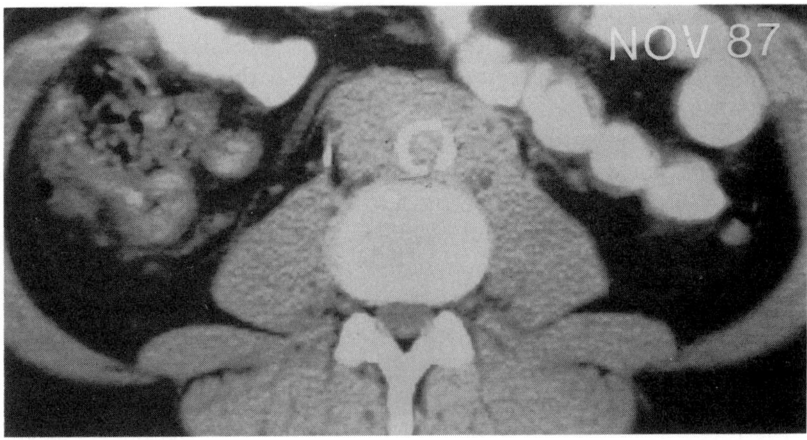

Fig. 10.1 Computerized tomography scan showing periaortic mass in idiopathic retroperitoneal fibrosis.

polymer of oxidized lipid and lipoprotein, and may be synthesized artificially by the oxidisation of low density lipoprotein. It is found in atheromatous plaques. The material is identified by staining with oil-red-O. Incubation of sections of aorta containing such plaques with mouse monoclonal antibody to human IgG reveals that antibody localizes in the region of the plaque where ceroid has been identified by oil-red-O staining. Identical findings are obtained by incubation with polyclonal rabbit anti-human IgG (Parums *et al.* 1986). Moreover, IgG and some IgM, but not IgA or IgE, can be identified in plasma cells in the fibrotic tissue where there are splits in the adjacent media. Parums and Mitchinson (1990) detected circulating antibodies to oxidized low density lipoprotein and to ceroid extracted from human atheroma using an ELISA technique in normals and various disease states, and compared their results with findings in normal individuals. They found no difference in the antibody titre between patients with ischaemic heart disease and normal controls, but highly significant increases in antibody titre in patients with chronic periaortitis. Using stored serum obtained from individuals subsequently shown at post mortem to have had subclinical periaortitis, these authors also demonstrated significantly increased antibody titres in these patients compared with normal subjects.

It seems a reasonable hypothesis that chronic periaortitis has an auto-allergic cause in which the allergen is a component of ceroid, probably oxidized low density lipoprotein, elaborated in human atheroma, and that a specific immune response involves T cells and plasma cells—but not polymorphonuclear leucocytes—which secrete IgG. It is well known that oxidized low density lipoprotein is highly immunogenic. This concept clarifies some issues which previously were difficult to explain. For example, the definite, though uncommon association between mediastinal fibrosis and idiopathic retroperitoneal fibrosis has always been difficult to understand. If one regards mediastinal fibrosis as a periaortitis, occuring in this instance around the thoracic aorta, the association becomes comprehensible. Surgeons who operate on aortic aneurysms quite often see fibrosis around the aneurysm. What is more, surgeons have described for many years what is termed inflammatory aneurysm. When operating upon an aortic aneurysm, technical difficulties may be encountered, in particular adhesion of the aorta and duodenum. Around the aorta is a dense fibrotic, chronic inflammatory infiltrate. The unifying hypothesis of an auto-allergic periaortitis clearly may account for this finding. Certainly, so-called idiopathic retroperitoneal fibrosis, mediastinal fibrosis, peri-aneurysmal fibrosis and inflammatory aneurysm have much in common (Table 10.2). The hypothesis accounts for the well known association between aortic disease, including aneurysm and aortic wall calcification, and retroperitoneal fibrosis. Finally, it may well be no coincidence that drugs such as methysergide and ergot derivatives, sometimes responsible for the condition, have effects upon the vasculature.

Clinical features

It is clear that periaortic fibrosis is more common than hitherto appreciated, if subclinical forms of the condition are taken into account. Even overt idiopathic retroperitoneal fibrosis is, in all probability, much more common than is generally recognized, a recent series from just two institutions having identified 60 such patients (Baker *et al.* 1988).

The condition was three times as common in men as in women, as is well known. On examination generally no abnormality was found other than hypertension, which was by no means universal and usually moderate in degree. Oedema, a palpable kidney and hydrocoele, features regularly described in standard texts, were found in less than 10 per cent of patients. About two-thirds of patients presented with pain, usually in the flank, abdomen or both.

Patients' ages ranged from the early 20s to 80 years, but the peak incidence was in the sixth and seventh decades of life, and the mean age of this group was

Table 10.2 Chronic periaortitis

	Idiopathic retroperitoneal fibrosis	Idiopathic mediastinal fibrosis	Perianeurysmal retroperitoneal fibrosis	Inflammatory aneurysm
Aortic dilatation	–	–	+	+
Abdominal	+	–	+	+
Thoracic	–	+	–	+
Ureteric	+	–	+	±

56 years. The majority of patients were anaemic and had a raised erythrocyte sedimentation rate, but a significant minority were normal in one or both of these respects. Significant bacteriuria and proteinuria on 'stix' testing of urine were uncommon. The typical patient with idiopathic retroperitoneal fibrosis is a middle-aged man with flank or abdominal pain, no abnormality on examination and no unusual associated conditions. Not a single patient had been exposed to methysergide. Diagnostic delay was the rule in this series. In many patients, 6–12 months or even longer had elapsed from the onset of symptoms to the time of diagnosis. Bilateral upper tract obstruction was common, being present in 49 of the 60 patients. Ten had unilateral obstruction and a single patient had obstruction of a single kidney.

Management

The management of periaortitis is empirical and controversial. Some patients receive corticosteroid therapy alone, others are managed surgically, and yet others are operated upon and subsequently receive steroid therapy with the aim of shrinking the periaortic mass and maintaining remission. In some patients with bilateral obstruction, bilateral ureterolysis is carried out. Sometimes one ureter is freed surgically and steroid therapy then relied upon to correct contralateral obstruction. All of these strategies were followed in some of the patients reported (Baker *et al.* 1988) and at least some conclusions may be drawn about optimal management.

Five patients with bilateral ureteric obstruction received steroid therapy alone, usually because they were elderly and unfit for operation. In two, obstruction was relieved. In two others, unilateral relief of obstruction occurred but the contralateral ureter remained obstructed. In one patient, steroid therapy failed entirely. Thus, corticosteroid therapy alone cannot be relied upon regularly to cure these patients. Sometimes, however, a dramatic response to such treatment occurs within 24 h of commencing treatment, particularly if high-dose intravenous methylprednisolone is employed. The reason for this is in some ways as mysterious as the occurrence of obstruction in the first place: although renal failure and upper tract dilatation are present in this condition, ureteric catheters can be passed in many cases with ease in a retrograde fashion up the ureters and into the renal pelves. Initial management of patients with urinary tract obstruction due to periaortitis by insertion of ureteric stents has sometimes yielded good results, although none of these patients were managed in this way.

Surgery alone was also not the complete answer. In 7 of 11 patients managed in this way, follow-up was sufficiently long to draw valid conclusions. Three did well but 4 did not. In 3 patients in whom unilateral obstruction was relieved surgically and steroid therapy was withheld, contralateral ureteric obstruction subsequently developed.

In 14 patients, bilateral obstruction was present and both kidneys were functioning, as judged by uptake of isotope on renographic examination or development of a nephrogram on high-dose or conventional urography. In 7 patients both ureters were freed surgically and in the remaining 7 unilateral ureterolysis was performed. Both groups received subsequent corticosteroid treatment. Typically, methylprednisolone 500 mg intravenously daily for 3 days followed by prednisolone 20 mg daily was administered, commencing when wound sutures were removed. All 7 patients in whom bilateral ureterolysis was performed did well in the long term. In 5 of the 7 patients in whom unilateral surgery was performed, the non-operated ureter remained obstructed. If a patient has periaortitis, bilateral obstruction and salvagable kidneys, and is fit for bilateral ureterolysis, that is the procedure of choice.

It is not uncommon for patients to present with an obstructed kidney on one side and a non-functioning kidney on the other, as judged by renography or urography. Operating on non-functioning kidneys in this situation is of little benefit. In only 2 patients did any function return, as judged isotopically or on subsequent urography; even in these severe impairment of function remained. Provided accurate imaging techniques are used to demonstrate non-function, patients with unilateral obstruction and a functioning kidney on one side, and a non-functioning kidney on the other, should be treated by unilateral ureterolysis.

Prognosis and follow-up

Actuarial survival of the patients studied is shown in Figure 10.2. Not surprisingly, those who died in this series were statistically significantly older and more uraemic at the time of presentation than the survivors.

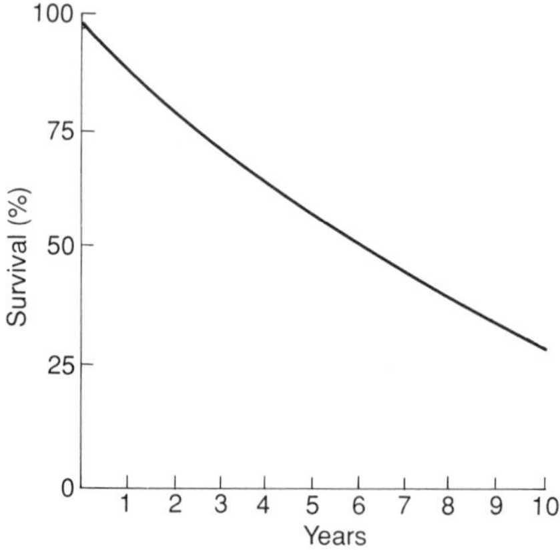

Fig. 10.2 Cumulative actuarial survival. Numbers available for analysis (followed up for stated period of time or dead) were 55 at 1 year, 53 at 2 years, 47 at 5 years and 43 at 10 years.

Of 60 patients, 10 relapsed more than 5 years after the time of diagnosis when steroid therapy had been stopped, in that erythrocyte sedimentation rate rose to an abnormal level and obstruction and diminished renal function redeveloped. Five patients relapsed even 10 years after the onset of the disease. In most patients, the dose of prednisolone can be reduced progressively and in some patients long-term remission occurs after complete withdrawal of corticosteroid therapy. Nevertheless, life-long follow-up is clearly mandatory. The best way in which to monitor such patients is not certain. Clinical assessment, serial measurement of erythrocyte sedimentation rate and assessment of renal function together with some imaging method aimed at detecting redevelopment of obstruction seem appropriate. Reduction in size of the periaortic mass can be detected on serial computerized tomography (CT) scanning, but residual periaortic tissue is seen frequently, even after steroid therapy, and the usefulness of CT in monitoring disease activity is limited (Brooks *et al.* 1987; Brooks 1990).

Brooks *et al.* (1987) described 15 patients with periaortitis who were scanned serially. In 12 a contrast-enhanced scan was performed before surgery was carried out and corticosteroid therapy was begun, and subsequent scans were performed at intervals. The antero-posterior diameter of the mass and its lateral extension on the left side were measured where the mass was largest. Assessment of the degree of extension of the mass on the right side was difficult since the inferior vena cava quite often failed to opacify. Variable reductions in the antero-posterior diameter and lateral extension of the mass occurred with time, and this was seen even in patients not treated with steroids. Whether corticosteroid release at the time of operation accounts for improvement in these patients—and also for reports of 'spontaneous' improvement in this condition—is unknown. Again, it is unclear whether in some patients the mass becomes progressively more fibrotic and smaller with time, even without treatment. Brooks *et al.* reported various outcomes in their patients. In some the mass shrank rapidly with steroid therapy, virtually disappearing in some cases. In others only a modest diminution in size occurred. In one, the mass shrank markedly after operation although corticosteroid therapy was withheld.

Periaortitis in the absence of ureteric obstruction

With the increased employment of CT scanning, it seems certain that more and more patients with periaortitis will present before the onset of urinary tract obstruction. Management of these cases is controversial.

References

Albarran, J. (1905). Retention renale per peri-ureterite. Liberation externe de l'uretere. *Compte Rendu de l'Association Francaise d'Urologie*, **9**, 511–7.

Baker, L.R.I., Mallinson, W.J.W., Gregory, M.C. *et al.* (1988). Idiopathic retroperitoneal fibrosis. A retrospective analysis of 60 cases. *Br. J. Urol.* **60**, 497–503.

Brooks, A.P. (1990). Computed tomography of idiopathic retroperitoneal fibrosis ('periaortitis'): variants, variations, patterns and pitfalls. *Clin. Radiol.* **42**, 75–9.

Brooks, A.P., Reznek, R.H., Webb, J.A.W., and Baker, L.R.I. (1987). Computed tomography in the follow-up of idiopathic retroperitoneal fibrosis. *Clin. Radiol.* **38**, 597–601.

Ormond, J.K. (1948). Bilateral ureteral obstruction due to envelopment and compression by an inflammatory process. *J. Urol.* **59**, 1072–9.

Parums, D.V., Chadwick, D.R., and Mitchinson, M.J. (1986). The localisation of immunoglobulin in chronic periaortitis. *Atherosclerosis,* **61**, 117–23.

Parums, D.V. and Mitchinson, M.J. (1990). Serum antibodies to oxidised LDL and ceroid in chronic periaortitis. *Arch. Pathol. Lab. Med.* **114**, 383–7.

Part 2 Diagnostic Techniques

11 Radionuclide imaging in nephrology

M.N. Maisey

Renal radiopharmaceuticals

Radioactively labelled compounds can provide a range of information about renal function and renal pathophysiology. This information may be related to total renal function, or it may provide images and measurements of the distribution of renal function between the two kidneys and within a single kidney. In almost all instances it is complementary to the predominantly structural information provided by other imaging methods, and should be interpreted with a full knowledge of the morphological abnormalities and the clinical context.

Radiopharmaceuticals for renal imaging

$^{99}Tc^{m}$-DTPA

$^{99}Tc^{m}$-DTPA is a chelating agent able to bind reduced technetium firmly. Pure $^{99}Tc^{m}$-DTPA is a glomerular filtered agent, which may be used for the measurement of glomerular filtration rate. The disadvantage of $^{99}Tc^{m}$-DTPA is its low extraction efficiency. It has a fourfold lower extraction efficiency than ^{123}I-O-iodohippurate (^{123}I-OIH), and the tissue and blood background is, therefore, always higher for $^{99}Tc^{m}$-DTPA than for ^{123}I-OIH. When renal function is poor, inferior quality images are obtained and data analysis is limited by the background levels and the poor statistical quality of the data. The advantages of $^{99}Tc^{m}$-DTPA are the ready availability of the $^{99}Tc^{m}$-generator and a robust kit that is simple to prepare.

^{131}I-O-iodohippurate

^{131}I-OIH has been an ideal tubule-secreted compound for probe renography for 30 years. It has the advantage of high extraction efficiency and cheapness, but ^{131}I is a poor radiolabel for the modern gamma-camera; its beta emission and long half-life limit its activity, so that image quality and the statistical quality of the data obtained is poor. ^{131}I is thus no longer used for routine renal gamma camera studies.

^{123}I-O-iodohippurate

^{123}I-OIH is a better renal imaging agent than ^{131}I-OIH and has a gamma-ray energy, 159 keV, suitable for the gamma-camera and a 13 h half life. The main problems with ^{123}I-OIH are related to its availability and cost. A 2.5 mCi

(100 MBq) amount of ^{123}I-OIH will give similar count rate images to 10 mCi (400 MBq) ^{99}Tc^{m}-DTPA, but because of its more rapid excretion, the background count of the ^{123}I-OIH image is much lower, giving a better kidney signal-to-noise ratio. Excellent quality images may be obtained even when renal function is poor.

Compounds secreted from the tubules are associated with physiological problems. Many drugs can interfere with their excretion, including the penicillins, probenecid and most diuretics.

^{99}Tc^{m}-mercaptoacetyltriglycine (MAG3)

The disadvantages of the current renal imaging agents have led many workers to search for a better alternative.

The three nitrogens and the sulphur of MAG3 are arranged in a ring structure, which is able to bind reduced technetium in a stable way. It contains a combination of polar and non-polar groups, which makes it suitable for proximal tubular uptake and secretion, and a potential technetium-labelled replacement for OIH.

The volume of distribution of MAG3 is 65 per cent smaller than that of ^{131}I-hippuran and the blood clearance half-lives are similar.

Clinical studies have shown that MAG3 is a successful radiopharmaceutical for routine renal work, combining the physiological advantages of OIH with the benefits of using ^{99}Tc labelling. It has been used in the whole range of clinical studies without side effects (Jafri *et al.* 1988).

^{99}Tc^{m}-dimercaptosuccinate (DMSA)

DMSA labelled with ^{99}Tc^{m} is taken up by the kidneys and retained in the proximal tubules, with less than 5 per cent being excreted. In order to avoid urinary excretion and liver uptake, careful preparation is required: poor preparations show high urinary loss which may interfere with the measurement of renal function. A dose of 100 MBq (2.7 mCi) is administered intravenously and static images are taken at 1 h if renal function is good, or 3–6 h if renal function is poor. The compound is filtered by the glomeruli and taken up by the kidney tubules.

Applications to clinical nephrology

The measurement of individual renal function

The relative contribution of each kidney to total function can be reliably determined using an excreted radiopharmaceutical. From the regions of interest over the two kidneys, activity-time curves, corrected for the background, are generated. From these, a normal peak occurs at about 150–180 s, at which time the tracer starts to leave the kidney. No analysis of percentage function can be taken after this time. During the first 90 s or so there are unstable conditions with mixing effects; it is best to limit the consideration of relative renal function to the period when uptake is unaffected by these factors, usually between 90 and 150 s in kidneys without outflow disorder. This period may occur later, for example 3–4 min in kidneys with an outflow disorder. The most important assumption is that the kidneys are at equal depth so that the attenuation of $^{99}Tc^{m}$ or radioiodine may be considered to be equal. In 75 per cent of kidneys there is less than 1 cm difference in depth and the count rate loss is about 10 per cent/cm for $^{99}Tc^{m}$ or ^{123}I. This leads to an error of about ± 3.5 per cent for the measured percentage relative function. Corrections for depth may be undertaken using true lateral views of the kidney, ultrasound or from a height/weight formula.

The normal range for relative uptake function determined in practice, and thus incorporating both biological variation and physical sources of error, is 42.5 per cent to 57.5 per cent for the contribution of one kidney. The accuracy of a particular figure depends on the overall renal function and the number of nephrons in a kidney but ranges from ± 4.5 per cent at 40 per cent of total uptake with a creatinine clearance of 90 ml/min to ± 7.5 per cent at 20 per cent of total uptake at a creatinine clearance of 20 ml/min. This accuracy is sufficient for most clinical applications.

The same considerations apply to the relative function determined by $^{99}Tc^{m}$-MAG3 or by the use of $^{99}Tc^{m}$-DMSA. The preferred method for determining relative function with $^{99}Tc^{m}$-DMSA is to use the geometric mean of background corrected anterior and posterior, left and right, renal counts taken between 2 and 3 h after injection.

Some of the clinical situations in which knowledge of the distribution of renal function is helpful in management will be considered.

Reflux nephropathy

The measurement of divided renal function is an essential part of the investigation and follow-up of the patient with ureteric reflux, as decisions about surgical management may depend on serial measurements of renal function. Urinary reflux may differentially affect parts of a single kidney, for example in a duplex system with reflux up one ureter only, damage may occur only to that part (usually the lower moiety) of a duplex system. In this situation it is essential to know the proportion of function in each of the two moieties in order to make rational decisions about the need for partial nephrectomy, total nephrectomy or other surgery. A $^{99}Tc^{m}$-DMSA scan performed at 3 h after injection of the radiolabel will clearly display the regional variations in functional damage brought about by reflux and associated infection.

Nephrolithiasis

The contribution of each kidney to total function should always be measured in the assessment and follow-up of patients with intrarenal stone formation.

Knowledge of the distribution of function within the kidney will also enable the surgical approach to be planned so as to minimize surgical damage to the residual renal tissue, and also contribute to the decision whether to carry out total or partial nephrectomy, or simple stone removal. Postoperative repeat assessment will be used to determine how successful the operation has been in preserving renal function and to monitor subsequent progress. Large masses of calcium lying between the kidney and the gamma-camera may result in misinterpretation of the distribution of function, so it is advisable to make anterior as well as routine posterior images.

Hypertension due to local ischaemia

Renal scanning with DMSA, DTPA, OIH or MAG3 may be helpful in defining an abnormal renal segment with poor function which is responsible for hypertension and which can be treated by partial nephrectomy or segmental angioplasty. The delayed DMSA scan will show a focal area of decreased function and the DTPA scan may show an early uptake defect at 2 min, with an increase later due to local delayed transit times secondary to local ischaemia and water reabsorption.

Renal localization

Although renal localization prior to biopsy is routinely performed under X-ray or ultrasound control, localization using $^{99}Tc^{m}$-DMSA may occasionally be preferable, for example when renal function is impaired so that localization of contrast media is poor or when there is a particular region of the kidney which is functionally abnormal. It may also be valuable in localizing the renal outline on the skin surface when planning radiotherapy of the abdomen in order to avoid unnecessary renal radiation.

Pyelonephritic scarring

Chronic pyelonephritis is one of the commonest causes of end-stage renal failure and in children is often associated with reflux. To follow up and treat children and young adults appropriately with long-term antibiotics or ureteric reimplantation it is necessary to identify the presence or absence of renal parenchymal scarring, and its progression. The identification of scars using urography is possible when there is impeccable technique and good preparation, especially with nephrotomography. However, bowel gas frequently complicates the picture, too low a dose of contrast is used and nephrotomography is not employed. Merrick *et al.* (1980) have shown that excretion urography has a sensitivity of 86 per cent and a specificity of 92 per cent whereas radionuclide imaging with $^{99}Tc^{m}$-DMSA has a sensitivity of 96 per cent and a specificity of 98 per cent. Renal scanning is, therefore, an important adjunct in the identification of scars as well as in the measurement of divided function in children and young adults with reflux or urinary infections. Care must be taken in the presence of current infection as focal 'scars' may be due to focal nephritis and resolve with time, whereas true scars cannot resolve. Recent advances in single photon emission computerized tomography (SPECT) are increasing the accuracy of the detection of focal abnormalities.

Congenital abnormalities

Renal imaging with $^{99}Tc^{m}$-DMSA is valuable for the proper assessment of many congenital abnormalities of the renal tract. Reflux and duplex kidneys have already been mentioned; other examples include the assessment of horseshoe kidneys where the function of the 'bridge' can often be assessed very much more easily than with urography. Ectopic kidneys, for example pelvic kidney, can usually be easily identified and investigated since once the radiopharmaceutical has been given a whole-body search can be undertaken if necessary, whereas a small pelvic kidney, especially if it is poorly functioning, cannot always be seen against the background of the pelvic bone. Renal abnormalities and the assessment of divided function associated with neurological abnormalities, such as meningomyelocele, can easily be documented.

Absolute measurement of each kidney's individual function

The difficulties in measuring absolute individual kidney function include the rates of change of the renal input, the blood and tissue background and renal uptake, the effects of different kidney depths on the attenuation of the renal count rate, the problem of relating the detected activity to the injected radiolabel, the sensitivity of the gamma-camera, and the dead time losses in count rate. Many attempts have been made to overcome these problems so that an absolute measurement of the uptake of an intravenously injected radiopharmaceutical by each kidney may be made. Most published methods correct for some, but not all, of the above problems (Fleming *et al.* 1987; Russell *et al.* 1985). Some methods require one or more blood samples to calibrate the activity time curves, but measurements with or without blood sampling tend to be made when the rate of change of activity in the blood is high. While these apparently direct methods may be quite practical, it is more accurate and reliable to measure the blood clearance properly in ml/min and apply the relative uptake function percentage measurements to it to give the individual renal clearance.

The use of $^{99}Tc^{m}$-DMSA is an alternative to a direct measure of absolute individual kidney function, since the conditions are much closer to a steady state at 2 or 3 h than they are during the first 30 min. The geometrical mean method overcomes some of the problems of count attenuation due to different kidney depths. In order to determine the absolute uptake the amount injected must be known either by preparing a proper standard, or by counting the active syringe before and after injection. Kidney depth may be measured from true lateral images of the kidneys or by ultrasound. Kidney phantom studies using different thicknesses of tissue equivalent material give the relationship between depth, count rate, and activity in the kidney, or a linear attenuation coefficient taking scatter into account may be incorporated into the calculation (Groshar *et al.* 1989).

The normal value for absolute uptake at 2 h is 27.0 ± 6.0 per cent, at 3 h 25.4 ± 8.9 per cent, and at 6 h 30.0 ± 9.2 per cent (mean ± SD) injected activity in each kidney.

Renovascular hypertension

Uptake function and transit times

The effect of renal artery stenosis on renal function has two features relevant to radionuclide studies. Because there is an increased proximal tubular water and salt reabsorption secondary to reduced renal perfusion, a non-reabsorbable solute, such as ^{131}I-hippuran, will become relatively concentrated within a pool of fluid that travels more slowly along the nephron. The time to peak of the renogram, which represents a crude measure of the mean transit time of hippuran entering and leaving the kidney, will be prolonged compared with a normal kidney. In the absence of outflow system disorder, and in a normally hydrated person, a difference of over a minute between peak times in the two renograms signifies prolonged hippuran transit. The blood supply, and therefore delivery, of the radiopharmaceutical to a kidney with renovascular disorder will be reduced below normal due to the occlusive arterial lesions. An uptake function less than 42 per cent of the total is abnormal, assuming that the function of the contralateral kidney is normal.

Functionally significant branch artery stenosis may be identified as delayed parenchymal uptake at one pole of a kidney. Such appearances are due to local prolongation of the tracer's parenchymal transit time.

The separation of the mean parenchymal transit time from the whole kidney transit time is important, since it obviates the effect of changes in pelvic transit time by which the specificity of the whole kidney time or activity time curve is reduced. In a normally hydrated hypertensive patient, a mean parenchymal transit time over 240 s or, in a small kidney, more than 60 s longer than that of a normal contralateral kidney, together with an uptake function of less than 42 per cent, is strongly suggestive of a functionally significant renovascular disorder. Such a finding does not distinguish between large-vessel and small-vessel disease. Bilateral functionally significant renal artery stenosis is rare and causes bilateral prolongation of mean parenchymal transit time (MPTT). Renal activity time curves and MPTT findings often demonstrate that only one stenosis is functionally significant at the time of the study when arteriography shows bilateral stenoses.

Congenitally small kidneys and the small kidney due to unilateral pyelonephritis will have normal parenchymal transit times if they occur incidentally in a hypertensive patient.

Captopril test

The 'captopril test' has been introduced as an alternative approach to improving the specificity of the changes in renal activity-time curves associated with renovascular disorder (Wenting *et al.* 1984). Captopril reduces angiotensin II formation, and thus relaxes the efferent arteriole, causing a fall in glomerular capillary pressure and filtration. For this test a baseline dynamic scan is performed using either ^{123}I-hippuran, ^{99}Tc^{m}-DTPA or ^{99}Tc^{m}-MAG3. After 1 h, 25 mg of captopril is given orally, followed 30 min later by a second injection of radiopharmaceutical and a second dynamic scan. Normally there should be no effect on the relative glomerular filtration rate or transit times; in renal artery stenosis the captopril will cause a drop in glomerular filtration rate and prolongation of transit times in the affected kidney. A single dose of captopril can cause potentially dangerous hypotension when renin levels are high and in salt-depleted patients. The latter is typically related to diuretic therapy, which must be stopped for at least 2 days before the use of captopril. Good hydration is important and blood pressure must be monitored during the test.

Use of the mean parenchymal transit time measurement in predicting the outcome of angioplasty of a renal artery stenosis has been recently demonstrated by Gruenewald *et al.* (1988). The importance of using MPTT in follow-up to detect restenosis was also demonstrated. In a study of 60 patients Geyskes *et al.* (1986) found that the uptake of hippuran was a good predictor of response to angioplasty. These radionuclide studies are complementary to angiography in the detection of renovascular hypertension, provide invaluable baseline data for follow-up and allow prediction of response and early detection of restenosis.

Vesicoureteral reflux

The micturating cystogram remains the preferred method and is the 'gold standard' for the diagnosis of vesicoureteral reflux. This investigation is usually necessary to establish the diagnosis and to achieve the anatomical information necessary for patient management. The micturating cystogram does, however, have the disadvantages of being an unpleasant investigation to perform, increasing the risk of infection due to bladder catheterization, and requiring a relatively high radiation dose. Radionuclide methods are used therefore, particularly in the follow-up of children with established reflux in whom the anatomical information has been obtained from a micturating cystogram.

There are two methods in routine use. The indirect method uses a standard dynamic ^{99}Tc^{m}-DTPA renal scan, with the generation of appropriate images, renogram curves and measurements of total and individual renal function. At the end of the study, instead of emptying the bladder, the child is encouraged to drink until the bladder is full and there is a desire to micturate. At this point the child stands (boys) or sits (girls) with the back against the gamma camera with the field of view including both kidneys and the bladder. After a baseline period of data acquisition the data are obtained while the bladder is emptied and for several minutes afterwards. Regions of interest are placed over each kidney and over the bladder. The time-activity curves from the renal regions of interest will show peaks in the renal area if radiolabelled urine refluxes up as far as the kidney. The bladder curve will show the rapid emptying phase if it is a good study and a partial refilling phase if the refluxed volume is large enough. The time-activity curves should always be correlated with the images during reflux to avoid false positive reports.

This method has been shown to be sensitive and accurate and can be used as a follow-up procedure. In addition to being less traumatic than the micturating cystogram, avoiding catheterization and having a lower radiation dose, it has the added advantage of providing functional renal information from the early part of the DTPA scan. The disadvantage, apart from lack of anatomical detail is that a true filling phase reflux is not seen unless it occurs spontaneously during the renogram phase, when it shows up as multiple spikes in the third phase. This method can also be used in the screening of siblings with reflux and the screening of patients presenting with urinary tract infection.

The retrograde (direct) method is used for children less than 3 or 4 years of age, who are usually not able to co-operate with the indirect method. $^{99}Tc^{m}$-DTPA diluted in saline is run into the catheterized bladder during imaging with the gamma camera. Although similar to a micturating cystogram, this investigation does not yield anatomical detail; on the other hand there is a much lower radiation dose and it can be made more quantitative.

Obstruction to outflow

An obstructing process may affect structure, as demonstrated by intravenous urography or ultrasound, renal function, as demonstrated by radionuclide studies; and pressure/flow relationships, as demonstrated by antegrade perfusion pressure measurements in the upper urinary tract and by urodynamic studies of the lower urinary tract. The passive component of salt and water reabsorption is increased, and the non-reabsorbable tracer is concentrated in the tubular luminal fluid, reducing its flow rate and prolonging its transit time along the nephron. Thus DTPA, MAG3 and I-OIH are concentrated in a smaller volume of fluid in the lumen of the nephron, the intranephron fluid flow rate is reduced, and the parenchymal transit time of these tracers is increased.

Clinical tests based on the use of the diuretic frusemide (diuretic renography) may help to determine the strength of the resistance to outflow. In order for frusemide to work, there must be a sufficient number of nephrons to produce a diuresis and the patient should not be chloride or sodium depleted. Thus frusemide diuresis is an unreliable test of outflow resistance when renal function is poor and when the patient has major electrolyte disturbancies. Conversely, if renal function is good, there may be such a diuretic response that a nephrologically important, but relatively slight outflow resistance may be overcome, leading to a false diagnosis of lack of obstruction. The response may also be lost in a grossly dilated pelvis and an inappropriately good response obtained if the pelvis is rigid. Variable correlations have been shown between the frusemide diuresis technique and antegrade perfusion pressure measurements.

The frusemide test is usually performed using $^{99}Tc^{m}$-DTPA, but ^{123}I-OIH or $^{99}Tc^{m}$-MAG3 may be used. If there is significant delay of tracer in the collecting system, intravenous frusemide is given 30 minutes after the tracer, at a dose of 0.5 mg/kg body weight. Data are collected for a further 15 min and time-activity curves are obtained from regions of interest over the renal areas and may be divided into three categories: (1) normal, when there is a steepening of the rate of fall with a concave washout curve; (2) obstructed, when there is no washout, and (3) intermediate, which is more difficult to interpret. Results are expressed as the percentage washout in 15 min and the washout rate (measured as half-life). Results are considered normal if washout is greater than 50 per cent, with a half-life faster than 10 min. In obstruction less than 25 per cent of the tracer washes out over 15 min, and half life is slower than 20 min; results between the two are considered intermediate.

Since the measured change in activity with time is the basis of determining the response to frusemide, this response is also crucially dependent on the amount of radiolabel taken up by the kidney and on the rate of uptake. Thus the rate of fall of the activity-time curve of the kidney in response to frusemide is dependent on its previous rate of rise and one has to judge whether the rate of fall is appropriate for a given rate of rise. A moderately poorly functioning kidney would have a moderately impaired rate of rise and a moderately impaired, but appropriate, rate of fall in response to frusemide in the absence of obstructing uropathy. An inappropriately slow rate of fall in response to frusemide would support a diagnosis of obstructing uropathy.

Poorly functioning kidneys often provide a real problem to the surgeon evaluating possible outflow obstruction and it is difficult to rely on the frusemide diuresis either visually, graphically or on numerical indices applied to the third phase unless they show unequivocal absence of obstruction. Recovery of renal function is often much greater than expected, particularly when the appearance of the kidney or an estimate of cortical thickness was determined from intravenous urography. Prolongation of the parenchymal transit time occurs before reduction in uptake function and shortening of the parenchymal transit time towards normal precedes an increase in uptake function after relief of the obstruction.

Renal failure

Renal failure may be chronic, resulting in progressive renal insufficiency, or may be an acute clinical presentation. The possible role of radionuclide studies in these conditions will be reviewed.

Acute renal failure

The mechanism of acute tubular necrosis is controversial, since it is now well established that necrosis is by no means always present. The initial insult is probably decreased renal perfusion, which results in a decrease in renal blood flow and renal ischaemia. As the glomerular filtration rate is acutely dependent on renal blood flow, the rate falls, resulting in a decrease in urine flow (oliguria) the renal ischaemia then becomes self-perpetuating. If the ischaemia is severe or nephrotoxins are present actual tubular cell necrosis will occur, with consequent back leakage of filtrate and tubule obstruction by debris.

The first problem is to decide upon initial investigations. Ultrasound examination combined with a radionuclide functional study will usually provide most of the information necessary for clinical management, with no risk of the toxicity and further deterioration of renal function which might be associated with the use of contrast agents. A dynamic $^{99}Tc^{m}$-DTPA or $^{99}Tc^{m}$-MAG3 scan is the method of choice, occasionally with the addition of a $^{99}Tc^{m}$-complex radiopharmaceutical, such as DMSA.

The radionuclide study provides information about first-pass renal perfusion, the handling of the glomerular filtered agent and whether there is sufficient urine flow through the collecting systems. The single most useful role of the dynamic radionuclide scan is to make a firm diagnosis of acute tubular necrosis which, if treated with adequate dialysis, has a good prognosis, compared with other causes of acute renal failure. It may also be possible to differentiate the early onset phase of acute tubular necrosis, which may be reversed by fluid and electrolye correction, from the established phase.

A practically normal perfusion phase during the first transit is followed by a moderately good visualization of the kidneys at 90 to 180 s which represents a blood pool image of the kidneys; as the tracer diffuses into the larger extracellular space from the vascular space the renal image diminishes without the appearance of tracer into the collecting system. Early signs of recovery are increasing retention of tracer in the kidney, as the glomerular filtration rate returns to normal and is superimposed on the blood pool image; progressive concentration as glomerular filtration continues to improve but intrarenal transit remains grossly prolonged; and excretion as at the onset of the diuretic phase of acute tubular necrosis.

Acute obstruction is characterized by a decreased perfusion image during first transit and poor early uptake image with progressive parenchymal accumulation, due to the grossly delayed intrarenal transit. Dilated calyces are frequently seen on the early images as negative photon-deficient areas which progressively accumulate tracer over several hours. These findings, although characteristic, should be confirmed with ultrasound prior to surgical treatment and the absence of evidence of obstruction on a $^{99}Tc^{m}$-DTPA scan should never be used to exclude the diagnosis without ultrasound confirmation.

Pre-renal failure associated with acute oliguria due to, for example, dehydration but before the onset of established acute tubular necrosis is identical to phase 2 or 4 of the acute tubular necrosis recovery pattern. Radionuclide imaging shows well-perfused kidneys with a significant secretory peak representing glomerular filtration, markedly delayed parenchymal transit, but with some excretion of tracer which may only appear in the calyces at the end of the normal 20 to 30 min period of imaging.

Acute nephritis and most other renal parenchymal diseases usually show significantly worse perfusion than acute tubular necrosis, with markedly decreased uptake at 2 min with either no accumulation or slow progressive accumulation in the renal parenchyma but without significant excretion.

These patterns represent the majority of cases of acute renal failure, but other causes, such as acute loss of perfusion unilaterally associated with renal artery embolus, may also be diagnosed. In aortic obstruction failure to visualize the aorta accompanies grossly diminished renal perfusion; aneurysm, if associated with a large lumen, will be identified as a large abdominal blood pool; venous thrombosis decreases perfusion but to a lesser extent than arterial occlusion.

When the visualization is very poor or absent using $^{99}Tc^{m}$-DTPA or MAG3 a repeat image following administration of $^{99}Tc^{m}$-DMSA may give better results, and be a better guide to eventual prognosis.

Chronic renal failure

Ultrasound examination remains the initial investigation in chronic renal failure and may reveal the aetiology. This may include irregularly scarred kidneys due to pyelonephritis or reflux; the enlarged and cystic kidneys of polycystic disease; the small, uniformly contracted kidneys of chronic glomerulonephritis, or dilated ureters, collecting system and enlarged kidneys arising from chronic obstruction. Although radionuclide investigations certainly show chronically damaged kidneys in this situation and, to a very limited extent, predict the degree of recoverable function, it is only occasionally of any clinical value and then is best performed with ^{123}I-hippuran, $^{99}Tc^{m}$-DTPA or MAG3, or $^{99}Tc^{m}$-DMSA.

Renal mass lesions

Renal parenchyma that has been replaced by a space-occupying lesion, whether tumour, cyst, infarct or scar, can be identified by a loss of functioning tissue compared with the surrounding normal parenchyma, with the presence or absence of changes in local perfusion during the angiographic phase. Early studies with $^{99}Tc^{m}$-DTPA and $^{99}Tc^{m}$-DMSA concentrated on the detection of tumours and cysts of the kidney but with the development of improved structural imaging techniques (high dose nephrotomography, ultrasonography and computerized tomography scanning) and more widespread use of diagnostic cyst puncture with cytological examination, the role of nuclear medicine techniques has changed markedly.

Radionuclide scans, which for the purpose of this particular problem are

$^{99}Tc^m$-DMSA scans, usually with first-pass radionuclide angiography, are used in three clinical situations. The first is when the ultrasound examination does not confirm a mass lesion or is equivocal. This is not often a problem with quality ultrasound examination, but when it does arise the renal scan is the next best way of confirming or excluding a tumour. The scan may also help to determine whether there is a single or multiple lesion. Lateral and oblique views of the kidney should always be performed, and the results should be reviewed in conjunction with the intravenous urogram, in order to correlate abnormalities seen on the two examinations and to identify the site and size of the calyces on the renal scan, because, with current high-resolution images of the renal parenchyma, normal calyces may be confused with a space-occupying lesion.

Renal radionuclide scanning may also be helpful in the further evaluation of a lesion seen on intravenous urography which is solid on ultrasound but could possibly be a pseudotumour, such as a lump from splenic pressure, fetal lobulation, compensatory hypertrophy, and prominent columns of Bertin. It is an advantage to be able to demonstrate that the 'tumour' is functionally normal renal tissue without resorting to arteriography.

When arteriography and surgery are contraindicated the renal scan, with particular emphasis on the first-pass arteriographic phase, may help to establish a diagnosis. Increased blood flow in the lesion indicates a high probability of tumour; no flow in a large lesion increases the likelihood of a cyst, but there are a significant number of cases in which the tumour has a relatively poor supply and differentiation of benign and malignant lesions is not possible.

Renal transplantation

Radionuclide investigations can contribute significantly to the management of renal transplant patients. Most patients who receive a cadaveric graft will have some degree of post-operative acute tubular necrosis, which may cause prolonged primary non-function. Acute rejection remains important and may be superimposed on primary non-function; the use of cyclosporin complicates the picture since it may have functionally similar effects to those of acute rejection.

Most of the important and common complications result in changes in renal function without morphological abnormalities; consequently techniques which essentially provide images of renal function are well suited to the monitoring of the graft. They are most effective when performed in a very reproducible way and with accurate quantification early after surgery (within 3 days), and repeated regularly until function is normal.

The dynamic $^{99}Tc^m$-DTPA scan is most widely used for monitoring graft function. A series of images of the first pass of tracer through the kidney and vessels (vascular phase) is obtained, followed by 1 min images for a total of 20 min as the tracer concentrates in the parenchyma and is excreted into the collecting system. Delayed images may be necessary if there is delayed excretion, suspected obstruction or a urinary leak. Data are stored on the computer and time-activity curves are generated from the regions of interest over the kidney, the iliac artery and background. Quantitative measurement may be made according to a number of published protocols (Baillet *et al.* 1986), one of which generates a flow index, which is a measure of perfusion, and an uptake index, which is a measure of glomerular filtration rate (Hilson *et al.* 1978). This method will diagnose acute tubular necrosis, and demonstrate decreased perfusion and function due to a number of causes, for example acute rejection, cyclosporin A toxicity and renal arterial disorders. In most cases, unless the results are unequivocal, changes in the scan over a period of days and the patient's clinical circumstances are needed for accurate diagnosis. Cyclosporin A toxicity can be particularly difficult to distinguish from acute rejection (Gedroyc *et al.* 1986). Many other complications (lymphocoele, leaks, obstruction etc.) can also be detected with a radionuclide scan, often before they are clinically apparent.

Although $^{99}Tc^{m}$ is the most common radiopharmaceutical used for transplant evaluation, others have been introduced and are used in some centres. These include ^{131}I-OIH or ^{123}I-OIH, which are used with imaging protocols similar to DTPA and a variety of different quantitative procedures. They may also be combined with serum sampling to measure effective renal plasma flow. Recently, $^{99}Tc^{m}$-MAG3 has been used as a DTPA replacement with encouraging results, the higher clearance giving better quality images, especially in patients with impaired function and in children. Some radiopharmaceuticals have been directed at the diagnosis of rejection, and these include ^{67}Ga-citrate, $^{99}Tc^{m}$-sulphur colloid, ^{111}In-labelled platelets and white cells. However, none of these has been widely accepted as a routine clinical tool.

References

Baillet, G. *et al.* (1986). Evaluation of allograft perfusion by radionuclide first-pass study in renal failure following renal transplantation. *European Journal of Nuclear Medicine*, **11**, 463–9.

Fleming, J.S., Keast, C.M., Waller, D.G., and Ackery, D. (1987). Measurement of glomerular filtration rate with 99mTc-DTPA: a comparison of gamma camera methods. *Journal of Nuclear Medicine*, **13**, 250–3.

Gedroyc, W., Taupe, D., Fogelman, I., Neild, G., Cameron, J.S., and Maisey, M.N. (1986). Tc-99m DTPA scans in renal allograft rejection and cyclosporine nephrotoxicity. *Transplantation*, **42**, 494–7.

Geyskes, G.G., Oei, H.Y., and Faber, J.A. (1986). Renography: prediction of blood pressure after dilatation of renal artery stenosis. *Nephron*, **44**, 54–9.

Groshar, D. *et al.* (1989). Quantitation of renal uptake of technetium-99m DMSA using SPECT. *Journal of Nuclear Medicine*, **30**, 246–50.

Gruenewald, S.M. *et al.* (1988). Quantitative renography in patient follow-up following treatment of renal artery stenosis. *Proceedings of the Royal Australian College of Physicians Golden Jubilee meeting*, **A**, 156.

Hilson, A.J., Maisey, M.N., Brown, C.B., Ogg, C.S., and Bewick, M.S. (1978). Dynamic renal transplant imaging with Tc-99m-DTPA (Sn) supplemented by a transplant perfusion index in the management of renal transplants. *Journal of Nuclear Medicine*, **19**, 994–1000.

Jafri, R.A. *et al.* (1988). 99m-Tc MAG3: a comparison with I-123 and I-131 orthoiodohippurate in patients with renal disorder. *Journal of Nuclear Medicine*, **29**, 147–58.

Merrick, M.V., Uttley, W.S., and Wild, S.R. (1980). Detection of pyelonephritic scarring in children by radioisotope imaging. *Br. Journal of Radiology*, **53**, 544–6.

Russell, C.D. *et al.* (1985). Measurement of glomerular filtration rate using 99mTc-DTPA and the gamma camera: a comparison of methods. *European Journal of Nuclear Medicine*, **10**, 519–21.

Wenting, G.J., Tan-Tjiong, H.L., Derkx, F.H.M., de Bruyn, J.H.B., Man in t'Veld, A.J., and Schalekamp, M.A.D.H. (1984). Split renal function after captopril in unilateral renal artery stenosis. *British Medical Journal*, **288**, 886–90.

Bibliography

Bischof-Delaloye, A and Blaufox, M.D. (1987). Radionuclides in Nephrology. *Contributions to Nephrology*, **86**, 77–81.

Blaufox, M.D. (1987). The current status of renal radiopharmaceuticals. *Contributions to Nephrology*, **56**, 31–7.

Dubovsky, E.V. and Russell, C.D. (1988). Radionuclide evaluation of renal transplants. *Seminars in Nuclear Medicine*, **18**, 181–98.

Fine, E.J. and Sarkar, S. (1989). Differential diagnosis and management of renovascular hypertension through nuclear medicine techniques. *Seminars in Nuclear Medicine*, **19**, 101–15.

Fogelman, I. and Maisey, M.N. (1988). *An atlas of clinical nuclear medicine*. Martin Dunitz, London.

Gordon, I. (1986). Use of TC-99m DMSA and Tc-99m DTPA in reflux. *Seminars in Urology*, **4**, 99–108.

Heyman, S. (1989). An update of radionuclide renal studies in paediatrics. In *Nuclear medicine annual*, p. 179.

O'Reilly, P.H., Shields, R.A., and Testa, H.J. (1987). In *Nuclear medicine in urology and nephrology*, (ed. H.J. Testa, R.A. Shields, and P.H. O'Reilly), Butterworths, London.

Sfakianakis, G.N. and Sfakianaki, E.D. (1988). Nuclear-medicine in paediatric urology and nephrology. *Journal of Nuclear Medicine*, **29**, 1287–300.

Part 3 Hypertension

12 The role of the kidney in the pathogenesis of hypertension

A.E.G. Raine

It has long been known that blood pressure and the kidney are intimately and mutually related, each being capable of adversely affecting the other. Hypertension is very common in patients with chronic renal failure, and occurs in 90 per cent or more of patients as end stage renal disease approaches (Raine and Ledingham 1990). Furthermore, hypertension may itself lead to renal impairment. Severe and often irreversible renal dysfunction is characteristic of malignant hypertension (Kincaid Smith *et al.* 1958). Benign hypertension, in contrast, is a rare primary cause of end stage renal disease in Europe (Broyer *et al.* 1986), though it is considerably more common in blacks in the USA (Rostand *et al.* 1989). It is also clear that in some acquired forms of hypertension other than chronic renal failure, such as renovascular hypertension, the kidney plays a central role.

However, whether or not the kidney is responsible for the development of essential hypertension remains a contentious issue. Cross-transplantation experiments have provided persuasive evidence of the primary role of the kidney in animal models of hypertension, yet have raised many new questions. Are the alterations in renal haemodynamics and sodium handling in human essential hypertension and in animal models a primary defect, causing hypertension, or are they purely a secondary adaptation to raised blood pressure? What is the evidence that membrane transport abnormalities demonstrated in blood cells of patients with hypertension are present in the kidney? If they are present, what role do they play in the pathogenesis of hypertension? Lastly, is insulin resistance a cause of abnormal sodium handling in hypertension?

Renal function in essential hypertension

In young subjects with borderline essential hypertension, renal blood flow has been reported to be normal or increased (Hollenberg *et al.* 1978; Bianchi *et al.* 1979). Once hypertension becomes established, renal blood flow is usually reduced, and renal vascular resistance is increased, a change consistent with the generalized increase in total peripheral resistance which is present in hypertension. The cause of the renal vasoconstriction remains disputed, though increased renal sympathetic tone may be a factor. In keeping with this possibility, de Leeuw and Birkenhager (1983) observed an inverse relationship between outer renal cortical blood flow and plasma noradrenaline concentration in hypertensive subjects.

Recent studies in a carefully-defined population (Van Hooft *et al.* 1991) have indicated that in young normotensive subjects with two hypertensive parents, and thus at risk for hypertension, renal vasoconstriction is increased, and renin and aldosterone secretion decreased. These findings are in contrast with earlier reports (Bianchi *et al.* 1979), and they are in keeping with the hypotheses of Hollenberg and coworkers (Hollenberg *et al.* 1986; Dluhy *et al.* 1988) that some normotensive offspring of hypertensive families are 'non-modulators', who have an impaired ability to reduce renal vascular tone in response to increases in dietary sodium intake and the resulting reduction in renal angiotensin II concentrations.

In established uncomplicated essential hypertension the glomerular filtration rate remains normal. Consequently, filtration fraction, the ratio of glomerular filtration rate to renal plasma flow, is increased. Micro-puncture studies in genetic models of hypertension, such as the spontaneously hypertensive rat, have indicated that these alterations in renal haemodynamics are achieved through an increase in resistance of both the afferent and efferent arterioles, so that glomerular capillary hydraulic pressure is maintained at a normal level (Arendhorst *et al.* 1979; Azar *et al.* 1979). Thus, in these models of spontaneous hypertension the kidney is 'protected' from elevated systemic arterial pressure, in contrast to animal models of diabetic nephropathy (Zatz *et al.* 1986) and renal ablation nephropathy (Anderson *et al.* 1986). In these models, an increase in systemic blood pressure is transmitted to the glomerulus and is associated with increased transcapillary hydraulic pressure and glomerular hyperfiltration. Brenner has argued that these latter renal haemodynamic changes are a major factor in progression of chronic renal disease (Brenner *et al.* 1982).

The rarity of development of significant renal impairment in patients with essential hypertension is consistent with these hypotheses. It implies that a balanced increase in afferent and efferent arteriolar resistance is present also in human hypertension, shielding the kidney from high systemic arterial pressure, while simultaneously enabling the maintenance of a near-normal glomerular filtration rate.

The apparent normality of renal function in essential hypertension has in some respects obscured the accumulation of evidence that the kidney plays a central part in the pathogenesis of spontaneous hypertension in animal models and in human hypertension. The most compelling proof so far that hypertension is linked with the kidney has come from cross-transplantation experiments.

Cross-transplantation studies in hypertension

Dahl and colleagues (Tobian *et al.* 1966) were the first to show that transplantation of a kidney from a salt-sensitive hypertensive rat strain into a normotensive rat could confer hypertension on the recipient. This concept was extended by

Bianchi *et al.* (1974), who demonstrated unequivocally that the kidney played a central role in promoting or suppressing development of hypertension in the Milan strain of genetically hypertensive rats. They showed that transplantation of a kidney from a hypertensive donor into a normotensive recipient resulted in development of hypertension within 2–3 weeks. Transplantation of a kidney from a young, pre-hypertensive rat into a normotensive animal had exactly the same effect, whereas receipt of a kidney from a normotensive rat prevented the increase in blood pressure in young animals belonging to the hypertensive strain.

Experiments such as these provide powerful evidence that the kidney itself contains the message conferring development of hypertension. Similar observations have been subsequently made in other models of hypertension, confirming the generality of the observations of Dahl and Bianchi. For example, in the Okamoto spontaneously hypertensive rat (SHR), in which development of hypertension is independent of salt intake, donation of a kidney from a normotensive (WKY) control to an SHR/WKY F1 hybrid caused blood pressure to fall. In contrast, transplantation of an SHR kidney into an F1 hybrid produced an increase in blood pressure (Kawabe *et al.* 1978). More recent studies utilizing SHR have extended these observations. Transplantation from 2-week old stroke-prone SHR (systolic blood pressure 186 mmHg) to SHR/WKYF1 hybrids caused the systolic blood pressure to increase from under 140 to nearly 240 mmHg over 8 weeks (Rettig *et al.* 1989). In further studies, blood pressure of young SHR was prevented from increasing by administration of the converting enzyme inhibitor ramipril. Despite this, subsequent kidney donation to an F1 hybrid still resulted in a 50 mmHg increase in systolic pressure (Rettig *et al.* 1990).

The conclusion from these transplantation studies in a number of different genetic models of hypertension is that the presence or absence of hypertension may be transmitted with the kidney. The relevance of these models to human essential hypertension remains a matter of debate, nevertheless. It is difficult to obtain evidence in humans to confirm or refute the primary role of the kidney, but observations in renal transplant recipients made by Curtis *et al.* (1983) do offer a counterpart to the experimental studies. These workers studied six black patients with end stage renal failure believed to be due to hypertensive nephrosclerosis, who received cadaveric renal allografts from a normotensive donor. The recipients' blood pressure prior to transplantation was 201/133 mmHg. After a mean follow-up period of 4.5 years, the blood pressure in all six recipients was normal (mean blood pressure 92 mmHg), and their left ventricular hypertrophy and hypertensive retinopathy had resolved.

Bianchi and colleagues have performed an analysis of the requirements for anti-hypertensive therapy in renal transplant recipients. This showed that when a recipient from a family with no history of hypertension received a kidney from a donor who was normotensive, but with a family history of hypertension, mean blood pressure was increased and requirement for therapy was doubled, in comparison with donors who had no such family history (Guidi *et al.* 1985).

These observations in animal models of hypertension and in patients undergoing renal transplantation support the possibility that an abnormality of some aspect of renal function may both initiate and maintain hypertension. The nature of this abnormality remains unknown, as does its presumed genetic basis. Most attempts so far to approach this question have concerned the role of the kidney in sodium and fluid homoeostasis.

Pressure natriuresis

An increase in renal perfusion pressure results in an increase in renal sodium excretion, the phenomenon of pressure natriuresis (Firth *et al.* 1990). Guyton has emphasized the fundamental importance of the renal sodium excretion mechanism in long-term blood pressure control (Guyton *et al.* 1972), pointing out that unless the pressure natriuresis mechanism were reset in established hypertension a state of perpetual negative sodium balance and volume depletion would ensue.

Many studies have confirmed that resetting of pressure natriuresis does occur when blood pressure is persistently increased, so that a higher perfusion pressure is required to excrete a given sodium load. Moreover, this resetting appears to be very precise, and is demonstrable in the isolated kidney, devoid of neural or humoral influence. Studies employing the isolated perfused kidneys of SHR and WKY controls showed that, as anticipated, the relationship between perfusion pressure and sodium excretion was shifted markedly to the right in SHR (Fig. 12.1). In addition, sodium excretion of the isolated SHR

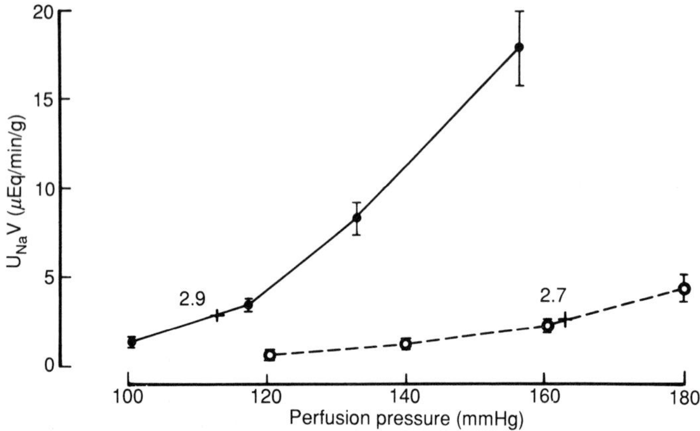

Fig. 12.1 Effect of increasing renal perfusion pressure on urinary sodium excretion of isolated kidneys from Okamoto spontaneously hypertensive rats (○) and age-matched Wistar-Kyoto controls (●). Crosses show interpolated sodium excretion corresponding to mean blood pressure *in vivo* (163 mmHg, SHR; 113 mmHg, WKY).

and WKY kidneys was virtually identical at the *in vivo* mean blood pressure previously determined for each strain (Raine *et al.* 1984). Thus, whatever the cause of pressure natriuresis resetting in this model of hypertension, it is demonstrable in vitro as an intrinsic adaptation of the kidney.

A second, more fundamental question is whether pressure natriuresis resetting is purely a secondary response, part of a renal adaption to maintain sodium balance in the face of increased systemic blood pressure, or whether it is a primary cause of hypertension. According to the latter hypothesis, a primary impairment of pressure natriuresis would lead to sodium retention, and the consequent volume expansion would increase blood pressure. This in turn would increase sodium excretion, and blood pressure would become re-established at the new, hypertensive, level which enabled restoration of sodium balance. These two alternative explanations of pressure natriuresis resetting are summarized in Fig. 12.2.

Guyton and Hall have argued that pressure natriuresis impairment may be a primary mechanism in hypertension, (Guyton 1987; Hall *et al.* 1990), although in practice it has proved very difficult to establish whether or not this is true. One approach is to study renal sodium handling in young pre-hypertensive animal models, and here the results obtained have been conflicting. While two studies have suggested that in the Dahl salt-sensitive hypertensive rat there is a rightward shift of the pressure natriuresis relationship before the development of hypertension (Tobian *et al.* 1978; Roman 1986), a third study (Girardin *et al.* 1980) showed that pressure natriuresis resetting was established in Dahl rats after 7 weeks of high-salt diet, when hypertension had

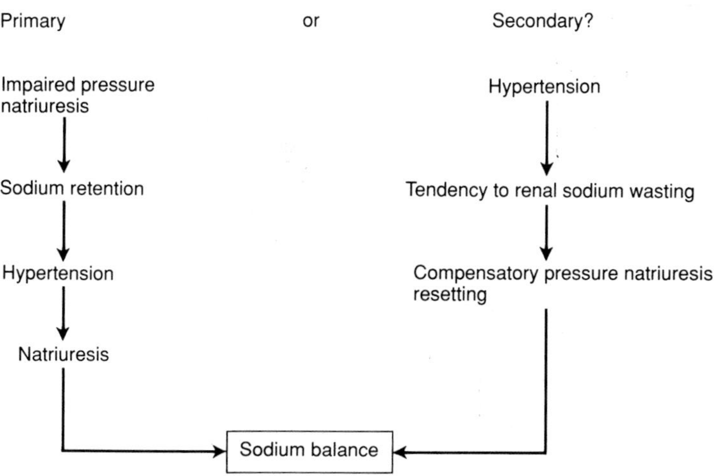

Fig. 12.2 Alternative explanations for the mechanism of resetting of pressure natriuresis in hypertension.

developed, but not after 3 weeks of high-salt diet, at a time when blood pressure was still normal.

A number of factors may alter the pressure natriuresis relationship. These include dietary sodium intake, an increase in which causes a leftward shift of the relationship (Harrap *et al.* 1988), and angiotension II (Hall *et al.* 1980) and aldosterone (Harrap *et al.* 1988). Chronic infusion of either hormone results in increased renal sodium reabsorption, with a rightward shift of pressure natriuresis demonstrable in the isolated kidney, and increased blood pressure. Renal haemodynamic changes, especially afferent or efferent arteriolar vasoconstriction may also blunten the sensitivity of pressure natriuresis (Guyton 1987). Hormonal systems which are primarily natriuretic may have the opposite effect. Atrial natriuretic factor, for example, causes an increase in sensitivity (left shift) of the pressure natriuresis relationship in the isolated perfused rat kidney (Fig. 12.3).

Thus, many of the neurohumoral and haemodynamic changes postulated to be implicated in the pathogenesis or maintenance of hypertension may themselves markedly affect renal sodium handling, and could account for the resetting of pressure natriuresis which occurs during the development of hyper-

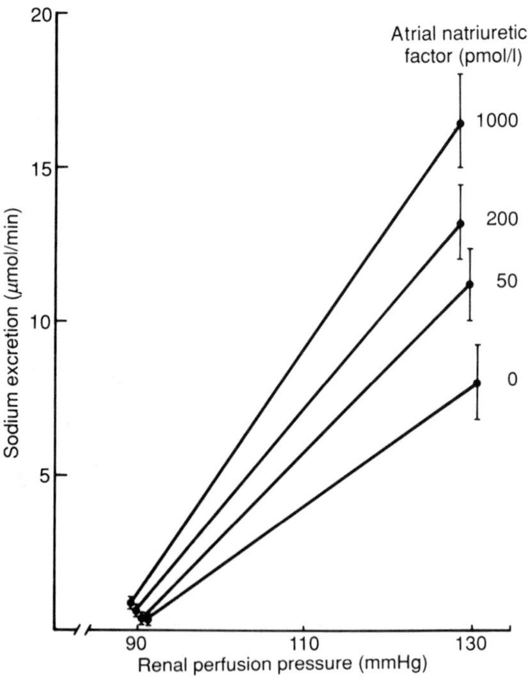

Fig. 12.3 Effect of increase in perfusion pressure on sodium excretion of isolated rat kidneys in response to atrial natriuretic factor. (Modified from Firth *et al.* 1988.)

tension. In established hypertension, it is clear that the kidney has adapted fully to maintain sodium balance at a higher perfusion pressure, and this adaptation is demonstrable *in vitro*. It is mediated primarily by increased proximal tubular reabsorption (Firth *et al.* 1989). However, whether altered pressure-natriuresis in hypertension is in part the expression of a 'fundamental' renal abnormality, or whether it is purely a renal response to primary abnormalities of hormonal control of sodium balance remains unknown.

Blood pressure, plasma volume and sodium excretion

Many of the neurohumoral systems which modify the pressure natriuresis relationship, such as angiotensin II, aldosterone, catecholamines and atrial natriuretic factor, are activated in response to changes in intravascular volume. It must be emphasized that increase in plasma volume is a major stimulus to renal sodium excretion, independent of changes in arterial pressure (de Wardener *et al.* 1961). The natriuresis is mediated by suppression of the renin-angiotensin-aldosterone system, suppression of antidiuretic hormone, withdrawal of renal sympathetic tone, increased cardiac release of atrial natriuretic factor, and possibly enhanced release of ouabain-like natriuretic hormone (see Chapter 13). The sensitivity of these volume-responsive mechanisms explains an important difference. In both the isolated kidney (Fig. 12.1) and in humans *in vivo* (Omvik *et al.* 1980), the relationship between changes in renal perfusion pressure and change in sodium excretion has appreciable slope. In contrast, in animal studies *in vivo* and in humans, sodium loading results in increased natriuresis with little or no increase in systemic blood pressure, and the 'pressure natriuresis' relationship is virtually vertical (Guyton 1987).

As a corollary, it is possible in humans to define the pressure natriuresis relationship by acute reduction of blood pressure with vasodilators such as nitroprusside (Omvik *et al.* 1980), but studies which employ sodium loading in humans are in general documenting volume natriuresis rather than pressure natriuresis. Even so, it is possible they may induce subtle changes in renal perfusion pressure, while systemic blood pressure remains unchanged. Thus, in practice, changes in both arterial pressure and volume status interact closely to determine ultimate sodium excretion; as Omvik *et al.* (1980) emphasized, a given increase in arterial pressure will cause greatly accentuated sodium excretion when intra-vascular volume is expanded. The enhanced natriuresis may be explained by the potentiating effect of the neurohumoral changes, such as an increase in atrial natriuretic factor, outlined above.

In summary, abnormal pressure natriuresis could, in theory, be a cause of hypertension, and in certain models such as chronic infusion of angiotensin II (Hall *et al.* 1980), this mechanism appears to operate. However, evidence that a shift in pressure natriuresis precedes the development of hypertension is conflicting in salt-sensitive Dahl models, and is not available for other models such as the SHR or in human hypertension. Pressure natriuresis is always reset in any form of established hypertension. Once resetting has occurred, it will act to maintain hypertension, as a fall in blood pressure will result in positive sodium balance. A corollary is that drugs acting to lower blood pressure may cause initial sodium retention, unless the agent in question also directly promotes renal sodium excretion.

Membrane transport abnormalities and the kidney in hypertension

Knowledge from cross-transplantation studies that hypertension is transferred with the kidney has emphasized the need to define the renal abnormality responsible. In the past decade a number of abnormalities of membrane ion transport in hypertension have been described, and it has been tempting to speculate that, if present in the kidney, these abnormalities could result in impaired renal sodium excretion. This, it is predicted, would cause plasma volume expansion, which could then result in hypertension, either as a direct result of the pressure natriuresis mechanism acting to let blood pressure rise until sodium balance is restored, or through an additional mechanism. De Wardener and MacGregor (1980) postulated, for example, that volume expansion secondary to renal sodium retention resulted in release of a circulating $Na^{+}K^{+}$ATPase sodium transport inhibitor, which restored sodium balance but also caused vasoconstriction and hypertension.

A difficulty with all these hypotheses, which is as yet unanswered, is that increased extracellular volume resulting from impaired renal sodium excretion is an essential causal link in the development of hypertension. However, the evidence for volume expansion is conflicting, and often lacking (De Wardener (1990)). Exchangeable sodium is not increased in young and adult Milan hypertensive rats, and plasma volume but not extracellular fluid volume is increased in Dahl salt-sensitive rats. In SHR there are reports of both increased (Mullins 1983) and decreased (Rippe *et al.* 1978) plasma volume in young SHR, at the time hypertension is developing. In established human essential hypertension both plasma and total blood volume are normal or low, although central blood volume is increased, with a resulting increase in central venous pressure (Tarazi 1983). There is little information on whether plasma and extracellular volume are abnormal in young pre-hypertensive humans, and such studies pose practical difficulties.

It may be argued that in animal models of hypertension and in humans any tendency to volume expansion is immediately counteracted by activation of neurohumoral volume defence mechanisms, or by an increase in blood pressure, resulting in volume or pressure natriuresis. Nevertheless, the lack of unequivocal demonstration of a phase of volume expansion in early hypertension emphasizes the need for caution when assessing current hypotheses which invoke inherited or acquired impairment of renal sodium excretion as a primary cause of raised blood pressure.

Several abnormalities of membrane sodium transport have been described in the past decade in blood cells of subjects with essential hypertension. These include a reduction in leucocyte Na^+K^+ATPase activity, a reduction in erythrocyte sodium–potassium cotransport and an increase in sodium–lithium countertransport in erythrocytes (Hilton 1986). This latter abnormality has been of particular interest, as sodium–lithium countertransport is also increased in first-degree relatives of hypertensives (Canessa *et al.* 1980) and activity of this membrane transport process has a major genetic component (Motulsky *et al.* 1987). Mahnensmith and Aronson (1985) postulated that sodium–lithium countertransport may also mark activity of the sodium–hydrogen antiporter, and thus the finding of Weder (1986) that renal lithium clearance was reduced in subjects with essential hypertension aroused great interest. It suggested that an inherited predisposition to enhanced proximal tubular sodium–hydrogen exchange and sodium reabsorption might represent a primary genetic membrane transport defect, whose expression in the kidney could account for sodium retention and hence development of hypertension. Consistent with this, proximal tubular sodium reabsorption is increased in young, six-week old SHR (Dilley *et al.* 1984) and recent studies have demonstrated that sodium–hydrogen exchange is increased by 60 per cent in isolated brush border vesicles of SHR proximal tubules (Morduchowicz *et al.* 1989).

Attractive as these hypotheses are, conflicting evidence has now arisen in relation to the role of the sodium–hydrogen antiporter. Further studies have failed to confirm reduction in renal lithium clearance in patients with essential hypertension (Hla-Yee-Yee *et al.* 1987) or their offspring (Stadler *et al.* 1987). It is also very unlikely that the sodium–lithium exchanger present in red blood cells and sodium–hydrogen antiporter in the proximal tubule are one and the same (Postnov *et al.* 1988). Lastly, family studies in essential hypertension have now shown conclusively that there is no linkage of the sodium–hydrogen antiporter gene to presence of hypertension (Lifton *et al.* 1991; Dudley *et al.* 1991).

A great number of other candidate genes expressed in the kidney and elsewhere may be relevant to the pathogenesis of hypertension, and many of these are being actively investigated. The recent demonstration (Mullins *et al.* 1990) that introduction of the mouse Ren-2 gene into transgenic rats results in fulminant hypertension, yet with suppressed plasma and renal renin activity and no demonstrable abnormality of renal function, emphasises that experimental genetic hypertension may be 'transplanted' without any apparent involvement of the kidney.

Insulin resistance and renal sodium reabsorption

A further hypothesis of current interest is that hyperinsulinaemia and insulin resistance present in patients with essential hypertension (Ferrannini *et al.* 1987) might be a cause of renal sodium retention. Insulin undoubtedly may augment renal sodium reabsorption. For example, in a euglycaemic clamp study, de Fronzo *et al.* (1975) infused exogenous insulin for two hours in normal subjects to achieve an increase in plasma insulin concentration from 12–149 mU/ml and observed a fall in sodium excretion from 400 to 200 mEq/min. Thus a 5 mU/ml increase in plasma insulin might be approximated to a 10 mmol/24 h fall in sodium excretion. *In vitro* studies employing perfusion of isolated rabbit proximal tubules with insulin have shown that the threshold for enhanced fluid reabsorption is 14 mU/ml (Baum 1987). While basal plasma insulin in subjects with essential hypertension and controls is equal (8 mU/ml, insulin resistance in the hypertensive subjects is typified by an enhanced increase in plasma insulin, 5–20 mU/ml more than controls, in response to a glucose load (Ferrannini *et al.* 1987).

When these observations are considered quantitatively, it is conceivable that intermittent hyperinsulinaemia in hypertensive subjects might cause a mild increase in renal sodium retention. However, it seems unlikely to be the sole explanation of the rather marked enhancement of proximal tubular reabsorption observed in *in vitro* studies (Morduchowicz *et al.* 1989). For example, long-term insulin infusion in dogs leads to transient sodium retention only, and no change in blood pressure (Hall *et al.* 1989). Moreover, it is difficult to see how an external stimulus, such as hyperinsulinaemia, could cause a renal sodium-retaining defect which is transmissable by cross-transplantation. The relationship between insulin resistance and hypertension appears also to be confined to whites, and is not demonstrable in Pima Indians or blacks (Saad *et al.* 1991).

Conclusions

The evidence is compelling that, if a primary cause of essential hypertension exists, it resides in the kidneys. The observations that cross-transplantation of kidneys of hypertensive and normotensive rat strains can induce or reverse hypertension are difficult to refute, and are supported, however imprecisely, by clinical observations in renal transplant recipients. Guyton has also argued persuasively that altered renal sodium handling, and in particular a resetting of pressure natriuresis is mandatory for the maintenance of hypertension.

The precise nature of the fundamental renal abnormality in question remains tantalizingly elusive. So, too, does the basis of the enhanced renal sodium reabsorption which characterizes the resetting of pressure natriuresis in hypertension. It remains unknown whether the resetting of the relationship between perfusion pressure and sodium excretion precedes development of hypertension, and is a cause of it, or whether it is merely a homoeostatic adjustment necessary to maintain sodium balance in a kidney exposed to systemic hypertension.

Lastly, the role of membrane sodium transport abnormalities in determining altered renal sodium handling in hypertension raises many questions, with few clear answers. The ability of techniques of molecular biology to define specific transporter proteins and to identify genes responsible for their inheritance should allow both the possible pathophysiological role of candidate sodium transporting systems and their association with the hypertension phenotype to be examined, and confirmed or refuted.

References

Anderson, S., Rennke, H.G., and Brenner, B.M. (1986). Therapeutic advantage of converting enzyme inhibitors in arresting progressive renal disease associated with systemic hypertension in the rat. *Journal of Clinical Investigation*, **77**, 1993–7.

Arendhorst, W.J. and Beierwaltes W.H. (1979). Renal and nephron hemodynamics in spontaneously hypertensive rats. *American Journal of Physiology*, **236**, F246.

Azar, S., Johnson, M.A., Scheinman, J. *et al.* (1979). Regulation of glomerular capillary pressure and filtration rate in young Kyoto hypertensive rats. *Clinical Science*, **56**, 203.

Baum, M. (1987). Insulin stimulates volume absorption in the rabbit proximal convoluted tubule. *Journal of Clinical Investigation*, **79**, 1104–9.

Bianchi, G., Fox, U., Di Francesco, G.F., Giovanetti, A.M., and Pagetti, D. (1974). Blood pressure changes produced by kidney cross-transplantation between spontaneously hypertensive rats and normotensive rats. *Clinical Science*, **47**, 435–48.

Bianchi, G., Cusi, D., Gatti, M., *et al.* (1979). A renal abnormality as a possible cause of 'essential' hypertension. *Lancet*, **1**, 173–7.

Brenner, B.M., Meyer, T.W., and Hostetter T.H. (1982). Dietary protein intake and the progressive nature of kidney disease: the role of hemodynamically mediated glomerular injury in the pathogenesis of progressive glomerular sclerosis in aging, renal ablation, and intrinsic renal disease. *New England Journal of Medicine*, **307**, 652.

Broyer, M., Brunner, F.P., Brynger, H. *et al.* (1986). Demography of dialysis and transplantation in Europe, (1984). *Nephrology Dialysis and Transplantation*, **1**, 1–8.

Canessa, M., Adragna, N., Solomon, H.S., Connolly, T.M., and Tosteston, D.C. (1980). Increased sodium-lithium countertransport in red cells of patients with essential hypertension. *New England Journal of Medicine*, **302**, 772.

Curtis, J.J., Luke, R.G., Dustan, H.P. *et al.* (1983). Remission of essential hypertension after renal transplantation. *New England Journal of Medicine*, **309**, 1009.

De Fronzo, R.A., Cooke, C.R., Andres, R., Faloona, G.R., Davis, P.J. (1975). The effect of insulin on renal handling of sodium, potassium, calcium, and phosphate in man. *Journal of Clinical Investigation*, **55**, 845–55.

De Leeuw, P.W. and Birkenhager, W.H. (1983). The renal circulation in essential hypertension (Editorial Review). *Journal of Hypertension*, **1**, 321–31.

De Wardener, H.E., Mills, I.H., Clapham, W.F. and Hayter, C.J. (1961). Studies on the efferent mechanism of the sodium diuresis which follows the administration of intravenous saline in the dog. *Clinical Science*, **21**, 249–258.

De Wardener, H.E. and MacGregor, G.A. (1980). Dahl's hypothesis that a saluretic substance may be responsible for a sustained rise in arterial pressure. Its possible role in essential hypertension. *Kidney International*, **18**, 1–9.

De Wardener, H.E. (1990). The primary role of the kidney and salt intake in the aetiology of essential hypertension. *Clinical Science*, **79**, 193–200.

Dilley, J.R., Stier, C.T., and Arendshorst, W.J. (1984). Abnormalities in glomerular function in rats developing spontaneous hypertension. *American Journal of Physiology*, **246**, F12–20.

Dluhy, R.G., Hopkins, P., Hollenberg, N.K., Williams, G.H., and Williams, R.R. (1988). Heritable abnormalities of the renin-angiotensin-aldosterone system in essential hypertension. *Journal of Cardiovascular Pharmacology*, **12**, (3) 149–54.

Dudley, C.R.K., Giuffra, L.A., Raine, A.E.G. and Reeders, S.T. (1991). Assessing the role of APNH, a gene encoding for a human amiloride-sensitive Na^+/H^+ antiporter, on the interindividual variation in red cell Na^+/Li^+ countertransport. *Journal of the American Society of Nephrology*. In press.

Ferrannini, E., Buzzigoli, G., Bonadonna, R., *et al.* (1987). Insulin resistance in essential hypertension. *New England Journal of Medicine*, **317**, 350–7.

Firth, J.D., Raine, A.E.G., and Ledingham, J.G.G. (1988). Low concentrations of ANP cause pressure-dependent natriuresis in the isolated kidney. *American Journal of Physiology*, **255**, F391–6.

Firth, J.D., Raine, A.E.G., and Ledingham, J.G.G. (1989). Sodium and lithium handling in the isolated hypertensive rat kidney. *Clinical Science*, **67**, 335–41.

Firth, J.D., Raine, A.E.G., and Ledingham, J.G.G. (1990). The mechanism of pressure-natriuresis. *Journal of Hypertension*, **8**, 97–104.

Girardin, E., Caverzasio, J., Iwai, J., Bonjour, J.P., Muller, A., and Grandchamp, A. (1980). Pressure natriuresis in isolated kidneys from hypertension-prone and hypertension-resistant rats (Dahl rats). *Kidney International*, **18**, 10–19.

Guidi, E., Bianchi, G., Rivolta, E. *et al.* (1985). Hypertension in man with a kidney transplant: role of familial versus other factors. *Nephron*, **41**, 14–21.

Guyton, A.C., Coleman, T.G., Cowley, A.W. *et al.* (1972). Arterial pressure regulation; over-riding dominance of the kidneys in long-term regulation and in hypertension. *American Journal of Medicine*, **52**, 584.

Guyton, A.C. (1987). Renal function curve—a key to understanding the pathogenesis of hypertension. *Hypertension*, **10**, 1–6.

Hall, J.E., Guyton, A.C., Smith, M.J.J. Jr., and Coleman T.G. (1980). Blood pressure and renal function during chronic changes in sodium intake: Role of angiotensin. *American Journal of Physiology*, **239**, F271–80.

Hall, J.E., Coleman, T.G., and Mizelle, H.L. (1989). Does chronic hyperinsulinaemia cause hypertension? *American Journal of Hypertension*, **2**, 171–3.

Hall, J.E., Mizelle, H.L., Hildebrandt, D.A., and Brands, M.W. (1990). Abnormal pressure natriuresis; a cause or a consequence of hypertension? *Hypertension*, **15**, 547–59.

Harrap, S.B., Clark, S.A., Fraser, R., Towrie, A., Browne, A.J., and Lever A.F. (1988). Effects of sodium intake and aldosterone on the renal pressure-natriuresis. *American Journal of Physiology*, **254**, F697–703.

Hilton, P.J. (1986). Cellular sodium transport in essential hypertension. *New England Journal of Medicine*, **314**, 222.

Hla-Yee-Yee, Shirley, D.G., Singer, D.R.S., Markandu, N.D., Jones, B.E. and Macgregor, G.A. (1987). Is renal lithium clearance altered in essential hypertension. *Journal of Hypertension*, **7**, 955–60.

Hollenberg, N.K., Borucki, L.J., and Adams, D.F. (1978). The renal vasculature in early essential hypertension; evidence for a pathogenetic role. *Medicine* (*Baltimore*), **57**, 167–78.

Hollenberg, N.K., Moore, T., Shoback, D., Redgrave, J., Rabinowe, S., and Williams, G.H. (1986). Abnormal renal sodium handling in essential hypertension: relation to failure of renal and adrenal modulation of responses to angiotensin II. *American Journal of Medicine*, **81**, 412–8.

Kawabe, K., Watanabe, T.X., Shionos, K., and Sokabe, H. (1979). Influence on blood pressure of renal isografts between spontaneously hypertensive and normotensive rats, utilizing the F_1 hybrids. *Jpn. Heart Journal*, **20**, 886–94.

Kincaid-Smith, P., McMichael, J. and Murphy, E.A. (1958). Clinical course and pathology of hypertension with papilloedema. *Quarterly Journal of Medicine*, **37**, 117–52.

Lifton, R.P., Hunt, S.C., Williams, R.R., Pouyssegur, J. amd Lalouel, J.M. (1991). Exclusion of the Na^+-H^+ antiporter as a candidate gene in human essential hypertension. *Hypertension*, **17**, 8–14.

Mahnensmith, R.L. and Aronson, P.S. (1985). The plasma membrane sodium-hydrogen exchanger and its role in physiological and pathophysoological processes. *Circulation Research*, **57**, 773–88.

Morduchowicz, G.A., Sheikh-Hamad, D., Ok, D.J., Nord, E.P., Lee, D.B.N., and Yahagawa, N. (1989). Increased Na/H antiport activity in the renal brush border membrane of SHR. *Kidney International*, **36**, 576–81.

Motulsky, A.G., Burke, W., Billings, P.R., *et al.* (1987). Hypertension and the genetics of red cell membrane abnormalities. In *Molecular approaches to human polygenic disease*, pp. 150–66. Wiley, Chichester (Ciba Foundation Symposium 130).

Mullins, M.M. (1983). Body fluid volumes in pre-hypertensive spontaneously hypertensive rats. *American Journal of Physiology*, **244**, H652–5.

Mullins, J.J., Peters, J., and Ganten, D. (1990). Fulminant hypertension in transgenic rats harbouring the mouse Ren-2 gene. *Nature*, **344**, 541–4.

Omvik, P., Tarazi, R.C., and Bravo, E.L. (1980). Regulation of sodium balance in hypertension. *Hypertension*, **2**, 515–23.

Postnov, Y.V., Kravtsov, G., Orlov, S. *et al.* (1988). Effect of protein kinase C activation on cytoskeleton and cation transport in human erythrocytes. *Hypertension*, **12**, 267–73.

Raine, A.E.G., Roberts, A.F.C., and Ledingham, J.G.G. (1984). Resetting of pressure-natriuresis and frusemide sensitivity in spontabeously hypertensive rats. *Journal of Hypertension*, **2**, (3), 359–61.

Raine, A.E.G., and Ledingham, J.G.G. (1990). Treatment of hypertension in renal disease. In *Handbook of hypertension*, vol. 13, *The management of hypertension*. (eds. F.R. Buhler and J.H. Laragh). pp. 466–82. Elsevier, Amsterdam.

Rettig, R., Stauss, H., Folberth, C., Ganten, D., Waldherr, R., and Unger, T. (1989). Hypertension transmitted by kidneys from stroke-prone spontaneously hypertensive rats. *American Journal of Physiology*, **257**, F197–203.

Rettig, R., Folberth, C., Stauss, H., Kopf, D., Waldherr, R., and Unger, T. (1990). Role of the kidney in primary hypertension. A renal transplantation study in rats. *American Journal of Physiology*, **258**, F606–11.

Rippe, B., Lundin, S., and Folkow, B. (1978). Plasma volume, blood volume and transcapillary escape rate (TER) of albumin in young spontaneously hypertensive rats (SHR) as compared with normal controls (NCR). *Clinical and Experimental Hypertension*. Part A. **1**, 39–50.

Roman, R.J. (1986). Abnormal renal hemodynamics and pressure-natriuresis relationship in Dahl salt-sensitive rats. *American Journal of Physiology*, **251**, F57–65.

Rostand, S.G., Brown, G., Kirk, K.A., Rutsky, E.A., Dustan, H.P. (1989). Renal insufficiency in treated essential hypertension. *New England Journal of Medicine*, **320**, 684.

Saad, M.F., Lillioha, S., Nyomba, B.L. *et al.* (1991). Racial differences in the relation between blood pressure and insulin resistance. *New England Journal of Medicine*, **324**, 733–9.

Stadler, P., Pusterla, C., and Beretta-Piccoli, C. (1987). Renal tubular handling of sodium and familial pre-disposition to essential hypertension. *Journal of Hypertension*, **5**, 727–32.

Tarazi, R.C. (1984). The hemodynamics of hypertension. In *Hypertension, physiopathology and treatment*. (ed. J. Genest, O. Kuchel, P. Hamet and M. Cantin) pp. 17–18. McGraw-Hill, New York.

Tobian, L., Coffee, K., McCrea, P. and Dahl, L. (1966). A comparison of the antihypertensive potency of kidneys from one strain of rats susceptible to salt hypertension and kidneys from another strain resistant to it. *Journal of Clinical Investigation*, **45**, 1080.

Tobian, L., Lange, J., Azar, S. *et al.* (1978). Reduction of natriuretic capacity and renin release in isolated, blood-perfused kidneys of Dahl hypertension-prone rats. *Circ. Res.* **43**, (1), 192–8.

Van Hooft, I,M.S., Grobbee, D.E. Derkx, F.H.M., de Leeuw, P.W., Schalekamp, M.A.D.H. and Hofman, A.H. (1991). Renal haemodynamics and the renin-angiotensin-aldosterone system in normotensive subjects with hypertensive and normotensive parents. *New England Journal of Medicine*, **324**, 1305–11.

Weder, A.B. (1986). Red-cell lithium-sodium countertransport and renal lithium clearance in hypertension. *New England Journal of Medicine*, **314**, 198–201.

Zatz, R., Dunn, B.R., Meyer, T.W., Anderson, S., Rennke, H.G., and Brenner, B.M. (1986). Prevention of diabetic glomerulopathy by pharmacological amelioration of glomerular capillary hypertension. *Journal of Clinical Investigation*, **77**, 1925.

13 The natriuretic hormone and hypertension

L. Poston

In 1961 De Wardener and colleagues suggested that a natriuretic hormone was involved in the control of sodium balance.[1] Subsequently, a role for such a hormone has been implicated in the pathophysiology of chronic renal failure and essential hypertension. The hypothesis involving a natriuretic hormone with the causation of essential hypertension was first suggested more than ten years ago.[2,3] The theory, summarized in Fig. 13.1, encompasses a primary role for the kidney in the aetiology of essential hypertension, and also a relationship between sodium balance and the hypothesized natriuretic hormone. The final pathway, relating the presence of a natriuretic hormone to an increase in peripheral vascular tone (and so to hypertension), rests on the assumption that the natriuretic hormone would inhibit the sodium pump in both kidney and vasculature, the latter leading to an increase in vascular smooth muscle constriction. It is beyond the remit of this short chapter to present detailed evidence for all the limbs of this hypothesis.

Historical background

Full appreciation of this theory is best achieved by summarizing the chronological events leading to the evidence for a natriuretic hormone and its role in the control of sodium balance. The awareness that a 'third factor', other than glomerular filtration rate and aldosterone, might be involved in sodium balance originated in the 1940s. It was found that a reduced glomerular filtration rate induced in rats by hypophysectomy did not lead to sodium retention despite the presence of exogenously administered mineralocorticoids. Furthermore, there was recognition that patients with renal failure, and no fall in plasma aldosterone, were able to maintain normal sodium balance despite an extremely low glomerular filtration rate.

Mineralocorticoid 'escape', first described in the 1950s, also provided indirect support for the presence of a third factor. This now well-recognized phenomenon occurs when a normal subject is given a mineralocorticoid over several days. Initially, as expected, this leads to profound sodium retention. After a few days, however, 'escape' from sodium retention occurs and sodium excretion returns to normal and the subject to sodium balance. This happens despite no significant change in the glomerular filtration rate.

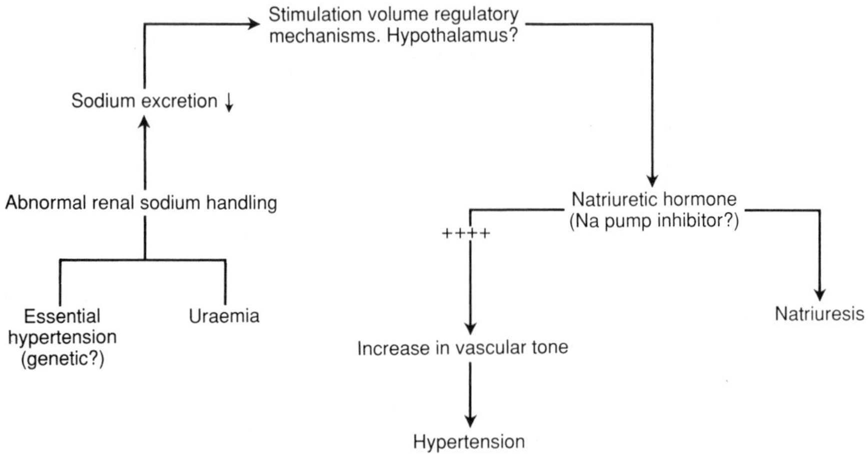

Fig. 13.1 The natriuretic hormone and hypertension.

In order to provide more direct evidence for a 'third factor', De Wardener and colleagues designed experiments which clearly demonstrated that a natriuresis occurs in response to volume expansion without a change in glomerular filtration rate and when mineralocorticoids and vasopressin were controlled artificially. The classic experiment, often described, involved the cross-circulation of two dogs via the femoral arteries and veins.[1] By carefully maintaining the blood volume of one animal, whilst expanding that of the other with saline, it was found that a natriuresis occurred in the non-expanded animal, thus suggesting the existence of a humoral mediator of natriuresis. Further studies which avoided the dilutional effect of infused saline, by using a reservoir of equilibrated blood, substantiated de Wardener's early work. There is a large literature on this topic, reviewed comprehensively by De Wardener and Clarkson.[4]

The demonstration of a humorally mediated natriuresis led to an extensive search for the natriuretic hormone. In summary, the natriuresis of sodium loading was found to be associated with the appearance of an inhibitor of active sodium transport in the serum and urine. The methods of extraction and characterization will be considered in more detail after a discussion of the possible role of an endogenous natriuretic hormone in renal failure and essential hypertension.

Renal failure

Patients with a reduced nephron population maintain sodium balance by increasing the fraction of filtered sodium excreted (FeNa) by each remaining functional nephron. The possibility that this results from the same factor as that involved in the natriuresis of volume expansion provided the basis of several studies, particularly as it appeared that the raised FeNa of uraemia was not the result of hyperfiltration or of abnormal nephron structure. Data were produced indicating that the increase in FeNa in uraemia was determined by salt intake and was not due to the accumulation of toxic waste products. These experiments showed that in animals with a similar degree of uraemia, a high salt intake was associated with a much greater FeNa, compared with those whose diet included little salt.[5]

Further experiments demonstrated that infusion of serum and urine extracts from uraemic patients into unilaterally nephrectomized animals led to an increase in the FeNa of the remnant kidney.[6] It also became apparent that the erythrocytes and leucocytes of patients with chronic renal failure demonstrate abnormal sodium transport and high intracellular sodium, and that the defect could be conferred on normal cells by incubation in uraemic serum (for a recent review see Kaji *et al.*[7]). This evidence therefore favoured, but did not prove, a role for the proposed natriuretic hormone in renal failure.

Essential hypertension

Until ten years ago the presumed natriuretic hormone was considered only to play a role in sodium balance and in uraemia. It now attracts more interest in association with hypertension. Strictly speaking it is the association of hypertension with an inhibitor of sodium transport rather than a natriuretic compound *per se* which has provided the stimulus for a large number of studies. Several observations led to this association. The first of these, by Tobian and Binion,[8] found that intracellular sodium was elevated in the vasculature of patients with essential hypertension and was followed by a number of experiments in which the electrolyte content and sodium transport rates were quantified in more easily accessible cells. In 1975, Edmondson *et al.* observed that active sodium transport was reduced in the peripheral blood leucocytes of patients with essential hypertension, a finding since repeated in many reports, but with equivocal results in erythrocytes (for review see Hilton 1986[9]). Studies in animals also demonstrated abnormal sodium pump activity in the vasculature of rats with low renin type experimental hypertension (G1K1C).[10]

Together with a number of observations that the plasma of patients with essential hypertension demonstrated enhanced pressor activity this evidence led to the elaboration of the hypothesis shown in Fig. 13.1. The pivotal role of the kidney in the scheme has gained considerable support since the hypothesis was first suggested. Experiments in hypertensive strains of rat have suggested that hypertension follows the kidney (for a recent review see De Wardener *et al.* 1990[11]). Cross transplantation of kidneys from the hypertensive rat to the relevant control rat have led to the development of hypertension in the control animal. This has been established in four strains of hypertensive rats.

Although obviously less easy to demonstrate in humans, there are indications that in essential hypertension blood pressure follows the kidney. The blood pressure of the recipients of cadaver kidneys has been found to relate to the blood pressure of the donor or the donor's parents. In another study, the blood pressure of patients who had received kidney transplants because of renal failure due to essential hypertension returned to normal and remained normal after transplantation. That the renal abnormality is a defect in the sodium excretory capacity of the kidney has also gained support from a number of well-controlled trials in which the effect of dietary manipulation clearly showed a hypertensive effect of a high salt diet in patients with essential hypertension. Amongst population studies, the evidence gathered from the recent Intersalt study[12] carried out in a total of 10 079 people from 52 world-wide centres also favours a contribution of salt intake in the hypertensive process.

Salt loading in hypertensive or normotensive animals or humans does not lead to prolonged changes in the extracellular fluid volume, although volume regulatory mechanisms must be mobilized. Neither is there volume expansion in essential hypertension, although it has been reported that there is an increase in central volume. It is not clear therefore what the stimulus is that promotes the postulated prolonged secretion of a sodium transport inhibitor in essential hypertension. The hypothalamus is thought to play an important role in volume control and may act as a detector or integrative centre.

The hypothesis relating a natriuretic hormone (or sodium transport inhibitor) to the aetiology of essential hypertension has been tested many times and in a variety of ways. In order to establish the presence of a sodium transport inhibitor experiments were carried out to determine the effect of serum from patients with essential hypertension on Na^+K^+ATPase. Using normal peripheral blood leucocytes, inhibition of the glycoside-sensitive sodium efflux rate constant was demonstrated after incubation in essential hypertensive patients' serum.[13] Hamlyn *et al.* also found serum induced inhibition of soluble Na^+K^+ ATPase.[14] In that investigation and several subsequent reports, a relationship between the Na^+K^+ATPase activity and the blood pressure of the patient was found. These results have been repeated by several investigators, although a few found no evidence for a sodium transport inhibitor.

There has been considerable contention as to whether impaired sodium transport will lead to vasoconstriction—a fundamental assumption in the hypothesis. In large vessels it is certain that elevated intracellular sodium leads to an increase in tone.[15] In small resistance arteries, however, a number of studies using animal tissue have clearly demonstrated no vascular constriction in response to inhibition of the sodium pump.[16] These observations certainly reduced the strength of the hypothesis. Recently, however, work carried out using human resistance arteries (obtained from subcutaneous fat of patients undergoing elective surgery) has shown that low concentrations of the cardiac glycoside, ouabain, can lead to a prolonged increase in resting tone and can cause potentiation of noradrenaline induced tone.[17] Moreover, the same laboratory has also found that very small concentrations of ouabain cause profound impairment of endothelium-dependent relaxation in human resistance arteries, apparently by inhibition of release of endothelium derived relaxing factor. The disparity of these results and those from rat arteries may reflect the relative cardiac glycoside insensitivity of rat tissue compared with that of humans. Determination of the circulating concentration of a sodium transport inhibitor remains, however, an important factor in relation to the degree of vasoconstriction or reduced relaxation that would be anticipated in human resistance arteries.

Characterization of the natriuretic hormone

The evidence for a natriuretic hormone outlined above led to an intensive search for the substance or substances responsible. In the search for a natriuretic hormone produced in response to volume expansion or salt loading, the detection of sodium transport inhibitory activity in serum and urine led to the general assumption that the substance responsible for the natriuresis was an inhibitor of active sodium transport. This would then presumably lead to decreased renal tubular sodium reabsorption by inhibition of sodium transport in the nephron. Inhibition of the sodium pump however, does not necessarily lead to a natriuresis; this is demonstrated by the lack of natriuresis following an infusion of cardiac glycosides (specific inhibitors of Na^+K^+ATPase).

None the less, volume expansion is associated with reduced tubular sodium reabsorption, particularly in the distal tubule and collecting duct. Recent evidence[18] that the kidney expresses different isoenzymes of Na^+K^+ATPase may be of importance in elaborating how inhibition of active sodium transport affects the kidney, particularly as the different isoenzymes show differing sensitivities to cardiac glycosides. In the rabbit, Na^+K^+ATPase in the collecting duct has been found to be 10–30 times as sensitive to ouabain as more proximal segments and may represent the preferential expression of the glycoside-sensitive isoform. Moreover, it has been suggested that expression of the different isoforms can be influenced by the sodium balance of the animal. It is possible, therefore, that the kidney may develop cardiac glycoside sensitivity in response to sodium loading.

Despite the expectation that a natriuretic hormone will be able to induce a natriuresis, the major research effort has centred on the search for a Na^+K^+ATPase inhibitor. The atrial natriuretic peptide has been identified, but fails to fulfil the criteria of an inhibitor of active sodium transport. The atrial natriuretic peptide is short-acting and may explain some of the immediate but probably not the prolonged natriuresis which follows volume expansion. Other than the atrial natriuretic peptide, approximately 20 different substances have been partially characterized in more than 30 different laboratories and their suitability for the role of the 'natriuretic' hormone assessed using one or more different assays of sodium transport (for review see Wechter *et al.*[19]

Assays for sodium transport inhibitor

The assays used in the detection of sodium transport inhibitors might at first appear to be one of the more straightforward elements of the natriuretic hormone story. The reverse is true; the diversity and non-specificity of the assays used is likely to have been the major cause of the delay in identification of the candidate substance or substances.[19] The commonly-used assays are listed in Table 13.1. Unfortunately, very few investigators have followed the detection of an inhibitor of sodium transport with a test for natriuresis, neither have many accompanied the original assay with a back-up assay for sodium transport.

Table 13.1 Commonly used assays for detection of sodium pump inhibitor

Isolated $Na^+ K^+$ ATPase
Glycoside sensitive ^{86}Rb uptake (equivalent to potassium uptake)
Leucocyte/erythrocyte glycoside sensitive ^{22}Na efflux
^{3}H Ouabain binding
Digoxin-like immunoreactivity

The major problem with these assays for active sodium transport lie in non-specificity. Many endogenous compounds in sufficiently high concentration will inhibit Na^+K^+ATPase, particularly if the enzyme is in the soluble form and not *in situ* in a cell membrane.[19] Sufficiently high concentrations of these compounds may well occur as contaminants in preparation of extracts from tissue or fluid. The radioimmunoassay for digoxin as a detection system for a cardiac glycoside-like material has probably led to many erroneous results. Digoxin antibodies will bind non-specifically to many steroids and to free fatty acids. Undoubtedly the most reliable assay for a sodium transport inhibitor is an estimate of the glycoside-sensitive transport of sodium out of a cell, or the glycoside-sensitive component of potassium influx into the cell. The latter is often achieved using ^{86} Rb uptake into erythrocytes as a marker for potassium, since the isotope of potassium has a short half life.

Methods of extraction of sodium pump inhibitors have included gel chromatography, high pressure liquid chromatography, flash chromatography, ion exchange chromatography and enzyme (Na^+K^+ATPase) affinity. Many involve an intitial preparative stage using reverse phase chromatography and methanol extraction. Several very polar fractions demonstrate sodium transport inhibitory activity, as do some more non-polar ones. Active fractions have been isolated from serum and urine of normal, salt loaded and uremic subjects and animals, and from adrenals, kidneys and brain. Amniotic fluid and cord serum are also rich sources. Molecular weights of an active compound reportedly range from 280–1468. Amongst the organs in which an active substance has been found, the hypothalamus has received the greatest attention. There is undoubtedly a very active inhibitor of sodium transport in the hypothalamus but, as yet, the same substance has not been shown to be present in the serum and it cannot be classified as a circulating hormone. There is also disagreement as to whether the hypothalamic substance is steroidal or peptidic in nature.

Two groups have recently identified the structure of circulating inhibitors of sodium transport which have considerable specificity in more than one Na^+K^+ATPase assay. Lichstein *et al.* have extracted 1500 litres of bovine plasma and, using the technique of high pressure liquid chromatography and characterization by mass spectroscopy have identified one inhibitory compound as 11, 13, dihydroxy-14-octadecaenoic acid, a fatty acid derivative.[20] Very recently, and as the result of eight years work, Hamlyn *et al.* in association with the Upjohn Company have isolated and completely characterized an active fraction obtained from human serum using high pressure liquid chromatography and an enzyme affinity stage in the preparation.[21, 22] This compound has been identified in a number of animal species and in several organs. To the surprise of a great many investigators in this field, the compound has been identified as a cardiac glycoside. Moreover, the substance is identical in all respects, structurally and biologically to the cardiac glycoside, ouabain, a cardiac glycoside more generally associated with the African Ouabaio tree than with mammalian blood.

It is unlikely that the cardiac glycoside originates in the diet since ouabain was found to increase in DOCA-salt induced hypertension in pigs and to be present in high concentration in patients undergoing parenteral nutrition. In common with ouabain, the endogenous cardiac glycoside demonstrates positive inotropy and causes constriction of pre-activated conduit arteries. The circulating concentration of the circulating compound is said to be of the order of 0.14 nM/l. Ouabain, as mentioned earlier, is not natriuretic, an observation repeated with endogenous ouabain.

In summary, there is ample evidence for the presence of hitherto unidentified natriuretic factors produced in response to sodium loading and in patients with uraemia; the function of these factors is presumed to be to increase FeNa and so contribute to re-establishment of sodium balance. There is no evidence which would confirm that the same compound is involved in both instances. There is also good reason to believe that endogenous sodium transport inhibitors exist, and that they may be related to sodium balance and to hypertension. To date, many active fractions have been isolated from animal and human tissues and fluids and to varying degrees of purity; in two laboratories this has culminated in positive identification. There is no suggestion, however, as to the relationship between these substances and those involved in the natriuresis of volume expansion or uraemia.

References

1. De Wardener, H.E., Mills, I.H., Clapham, W.F., and Hayter, C.J. (1961). Studies on the efferent mechanism of the sodium diuresis which follows the administration of intravenous saline in the dog. *Clinical Science*, **21**, 249–58.
2. Haddy, F.J. and Overbeck, H.W. (1976). The role of humoral substances in volume expanded hypertension. *Life Sciences*, **19**, 935–48.
3. De Wardener, H.E. and MacGregor, G.A. (1980). Dahl's hypothesis that a saliuretic substance may be responsible for a sustained rise in arterial pressure: its possible role in essential hypertension *Kidney International*, **18**, 1–9.
4. De Wardener, H.E. and Clarkson, E.M. (1985). Concept of natriuretic hormone. *Physiological Reviews*, **65**, 658–759.
5. Schultze, R.G., Shapiro, H.S., and Bricker, N.S. (1969). Studies on the control of sodium excretion in experimental uraemia. *Journal of Clinical Investigation*, **48**, 869–77.
6. Bourgoignie, J.J., Hwang, K.H., Ipakchi, E., and Bricker, N.S. (1974). The presence of a natriuretic factor in the urine of patients with chronic uraemia. *Journal of Clinical Investigation*, 1559–67.
7. Kaji, D. and Kahn, T. (1987). $Na^+ - K^+$ pump in chronic renal failure. *American Journal of Physiology*, **252**, F785–93.
8. Tobian, L. and Binion, J.T. (1952). Tissue cations and water in arterial hypertension. *Circulation*, **5**, 754–58.
9. Hilton, P.J. (1986). Cellular sodium transport in essential hypertension. *New England Journal of Medicine*, **314**, 222–9.
10. Haddy, F.J. and Pamnani, M.B. (1985). Evidence for a circulating endogenous $Na^+ - K^+$ inhibitors in low renin hypertension. *Federation Proceedings*, **44**, 2789–94.
11. de Wardener, H.E. (1990). The primary role of the kidney and salt intake in the aetiology of essential hypertension: *Clinical Science*, **79**, 193–200.
12. The Intersalt study. (1989). *Journal of Human Hypertension*, **3**, 279–407.
13. Poston, L., Sewell, R.B., Wilkinson, S.P., Richardson, P.J., Williams, R., and Clarkson, E.M., MacGregor, G.A., and DeWardener, H.E. (1981). Evidence for a circulating sodium transport inhibitor in essential hypertension. *Br. Med. J.* **282**, 847–9.
14. Hamlyn, J.M., Ringel, R., Schaeffer, J., Levinson, P.D., Hamilton, B.P., Kowarski, A.A., and Blaustein, M.P. (1982). A circulating inhibitor of $(Na^+ + K^+)$ ATPase associated with essential hypertension. *Nature*, **300**, 650–2.
15. Blaustein, M.P. (1977). Sodium ions, calcium ions, blood pressure regulation and hypertension: a reassessment and a hypothesis. *Am. J. Physiol.* **232**, C165–73.
16. Mulvany, M.J. (1985). Changes in sodium pump activity and vascular contraction. *J. Hypertens.* **3**, 429–36.
17. Woolfson, R.G., Hilton, P.J., and Poston, L. (1990). Effects of ouabain and low sodium on contractility of human resistance arteries. *Hypertension*, **15**, 583–90.
18. Sweadner, K.J. (1989). Isoenzymes of the Na^+/K^+-ATPase. *Biochim. Biophys. Acta*, **988**, 185–220.
19. Wechter, W.J. and Benaksas, E.J. (1990). Natriuretic Hormones. *Progress in Drug Research*, **34**, 231–57.
20. Lichstein, D., Samuelov, S., Gati, I., Felix, A.M., Gabriel, T.F., and Deutsch, J. (1991). Identification of 11,13,dihdroxy-14-octadecanoic acid as a circulating Na/K ATPase inhibitor. *J. Endocrinol.*, **128**, 71–8.

21. Hamlyn, J.M., Harris, D.W., and Ludens, J.H. (1989). Digitalis-like acitivity in human plasma. *Journal of Biological Chemistry*, **264**, 7395–404.
22. Hamlyn, J.M., Blaustein, M.P., Bora, S., DuCharme, D.W., Harris, D.W., Mandel, F., Matthews, W.R., and Ludens, J.H. (1991). Identification and characterization of a ouabain-like compound from human plasma. *Proc. Natl. Acad. Sci. USA*, **88**, 6259–63.

14 Hypertension in parenchymatous renal disease and chronic renal failure

R. Wilkinson

Parenchymatous renal disease accounts for approximately 4 per cent of all cases of hypertension. The prevalence of hypertension is increased in all types parenchymal renal disease, although the proportion of patients affected varies from around 15–20 per cent in patients with minimal change nephropathy to around 70–80 per cent in patients with focal glomerulosclerosis and mesangiocapillary glomerulonephritis.[1] The prevalence of hypertension in a particular type of renal disease depends on the stage of the disease, thus in children with chronic pyelonephritis approximately 10 per cent are hypertensive, in adults at presentation 33 per cent have been found to be hypertensive, at six year follow-up 45 per cent are hypertensive and by the time end stage renal failure is reached approximately 80 per cent are hypertensive.[2]

End stage renal failure

The mechanism of hypertension in end stage renal failure is reasonably well established and a combination of raised plasma renin activity and sodium retention seems sufficient to account for the hypertension. A positive correlation has been found between plasma renin concentration and mean arterial pressure in patients with chronic renal failure studied on the day following dialysis.[3] In support of the importance of renin is the further observation that bilateral nephrectomy resulted in a reduction of blood pressure in patients with severe hypertension on dialysis[3] and that in renal transplant recipients who had been subjected to bilateral nephrectomy hypertension was very uncommon. Furthermore in patients not subjected to bilateral nephrectomy, the blood pressure following transplantation was significantly related to the pre-transplant blood pressure. This observation suggests the presence of a factor, probably renin, which was present before transplantation as a major determinant of the post-transplant blood pressure.[4]

There is little doubt that sodium retention is also a major factor in determining the level of blood pressure in patients with end stage renal failure. The hypertension which affects 80 per cent of patients with end stage renal failure can be controlled in the vast majority by adequate dialysis, with control of sodium and water balance. In addition, in patients who have undergone bilateral nephrectomy a significant correlation can be demonstrated between mean arterial blood pressure and exchangeable sodium.[3]

There is some uncertainty as to the interpretation of measured exchangeable sodium, since total body exchangeable sodium depends on lean body mass, the determination of which is difficult. The formulae used to derive lean body mass from anthropometric measurements tend to be inaccurate at extremes of body weight. The estimation of lean body mass from total body potassium, probably the most reliable method in normal subjects, may be unreliable in conditions, such as chronic renal failure, where there is likely to be a disturbance of total body potassium. Although the exchangeable sodium calculated by a variety of different formulae is in close agreement, it is not independent of body build; there is a progressive increase in exchangeable sodium when expressed per kilogram lean body mass from fat to lean subjects, indicating that the formulae for calculation of lean body mass are not reliable.[5] It seems likely, however, that both sodium retention and hyperreninaemia are of major importance in the hypertension of end stage renal failure, since blood pressure is more closely correlated to the product of exchangeable sodium and the log of plasma renin concentration than it is to either measurement alone.[5]

Bilateral nephrectomy in end stage renal failure

In approximately 5 per cent of patients with end stage renal failure blood pressure control is difficult despite adequate volume control[3,6], and in these patients plasma renin is inappropriately high in relation to body sodium. Bilateral nephrectomy results in a reduction in blood pressure in these patients and reduces the risk of hypertension following renal transplantation.[4] Although bilateral nephrectomy was undertaken quite regularly until about 15 years ago it has always been embarked upon reluctantly because of the problems associated with it. There is an increase in morbidity and mortality arising from the operation itself, and increased problems with anaemia and bone disease during dialysis following bilateral nephrectomy. Furthermore, it is impossible to predict the blood pressure response to the operation, since neither the measurement of peripheral plasma renin activity, the hypotensive response to angiotensin converting enzyme inhibition, or renal vein renin levels are reliable in predicting response.

The requirement for bilateral nephrectomy has fallen over the past 15 years or so because of the introduction of more potent antihypertensive agents, and it is now rarely necessary to proceed to bilateral nephrectomy for the control of blood pressure. Percutaneous renal ablation has been undertaken to reduce the risk arising from bilateral nephrectomy and did appear to lower pressure,[7] but no controlled study of this treatment has been undertaken. It seems unlikely that it will prove possible totally to ablate the kidneys by embolization of the renal arteries, since capsular vessels will remain intact and renin secretion will probably continue. The availability of 1-alpha-hydroxycholecalciferol and of erythropoietin have made the prospect of bilateral nephrectomy less daunting, but the risks of the operation itself persist and it seems unlikely that it will become widely practised.

Hypertension in parenchymatous renal disease without renal failure

There is an increased prevalence of hypertension in parenchymatous renal disease even before there is detectable impairment of renal function. The mechanism of this hypertension has not yet been fully elucidated. Zucchelli *et al.*[8] studied 85 patients with glomerulonephritis and 24 normal subjects and found that plasma renin activity, exchangeable sodium, plasma adrenaline, urinary prostaglandin E and autonomic function were normal in all of the hypertensive patients, and the only abnormality detected was an increase in plasma renin activity in 9 of the 30 normotensive patients with glomerulonephritis.

In a study of 25 patients with chronic pyelonephritis who were carefully matched for age and sex with a group of normal control subjects mean plasma renin activity, plasma angiotensin II, plasma aldosterone and exchangeable sodium were all increased in the pyelonephritic group in comparison with the controls, but there was no significant difference between the hypertensive and normotensive patients with chronic pyelonephritis in any of these parameters.[9] These data seemed to indicate that the renal disease had resulted in an increase in renin secretion either through areas of ischaemia within the kidney, or possibly because of an increased sympathetic activity in parenchymatous renal disease. The latter possibility is supported by the observation of an enhanced response to infused noradrenaline in renal disease.[10] and of a raised plasma renin activity in normotensive patients with renal disease[8], which was reduced by treatment with adrenergic receptor blocking drugs.

It is likely that the increased levels of renin, by increasing levels of angiotensin II and subsequently aldosterone had led to sodium retention, with the subsequent development of hypertension. The hypertension may have arisen with the passage of time, since Brod *et al.* described sodium retention with hypervolaemia and increased cardiac output, but normal blood pressure, when they first investigated a group of patients with parenchymatous renal disease. At subsequent investigation these patients had gone on to develop hypertension with an increase in peripheral resistance and a return of hypervolaemia and increased cardiac output towards normal.[11,12] Brod's suggestion was that the hypervolaemia may lead eventually to hypertension through the stimulation of an inhibitor of sodium potassium ATPase, which, via an increase in intracellular sodium and intracellular calcium, may be responsible for increased arteriolar tone and hypertension.

The fact that a difference could not be demonstrated in renin, angiotensin, aldosterone or sodium between hypertensive and normotensive patients with chronic pyelonephritis may be explained purely by a shorter duration of hypervolaemia in the normotensives, so that if they had been followed for longer they would have become hypertensive. Alternatively, when faced with the same stress of sodium retention and hyper-reninaemia, patients with a predisposition to hypertension may become hypertensive, whereas others do not. In support of this suggestion is the observation that the proportion of patients with renal hypertension who have abnormally raised sodium–lithium counter-transport (thought to be a marker of an inherited form of hypertension) is similiar to that seen in a group of essential hypertensives.

The suggestion is therefore that in parenchymatous renal disease an initial rise in renin secretion leads to an increase in plasma angiotensin II and thence aldosterone, which in turn leads to sodium retention and hypervolaemia with an increase in cardiac output. With time, this increased cardiac output leads to an increase in peripheral resistance, possibly through hormonal inhibition of sodium potassium ATPase, and susceptible patients become hypertensive. This may be an over-simplistic view, since it does not take into account possible disturbances of renal hypotensive factors, or the recently described endothelial factors (endothelin and endothelium derived relaxing factor) or atrial natriuretic peptide. Indeed there may be other hormonal factors influencing blood pressure in this situation.

The probable role of renin in the hypertension of parenchymatous renal disease has been supported by a recent study in patients with adult polycystic kidney disease.[14] Plasma renin activity was higher in hypertensive patients with polycystic kidney disease than in patients with essential hypertension matched for blood pressure and other parameters, although it did not differ from normal control subjects or normotensive patients with polycystic kidney disease. This suggested that renin was not normally suppressible by raised blood pressure in polycystic kidney disease and its hypersecretion might therefore have been responsible for the development of hypertension. Interestingly, as in the earlier studies in chronic pyelonephritis, plasma renin activity did not differ in normotensive and hypertensive patients with polycystic kidney disease although in both groups it exceeded the levels in essential hypertensives.

Unilateral pyelonephritis

In most cases of hypertension due to parenchymatous renal disease there is no prospect of cure of the hypertension. However, when the renal disease is predominantly unilateral there is the prospect of cure or improvement by removing the affected kidney. There have been many reports over the years of the effectiveness of this procedure, but there have been no controlled trials comparing medical with surgical treatment. In a retrospective review of experience of nephrectomy for predominantely unilateral chronic pyelonephritis,[15] 15 patients treated by nephrectomy were compared with 16 treated medically.

The choice of treatment had not been random; those patients with more severe hypertension, smaller kidneys, younger age and having higher renal vein renin ratios tended to be offered surgery and the others were treated medically.

The control of blood pressure was found to be similar in the two groups. Control was achieved in the medical group by increased antihypertensive therapy in 64 per cent of cases. In 30 per cent there was some improvement in blood pressure without a change or with a reduction in dosage of drugs. In one patient only (6 per cent) the blood pressure remitted spontaneously. By contrast, in the surgical group 4 patients were cured, 10 showed an improvement, in that there was a reduction in systolic blood pressure of more than 20 mmHg and in diastolic blood pressure 10 mmHg without increased drug dosage, and in only one patient was there no improvement. Perhaps surprisingly, renal function at follow-up had remained stable in the nephrectomy group. Plasma creatinine was 119 ± 61 μmol/l prior to surgery and 117 ± 67 μmol/l following surgery (not significant). In the patients treated medically there was again no significant change, but there was a tendency for plasma creatinine to rise from 89 ± 20 μmol/l at the time of entry to the study to 102 ± 32 μmol/l at follow-up. Unfortunately, renal vein renin ratios did not distinguish between patients in whom hypertension was cured by surgery and those with improvement only, and there is no test which can accurately predict the response to nephrectomy.

Our present policy is to consider nephrectomy in the following patients: those with unilateral renal disease in whom the diseased kidney is very small and certainly contributes less than 20 per cent of overall function; those patients with hypertension which is difficult to control medically or who are suffering side effects from medical treatment; and young women planning further pregnancy where there have been problems with hypertension during their first pregnancy. We have performed nephrectomy less in recent years than formerly, probably because of the availability of a wider range of antihypertensive drugs which make control of hypertension easier.

The effects of control of blood pressure on the progression of renal failure

In patients with malignant hypertension, treatment is obligatory regardless of any possible effect on renal function, since without treatment death ensues in the majority of patients within a year, usually from a vascular cause. The treatment of malignant hypertension may result in an initial deterioration in renal function, followed later in some cases by improvement. Some patients who have become dialysis-dependent during treatment of hypertension subsequently regain some function.[16]

Progressive renal failure is uncommon in patients with essential hypertension and does not seem closely related to the degree of control of blood pressure achieved. Rostand *et al.* found no difference in the degree of blood pressure control achieved in those patients with essential hypertension who progressed to renal failure and those whose renal function remained stable.[17]

In patients with renal disease there is no doubt that the presence of hypertension is associated with a high risk of progression of renal failure.[2,18,19] Perhaps surprisingly there is some dispute as to whether control of blood pressure delays the progression of renal failure. No relationship was found between deterioration of renal function and blood pressure.[20–22] However, a preliminary report has shown a correlation between the rate of fall in renal function and blood pressure control in patients with progressive renal disease.[23]

If it is established that control of blood pressure delays progression of renal failure in patients with parenchymatous renal disease but not in those with essential hypertension, why may this be so? In essential hypertension there is evidence that afferent arteriolor tone is increased since the glomerular filtration rate and renal plasma flow are reduced in parallel.[24] If this is the case, then it would account for auto-regulation of renal blood flow, since as systemic blood pressure rises the increased pressure will not be transmitted to the glomerular capillaries, except at extremes of pressure. In contrast, in renal hypertension there is some evidence that the efferent arteriole is perferentially constricted, and the afferent arteriole relatively dilated, so that the glomerular capillaries are exposed to any increase in systemic pressure. Reduction of systemic pressure would therefore be expected to delay progression of renal damage by reducing glomerular capillary pressure. One piece of evidence supporting this concept is an old observation that in patients with renal failure subjected to sodium loading creatinine clearance was directly proportional to mean arterial pressure, suggesting direct transmission of systemic arterial pressure to the glomerular capillaries.[25]

Possible specific effects of angiotensin converting enzyme inhibitors in parenchymatous renal disease

If there is relative afferent arteriolar vasodilatation and efferent arteriolar vasoconstriction in patients with parenchymatous renal disease, maintaining an increase in filtration fraction at the price of raised glomerular capillary pressure, then this has important implications. Since efferent arteriolar tone is thought to be maintained mainly by angiotensin II, converting enzyme (ACE) inhibitors may be specifically beneficial by inducing efferent vasodilatation with a resulting fall in glomerular capillary pressure and a possible delay in progression of renal damage. In experimental animals there seems little doubt that ACE inhibitors can protect against glomerular injury, both in animals with primary parenchymatous renal disease and with diabetes.[26] In humans there have been no controlled studies, although these are in progress.[27,28] There have been reports of retrospective studies which have shown a flattening of the slope of the reciprocal of plasma creatinine against time curve during treatment with ACE inhibitors, suggesting delay in the progression of renal failure. Although reciprocal plasma creatinine is not an accurate measure of changes in renal function, it does probably give a reasonable estimate of such changes. Using only historical conrols Rodicio *et al.* described a flattening of the reciprocal plasma creatinine curve, stabilization of inulin clearance and reduced proteinuria during treatment with ACE inhibitor in patients with parenchymatous renal disease.[29]

Although it has been difficult to establish an effect of ACE inhibitors on the progression of renal failure it has been much easier to demonstrate an effect on proteinuria. ACE inhibitors have been shown to reduce proteinuria in diabetic animals and in animals with reduced renal mass.[30,31] Their effect has also been clearly demonstrated in humans.[32] The mechanism of the reduction in proteinuria is not certain, and in particular it is not clear whether it is due to a reduction in systemic blood pressure or a specific effect of ACE inhibitors in bringing about efferent arteriolar dilatation with a consequent reduction in glomerular capillary pressure. There is some evidence to suggest that at least the effect in reducing proteinuria is brought about by a reduction in efferent arteriolar tone.[32]

Before the renal haemodynamic changes thought to occur in renal artery stenosis (i.e. angiotensin II-induced efferent arteriolar constriction with relative afferent dilatation) are extrapolated to the situation in parenchymatous renal disease it should be pointed out that in patients with chronic renal failure ACE inhibitors have not yet been convincingly shown to influence glomerular filtration rate, renal perfusion factor or filtration fraction.[33] A further word of warning comes from the observation of Walser that in non-diabetic patients with parenchymatous renal disease who were treated with an ACE inhibitor progression of renal failure was more rapid than in those not taking the drug.[34] This of course confirms observations from in clinical practice that patients may deteriorate during treatment with ACE inhibitors despite the absence of renal artery stenosis, although fortunately this change is usually reversible.

Implications for treatment of hypertension in patients with parenchymal renal disease

Although no controlled studies have been undertaken in patients with renal disease to demonstrate the reductions in cardiovascular morbidity and mortality which have been demonstrated in essential hypertensive patients with control of hypertension, there seems no reason to believe that equivalent benefit will not be gained. Hence whatever the effect of treatment of hypertension on the progression of renal disease, treatment is indicated.

In patients with proteinuria and hypoproteinaemia not responsive to specific treatment blood pressure should be controlled with conventional drugs; if proteinuria persists then a trial of ACE inhibitor is worthwhile, but it may be no more effective than other antihypertensive agents.

The drugs used for the control of hypertension in parenchymatous renal disease should not differ from those used in the treatment of essential hypertension until clear evidence emergence regarding the possible specific benefit of ACE inhibitors. In patients with imparied renal function in whom plasma creatinine exceeds 200 μmol/l a loop diuretic rather than an thiazide diuretic should be used.

References

1. Smith, M.C. and Dunn, M.J. (1990). Hypertension in renal parenchymal disease. In *Hypertension: pathophysiology, diagnosis, and management*, (ed. J.H. Laragh and B.M. Brenner), Raven Press, New York.
2. Arze, R.S., Ramos, J.M., Owen, J.P., Morley, A.R., Elliott, R.W., Wilkinson, R., Ward, M.K., and Kerr, D.N.S. (1982). The natural history of chronic pyelonephritis in the adult. *Quarterly Journal of Medicine*, **51**, 396–410.
3. Wilkinson, R., Scott, D.R., Uldall, P.R., Kerr, D.N.S., and Swinney, J. (1970). Plasma renin and exchangeable sodium in the hypertension of chronic renal failure. *Quarterly Journal of Medicine*, **39**, 377–94.
4. McHugh, M.I., Tanboga, H., Marcen, R., Liano, R., Robson, V., and Wilkinson, R. (1980). Hypertension following renal transplantation: the role of the host's kidneys. *Quarterly Journal of Medicine*, **49**, 395–403.
5. Wilkinson, R. (1978). Studies on the mechanism of hypertension in renal disease. *MD Thesis*, University of Newcastle upon Tyne.
6. Vertes, V., Cangiano, J.L., Berman, L.B., and Gould, A. (1969). Hypertension in end-stage renal disease. *New England J. Med.*, **280**, 978–81.
7. Thompson, J.F., Fletcher, E.W.L., Wood, R.F.M., Chalmers, D.H.K., Taylor, H.M., Benjamin, I.S., and Morris, P.J. (1984). Control of hypertension after renal transplantation by embolisation of host kidneys. *Lancet*, **2**, 424–7.
8. Zuchelli, P., Zuccala, A., and Mancini, E. (1989). Hypertension in primary glomerulonephritis without renal insufficiency. *Nephrology Dialysis Transplantation*, **4**, 605–10.
9. Siamopoulos, K.C. and Wilkinson, R. (1987). Hypertension in chronic pyelonephritis. *Contr. Nephrol.*, **54**, 119–23.
10. Beretta-Piccoli, C., Weidmann, P., Schiffl, H., Cottier, C., and Reubi, F.C. (1982). Enhanced cardiovascular pressor reactivity to norepinephrine in mild renal parenchymal disease. *Kidney International*, **22**, 297–303.
11. Brod, J., Bahlmann, J., Cachovan, M., and Pretschner, P. (1983). Development of hypertension in renal disease. *Clinical Science*, **64**, 141–52.
12. Brod, J., Schaeffer, J., Hengstenberg, J.H., and Kleinschmidt, T.G. (1984). Investigations on the Na^+, K^+ –pump in erythrocytes of patients with renal hypertension. *Clinical Science*, **66**, 351–5.
13. Carr, S.J., Thomas, T.H., and Wilkinson, R. (1989). Erythrocyte sodium-lithium countertransport in primary and renal hypertension: relation to family history. *European Journal of Clinical Investigation*, **19**, 101–6.
14. Chapman, A.B., Johnson, A., Gabow, P.A., and Schrier, R.W. (1990). The renin-angiotensin-aldosterone system and autosomal dominant polycystic kidney disease. *New England Journal of Medicine*, **323**, 1091–6.
15. Siamopoulos, K.C., Sellars, L., Mishra, S.C., Essenhigh, D.M., Robson, V., and

Wilkinson, R. (1983). Experience in the management of hypertension with unilateral chronic pyelonephritis: results of nephrectomy in selected patients. *Quarterly Journal of Medicine*, **52**, 349–62.

16. Cordingley, F.T., Jones, N.F., Wing, A.J., and Hilton, P.J. (1980). Reversible renal failure in malignant hypertension. *Clinical Nephrology*, **14**, 98–103.
17. Rostand, S.G., Brown, G., and Kirk, K.A. (1989). Renal insufficiency in treated hypertension. *New England Journal of Medicine*, **320**, 684–8.
18. Baldwin, D.S., Neugarten, J. (1986). Blood pressure control and progression of renal insufficiency. *Contemporary Issues in Nephrology*, **14**, 81–110.
19. Brazy, P.C., Stead, W.N., and Fitzwilliam, J.F. (1989). Progression of renal insufficiency. Role of blood pressure. *Kidney International*, **35**, 670–4.
20. Steinvinkel, P., Alvestrand, A., and Bergstrom, J. (1989). Factors influencing progression in patients with chronic renal failure. *J. Intern. Med.*, **226**, 183–8.
21. Hannedouche, T., Chauveau, P., Fehrat, A., Albouze, G., and Jungers, P. (1989). Effect of moderate protein restriction on the rate of progression of chronic renal failure. *Kidney International*, **36**, (Suppl. 27) S91–5.
22. Williams, P.S., Fass, G., and Bone, J.M. (1988). Renal pathology and proteinuria determine progression in mild/moderate chronic renal failure. *Quarterly Journal of Medicine*, **67**, 343–54.
23. Klahr, S., Levey, A.S., Sandberg, A.M., and Williams, G.W. (1989). Major results of the feasibility study of the modification of diet in renal disease (MORD). *Kidney International*, **35**, 195. (abstr.)
24. London, G.M., Safar, M.E., Sascard, J.E., Levinson, J.A., and Simon, A.C. (1984). Renal and systemic haemodynamics on sustained essential hypertension. *Hypertension*, **6**, 743–54.
25. Wilkinson, R., Luetscher, J.A., Dowdy, A.J., Gonzales, C., and Nokes, G.W. (1972). Studies on the mechanism of sodium excretion in uraemia. *Clinical Science*, **42**, 711–23.
26. Anderson, S. and Brenner, B.M. (1987). Therapeutic implications of converting enzyme inhibitors in renal diseases. *American Journal of Kidney Disease*, **10**, (1), 81–7.
27. Keane, W.F., Anderson, S., Aurell, M., de Zeeuw, D., Narins, R.G., and Povar, G. (1989). Angiotensin converting enzyme inhibitors and progressive renal insufficiency: current experience. *Annals Internal Medicine*, **111**, 503–16.
28. Feig, P.U. and Rutan, G.H. (1989). Angiotensin converting enzyme inhibitors: the end of the end-stage renal disease. *Annals Internal Medicine*, **111**, 451–4.
29. Rodicio, J.L., Alcazar, J.M., and Ruilope, L.M. (1990). Influence of converting enzyme inhibition on glomerular filtration rate and proteinuria. *Kidney International*, **38**, 590–4.
30. Anderson, S., Meyer, T.W., Rennke, H.G., and Brenner, B.M. (1985). Control of glomerular hypertension limits glomerular injury in rats with reduced renal mass. *Journal of Clinical Investigation*, **76**, 612–9.
31. Zatz, R., Dunn, B.R., Meyer, T.W., Anderson, S., Rennke, H.G., and Brenner, B.M. (1986). Prevention of diabetic glomerulopathy by pharmacological amelioration of glomerular capillary hypertension. *Journal of Clinical Investigation*, **77**, 1925–30.
32. Bedogna, V., Valvo, E., Casagrande, P., Braggio, P., Fontanarosa, C., Dal Santo, F., Alberti, D., and Maschio, G. (1990). Effects of ACE inhibition in normotensive patients with chronic glomerular disease and normal renal function. *Kidney International*, **38**, 101–7.

33. Smith, W.G.J., Dharmasena, A.D., El Nahas, A.M., Thomas, D.M., and Coles, G.A. (1989). Short-term effect of captopril on renal haemodynamics in chronic renal failure. *Nephrology Dialysis Transplantation*, **4**, 696–700.
34. Walser, M. (1990). Progression of chronic renal failure in man. *Kidney International*, **37**, 1195–210.

15 Role of transluminal angioplasty in renovascular disease

W.W. Yeo, L.A. Brawn, P.C. Waller, and L.E. Ramsay

In 1975, surgical treatment of renovascular disease was reported to result in cure of hypertension in 51 per cent of cases, improvement in 15 per cent, and failure in 34 per cent. Those with atherosclerotic renal artery stenosis responded less well than those with fibromuscular disease (Foster *et al.* 1975). However the overall surgical mortality was 5.9 per cent, and 13.1 per cent had major complications related to surgery (Franklin *et al.* 1975). Mortality was particularly high in those with atherosclerotic renovascular disease (9.3 per cent), in older patients and in those with renal impairment or generalized vascular disease. Nephrectomy had to be performed as a primary or secondary procedure in about 50 per cent of patients (Foster *et al.* 1975; Sellars *et al.* 1985*a*,*b*; Mackay *et al.* 1983). Considering these figures for mortality, morbidity, and loss of functioning renal tissue, and the improved efficacy and tolerability of modern antihypertensive drugs, many physicians abandoned the search for renovascular disease except in very young patients or very resistant hypertension.

Interest in renovascular disease was reawakened by a report of dilatation of renal artery stenosis with a balloon catheter (Gruntzig *et al.* 1978) and reports soon followed of large series of patients treated by percutaneous transluminal angioplasty (Martin *et al.* 1981; Colapinto *et al.* 1982; Geyskes *et al.* 1983; Sos *et al.* 1983; Tegtmeyer *et al.* 1984*a*; Miller *et al.* 1985; Martin *et al.* 1985; Kaplan-Pavlobcic *et al.* 1985; Kuhlmann *et al.* 1985; Bell *et al.* 1987). The outcome reported appeared to be similar to that for surgery as regards blood pressure response, but angioplasty was simpler, safer and rarely resulted in loss of functioning renal tissue. Physicians have had to rethink completely their policies on detection, investigation and management of renovascular disease. In this chapter current practice is reviewed, with emphasis on the role of transluminal angioplasty in management.

Clinical features suggesting renovascular disease

The prevalence of renovascular disease in unselected hypertensive subjects is probably less than 1 per cent (Berglund *et al.* 1976) and routine screening of all patients is not justified (Atkinson and Kellett, 1974). The condition

Table 15.1 Features of hypertension which should prompt investigation for underlying renovascular disease

Recent onset or recent worsening of hypertension
Accelerated (malignant) phase
Treatment resistance: uncontrolled by a three-drug regimen
Bruit in abdomen or loin
Young age: moderate or severe hypertension, age less than 30 years
Proteinuria, haematuria or elevated serum creatinine
Hypokalaemia not otherwise explained, for example by diuretics
Renal failure caused by ACE inhibitor
Severe peripheral vascular disease
Pulmonary oedema not otherwise explained

should be sought only in the minority of hypertensive patients who have clinical features which point to the possibility of its presence (Simon *et al.* 1972). These clinical features are shown in Table 15.1. Approximately 20 per cent of patients with one or more of these features will have renovascular disease.

Diagnosis of renovascular disease

The choice of screening test for renovascular disease has been the subject of considerable debate. The rapid sequence intravenous urogram (IVU) has been used widely for many years, but some have abandoned its use in favour of other screening methods such as digital-subtraction angiography, captopril isotope renography, or Doppler ultrasound scanning (Thornbury *et al.* 1982; Pickering, 1990). Others have suggested that screening tests should be abandoned altogether, and that renal arteriography should be the first and only test in all patients suspect of having renovascular disease (Carmichael *et al.* 1986). We still favour the rapid sequence IVU as the screening method of choice because it is widely available, and because its predictive value has been thoroughly investigated (Havey *et al.* 1985). Moreover, it is the only screening test which will also detect renal causes of hypertension other than renovascular disease, for example renal scarring, obstructive uropathy, renal tuberculosis or mass lesions. This important advantage of the IVU over other screening methods is generally overlooked.

The rapid sequence IVU is, however, an imperfect screening test for renovascular disease, with a sensitivity of 80 per cent and specificity of 85 per cent (Harvey *et al.* 1985). Renal digital-subtraction angiography is the only other test which has been thoroughly validated and can match these figures, but it has no clear advantage over the IVU (Havey *et al.* 1985) and is not in any event universally available. The predictive values of captopril renography (Fommei *et al.* 1987) and Doppler ultrasound scanning to record velocity profiles from the renal arteries (Kohler *et al.* 1986) have not been examined in sufficiently large series of patients.

Is it necessary to screen patients by rapid sequence IVU, or should renal angiography be performed in all patients (Carmichael *et al.* 1986)? The rapid sequence IVU has a sensitivity of 80 per cent and specificity 85 per cent, and the prevalence of renovascular disease in patients appropriately selected (Table 15.1) is 20 per cent. One hundred consecutive IVUs will diagnose correctly 16 cases (80 per cent) and miss 4 cases (20 per cent) of renovascular disease, and identify incorrectly 12 patients who do not have renovascular disease. Of the 100 patients screened only 28 would proceed to angiography. When compared with a policy of renal arteriography for all, the benefit of screening is avoidance of the morbidity, inconvenience and expense of 72 renal arteriograms. The price paid for this is the failure to diagnose four (20 per cent) patients with renovascular disease. Because screening will fail to diagnose 20 per cent of patients it is important to proceed to renal arteriography despite a normal IVU when the clinical suspicion of renal artery stenosis is very high, or when it is essential to exclude renovascular disease. This is the case when there is severe and resistant hypertension, declining renal function, or pulmonary oedema which is not readily explained (Pickering *et al.* 1988).

In summary we favour the rapid sequence IVU as a routine screening procedure for those with suggestive clinical features (Table 15.1), but perform renal arteriography regardless of the result of the IVU in a minority of patients with the particular indications described above.

Predicting cure of blood pressure

Many tests have been used to attempt to predict the blood pressure response to angioplasty or surgery in renovascular disease, for example measurement of renal vein renins, divided renal function tests, and the response to saralasin. Of these, the renal vein renin ratio proved the most satisfactory (Mackay *et al.* 1983), but its predictive power is too low to be of practical value (Mackay *et al.* 1983; Geyskes *et al.* 1983; Sellars *et al.* 1985*a*,*b*). Because of this, and because of the low morbidity associated with angioplasty, use of the renal vein renin ratio and other predictive tests has largely been abandoned. The usual policy is to dilate any significant renal artery stenosis (>50 per cent) without further investigation (Colapinto *et al.* 1982; Geyskes *et al.* 1983). The best predictors of the blood pressure response to angioplasty or surgical intervention of the blood pressure response to angioplasty or surgical intervention are in fact two simple variables—age and renal function (Foster *et al.* 1975; Brawn and Ramsay, 1987; Marshall *et al.* 1990). Patients over the age of 60 years, or those with elevated serum creatinine, are less likely to have satisfactory blood pressure responses to angioplasty (Brawn and Ramsay 1987).

Performing angioplasty

Whenever possible angioplasty should be performed at the same time as the diagnostic arteriogram, otherwise the patient is subjected to two invasive procedures. Severe hypotension occasionally occurs immediately after the stenosis is dilated (Schwarten 1980; Colapinto *et al.* 1982), particularly in patients taking potent diuretic and antihypertensive therapy. Patients on such treatment should be admitted 48 h before the procedure to withdraw diuretics and reduce other antihypertensive drugs as far as possible, aiming for acceptable blood pressure control with minimum treatment at the time of angioplasty. Routine procedures before angioplasty should include written consent; a record of the presence or absence of all peripheral pulses; measurement of serum creatinine, haemoglobin, urinalysis, and blood group; and ensuring that a vascular surgeon is available at short notice when the procedure is to be performed.

The technical aspects of angioplasty have been described elsewhere (Tegtmeyer *et al.* 1980). The stenotic lesion is crossed by a balloon-tipped catheter and the balloon is then inflated to compress and abolish the stenosis. The procedure is performed using local anaethesia, usually through the femoral artery and under cover of infused heparin. It is not clear whether bilateral lesions are best dealt with at a single session or two separate sessions. There is a small risk of causing bilateral renal artery thrombosis or dissection which could leave the patient functionally anephric if both sides are dilated. It is probably safer to dilate the more severely affected artery and leave the contralateral artery to a later date, but some groups perform bilateral angioplasties at a single session, apparently without serious complications (Tegtmeyer *et al.* 1984*a*,*b*).

Complications of angioplasty

An overview of ten large series of renal artery angioplasty in 691 patients reported a complication rate of 9.1 per cent, and a mortality of 0.43 per cent (Ramsay and Waller 1990). According to the classification of Mahler *et al.* (1986) 89 per cent of complications were a direct consequence of the procedure, and 11 per cent an indirect consequence. It was difficult to determine from the published reports what proportions of complications were major or minor. The most common direct complications of angioplasty are renal artery dissection, renal artery thrombosis, segmental renal infarction, haematoma formation or puncture trauma, retroperitoneal haemorrhage, profound hypotension, and acute tubular necrosis. The most feared direct complication is cholesterol embolism (athero-embolic disease). Serious indirect complications include stroke, myocardial infarction, spinal artery thrombosis, and bowel infarction. The services of a vascular surgeon should be available immediately if required. Patients should be observed in hospital for about five days after the procedure, monitoring particularly the urine output, urinalysis, serum creatinine, haemoglobin, puncture site and peripheral pulses.

Technical failure of angioplasty

In large series of renal angioplasties the procedure has been technically unsuccessful in 12 per cent of patients, but the rate of failure has varied widely between series, from 3 to 24 per cent (Ramsay and Waller 1990). This is probably a consequence of patient selection in tertiary referral centres. These figures are for all forms of renovascular disease, but there is an important difference in the success rate with angioplasty between fibromuscular and atherosclerotic forms of renovascular disease. Renal artery lesions can be dilated satisfactorily in 80–90 per cent of patients with fibromuscular dysplasia. However, in unselected patients with atherosclerotic renovascular disease complete correction can be achieved in only 40 per cent of cases (Sos *et al.* 1983; Brawn and Ramsay 1987). Failure in atherosclerotic disease is usually due to renal artery occlusion, long atherosclerotic lesions, or ostial stenosis, i.e. encroachment of an aortic plaque upon the origin of the renal artery. Short proximal atherosclerotic stenoses which do not involve the ostium are generally amenable to dilatation (Cicuto *et al.* 1981; Sos *et al.* 1983; Tegtmeyer *et al.* 1984*a*,*b*; Miller *et al.* 1985).

Blood pressure response after angioplasty

The initial blood pressure response to angioplasty can be quite misleading. A response may be delayed for as long as one month, and on the other hand hypertension may recur after an apparent early response (Colapinto *et al.* 1982). Blood pressure must therefore be monitored closely after discharge from hospital, as these patients have often had severe hypertension, may be discharged on no treatment, and may have a rapid and dangerous rise in blood pressure. The ultimate blood pressure response is intimately related to the technical success of angioplasty (Sos *et al.* 1983) and improvement should not be anticipated if a stenosis of 50 per cent or more persists after the procedure.

The long-term blood pressure responses to angioplasty have generally been classed as cure, ‘improvement’, or no change, a convention previously used to assess the outcome of surgical treatment of renovascular disease. This classification is unsatisfactory in many respects, and the results cited for angioplasty (and indeed surgery) have to be viewed with considerable circumspection. The definition of ‘improvement’ has differed between studies (Ramsay and Waller 1990). The series reported have all been uncontrolled, and none has taken adequate account of factors such as altered compliance, regression to the mean, placebo response, and observer bias, all of which confound the assessment of response in hypertension (Ramsay and Waller 1990). Furthermore, an ‘improvement’ rate as high as 33 per cent has been observed in patients with a similar degree of hypertension who had no intervention (Brawn and Ramsay 1987).

The figures for 'improvement' of blood pressure cited in studies of angioplasty and surgery are therefore unreliable at the very least, and may indeed be entirely spurious (Brawn and Ramsay 1987). These strictures do not apply to figures cited for cure of hypertension, as the criteria have been reasonably uniform between studies (Ramsay and Waller 1990), and spontaneous 'cure' is uncommon in hypertension of this severity. Figures for cure should therefore be given most weight when assessing the published results (Brawn and Ramsay 1987).

The blood pressure responses to angioplasty in ten large series (Ramsay and Waller 1990) are summarized in Fig. 15.1. It can be seen that the outcome differs markedly in fibromuscular disease and atheosclerotic renal artery stenosis. The cure rate for hypertension in fibromuscular disease is approximately 50 per cent (Fig. 15.1). In atherosclerotic disease, however, the cure rate for hypertension after technically succesful angioplasty was only 19 per cent. A further 52 per cent were classed as 'improved', but the reservations mentioned above about the validity of this 'improvement' must be born in mind. In completely unselected patients the technical failure rate is as high as 60 per cent (Sos *et al.* 1983; Brawn and Ramway 1987), and the 'intention to treat' cure rate for rate for hypertension is therefore as low as 7 per cent (Fig. 15.1) (Ramsay and Waller 1990). This disappointing outcome is not really surprising, as atheroma of the renal arteries may often be a consequence of longstanding hypertension, and not its cause (Pickering 1990). Furthermore renal artery stenosis is commonly observed in normotensive subjects who have extensive vascular disease (Choudhri *et al.* 1990).

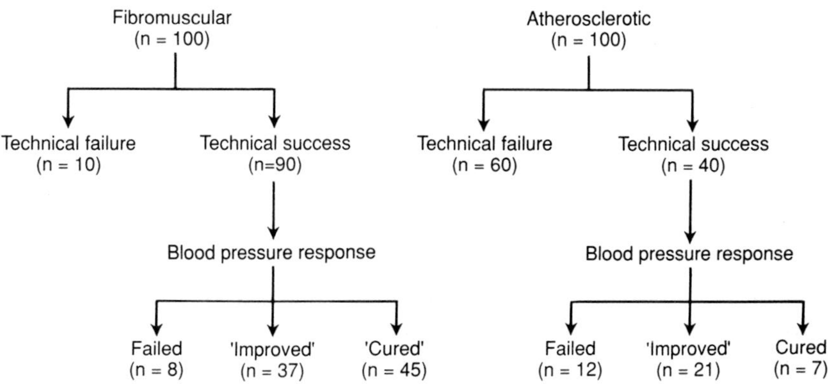

Fig. 15.1 Response of blood pressure to percutaneous transluminal angioplasty of fibromuscular and atherosclerotic renovascular disease in ten large series (Ramsay and Waller 1990). Note that 'improvement' of blood pressure is an unreliable endpoint (Brawn and Ramsay 1987).

Effect of angioplasty on renal function

Data on the effects of angioplasty on renal function are sparse and inconsistent. It can improve renal function substantially in patients with critical stenosis of the artery supplying a single functioning kidney (Sutters *et al.* 1987), and some small series have suggested a more general improvement in renal function after angioplasty (Bell *et al.* 1987; Pickering *et al.* 1986). However, this has not been observed in other series (Luft *et al.* 1983; Brawn and Ramsay, 1987; Marshall *et al.* 1990), and detailed study of the effect of angioplasty on renal function is needed.

Recurrence of stenosis

Restenosis occurs in about 15 per cent of patients (Tegtmeyer *et al.* 1984*a*,*b*). It is more common, and occurs earlier, in atherosclerotic disease (Grim *et al.* 1981) and is more likely to occur if there is residual stenosis of 30 per cent or more at the original angioplasty. Those free of recurrence after 6 months are apparently likely to remain so in the long-term (Tegtymeyer *et al.* 1984*a*,*b*). Recurrent stenotic lesions can be dilated readily and repeatedly if necessary (Tegtmeyer *et al.* 1984*a*,*b*; Grim *et al.* 1981). Various measures have been employed to prevent thrombosis or restenosis after angioplasty (Ramsay and Waller 1990). Aspirin is recommended because it is simple, economic and acceptably safe. Quite apart from restenosis, patients with hypertension which has been 'cured' by angioplasty need to be followed-up long-term, as they may develop new stenosing lesions (Jones *et al.* 1978).

Role of angioplasty in management

Angioplasty is clearly the treatment of choice for fibromuscular renal artery stenosis. The chance of technically successful dilatation is high, complications of angioplasty in this group are relatively infrequent, and the cure rate of hypertension is approximately 50 per cent. The role of transluminal angioplasty in the treatment of atherosclerotic renovascular disease is much less clear. The technical failure rate is high, the risk of complications is substantial, and the cure rate for hypertension is a disappointing 7–8 per cent on the intention to treat principle. Benefit in this group hinges largely on 'improvement' of hypertension, and this is distinctly suspect. In any event those with 'improved' blood pressure have to continue on antihypertensive therapy (by definition). It is generally possible to control blood pressure adequately with modern antihypertensive therapy, without intervention, and it might be argued that those who have 'improved' are little better off. There is plainly a need for a prospective controlled trial comparing angioplasty to conservative medical management in atherosclerotic renovascular disease (Brawn and Ramway 1987; Ramsay and Waller 1990). There are however clear indications for attempting angioplasty in certain groups of patients with atherosclerotic renovascular disease, for example those with severe and uncontrollable hypertension, those with declining renal function despite adequate blood pressure control and those with pulmonary oedema caused by advanced bilateral renovascular disease (Pickering *et al.* 1988). Although angioplasty should be attempted in these circumstances the outcome has been disappointing in our experience. A case can also be made for dilating discrete proximal non-ostial atherosclerotic lesions.

Role of surgery

Surgical techniques used in renovascular disease have been reviewed by Novick (1990). Surgical intervention should be considered in the 10–20 per cent of patients with fibromuscular disease which is not amenable to angioplasty, for example those with distal aneurysms or intra-renal lesions. In atherosclerotic renovascular disease surgery might be considered in the 60 per cent so of patients whose disease cannot be corrected by angioplasty, most often because of ostial stenosis, renal artery thrombosis, or long atherosclerotic lesions. However, such lesions are generally found in older subjects with renal impairment and generalized vascular disease, precisely the patients who have the highest operative mortality (Franklin *et al.* 1975) and least satisfactory blood pressure responses (Foster *et al.* 1975) to surgery. When angioplasty cannot be performed in such patients we generally advise conservative medical management, and recommend surgery only when hypertension is severe and uncontrollable, when renal function declines, or in patients with pulmonary oedema.

Summary

Efforts to diagnose renovascular disease should be directed particularly at hypertensive patients aged 60 years or less who have the suggestive clinical features listed in Table 15.1. The diagnosis should also be considered in older patients if there are special circumstances, such as severe and uncontrollable hypertension, declining renal function, or pulmonary oedema. The rapid sequence IVU remains at present the best screening test which is generally available, and also provides information on other renal causes of hypertension. It is however an imperfect screening test, and renal arteriography should be performed when it is essential to exclude renovascular disease. In fibromuscular renal artery stenosis the lesion can usually be dilated by transluminal angioplasty, with few complications and a 50 per cent cure rate of hypertension. In atherosclerotic renovascular disease technical failure is common, complications of angioplasty are frequent, and the cure rate of hypertension is very low. Conservative management may be preferable unless there is severe and uncontrollable hypertension, declining renal function, or pulmonary oedema. Those who are unsuitable for angioplasty are generally the patients who also do least well with surgical management.

References

Atkinson, A.B. and Kellett, R.J. (1974). Value of intravenous urography in investigating hypertension. *J. Roy. Coll. Phys. Lond.* **8**, 175–81.

Bell, G.M., Reid, J., and Buist, T.A.S. (1987). Percutaneous transluminal angioplasty improves blood pressure and renal function in renovascular hypertension. *Q. J. Med. (new Series)*, **63**, 393–403.

Berglund, G., Andersson, O., and Wilmhelmsen, L. (1976). Prevalence of primary and secondary hypertension: studies in a random population sample. *Br. Med. J.* **3**, 554–6.

Brawn, L.A. and Ramsay, L.E. (1987). Is 'improvement' real with percutaneous transluminal angioplasty in the management of renovascular hypertension? *Lancet*, **2**, 1313–6.

Carmichael, D.J.S., Mathias, C.J., Snell, M.E., and Peart, S. (1986). Detection and investigation of renal artery stenosis. *Lancet*, **1**, 667–70.

Choudhri, A.H., Cleland, J.G.F., Rowlands, P.C., Tran, T.L., McCarty, M., Al-Kutoubi, M.A.O. (1990). Unsuspected renal artery stenosis in peripheral vascular disease. *Br. Med. J.* **301**, 1197–8.

Cicuto, K.P., McLean, G.K., Oleaga, J.A., Freiman, D.B., Grossman, R.A., and Ring, E.J. (1981). Renal artery stenosis: anatomic classification for percutaneous transluminal angioplasty. *Am. J. Roentgenol.* **137**, 599–601.

Colapinto, R.F., Stronell, R.D., Harries-Jones, E.P. *et al.* (1882). Percutaneous transluminal dilatation of the renal artery. Follow-up studies on renovascular hypertension. *Am. J. Roentgenol.* **139**, 722–32.

Fommei, E., Ghione, S., Palla, L. *et al.* (1987). Renal scintigraphic captopril test in the diagnosis of renovascular hypertension. *Hypertension*, **10**, 212–20.

Foster, G.H., Maxwell, M.H., Franklin, S.S. *et al.* (1975). Renovascular occlusive disease. Result of operative treatment. *J.A.M.A.* **231**, 1043–8.

Franklin, S.S., Young, J.D., Maxwell, M.H. *et al.* (1975). Operative morbidity and mortality in renovascular disease. *J.A.M.A.* **231**, 1148–53.

Geyskes, G.G., Puyleart, C.B.A., Oei, H.Y., and Dorhout Mess, E.J. (1983). Follow up study of 70 patients with renal artery stenosis treated by percutaneous transluminal dilatation. *Br. Med. J.* **287**, 333–6.

Grim, C.E., Luft, F.C., Yune, H.Y., Klatte, E.C., and Weinberger, M.H. (1981). Percutaneous transluminal dilatation in the treatment of renal vascular hypertension. *Ann. Intern. Med.* **95**, 439–42.

Gruntzig, A., Kuhlmann, U., Vetter, W., Lutolf, U., Meier, B., and Siegenthaler, W. (1978). Treatment of renovascular hypertension with percutaneous transluminal dilatation of a renal-artery stenosis. *Lancet*, **1**, 801–2.

Havey, R.J., Krumlovsky, F., del Greco, F., and Martin, H.G. (1985). Screening for renovascular hypertension. Is renal digital-subtraction angiography the preferred noninvasive test? *J.A.M.A.* **254**, 388–93.

Jones, E.O.P., Wilkinson, P., and Taylor, R.M.R. (1978). Contralateral renal artery fibromuscular dysplasia after nephrectomy for renal artery stenosis. *Br. Med. J.* **2**, 825–6.

Kaplan-Pavlobcic, S., Koselj, M., Obrez, I. *et al.* (1985). Percutaneous transluminal renal angioplasty: follow up studies in renovascular hypertension. *Przegl. Lek.* **42**, 342–4.

Kohler, T.R., Zierler, E., Martin, R.L. *et al.* (1986). Noninvasive diagnosis of renal artery stenosis by ultrasonic duplex scanning. *J. Vasc. Surg.* **4**, 450–6.

Kuhlmann, U., Greminger, P., Gruntzig, A. *et al.* (1985). Long-term experience in percutaneous transluminal dilatation of renal artery stenosis. *Am. J. Med.* **79**, 692–8.

Luft, F.R., Grim, C.E., and Weinberger, N.J. (1983). Intervention in patients with renovascular hypertension and renal insufficiency. *J. Urol.* **130**, 645–56.

Mackay, A., Boyle, P., Brown, J.J. *et al.* (1983). The decision on surgery in renal artery stenosis. *Q. J. Med. (New Series)*, **52**, 363–81.

Mahler, F., Triller, J., Weidmann, P., and Nachbur, B. (1986). Complications in percutaneous transluminal dilatation of renal arteries. *Nephron*, **44**(1), 60–3.

Marshall, F.I., Hagen, S., Mahaffy, R.G. *et al.* (1990). Percutaneous transluminal angioplasty for atheromatous renal artery stenosis—blood pressure response and discriminant analysis of outcome predictors. *Q. J. Med. (New Series)*, **75**, 483–9.

Martin, E.C., Mattern, R.F., Baer, J., Fankuchen, E.I., and Casarella, W.JU. (1981). Renal angioplasty for hypertension: predictive factors for long-term success. *Am. J. Roentgenol.* **137**, 921–4.

Martin, L.G., Price, R.B., Casarella, W.J. *et al.* (1985). Percutaneous angioplasty in clinical management of renovascular hypertension: initial and long-term results. *Radiology*, **155**, 629–33.

Miller, G.A., Ford, K.K., Braun, S.D. *et al.* (1985). Percutaneous transluminal angioplasty vs surgery for renovascular hypertension. *Am. J. Roentgenol.* **144**, 447–50.

Novick, A.C. (1990). Renovascular hypertension. Surgical treatment. In *Hypertension: pathophysiology, diagnosis and management*, (ed. J.H. Laragh and B.M. Brenner) pp. 1561–71. Raven Press, New York.

Pickering, T.G. (1990). Renovascular hypertension. Medical evaluation and non-surgical treatment. In *Hypertension: pathophysiology, diagnosis and management.* (ed. J.H. Laragh and B.M. Brenner) pp. 1539–59. Raven Press, New York.

Pickering, T.G., Sos, T.A., Saddekni, S. *et al.* (1986). Renal angioplasty in patients with azotemia and renovascular hypertension. *J Hypertens*, **4**(6), S667–9.

Pickering, T.G., Herman, L., Devereux, R.B. *et al.* (1988). Recurrent pulmonary oedema in hypertension due to bilateral renal artery stenosis: treatment by angioplasty or surgical revascularisation. *Lancet*, **2** 551–2.

Ramsay, L.E. and Waller, P.C. (1990). Blood pressure response to percutaneous transluminal angioplasty for renovascular hypertension: an overview of published series. *Br. Med. J.* **300**, 569–72.

Schwarten, D.E. (1980). Transluminal angioplasty of renal artery stenosis: 70 experiences. *Am. J. Roentgenol.* **135**, 969–74.

Sellars, L., Siamopoulos, K., Hacking, P.M. *et al.* (1985*a*). Renovascular hypertension: ten years experience in a regional centre. *Q. J. Med. (New Series)*, **56**, 403–16.

Sellars, L., Shore, A.C., and Wilkinson, R. (1985*b*). Renal vein renin studies in renovascular hypertension—do they really help? *J. Hypertens.* **3**, 177–81.

Simon, N., Franklin, S.S., Bleifer, K.H., and Maxwell, M.H. (1972). Clinical characteristics of renovascular hypertension. *J.A.M.A.* **220**, 1209–18.

Sos, T.A., Pickering, T.G., Sniderman, K. *et al.* (1983). Percutaneous transluminal renal angioplasty in renovascular hypertension due to atheroma or fibromuscular dysplasia. *N. Engl. J. Med.* **309**, 274–9.

Sutters, M., Al-Kutoubi, M.A., Mathias, C.J., and Peart, S. (1987). Diuresis and syncope after renal angioplasty in a patient with one functioning kidney. *Br. Med. J.* **295**, 527–8.

Tegtmeyer, C.J., Dyer, R., and Teates, C.D. (1980). Percutaneous transluminal dilatation of the renal arteries. Techniques and results. *Radiology*, **135**, 589–99.

Tegtmeyer, C.J., Kellum, C.D., and Ayers, C. (1984*a*). Percutaneous transluminal angioplasty of the renal artery. *Radiology*, **153**, 77–84.
Tegtmeyer, C.J., Kofler, T.J., and Ayers, C.A. (1984*b*). Renal angioplasty: current status. *Am. J. Roentgenol.* **142**, 17–21.
Thornbury, J.R., Stanley, J.C., and Fryback, D.J. (1982). Hypertensive urogram: nondiscriminatory test for renovascular hypertension. *Am. J. Roentgenol.* **138**, 43–9.

Part 4 Molecular Genetics in Renal Disease

16 Inherited disorders of the kidney: the application of modern molecular genetics

S.H. Morgan

Inherited disorders of the kidney are becoming increasingly important, not only in paediatric nephrology, but also in adult practice. In almost 50 per cent of all children and 15 per cent of adults accepted for renal replacement therapy the underlying disorder has an hereditary basis. Hereditary disorders of the kidney may affect all parts of the nephron. Many of these diseases manifest at birth, but others do not become obvious until later in adult life. Within the past decade, the ability to define the genetic lesions responsible for some of these disorders has revolutionized our diagnostic approach, and in some instances it may in the future pave the way to rationalization of their management. In this chapter, discussion has been confined to developments in inherited renal disorders which usually present first in adult life (Table 16.1).

Table 16.1 Common heritable disorders which may present with renal failure in adult life

Cystic disorders
Autosomal dominant polycystic disease
Structural disorders of basement membrane
Alport-type hereditary nephritis
Thin membrane syndrome
Onycho-osteodysplasia (nail-patella syndrome)
Storage disorders
Alpha-galactosidate A deficiency (Anderson–Fabry disease)
Lecithin–cholesterol acyltransferase deficiency
Amyloidosis
Gaucher's disease
Von-Gierke's disease
Disorders predisposing to renal calculi
Primary hyperoxaluria
Cystinuria
Multisystem disease
Diabetes mellitus
Systemic lupus erythematosus

Investigation of genetic disorders

The strategies employed to study any disorder with a genetic basis will depend on several factors. In some diseases, expression of the mutant gene may be very specifically defined, as in Anderson–Fabry disease, where the disease gene results in a failure to synthesize the catalytically active form of an enzyme, alpha-galactosidase A. There are however other disorders in which the underlying biochemical defect is unknown, as is the case with adult polycystic kidney disease. Pedigree pattern analysis can often confirm that a disorder is inherited as a single gene, autosomal or sex-linked trait. The next step is to identify the locus or position on the chromosome of the mutant gene. Such mapping is a prerequisite for improving prenatal diagnostic and carrier state detection. Determining the exact nature of the mutation may provide a basis for the rational design of therapy.

The confirmation that a disorder is linked to a particular sex, automatically assigns or locates the causative gene to a particular sex chromosome as in Anderson–Fabry disease and classical X-linked Alport's syndrome. Chromosomal location is much more difficult for autosomal disorders but still remains the first step in gene mapping, and family linkage studies may provide the most useful information.

Family linkage studies

Two genes are said to be 'linked' if their loci are in close proximity on the same chromosome, so that alleles at these two loci are inherited together at meiosis. If the chromosomal location of one of the genes is known, then by inference the other can be mapped to that same area of the chromosome. In family linkage studies two loci are considered—one unknown for the disease being studied, and another, the known marker. Such markers include serum protein or enzyme polymorphisms, blood groups, and specific restriction fragment length polymorphisms (RFLPs).

Restriction fragment length polymorphisms

DNA can be extracted from any nucleated cell including peripheral blood lymphocytes, amniocytes and fetal tissue obtained at chorionic biopsy. The DNA can be cleaved into fragments using bacterially derived restriction endonucleases. The number and size of the DNA fragments depends on the number and site of specific points of cleavage or endonuclease recognition sites—these being determined genetically according to changes in base sequences (Fig. 16.1). The DNA fragments can be separated on the basis of size by agarose gel electrophoresis, and individual fragments identified by hybridization to radiolabelled complementary DNA sequences (Southern analysis) (Fig. 16.2). In some genetic diseases, the mutation causing the disease may cause the elimination or addition of an endonuclease recognition site, or the disease gene may be tracked by the presence of a neighbouring recognition site.

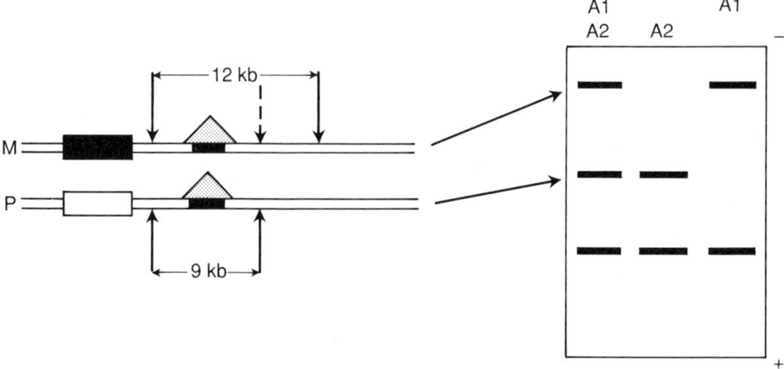

Fig. 16.1 Segment of a pair of chromosomes with the paternal chromosome (P) and maternal chromosome (M). The genetic locus under study is shown as a solid bar. DNA from P has three restriction endonuclease recognition sites, and so produces two fragments when cleaved. DNA from M however lacks one of these sites, so only produces one larger fragment. The presence or absence of this site is called a restriction fragment length polymorphism (RFLP).

There are a number of sophisticated computer programmes that can be used to analyse the statistical significance of RFLPs and other linkage data from pedigrees. The odds in favour of linkage of a marker locus to a disease locus are expressed as a LOD score (logarithm of the odds against linkage). A LOD score of 3 or more favours linkage with odds of 1000: 1. A negative LOD score usually implies non-linkage. Family studies using several marker loci allow the construction of linkage maps, ordering the position of markers in relation to the specific disease gene under investigation. The developmet of the polymerase chain reaction (PCR) has further facilitated the use of DNA technology as a valuable diagnostic tool as well as a research tool. It has allowed the intricate study of disease mutations localized by linkage studies and other methods of gene mapping within all areas of medicine. Some specific renal disorders are discussed below.

Anderson–Fabry disease (alpha-galactosidase A deficiency)

This disorder, was first described in 1898 as a dermatological curiosity by two physicians independently (Anderson 1898; Fabry 1898). Genetically, it represents a heterogeneous group of monogenic disorders where the gene product is known. A mutation wihtin the gene responsible for the synthesis of the mature alpha-galactosidase A (alpha-Gal A) produces deficiency or catalytic inactivity of this lysosomal hydrolase, resulting in the accumulation of an abnormal storage product – globotriaosyceramide, a glycosphingolipid. Deposition of this storage product within the lysosomes of vascular endo-

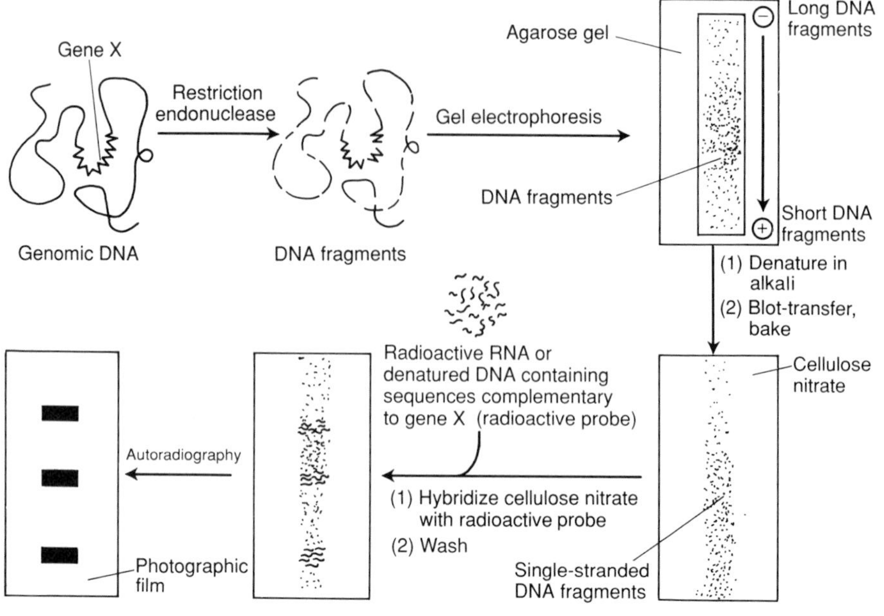

Fig. 16.2a DNA fragment identification by Southern analysis.

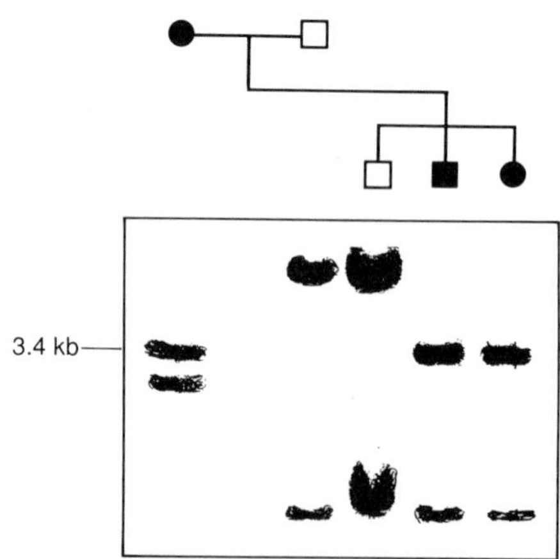

Fig. 16.2b A small part of an adult polycystic kidney disease pedigree is shown. Affected individuals are indicated by the filled symbols. (circles = female, squares = male). The autoradiograph is from the Southern analysis of *Pvu II* digested genomic DNA extracted from peripheral blood leucocytes (of each individual) hybridized with the 3′HVR probe is shown below. Each track contains two polymorphic fragments, one from each parental chromosome. The larger fragment (3.4 kb) segregates with the disease.

thelium and smooth muscle cells produces the protean clinical manifestations of the disease including angiokeratomata (Fig. 16.3), a small fibre neuropathy, dysrhythmias, cardiac failure, thrombotic cerebrovascular events and progressive renal failure.

Pedigree analysis suggested that the disorder was X-linked (Opitz *et al.* 1965) and the enzymatic defect was subsequently defined in 1970 (Kint 1970). The genetic locus responsible for alpha-Gal A production was physically assigned to the middle of the long arm of the X chromosome, between the banding regions Xq21 and 24, by somatic cell hybridization studies (Shows *et al.* 1978) (Fig. 16.4). In males, the diagnosis of alpha-Gal A deficiency is suggested by the presence of the cutaneous lesions and slit lamp ophthalmoscopy, which invariably demonstates an asymptomatic keratopathy (cornea verticillata). Absent or residual alpha-Gal A activity in peripheral blood leucoytes or cultured skin fibroblasts confirms the diagnosis.

Most female carriers of the disease are asymptomatic and may often be difficult to diagnose with certainty. Alpha-Gal A activity may be reduced, but often falls within the lower end of the normal range. Some female carriers may, however, have clinical features as severe as those seen in affected males. RFLP studies in alpha-Gal A deficiency have enabled more precise localization of the Anderson–Fabry disease gene (MacDermot *et al.* 1987, Morgan *et al.* 1987) and can in sufficiently extensive pedigrees be a useful guide to carrier

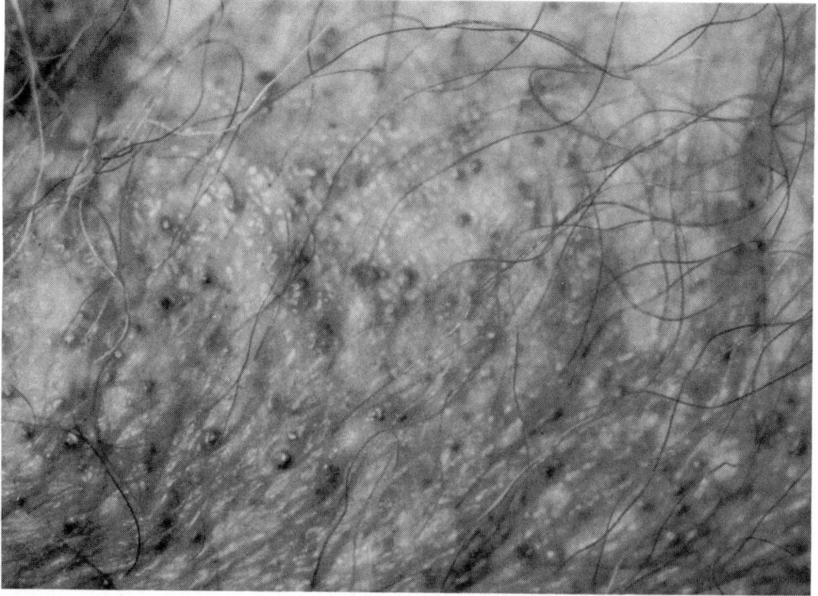

Fig. 16.3 The classical cutaneous lesions of alpha-galactosidose A deficiency—angiokeratoma.

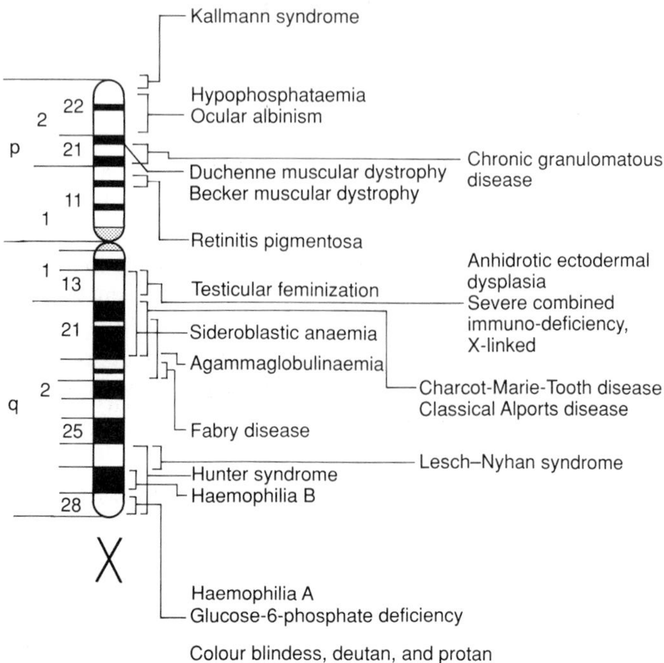

Fig. 16.4 Map of the X-chromosome showing physical location of common disease loci.

state. Pre natal diagnosis of affected males is achieved by enzymatic assay of amniocytics or direct analysis of chorionic villous tissue.

Within the past five years Desnick *et al.* (1989) have successfully sequenced the mature alpha-Gal A polypeptide and cloned a complementary cDNA enabling the characterization of the structure and organization of the alpha-Gal A gene. This gene specific probe has subsequently been used to explore and define mutations within the alpha-Gal A gene (Desnick *et al.* 1987) (Fig. 16.5). Anderson–Fabry disease however remains incurable and can only be managed symtomatically and limited by effective genetic counselling (Morgan and Crawfurd 1988).

Adult polycystic kidney disease

The mutant gene for adult type autosomal dominant polycystic kidney disease is carried by 1 in 1000 people and accounts for end-stage renal failure in 10 per cent of patients on replacement therapy within Europe. This is however the fate of only 50 per cent of patients with the disease. In adult polycystic kidney disease, the product of the disease gene is unknown and it appears

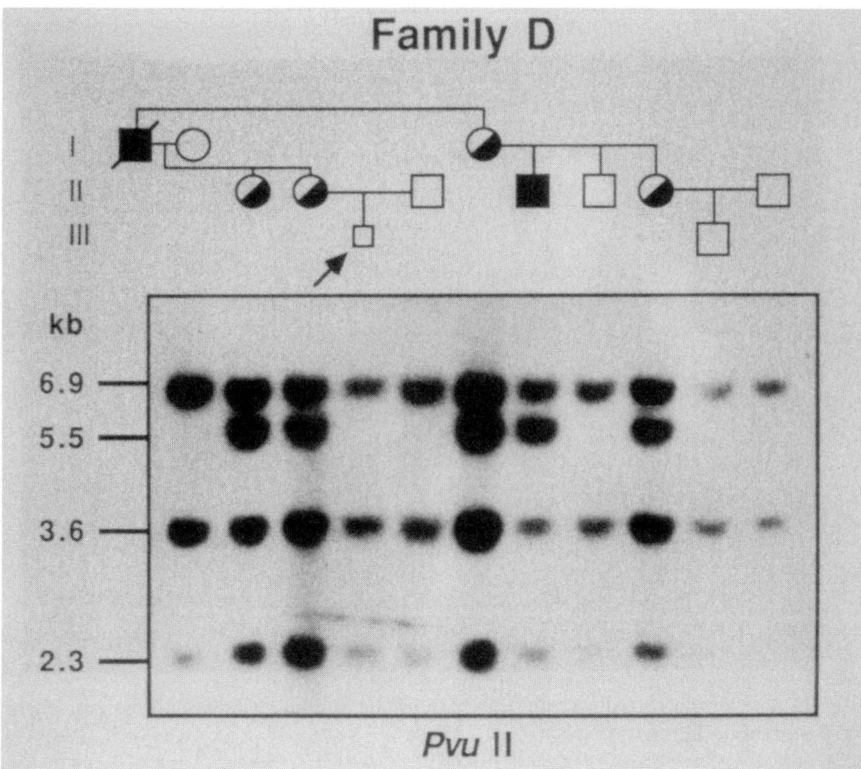

Fig. 16.5 Autogradiograph of a Southern blot of genomic DNA from a family with Fabry's disease digested with a restriction enzyme (PVUII) and hybridized with the gene-specific probe for α-galactosidase A. In normal family members and individuals marrying into the family there are three α-galactosidase A specific restriction fragments at 2.3, 3.6, and 6.9 kb. In affected family members there is an extra restriction fragment at 5.6 kb. Restriction fragment mapping suggests that in this family the gene mutation is due to a reduplication. (Reproduced with permission, from Desnick *et al.* 1987)

that there is considerable variability in terms of the clinical expression of the disease. The term adult polycystic kidney disease is somewhat ambiguous as there are may extrarenal disease features and associations (Gabow 1990) (Table 16.2). Indeed, these associations suggest that the basic defect in adult polycystic kidney disease is within the extracellular matrix and that the disease gene may produce a developmental abnormality of connective tissue.

Despite sophisticated imaging techniques (Fig. 16.6) it is often impossible to reassure 'at-risk' members of a family who carry the disease gene that they will not develop adult polycystic kidney disease. Even with a normal abdominal visceral scan (computerized tomography or ultrasound) at the age of 35 years, such an individual will still have a 10–15 per cent chance of developing cystic

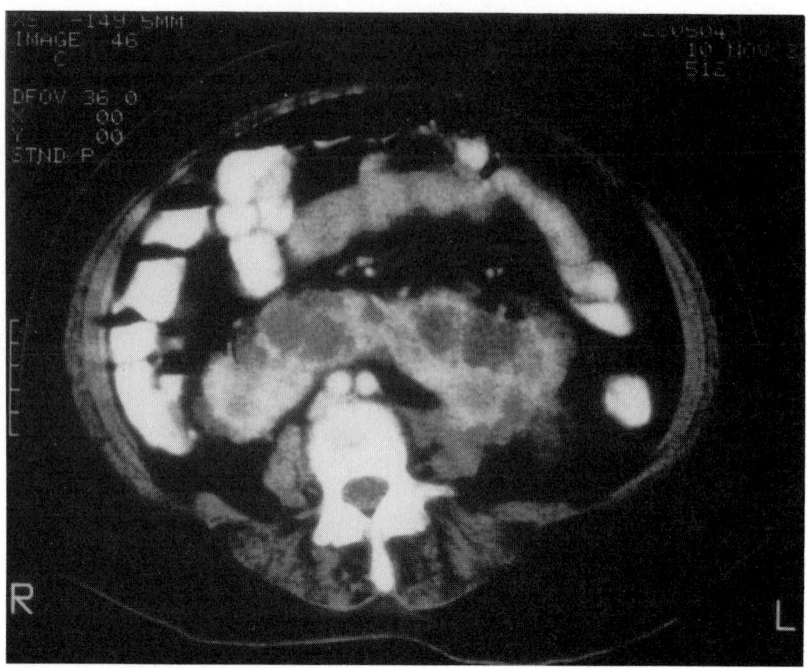

Fig. 16.6 Computerized tomography scan of the abdomen showing polycystic change in a horseshoe kidney.

change in later life. Over the last decade, the search for a candidate gene began. Since the gene product was chemically unknown, there were no clues, and a strategy of 'reverse genetics' was followed (Weatherall 1986) to identify the mutant gene and its role in the pathogenesis of adult polycystic kidney disease. In 1985 Reeders *et al.* using cloned DNA markers which recognized base sequences within the alpha globin gene cluster on the short arm of chromosome 16, found close linkage between a marker—3′HVR—and the disease

Table 16.2 Extra-renal manifestations of adult poylycystic kidney disease

Hepatic cysts
Pancreatic cysts
Ovarian cysts
Mitral valve prolapse
Intra-cranial aneurysms
Aortic aneurysms
Diverticular disease
Inguinal hernias

gene in pedigrees from the Oxford region. Subsequently additional linkage studies using a chromosome 16 assigned enzyme polymorphism—phosphoglycolate phosphatase, confirmed this (Reeders *et al.* 1986*a*). The restriction map at this end of chromosome 16 is now expanding and the genetic distance between flanking markers and the putative gene is closing. Even with such relatively crude localization it has still been possible, in informative families, to offer antenatal DNA diagnosis in adult polycystic kidney disease with high levels of confidence (Reeders *et al.* 1986*b*).

Genetic heterogeneity in adult polycystic kidney disease

In adult polycystic kidney disease (APKD) there may be marked heterogeneity between one individual and another from the same family and indeed between individuals from different families in terms of the pattern of extra-renal manifestations and the rate of progression to end stage renal failure. Initial linkage studies using pedigrees from several centres suggested that these was no evidence of genetic heterogeneity and that linkage with the alpha-globin gene cluster was found in all families studied. As these studies were extended to include individuals from other centres and racial groups, a few large pedigrees did not show linkage with chromosome 16 markers (Romeo *et al.* 1988; Kimberling *et al.* 1988). This second—or possibly multiple—alternative genetic locus has so far not been chromosomally assigned.

Clinical studies, comparing APKD1 (chromosome 16) and APKD2 do not suggest any clear differences although it appears that progression to end stage renal failure in patients with APKD2 may be much slower (Parfrey *et al.* 1990). The proportion of families carrying the APKD2 gene is not clear, but has been estimated at 4 per cent (Pieke *et al.* 1989). This may well be an underestimate, since selection of patients for many of the previous studies is biased towards patients presenting to nephrologists, and may not include a phenotypically distinct body of patients with adult polycystic kidney disease who are seen in hypertension clinics, by urologists, or by gastroenterologists. The study of population genetics by these techniques is often linited by an inability to generate linkage studies in the pedigrees of all the reference cases. In adult polycystic kidney disease we have very much found this to be the case as the family structure may be too limited to provide informative data (Jeffery *et al.* 1992).

Hereditary nephritis

In 1927, Alport described a pedigree with haematuria and progressive renal failure with deafness in affected males. Ocular defects (lenticonus) were also a feature, and occasionally more unusual associations, such as macrothromobocytosis and oesophogeal leiomyomatosis, were found. Classical X-linked (Alport's) hereditary nephritis is now part of a spectrum of hereditary renal disease in which the primary abnormality is a defect in the biochemical composition of renal basement membrane—tubular and glomerular. These abnormalites include benign familial haematuria (thin membrane syndrome), nail-patella syndrome (onycho-osteodysplasia) and nephronophthisis (cystic renal medulla complex).

Histological examination of renal biopsy specimens from patients with Alport type nephritis shows irregular thickening of the glomerular basement membrane, with splitting of the lamina densa. A number of biochemical abnormalities have been documented in glomerular basement membrane isolated from these patients; interestingly, serum from patients with Goodpasture's syndrome (containing antibodies against glomerular basement membrane) will not bind to it. The defect in Alport's syndrome has been defined as a polypeptide abnormality within the non-collagenous domain of type IV collagen (Noel *et al.* 1989). Localization of the disease gene in classical Alport's has been mapped to the long arm of the X chromosome (Brunner *et al.* 1989; Flinter *et al.* 1988). Nail-patella syndrome, which is associated with glomerular basement membrane abnormalities which progress to end stage renal failure, has been mapped to chromosome 9.

Management of hereditary renal disorders

Despite the rapid advances in molecular genetics and molecular biology in the past decade, specific therapy for all the disorders discussed here has not evolved. The only exception is perhaps the recognition of the enzymatic defect in primary hyperoxaluria type I, in which hepatic activity of a transaminase, alanine-glyoxylate aminotransference, is deficient or absent (Chapter 17). In this disorder, which usually presents in adolescence with renal calculi and progressive development of renal parenchymal oxalate deposition, liver transplantation appears curative (Morgan and Watts 1989). In adult polycystic kidney disease and the other conditions discussed in this chapter the goals are directed at the genetic counselling of affected families, supported by the genetic techniques which enable accurate antenatal and carrier state diagnosis. Several studies suggest that even this will not alter attitudes and planning in such families (Sujansky *et al.* 1990; MacNicol *et al.* 1991).

References

Anderson, W. (1898). A Case of 'angio-keratoma' *British Journal of Dermatology*, **10**, 113–7.

Brunner, H., Van Bennekun, C., Schroder, C. *et al.* (1987). X-linked Alport syndrome: localisation of the gene in three families. *Kidney International*, **31**, 1044–7.

Desnick, R.J., Bernstein, H.S., Astrin, K.H., and Bishop D.F. (1987). Fabry disease: molecular diagnosis of hemizygotes and heterosygotes. *Enzyme*, **38**, 54–64.

Desnick, R.J., Astrin, K.H., and Bishop, D.F. (1989). Fabry disease: molecular genetics of the inherited nephropathy. *Advances in Nephrology*, **18**, 113–28.

Fabry, J. (1988). Ein beitrag zur kenntniss der purpura haemorrhagica nodularis (Purpura papulosa haemorrhagica hebrae) *Archives for Dermatologie and Syphilistherapie*, **43**, 187–200.

Flinter, F.A., Cameron, J.S., Chantler, C., Houston, I., and Bobrow, M. (1988). Genetics of classic Alport's syndrome. *Lancet*, **2**, 1005–7.

Gabow, P.A. (1990). Autosomal dominant polycystic kidney disease—more than a renal disease. *American Journal of Kidney Diseases*, **16**, 403–13.

Jeffery, S., Morgan, S.H., and Patton, M. (1992). Genetic counselling in families with adult polycystic kidney disease: The problem of heterogeneity. *Nephrology Dialysis and Transplantation* (In press).

Kint, J.A. (1970). Fabry;s disease: alpha galactosidase deficiency. *Science*, **167**, 1268–9.

Kimberling, W.J., Fain, P.R., Kenyon, J.B., Goldgar, D., Sujansky, E., and Gabow, P.A. (1988). Linkage heterogeneity of autosomal dominant polycystic kidney disease. *New England Journal of Medicine*, **319**, 913–8.

MacDermot, K.D., Morgan, S.H., Cheshire, J.K., and Wilson, T.M. (1987). Anderson–Fabry Disease, a close linkage with highky polymorphic DNA markers DXS 17, DXS 87 and DXS 88. *Human Genetics*, **77**, 263–6.

MacNicol, A.M., Wright, A.F., and Watson, M.L. (1991). Education and attitudes in families with adult polycystic kidney disease. *Nephrology Dialysis and Transplantation*, **6**, 27–30.

Morgan, S.H., Cheshire, J.K., Wilson, T.M., MacDermot, K., and Crawfurd M. d'A (1987). Anderson–Fabry disease—family linkage studies using two polymorphic X-linked DNA probes. *Paediatric Nephrology*, **1**, 536–9.

Morgan, S.H., and Crawfurd, M. d'A (1988). Anderson–Fabry disease. *British Medical Journal*, **297**, 872–3.

Morgan, S.H., and Watts, R.W.E. (1989). Perspectives in the assessment and management of patients with primary hyperoxaluria type I. *Advances in Nephrology*, **18**, 95–106.

Noel, L.H., Gubler, M.B., Bobrie, G., Savage, C.O.S., Lockwood, C.M., and Grunfeld J.P. (1989). Inherited defects of renal basement membranes. *Advances in Nephrology*, **18**, 77–94.

Opitz, J.M. *et al.* (1965). The genetics of angiokeratoma corporis diffusum (Fabry's disease) and its linkage relations with the Xg locus. *American Journal of Human Genetics*, **17**, 325–42.

Parfrey, P.S., Bear, J.C., Morgan, J. *et al.* (1990). The diagnosis and prognosis of autosomal dominant polycystic kidney disease. *New England Journal of Medicine*, **323**, 1085–90.

Pieke, S.A., Kimberling, W.J., Kenyon, J.B., and Gabow, P. (1989). Genetic heterogeneity of polycystic kidney disease: an estimate of the proportion of families unlinked to chromosome 16. *American Journal of Human Genetics*, **45**, A58, (Abstr.).

Reeders, S.T., Breuning, M.H., Davies, K.E., Nicholls, R.D., Jarman, A.P., Higgs, D.R., Pearson, P.L., and Weatherall, D.J. (1985). A highly polymorphic DNA marker linked to adult polycystic kidney disease on chromosome 16. *Nature*, **317**, 542–4.

Reeders, S.T., Breuning, M.H., Corney, G. *et al.* (1986*a*). Two genetic markers closely linked to adult polycystic kidney disease on chromosome 16. *British Medical Journal*, **292**, 851–3.

Reeders, S.T., Zerres, K., Gal, A *et al.* (1986*b*). Prenatal diagnosis of autosomal dominant polycystic kidney disease with a DNA probe. *Lancet*, **2**, 6–8.

Romeo, G., Devoto, M., Costa, G *et al.* (1988). A second genetic locus for autosomal dominant polycystic kidney disease. *Lancet*, **2**, 8–11.

Shows, T.B., Brown , J.A., Haley, L.L., Groggin, A.P., Eddy, R.L., and Byers, M.G. (1978). Assignment of alphagalactosidase to the q22-qter region of the X chromosome in man. *Cytogenetics and Cell Genetics*, **22**, 541–4.

Sujansky, E., Kreutzer, S.B., Johnson, A.M., Lezotte, D.C., Schrier, R.W., and Gabow, P.A. (1990). Attitudes of at-risk and affected individuals regarding presymtomatic testing for autosomal dominant polycystic kidney disease. *American Journal of Medical Genetics*, **35**, 510–15.

Weatherall, D.J. (1986). *The new genetics and clinical practice*. (2nd edn). Oxford University Press.

17 Molecular aetiology and pathogenesis of primary hyperoxaluria: implications for clinical management

C.J. Danpure

Oxalate stones and hyperoxaluria

Oxalate stone disease is a common condition, which has been estimated to affect up to 2 per cent of the population (Resnick *et al.* 1968). Although clearly multifactorial in nature, the most important factor in the intrarenal crystallization of calcium oxalate is the urinary concentration of oxalate (Robertson and Peacock 1980). This is the case even though most patients with oxalate urolithiasis do not have overt hyperoxaluria. Excessive excretion of oxalate can result from a primary (genetic) disease or be a secondary consequence of some other abnormality. Secondary hyperoxaluria can be caused by a variety of conditions, including ethylene glycol poisoning and excessive oxalate absorption from the gut, such as that caused by extensive disease of the small intestine or major resections of the small bowel (Smith 1980). The term primary hyperoxaluria (PH) covers a small number of genetic diseases, of which only two, type 1 (PH1) and type 2, (PH2), have been characterized at the level of the basic biochemical defect. There might also be a third, much less well characterized PH, due to the primary hyperabsorption of oxalate (Yendt and Cohanim 1986).

The primary hyperoxalurias—clinical/metabolic description

PH1 and PH2 are both autosomal recessive diseases. Although both are rare, PH1 is considerably more common than PH2. They are both typically progressive disorders, in which the slow accumulation of calcium oxalate in the kidney eventually leads to renal failure. It is frequently assumed that PH2 is a somewhat milder disease than PH1, but the very small number of PH2 patients described in the literature, together with the considerable clinical heterogeneity found in PH1 (see below), make this an unsafe generalization. In PH1, at least, oxalate crystallization can take the form of various combinations of urolithiasis, nephrocalcinosis and systemic oxalosis (Williams and Smith 1983). Although the latter generally becomes significant only after renal failure, this may not always be the case.

Unlike the majority of PH1 patients, who have a chronically progressing disease, often starting in early childhood and only leading to renal failure in early adulthood, a small subgroup of PH1 patients suffer from an acute aggressive neonatal form of the disease, in which the first symptoms may appear in the first few months of life and death occurs within the first few years. In these patients, clinical overt systemic oxalosis (e.g. oxalotic cardiomyopathy) might feature relatively early in the disease history.

Purely on clinical grounds, PH1 and PH2 may be difficult to distinguish in many patients. In the absence of the necessary biochemical information, PH1 is often assumed, simply because it is very much more common. However it is not difficult to see the self-perpetuating consequences of such an assumption. Usually PH1 and PH2 are only distinguished unambiguously by urinary or, less commonly, plasma organic acid analysis. While PH1 is characterized by concomitant hyperoxaluria/aemia and hyperglycolic aciduria/aemia, PH2 is characterized by hyperoxaluria and hyper L-glyceric aciduria. Major problems can occur with the interpretation of organic acid levels when renal function deteriorates significantly. In an anuric patient with end-stage renal failure, urine is not available for assay and biochemical diagnosis has to allow for the fact that plasma oxalate levels can increase enormously in patients with renal failure from other causes (Morgan *et al.* 1987). The use of these procedures in the differential diagnosis of the PHs is also complicated by the recent observations that some patients, unambiguously diagnosed as PH1 by other means (see below), do not have hyperglycolic aciduria or may have only mild hyperoxaluria (Danpure 1990*a*).

The primary hyperoxalurias—enzymological description

Luckily much more precise definitions of the PHs can be made because, for both PH1 and PH2, the basic enzymatic defects are known. PH1 is caused by a deficiency of the enzyme alanine:glyoxylate aminotransferase (AGT) (Danpure and Jennings 1986), whereas PH2 is caused by a deficiency of the enzyme glyoxylate reductase/D-glycerate dehydrogenase (GR) (Williams and Smith, 1968; Mistry *et al.* 1988). AGT is a liver-specific enzyme which is located within the peroxisomes of hepatocytes (Noguchi and Takada 1979). GR is found in a wide variety of tissues, and is located in the cytosol (Mistry *et al.* 1988). These enzymological findings provide an explanation for the elevated production and excretion of oxalate and glycolate in PH1 and oxalate and L-glycerate in PH2.

The metabolic role of AGT is to detoxify glyoxylate by catalysing its transamination to glycine, using alanine as the amino donor. Failure to do this in PH1 allows glyoxylate to be oxidized to oxalate either within the peroxisome, catalysed by L-alpha-hydroxy acid oxidase, or in the cytosol, catalysed by lactate dehydrogenase (LDH). In the cytosol, glyoxylate can also be reduced to glycolate, catalysed by GR and possibly LDH (Danpure and

Jennings 1986; Danpure 1989). In PH2, a deficiency of GR allows glyoxylate in the cytosol to be oxidized to oxalate, again catalysed by LDH (Mistry *et al.* 1988; Danpure *et al.* 1989*a*). GR also catalyses the reduction of hydroxypyruvate to D-glycerate. A deficiency in PH2 allows the hydroxypyruvate to be reduced instead to L-glycerate, catalysed yet again by LDH (Williams and Smith 1968). Thus LDH appears to be able to catalyse both oxidation and reduction reactions depending on the particular substrates and presumably the relative concentrations of NAD and NADH.

In the light of the knowledge above, the PHs can be more logically defined as either AGT deficiency (PH1) or GR deficiency (PH2). Further molecular characterization of the PHs has been almost entirely confined to PH1, which in recent years has been studied to the exclusion of PH2.

Enzymatic and clinical heterogeneity of PH1

A study of some 80 liver biopsies taken from PH1 patients over the past few years, has led to the identification of considerable heterogeneity at both the enzymic and clinical levels (Danpure 1991*a*). The enzymic heterogeneity is demonstrated by the wide range of hepatic AGT enzyme activities found in such patients, which can vary from 0 per cent to nearly 50 per cent of the mean control level. Simply on the basis of AGT enzyme activity the PH1 patients at the upper end of this range can not be distinguished from asymptomatic obligate heterozygotes. Most of the patients with zero AGT activity also have no detectable immunoreactive AGT protein (Wise *et al.* 1987). All patients with enzyme activity, together with a minority of those with zero activity, have immunoreactive AGT protein.

In human liver, AGT is normally exclusively localized within the peroxisomes. However in all of the PH1 patients who have significant AGT levels, the residual enzyme is mislocalized to the mitochondria (Danpure *et al.* 1989*b*). This apparent trafficking defect is without precedent in human disease. AGT is unable to carry out its metabolic function, namely glyoxylate detoxification, efficiently when located in the mitochondria, probably because it is the peroxisome that is the major site of glyoxylate synthesis (see Danpure 1989). Detoxification needs to take place within the peroxisome, otherwise the glyoxylate has the opportunity to be oxidized to oxalate in the cytosol. It has been estimated that about one third of all PH1 patients have their disease, at least in part, due to the mistargeting of AGT (Danpure 1991*a*).

Attempts to relate the enzymic heterogeneity in PH1 to the clinical heterogeneity, for example age of onset, rate of progression, age of ESRF, or urinary or plasma metabolite levels have been without success (Danpure 1991*a*). Although this is partly due to the paucity of clinical information on such patients, it is also due to the likelihood that, although PH1 is clearly a monogenic disease, its clinical manifestation is obviously multifactorial, being influenced by a variety of genetic factors, such as other enzyme activities, and environmental factors, such as diet. The observations that a number of severely AGT-deficient patients were apparently asymptomatic until as late as the sixth decade of life, the absence of hyperglycolic aciduria and the occurence of only mild hyperoxaluria in some PH1 patients, leads to the conclusion that such patients would have been diagnosed only with difficulty using the classical criteria alone (see above).

Molecular biology of PH1

Full length human AGT cDNA has been cloned and sequenced (Takada *et al.* 1990)and a full length AGT genomic clone has been restriction mapped, partially sequenced and localized at the sub-chromosomal level (Purdue *et al.* 1990*a*). Recently the mutations associated with AGT mistargeting have been identified (Purdue *et al.* 1991). All such PH1 patients, which amount to one third of the total, possess at least one allele containing the same three mutations. A minority of patients with the trafficking defect are homozygous for this allele, while the majority are heterozygous, the other allele probably being a null (Purdue *et al.* 1991). Enough information is now available on the AGT gene sequence and structure to allow the identification of most, if not all, of the mutations that lead to PH1. Whether this turns out to be a feasible proposition depends, at least in part, on the frequency of specific mutations in the PH1 population.

Unfortunately the GR gene has not been cloned. Therefore the identification of the mutations causing GR deficiency in PH2 is not yet possible.

Implications for the clinical management of PH1

The recent characterization of the aetiology and pathogenesis of PH1 at the molecular level (see above) has had a major influence on all aspects of clinical management of PH1, namely diagnosis, prenatal diagnosis and treatment.

Diagnosis

Until recently, diagnosis of PH1 and PH2 relied on the clinical observation of primary urolithiasis, nephrocalcinosis and/or systemic oxalosis (i.e. after the exclusion of possible secondary factors), together with a variety of urinary/plasma organic acid measurements. Although adequate for many patients, as mentioned above, it is not adequate for patients who present in end state renal failure or those PH1 patients whose phenotype does not include hyperglycolic aciduria.

However, definitive unambiguous diagnosis of PH1 and PH2 can be made, irrespective of the patient's clinical condition, by AGT or GR enzyme assay. For PH1 this needs to be done on liver tissue (a percutaneous needle biopsy is suitable) (Danpure *et al.* 1987), whereas for PH2 peripheral blood leukocytes have been found suitable (Williams and Smith 1968; Chalmers *et al.* 1984), although a more clear cut result may probably be obtained from a liver biopsy (Mistry *et al.* 1988). As far as PH1 is concerned, if AGT enzyme assay is accompanied by immunoblotting (to detect the presence or absence of immunoreactive AGT protein) (Wise *et al.* 1987) and immunoelectron microscopy (to determine the intracellular localization of the immunoreactive protein) (Cooper *et al.* 1988), all the currently recognized phenotypes can be detected. Each technique used on its own runs the risk of missing a small number of patients. Although the minimum amount of liver required for PH2 diagnosis is uncertain, unambiguous diagnosis of PH1 has been accomplished on as little as 2 mg.

The recent characterization of AGT genomic DNA (Purdue *et al.* 1991) has allowed the use of exon-specific polymerase chain reaction (PCR) followed by restriction digestion or allele-specific oligonucleotide hybridization to identify the mutations responsible for disease in some PH1 patients (those with the trafficking defect) (Purdue *et al.* 1990). These specific patients could, in theory, be diagnosed solely by analysis of the DNA extracted from any nucleated cell, such as peripheral blood leukocytes. Whether this method of diagnosis becomes a practicable alternative to enzyme assay for the majority of PH1 patients depends largely on the identification of the causative mutations and how many mutations this involves. The reliability and versatility of enzymic diagnosis, but the inconvenience of liver biopsy, might have to be balanced against the inconvenience of having to screen for many mutations, but the ease of blood sampling.

Prenatal diagnosis

Various attempts have been made to diagnose PH1 prenatally by measuring amniotic fluid organic acid levels. The lack of success of such procedures may be because the disease is not metabolically manifested *in utero* or because the oxalate and glycolate are very efficiently cleared by the placenta to be excreted by the maternal kidneys.

Prenatal diagnosis of PH1 has been successfully accomplished by AGT assay of fetal liver biopsies (Danpure *et al.* 1989*c*). Notwithstanding problems associated with the development expression profile of AGT in fetal liver, unambiguous prenatal diagnosis can be made if a combination of enzyme assay, immunoblotting and immunoelectron microscopy is used. Although immunoreactive AGT can be detected within peroxisomes of fetal liver at 9–10 weeks (Cooper *et al.* 1989), enzymic diagnosis can only be made fairly late in pregnancy, as it is not a practicable proposition to attempt fetal liver biopsy before the 16th week of gestation.

Exon-specific PCR of DNA extracted from a chorionic villus biopsy in the first trimester should provide an alternative approach. However this suffers from the same disadvantages outlined above for postnatal diagnosis by DNA analysis.

Treatment

Conventional treatment for both PH1 and PH2 is largely symptomatic. Attempts to control diet and the urinary solubility of oxalate predominate. Despite such treatment, renal oxalate deposition will continue, eventually leading to end stage renal failure. Such patients will then receive a variety of dialysis regimes, haemodialysis generally being more successful than peritoneal dialysis. Even the most rigorous dialysis regimes will not remove oxalate efficiently enough to account for its excessive synthesis (Watts *et al.* 1984), leading to oxalate deposition in many extra-renal sites throughout the body. Kidney transplantation may provide temporary relief due to more efficient clearance of oxalate, but, because the basic defect is in the liver (in PH1), the new kidney will eventually suffer the same fate as the first. Repeated kidney transplantations may be necessary.

The only conventional treatment that, in retrospect, addresses the root cause of PH1 is the administration of pharmacological doses of pyridoxine. A minority of PH1 patients have their clinical condition improved and oxalate excretion lowered by large doses (Watts *et al.* 1985). Although the molecular basis of this is not known, it is clearly related to the fact that the deficient enzyme (AGT) uses pyridoxal phosphate as an essential cofactor.

The identification of the basic enzymic defect in PH1 has allowed the formulation of a new strategy for treatment. This consists of enzyme replacement therapy by liver transplantation (Watts *et al.* 1987; Danpure 1991*b*). Because AGT is liver-specific, such as approach has the advantage of replacing virtually all of the body's requirement for AGT and reintroduces it in the correct cell and subcellular location. Combined hepatorenal transplantation not only replaces the deficient enzyme, but also replaces the pathophysiologically affected organ. At least 30 such procedures have been carried out world-wide. Most have been very successful, with normalization of body fluid organic acid levels, and in a number of cases resolution of the systemic oxalosis. Assuming normal functioning of the grafted organs, the speed with which the oxalate levels normalize depends largely on the degree of systemic oxalosis (Danpure 1991*b*). This approximates to the length of time the patient has suffered from compromised renal function.

Liver transplantation has not been carried out in PH2. This might be due to the reputed mildness of the condition, but also due to the poorer scientific rationale of such a procedure. Because GR is not liver-specific, it is not clear how much of the body's requirement would actually be replaced by liver or liver-kidney transplantation.

Future

The future of PH1 patients, at least, is very much brighter than it was 5–10 years ago. Part of this is due to the much greater understanding of the molecular basis of its aetiology and pathogenesis. Identification of a greater number of mutations in the AGT gene will make definitive diagnosis of PH1 very much easier under all clinical conditions. It will also allow first trimester prenatal diagnosis. The best option for long-term treatment of severely-affected PH1 patients would still appear to be liver transplantation, especially as the hazards of such a procedure will inevitably decrease. In the very long term, gene therapy would appear to be a feasible proposition.

References

Chalmers, R.A., Tracey, B.M., Mistry, J., Griffiths, K.D., Green, A., and Winterborn, M.H. (1984). L-glyceric aciduria (primary hyperoxaluria type 2) in siblings in two unrelated families. *J. Inher. Metab. Dis.* **7**, (2), 133–4.

Cooper, P.J., Danpure, C.J., Wise, P.J., and Guttridge, K.M. (1988). Immunocytochemical localization of human hepatic alanine:glyoxylate aminotransferase in control subjects and patients with primary hyperoxaluria type 1. *J. Histochem. Cytochem.* **36**, 1285–94.

Cooper, P.J., Danpure, C.J., and Penketh, R.J. (1989). Prenatal differentiation of primary hyperoxaluria type 1 phenotypes in the first trimester using immuno-electron micrescopy. *Clin. Sci.* **76**, (20), 13.

Danpure, C.J. (1989). Recent advances in the understanding, diagnosis and treatment of primary hyperoxaluria type 1. *J. Inher. Metab. Dis.* **12**, 210–24.

Danpure, C.J. (1991*a*). Molecular and clinical heterogeneity in primary hyperoxaluria type 1. *Am. J. Kid. Dis.* **17**, 366–9.

Danpure, C.J. (1991*b*). Scientific rationale for hepatorenal transplantation in primary hyperoxaluria type 1. *Transplant. Clin. Immunol.* **22**, 91–8.

Danpure, C.J. and Jennings, P.R. (1986). Peroxisomal alanine:glyoxylate aminotransferase deficiency in primary hyperoxaluria type I. *FEBS Lett.* **201**, 20–4.

Danpure, C.J., Jennings, P.R., and Watts, R.W.E. (1987). Enzymological diagnosis of primary hyperoxaluria type I by measurement of hepatic alanine:glyoxylate aminotransferase activity. *Lancet*, **1**, 289–91.

Danpure, C.J., Jennings, P.R., Mistry, J., Chalmers, R.A., McKerrell, R.E., Blakemore, W.F., and Heath, M.F. (1989*a*). Enzymological characterization of a feline analogue of primary hyperoxaluria type 2: a model for the human disease. *J. Inher. Metab. Dis.* **12**, 403–14.

Danpure, C.J., Cooper, P.J., Wise, P.J., and Jennings, P.R. (1989*b*). An enzyme trafficking defect in two patients with primary hyperoxaluria type 1: peroxisomal alanine:glyoxylate aminotransferase rerouted to mitochondria. *J. Cell. Biol.* **108**, 1345–52.

Danpure, C.J., Jennings, P.R., Penketh, R.J., Wise, P.J., Cooper, P.J., and Rodeck, C.H. (1989*c*). Fetal liver alanine:glyoxylate aminotransferase and the prenatal diagnosis of primary hyperoxaluria type 1. *Prenat. Diag.* **9**, 271–81.

Mistry, J., Danpure, C.J., and Chalmers, R.A. (1988). Hepatic D-glycerate dehydrogenase and glyoxylate reductase deficiency in primary hyperoxaluria type 2. *Biochem. Soc. Trans.* **16**, 626–7.

Morgan, S.H., Purkiss, P., Watts, R.W.E., and Mansell, M.A. (1987). Oxalate dynamics in chronic renal failure. Comparison with normal subjects and patients with primary hyperoxaluria. *Nephron*, **46**, 253–7.

Noguchi, T. and Takada, Y. (1979). Peroxisomal localization of alanine:glyoxylate aminotransferase in human liver, *Arch. Biochem. Biophys.* **196**, 645–7.

Purdue, P.E., Lumb, M.J., Fox, M., Griffo, G., Hamon-Benais, C., Povey, S., and Danpure, C.J. (1991). Characterization and chromosomal mapping of a genomic clone encoding human alanine:glyoxylate aminotransferase. *Genomics*, **10**, 34–42.

Purdue, P.E., Takada, Y., and Danpure, C.J. (1990). Identification of mutations associated with peroxisome-to-mitochondrion mistargeting of alanine:glyoxylate aminotransferase in primary hyperoxaluria type 1. *J. Cell. Biol.* **111**, 2341–51.

Resnick, M., Pridgen, D.B., and Goodman, H.O. (1968). Genetic predisposition to formation of calcium oxalate renal calculi. *New. Engl. J. Med.* **278**, 1313–8.

Robertson, W.G. and Peacock, M. (1980). The cause of idiopathic calcium stone disease: hypercalciuria or hyperoxaluria. *Nephron*, **26**, 105–10.

Smith, L.H. (1980). Enteric hyperoxaluria and other hyperoxaluric states. In *Contemporary issues in nephrology*, Vol. 5, (ed. F.L. Coe, B.M., Brenner, and J.H. Stein) pp. 136–64, Churchill Livingstone, New York.

Takada, Y., Kaneko, N., Esumi, H., Purdue, P.E., and Danpure, C.J. (1990). Human peroxisomal L-alanine:glyoxylate aminotransferase. Evolutionary loss of a mitochondrial targetting signal by point mutation of the initiation codon. *Biochem. J.* **268**, 517–20.

Watts, R.W.E., Veall, N., and Purkiss, P. (1984). Oxalate dynamics and removal rates during haemodialysis and peritoneal dialysis in patients with primary hyperoxaluria and severe renal failure. *Clin. Sci.* **66**, 591–7.

Watts, R.W.E., Veall, N., Purkiss, P., Mansell, M.A., and Haywood, E.F. (1985). The effect of pyridoxine on oxalate dynamics in three cases of primary hyperoxaluria (with glycollic aciduria). *Clin. Sci.* **69**, 87–90.

Watts, R.W.E., Calne, R.Y., Rolles, K., Danpure, C.J., Morgan, S.H., Mansell, M.A., Williams, R., and Purkiss, P. (1987). Successful treatment of primary hyperoxaluria type 1 by combined hepatic and renal transplantation. *Lancet*, **2**, 474–5.

Williams, H.E. and Smith, L.H. (1968). L-glyceric aciduria: a new genetic variant of primary oxaluria. *New Engl. J. Med.* **278**, 233–9.

Williams, H.E. and Smith, L.H. (1983). Primary hyperoxaluria. In *The metabolic basis of inherited disease*, (ed.) J.B. Stanbury, J.B. Wyngaarden, D.S. Fredrickson, J.L. Goldstein, and M.S. Brown, (5th edn) pp. 204–228. McGraw-Hill, New York.

Wise, P.J., Danpure, C.J., and Jennings, P.R. (1987). Immunological heterogeneity of hepatic alanine:glyoxylate aminotransferase in primary hyperoxaluria type 1. *FEBS Lett.* **222**, 17–20.

Yendt, E.R. and Cohanim, M. (1986). Absorptive hyperoxaluria: a new clinical entity—sucfessful treatment with hydrochlorothiazide. *Clin. Invest. Med.* **9**, 44–50.

18 Genetic aspects of non-hereditary nephritis

A. Burns and A.J. Rees

Genetic factors have long been known to influence susceptibility to glomerulonephritis. In 1812 Wells noted that the siblings of a child who developed nephritis after an attack of scarlet fever were much more likely to develop nephritis than were the siblings of children who did not develop nephritis after scarletina.[1] This observation, which has been confirmed many times, implies that inherited factors influence susceptibility to nephritis in humans. Similarly, in animal models genetic factors have been shown to influence the severity of experimentally-induced nephritis.

Recent advances in the understanding of the pathogenesis of nephritis have led to the suggestions that most forms of nephritis may be autoimmune mediated. This important change in emphasis has focused attention on the control of individual immune responses.

The critical event in the development of an immune response is presentation of antigen to helper T lymphocytes, and it is this that is probably responsible for discrimination of self from foreign antigen. Antigen 'processing' is the first stage, and involves antigen being internalized by antigen presenting cells before being partially digested. Processed antigen is re-expressed on the cell surface bound to a major histocompatibility complex (MHC) molecule.[2] T cell receptors can only recognize antigenic peptides when bound to MHC molecules. Crystallographic studies of the three-dimensional structure of MHC molecules have shown that the outer domains, which are highly polymorphic, form an antigen binding groove.[3] Single amino acid substitutions can change the three-dimensional conformation of the groove and alter the range of antigenic peptides that it can bind.[4] This is thought to be the basis of the allele-specific effects, which define immune response genes that are believed to be responsible for many of the MHC associated influences on autoimmunity.

Antigen-bearing MHC molecules bind directly to T cell receptors and this tri-molecular complex (Fig. 18.1) is crucial to the development of immune responses. It may initiate the destruction of autoreactive T cells during ontogeny in the thymus,[5,6] it is essential for the generation of T cell help to delayed hypersensitivity and antibody response,[7] and it is possibly responsible for antigen specific suppression.[8] Thus, the tri-molecular complex provides important opportunities for directing immune responses and for eliminating potentially harmful responses during development, a process referred to as 'moulding of T cell repertoire'. Inherited differences in T cell receptor genes

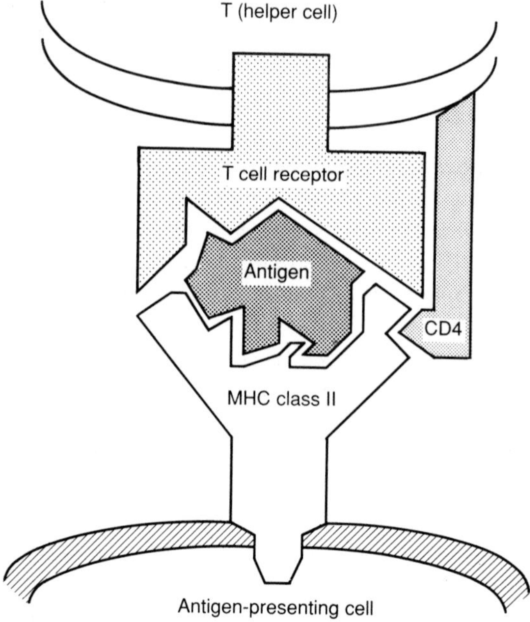

Fig. 18.1 Diagrammatic representation of the tri-molecular complex.

are likely to have equally important effects on the tri-molecular complex, and thus on the development of autoimmune responses. They have already been implicated in experimental autoimmune encephalomyelitis (EAE),[9-11] but, to date, no studies on experimental autoimmune nephritis have been reported. This chapter will concentrate mainly on the influence of the HLA complex on glomerulonephritis, but will start by considering the organization of human MHC.

The human major histocompatibility complex

One major focus of the efforts to understand the inheritance of autoimmune nephritis has been the attempt to identify genes within the human major histocompatibility complex (MHC) which confer susceptibility to nephritis. These studies have been based largely on the finding of associations between class II serological specificities and disease. The results have been difficult to interpret, and complicated further by the nomenclature that is used to describe the human leucocyte antigen (HLA) class II system. The advent of DNA recombinant technology, however, has greatly expanded knowledge of this system and has enabled more accurate assignment of HLA alleles. It will ultimately simplify the hitherto very difficult nomenclature, as alleles will be assigned definitive names on the basis of their nucleotide sequence.

Recent technical developments have facilitated more detailed studies of this region. First, restriction fragment length polymorphism (RFLP) analysis using cDNA probes to MHC genes can be used to assign MHC types from genomic DNA. Secondly, the nucleotide sequence of HLA class II genes has now been established relatively easily, and thirdly, analysis of DNA for the presence of particular polymorphisms can be readily accomplished by differential hybridization with sequence-specific oligonucleotide probes. These last two techniques have become possible with the development of the *in vitro* DNA amplification technique, the polymerase chain reaction (PCR), which greatly increases the number of copies of a target DNA sequence.

The genes that encode the MHC lie on the short arm of chromosome 6 in the distal portion of the 6p21.3 band (Fig. 18.2).[12,13] The complex contains three regions, designated classes II, III and I in order from the centromere. Their genetic organization has been elucidated using gene cloning and DNA sequencing. Class I and II regions are highly polymorphic,[14] the class III genes less so. The products of these classes have been characterized serologically and biochemically.

The class I region consists of *HLA-A*, *B* and *C* loci. Genes at these loci encode cell surface antigens which are expressed on most nucleated cells. At least 24 different HLA-A antigens, 50 HLA-B antigens, and 11 HLA-C antigens have been identified serologically. Class I molecules consist of two polypeptide chains, a 44 kDa heavy chain which spans the plasma membrane, and an extracellular portion divided into three domains ($\alpha 1$, $\alpha 2$ and $\alpha 3$), each of about 90 amino acids encoded by separate exons. The heavy chain is bound covalently to $\beta 2$-microglobulin ($\beta 2$m). The $\alpha 3$ domain and $\beta 2$m are relatively conserved, while the $\alpha 1$ and $\alpha 2$ domains are highly polymorphic.

The three-dimensional structures of two different class I molecules, HLA-A2 and HLA-A68 have been determined by X-ray crystallography.[15] The $\alpha 1$ and $\alpha 2$ domains form a platform composed of a single beta pleated sheet topped by alpha helices which together form a long antigen binding groove. Differences in amino acid sequence influence the three-dimensional shape of the antigen binding groove and presumably the peptides that they are able to bind.

The class II gene products form heterodimers consisting of heavy (α) and light (β) chains with molecular weights 30–34 kDa and 26–29 kDa, respectively, which fold together to form four external domains, two on each chain; $\alpha 1$ and $\alpha 2$ and $\beta 1$ and $\beta 2$ respectively. The $\alpha 2$ and $\beta 2$ domains lie closest to the plasma membrane and are highly conserved immunoglobulin-like molecules, while the two distal (*N*-terminal) domains $\alpha 1$ and $\beta 1$ carry the polymorphic determinants.[16] HLA-DR1 has been crystallized and shown to have a structure broadly similar to that of the class I molecules, though the exact three-dimensional structure has yet to be unravelled. Nevertheless, a hypothetical model of the foreign antigen binding site of class II histocompatibility molecules has been proposed by Brown *et al.* and is generally thought to be very accurate.[17]

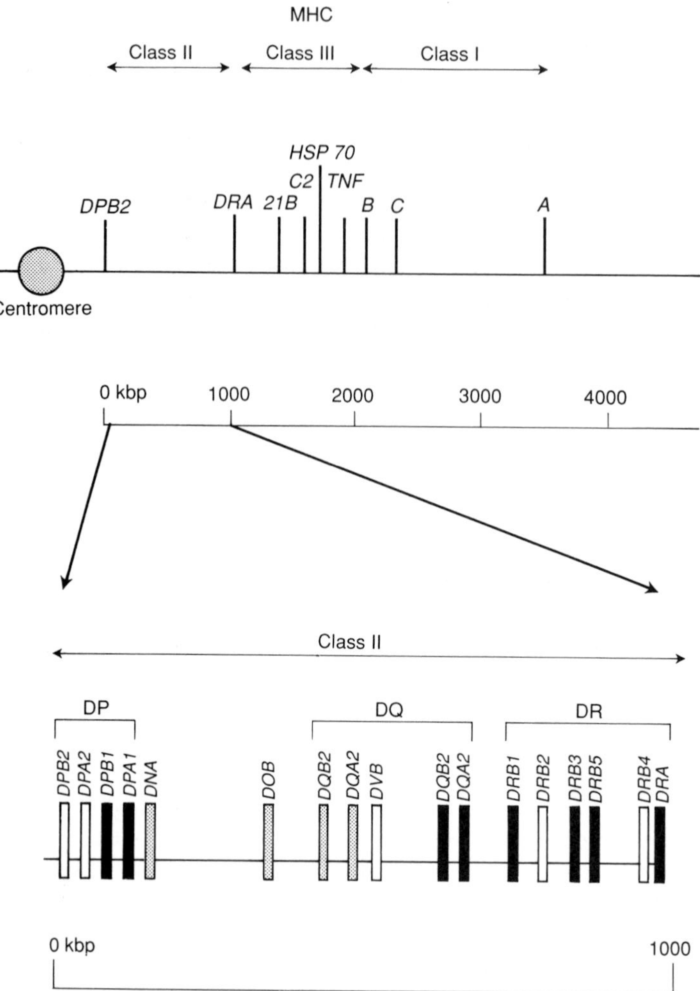

Fig. 18.2 Physical map of the human MHC. The number of expressed genes and pseudogenes varies according to the haplotype. For the class II region, expressed genes are depicted in black, pseudogenes in white, and undefined genes have cross-hatched markings.

The MHC class II region contains genes encoding at least three expressed class II molecules; DR, DQ, and DP. Each of these is a heterodimer consisting of an α and β chain. The α chains are encoded by single functional genes (*DPA1*, *DQA* and *DRA*) but the β chains are more complicated. DP and DQ β chains are also encoded by single functional genes (*DPB1*, *DQB1*) but the

DRA gene products can combine with products of at least two beta chains (DRβ1 and DRβ3, DRβ4 or DRβ5). The combination of the α1 and the β1 chain is responsible for DR specificities DR1–w18 (formerly DR1–10), whereas binding of α1 to the β3 or β4 chains results in DRw52 and DRw53 specificities, and combination with DRβ5 forms a unique second DR product in DR2 individuals. Finally, there are at least eight other genes in this region, which are not expressed, and are referred to as pseudogenes.

Genes within the MHC, and particularly those within the class II region, are inherited together more often than would be expected by chance. This phenomenon, known as linkage disequilibrium, has obvious implications when trying to interpret associations between HLA genes and disease. It is often impossible to decide which genetic locus within the MHC is more important in disease susceptibility, and indeed it may be that these alleles acting together confer disease susceptibility.

Analysis of data

Specific genetic loci can only be linked to a disease by studying families with more than one affected member.[18] The rarity of such affected kindred effectively precludes this approach in patients with glomerulonephritis. It follows that the involvement of the HLA complex must be inferred from the association with a particular HLA gene product. This can be ascertained from samples of patients and healthy controls, but only after certain preconditions have been satisfied. First, the population from which disease and control samples are drawn must be homogeneous, because the same disease may be associated with different HLA-molecules in different racial groups. This is illustrated in Table 18.1, which compares the HLA markers for nephritis in European caucasian and Japanese subjects. The need for homogeneity of the study population presents considerable difficulties when attempting to study populations such as those of the USA, where considerable racial admixture has occurred recently through immigration. The second precondition is that associations are much more likely to be found when the disease under consideration is a single entity, and when all patients are studied at the same stage of their disease. Lastly, appropriate statistical approaches must be used.[19]

Table 18.1 Differences in significant HLA associations between European caucasians and Japanese

	Associated HLA antigen	
Disease	Caucasians	Japanese
Membranous nephropathy	DR3	DR2
Minimal change nephropathy	DR7	DR8
Mesangial IgA disease	–	DR4

Associations between HLA antigens and disease are defined in terms of their statistical significance and their strength. The various methods to determine statistical significance have been reviewed by Thompson.[19] The usual approach is to calculate the probability from contingency tables and then make an appropriate correction for multiple comparisons.[20] Probability values from the first study of a particular disease are multiplied by the number of comparisons made because of the lack of a prior hypothesis. Subsequent studies to confirm a previously suspected association need no correction.

Two basic measures are commonly used to define the strength of an association (Table 18.2). The relative risk is a simple measure of the increased risk of developing the disease conferred by a genetic marker. It does not measure the closeness of association between marker and disease, and is strongly influenced by the frequency of the marker in the control sample. The closeness of associations is most easily assessed by the aetiological fraction. Essentially, it measures the proportion of the disease that can be attributed to the associated marker. It is independent of the frequency of the marker in the control sample and can be used to compare directly the strengths of association of different markers.[19] It can also be used to estimate the degree of linkage disequilibrium between marker and susceptibility genes, provided the mode of inheritance is known.[21]

Table 18.2 Measures of the strength of association between HLA antigens and disease

Test		Calculation
Relative risk (Scale: >0)	(RR)	$\frac{\text{Patients }(+)\text{ marker}}{\text{Patients }(-)\text{ marker}} \times \frac{\text{Control }(-)\text{ marker}}{\text{Control }(+)\text{ marker}}$
Etiological fraction (Scale: 0–1)	(EF)	$\frac{(\text{RR} - 1)}{\text{RR}} \times \frac{\text{Patients }(+)}{\text{Total patients}}$
Preventive fraction (Scale: 0–1)	(PF)	$\frac{(1 - \text{RR})}{\text{RR}} \times \frac{\text{Patients }(+)}{\text{Total patients}} \div 1 - \frac{\text{Patients }(+)}{\text{Total patients}} + \frac{\text{Patients }(+)}{\text{Total patients}}$

Association with disease

Most of the information from patients with glomerulonephritis comes from four diseases: Goodpasture's syndrome due to anti-glomerular basement membrane antibodies, membranous nephropathy, mesangial IgA disease and steroid responsive minimal change nephritis.

Anti-glomerular basement membrane disease

Human anti-glomerular basement membrane (anti-GBM) disease is an autoimmune disorder characterized by the presence of antibodies against a target restricted to the glomerular and a few other specialized basement membranes. It is unique amongst the nephritides, in that the antibodies that characterize the disease are known to be pathogenic.[22] Anti-GBM antibodies have highly restricted specificity and react with epitopes that reside in the alpha 3 chain of type four collagen.[23,24] Genetic factors undoubtedly play a role in susceptibility to this disease. A familial incidence has been described and the disease has been reported in four sibling pairs,[25,26] including two sets of identical twins.[27] However, the importance of non-genetic factors is emphasized by two sets of identical twins in our series, who are discordant for the disease, and another set reported by others.[28]

The first report of an HLA association with anti-GBM disease came from the UK, when the frequency of the class II allele *HLA-DR2* was found to be greatly increased.[29] It was present in 88 per cent of 17 patients diagnosed by immunohistology and radioimmunoassay, compared with 32 per cent of a panel of 100 controls. The strength of the association (relative risk 16) has meant that the association was statistically significant despite the small numbers. The incidence of *HLA-B7* was also increased in these studies, though to a lesser extent, and this was due to the high degree of linkage disequilibrium between *HLA-DR2* and *B7*. A later study extended this series to 38 patients, 34 of whom (88 per cent) had *DR2* compared with 19 per cent of 153 controls.[30] Currently 75 per cent of all 95 patients with anti-GBM disease on whom we have DR types are *DR2*. Thus, the findings are remarkably consistant and there is little doubt that, in the UK *HLA-DR2*, or genes in strong linkage disequilibrium with them, confer susceptibility to anti-GBM disease. Two studies elsewhere support this conclusion. Perle *et al.* in Australia reported that 7 of 8 patients with anti-GBM disease had *DR2*[31] and Garavoy in the USA found a lower, but still significantly increased, frequency of *DR2* which was present in 23 of 46 patients, but no clinical details were provided.[32]

The association of HLA alleles with anti-GBM disease is now known to be more complicated. The frequency of *DR4* was consistently increased in the early studies from the UK, but never reached statistical significance. This was partly because of the small number of patients studied, and also because of limitations in the serological typing procedures used at the time. In a more recent study of 71 patients who were allogenotyped using 'RFLP analysis'[33] the frequency of *DR2* was 78 per cent compared with 31 per cent in controls and *DR4* was 44 percent, compared with 37 per cent. However *DR4* was found in 8 of 13 (62 per cent) non *DR2* patients and 92 per cent of patients were found to have inherited *DR2*, or *DR4*, or both. There are precedents for the associations of more that one DR allele with a particular disease; previously, this have been described in diabetes (*DR3* and *DR4*) and in rheumatoid arthritis

(*DR1* and *DR4*). DR alleles can also confer protection against disease, as has been described for *DR2* in diabetes mellitus. This also appears to be the case in anti-GBM diseases as *DR1* was found very infrequently (2.7 per cent) in the patients compared with 21.7 per cent in the control group.

The linkage disequilibrium which exists between DR and DQ genes means that a given DR allele is almost invariably inherited together with a particular DQ allele. For example, almost all *DR2* caucasian individuals also inherit *DQw6*, and similarly Caucasian individuals who inherit *DR4* also have either *DQw7* or *DQw8* (in a ratio of 30 to 70 in samples of healthy individuals). This has considerable influence on the frequency of DQ allotypes in our patients with anti-GBM disease. *HLA-DQw6, 7* and *8* were found in 79 per cent, 34 per cent, and 25 per cent, respectively (n = 53), compared with control frequencies of 41 per cent, 27 per cent, and 36 per cent (n = 1103).

The obvious difficulty is to decide which genes within these haplotypes are responsible for susceptibility, and also why different haplotypes confer susceptibility to the same disease. These questions have been studied extensively in insulin-dependent diabetes. Here it appears that the absence of an aspartic acid residue at position 57 of the DQβ chain is important for susceptibility,[34] but that this has to be inherited in the context of particular DR alleles,[35] or by implication as part of an extended haplotype. This type of analysis of DQ and DR genes in anti-GBM disease suggests that the presence of a seven amino acid motif present on DRβ1 chains of *DR2* individuals and DRβ1 chains of *DR4* individuals may be important. It is easy to speculate that differences such as these could be involved in presentation of the autoantigen. Very soon it will be possible to test these predictions *in vitro* using purified autoantigen, mouse cells transfected with particular DR and DQ genes, and antigen specific T cells, to assess the ability of individual class II molecules to present autoantigen effectively.

Idiopathic membranous nephropathy

Idiopathic membranous nephropathy (IMN) is a frequent cause of nephrotic syndrome in adults. It is characterized by diffuse thickening of the glomerular capillary wall, with granular deposition of immunoglobulins and complement visible on immunofluorescence; sub-epithelial electron dense deposits are also evident on electron microscopy. There is little doubt that IMN is an immunologically mediated disease, although the mechanism is uncertain. Identical appearances can be produced experimentally either by autoantibodies to glomerular epithelial cells in rats with Heyman nephritis[36] or by 'serum sickness' after injection of cationic antigens into rabbits. In humans there is tentative evidence for the presence of autoantibodies to glomerular structures.

Familial membranous nephropathy has been described in five pairs of brothers[37,38] and two pairs of monozygotic twins.[38,39] Nevertheless, results from studies of the HLA complex show that heredity plays an important part in susceptibility. IMN has been repeatedly associated with certain HLA antigens, especially the class II allele *HLA-DR3*. In caucasians, 55–78 per cent of patients who develop IMN have the *DR3* allele, compared with 20–25 per cent of controls.[40] Similar studies in Japanese patients show an equally strong association between IMN *HLA-DR2*. *HLA-DR3* is very uncommon in Japanese but, even so, it is not clear why IMN should be associated with different DR alleles in caucasian and Japanese populations. There are at least three possible explanations: first, that both DR molecules share the properties needed to confer susceptibility to the disease; secondly, that the DR molecules themselves do not confer susceptibility, but are in linkage disequilibrium with the single true susceptibility gene and that the linkage is different in Japanese and caucasians; and thirdly, that the pathogenesis of MGN is different in different populations.

A recent report by Vaughan *et al.* adds support to the second hypothesis. Using RFLP analysis of genomic DNA they confirmed the serological association between *DR3* and IMN in a group of 31 patients. Only one of two subtypes of *DR3* was found to be increased and in addition, they observed a 4.5 kb *DQA* RFLP to occur with even greater frequency in the IMN patients than *DR3*. This RFLP pattern defines the *DQA1* allele, which is inherited exclusively with *DR3* and *DR5*. Using data from two other published studies Vaughan *et al.* made an indirect estimate of the likely frequency of this 4.5 kB *DQA* RFLP in the two groups of IMN patients and found them to be remarkably similar to their own findings. Thus, it is proposed that the DQ rather than the DR region is important in susceptibility to IMN, and that the *DR3* and *DR5 DQA1* allele, which are identical and different from other *DQa1* genes, are an important susceptibility factor in caucasian IMN.[41]

Mesangial IgA nephropathy

Mesangial IgA nephropathy (IgAN) is probably the most common chronic glomerulonephritis in the world[42] and a significant proportion of affected patients develop end stage renal failure.[43] Patients with IgA nephropathy have a generalized disease of cellular and humoral immunity.[44,45] An autoimmune model has been postulated for the disease recently because IgG autoantibodies against specific determinants on mesangial cells have been found in sera from some patients.[46]

The familial incidence of IgAN is well described in both caucasians and Japanese populations, and there are reports of disease in HLA identical siblings and in twins.[47–50] A number of studies, though not all,[52,53] have linked familial IgAN to particular alleles, but are not powerful enough to implicate the HLA system in susceptibility. The results of population studies attempting to correlate IgAN with specific HLA alleles have been even more confusing. In 1978 Berthoux *et al.* reported IgA nephropathy to be significantly associated with *HLA-Bw35*, but only a minority of subsequent studies have had similar findings. Attempts to explain the inconsistancies based on staging of the stage of the disease, or its tendency to progress to end stage renal failure have not been convincing.

Faucet (1980) reported that the frequency of DR4 was significantly increased in a group of 45 French patients with IgA disease, but subsequent studies have not confirmed this. The situation is quite different in Japanese, in which seven separate studies have shown an increased prevalence of *HLA-DR4*, with an incidence of about 60 per cent in patients compared with 32–44 per cent in healthy controls.

Two recent studies[54,55] may provide the explanation for the different association in caucasians and Japanese. Moore *et al.*[54] analysed genomic DNA from 35 caucasian patients with IgA nephropathy by RFLP techniques using probes to polymorphic DR and DQ genes. Although they found no significant differences between patients and controls using the DRβ probe, there was a significant increase in the frequency of 2 RFLPs using probes for the *DQB* gene. They did not assign specific alleles to these RFLPs but described two bands (2 kb and 6 kb respectively) which were significantly more frequent in patients with IgAN (50 per cent, n = 78) than in controls (15 per cent, n = 94 relative risk = 5.75 $p < 0.0001$). Li *et al.* in a study of 37 caucasian patients had similar findings using RFLPs but, in addition, used allele-specific oligotyping to assign DQ specificities to the patients. The patients had an increased prevalence of both *DR4* (51 per cent) and *DR5* (31 per cent) neither of which reached statistical significance. This contrasted with a highly significant increase in *DQw7* which may have been responsible for the 6 kb band described by Moore. *DQw7* was present in 71 per cent of patients (n = 32) compared with 28 per cent of controls (n = 1103 *RR6.2*).

This result could have been expected given the results of the DR studies, because both *DR4* and *DR5* are in linkage disequilibrium with *DQw7*. The Japanese results could also be explained by a primary association with a DQ allele, either *DQw7* itself or *DQw4*, which is the commonest DQ allele found in linkage disequilibrium with *DR4* in Japanese, and differs at only 4 polymorphic amino acid residues from *DQw7*. *DQw4* is found almost exclusively in Japanese populations. This means that *DR4* would be a good reporter gene for a *DQw7* or *DQw4* in Japanese, although a poor one for *DQw7* in caucasians. This suggests that susceptibility to *IgAN* is more closely associated with the DQ locus than the DR locus and identifies *DQw7* and its Japanese equivalent as the susceptibility gene.

Steroid responsive (minimal change) nephrotic syndrome of childhood

Minimal change nephrotic syndrome (MCNS) is the most common form of nephrotic syndrome in childhood, but can occur at any age. There are plenty of clues to suggest that MCNS is an immunological disease, but the exact pathogenesis is unknown. There have been a large number of studies to look for associations with MCNS. In 1976, Thompson *et al.* reported an association with *HLA-B12*;[56] of 71 affected children, 54 per cent inherited *HLA-B12*, compared with 15 per cent of 39 controls (relative risk 6.3, etiology fraction 0.45). Trompeter confirmed this association, in a larger group of patients from the same institution but the results were less striking.[57] Since then, at least seven other studies of caucasian children from various European countries and from Australia have failed to show a convincing association with *HLA-B12* but have almost all demonstrated an association with the class II specificity *HLA-DR7*. It is notable that *HLA-DR7* is in linkage disequilibrium with *B12*, which probably accounts for the original suggestion of a *B12* association. However the association of MCNS with *DR7* does not extend to other racial groups and Japanese children suffering from MCNS appear to inherit the class II antigen *DR8* more frequently than controls (13/28 compared with 9/114 respectively, (relative risk 10.1, etiology fraction 0.42)[58]). Clearly studies of the DQ loci will be needed in both populations before these differences can be interpreted.

Post-streptococcal glomerulonephritis

Post-streptococcal nephritis (PSGN) is an extremely common type of glomerulonephritis in many parts of the world, though no longer in Europe and the USA. It is often considered the model of immune complex mediated glomerulonephritis. The fact the not all patients who suffer streptococcal infections develop acute glomerulonephritis would suggest that genetic susceptibility may play at least some part in this condition. A familial incidence has been reported in epidemic outbreaks,[59-61] which may reflect the inheritance of a genetic susceptibility. Initial studies failed to show associations between class I antigens and PSGN, but Layrisse *et al.*,[62] though unable to show any evidence of linkage of susceptibility to the HLA complex in 18 Venezuelan families, did show a weak association between the class II antigen *DR4* and this condition in 42 unrelated patients.

Sasazuki *et al.*[63] have described similar findings in the Japanese population, in which patients with PSGN had a significantly increased prevalence of a class II molecule defined by mixed lymphocyte culture and designated D'EN', which is closely related to *HLA-DRw6*. More recently Naito[64] has suggested in another Japanese study an association between the class II antigen *DR1* and PSGN, which was present in 40 per cent of patients (n = 11) compared with 12 per cent of controls (n = 100) (relative risk = 4.9, $p < 0.02$). Uncorrected and corrected p values were not significant. These results are too preliminary to draw accurate conclusions but these class II genes, or immunoregulatory genes linked to them, may determine susceptibility to PSGN. Thus, a particular antigenic determinant of the nephritogenic streptococci may only lead to disease when presented together with particular immune response genes within the HLA complex or linked to it.

Rapidly progressive glomerulonephritis

Rapidly progressive glomerulonephritis (RPGN) is a clinical term used to describe patients with focal necrotizing glomerulonephritis which, if left untreated, progresses to end stage renal failure in weeks or months. Glomeruli from these patients have a common histological appearance with widespread crescent formation, but this can be due to a variety of different immunopathologies including anti-GBM disease. Nevertheless, there are a group of these patients who either have small vessel vasculitis, Wegener's granulomatosis and microscopic polyarteritis, or apparently isolated crescentic glomerulonephritis. This last group is often referred to as idiopathic RPGN, but there is increasing evidence that this is a more localized form of microscopic polyarteritis. The prevailing view is that these disease are immunologically mediated and possibly autoimmune, because of recent findings that the anti-neutrophil cytoplasmic antibodies (ANCA) are closely correlated with vasculitis and idiopathic RPGN.[65]

Early studies to look for immunogenetic factors associated with RPGN were flawed by the heterogeneity of disease leading to crescentic nephritis. Nevertheless, Muller *et al.*, in 1984, reported that idiopathic RPGN was associated with *HLA-DR2*, *MT3* and the complement allotype *BfF*, especially when inherited together as an extended haplotype.[66] None of his patients was said to have had evidence of systemic disease, which is surprising in view of the high incidence of systemic symptoms in most series of RPGN. This was the case in 45 patients presenting with vasculitis, many of whom also had RPGN, studied by Elkon *et al.*[67] The overall frequency of *DR2* was not increased,[67] but when the patients were segregated by the clinical diagnosis of Wegener's granulomatosis and microscopic polyarteritis, it was found that patients with Wegener's granulomatosis had a significantly increased frequency of *DR2* (65 per cent, n = 17), compared with normal controls (21 per cent, n = 113, $x^2 = 12.1$). These was no significant *HLA-DR* association with microscopic polyerteritis. It is too early to draw any conclusions from these studies and it will not be worthwhile doing so until the problem of heterogeneity has been resolved. New laboratory techniques, such as ANCA detection will undoubtedly facilitate more meaningful searches for immunogenetic influences on the development of RPGN.

Systemic lupus erythematosis

Systemic lupus erythematosis (SLE) is the prototype of a non-organ specific auto-immune disease, with immune complexes deposited in all pathological lesions and antibody mediated damage occurring in both lymphoid and non-lymphoid tissues. The occurrence of familial cases suggests a role for genetic factors which predispose to SLE. As many as 5 per cent of patients with SLE have a relative with the same disease and 57 per cent of monozygotic twins are concordent for SLE. The occurrence of SLE in dizygotic twins is similar to that of other first degree relatives. Thus, both genetic and environmental factors play a significant role in susceptibility to SLE.

Clinically detectable evidence of renal involvement in SLE is seen in about 50 per cent of patients, though electronmicroscopically detectable abnormalities are ubiquitous even when renal function and sediment are entirely normal. Thus, when considering the immunogenetics of lupus nephritis one must necessarily consider the genetic aspects of SLE in general and the wide variety of autoimmune systems involved.

Some patients developing lupus nephritis progress inexorably to end stage renal failure, while others suffer exacerbations and remissions. The diversity of the pathological and clinical features of this condition may reflect disease heterogeneity, which makes study of immunogenetics in this condition difficult. The strongest disease-susceptibility genes to be identified in humans to date are those responsible for deficiencies of the proteins of the classical pathway of complement, encoded within (*C2* and *C4*) and without (*C1q*) the MHC. Inherited complete deficiencies of these proteins account for only a tiny minority of patients suffering from SLE, and the prevalence of SLE in patients with hereditary deficiencies of the proteins of the classical pathway of complement is 68–88 per cent. Recognition of these facts have stimulated detailed studies of complement proteins to look for partial deficiencies.

The complement components C4 and C2 are encoded within the MHC and it has been suggested that the association of SLE with deficiency of these components occurs as a result of genetic linkage with other immune response genes, perhaps located within the class II region. This seems unlikely to be the explanation, as patients with complete deficiencies of these proteins have a variety of MHC haplotypes.[68] Many workers have looked for evidence of partial complement deficiency in patients with SLE, and Glass *et al.* have found an association between heterozygous C2 deficiency and SLE.[69] The C4A and C4B proteins exhibit extensive polymorphism, as discussed earlier in this chapter, and a number of groups have shown a markely raised prevalence of *C4AQ0* genes in patients with lupus nephritis.[70–74] However, the majority of these *C4AQ0* alleles were on *DR3*-bearing haplotypes.[71,73,74] The relative contributions of each of these alleles to disease susceptibility is difficult to assess, though Bachelor *et al.* have found an increased prevalence of the *C4A1Q* allele in *DR3* negative lupus patients.[75] Indeed, the *C4AQ0* allele appears to occur more commonly in lupus patients from other racial groups compared with non-lupus individuals of similar descent.

Initial studies of class II antigens demonstrated a raised prevalence of *HLA-DR2* and *HLA-DR3* in caucasian SLE patients. Subsequent studies have in general only confirmed the *DR3* association, though Woodrow, combining the data from 8 published reports, found that the presence of *DR2* and *DR3* is associated with an increased risk of developing SLE of 2 and 2.4 respectively.[76] The *DR2* association has also been seen in a study of SLE in southern Chinese (relative risk = 2.64).[77] The haplotype *HLA-A1 B8 DR3* is particularly associated with SLE as indeed it is with many other auto-immune conditions including Grave's disease, Addison's disease, type I diabetes mellitus and coeliac disease. One current hypothesis is that these class II molecules are particularly favourable restriction elements for the presentation of antigen to T-lympocytes in lupus as well as in these other auto-immune conditions.

The recent discovery that the genes for tumor necrosis factor are encoded within the MHC has led to the search for polymorphisms of these genes. Jacob and McDevitt have found reduced TNFα production in the offspring of lupus susceptible (NZW) mice crossed with non-susceptible mice (NZB) and regular injections of TNFα reduced the severity of nephritis and prolonged survival in these mice.[78] Recently, they have described analogous observations in patients with SLE.[79] These observations imply that inherited variation in TNF expression may be another disease-susceptibility gene for SLE.

It does not appear that distinct allotypic variants of antibodies, T-cell receptors or auto-antigens are involved in disease susceptibility. There is some evidence in mice for selective use of certain V gene families in the formation of autoantibodies, but these can also be present in non lupus-prone mice.[80] However, there is, as yet no evidence to support a role for genetic variation in antibody repertoire as a disease-susceptibility factor in humans. Stenezky *et al.* looked at *Gm* allotypes and related them to clinical features in 90 Hungarian patients with SLE and found that patients homozygous for *Gm3;5,13* are at special risk of developing SLE-related nephropathy.[81] The association of *Gm* with renal disease may be related to the fact that anti-DNA antibodies in *Gm3;5,13* homozygotes bind to anionic sites on glomerular basement membranes, thus initiating tissue injury.

Much of the work on genetic influences on the development of SLE has been done using animal models. Disease can be transmitted by bone marrow transplantation from susceptible to non-susceptible irradiated mice. Inter-breeding of different susceptible strains of mice, and cross-breeding of susceptible and non-susceptible mice strains has demonstrated that several genes are likely to be involved in disease susceptibility, at least in mice, and both dominant and recessive genes have been implicated. In addition, these experiments point to the existance of disease-protective genes which inhibit disease development. The onset of SLE, in susceptible mice, can be accelerated or delayed by viral and bacterial infection. One such virus, the lactate dehydrogenase (LDH) virus, which inhibits disease development has been shown to use the class II MHC molecules as its receptor and thereby may enter and inhibit the activities of antigen presenting cells.[82]

Thus, the causes of SLE in humans are likely to be multifactorial. Genetic factors certainly exist which, taken alone, are relatively weak but may act in consort to determine disease susceptibility. Alternatively, they may merely represent linkage to another true susceptibility gene.

HLA alleles and the severity of disease

Thus far, this chapter has been concerned only with the effects of HLA alleles on susceptibility to nephritis. There is also evidence that the HLA complex influences the severity of nephritis, its tendency to relapse, or to progress to end stage renal failure, though this evidence is considerably harder to collect. Often, it refers to associations between particular haplotypes rather than individual alleles, and there are even fewer precedents from experimental studies to guide interpretation of such associations. Nevertheless, the existence of extended haplotypes and the genes for complement, TNF and heat shock proteins within the class III region is tantalizing. There are no reports to date of the influence of HLA haplotypes on the intensity of inflammation but *DR2* has been reported to be associated with a reduced capacity for TNF secretion *in vitro*.[79]

The severity of anti-GBM disease has been reported to be influenced by whether *HLA-B7* is inherited together with *DR2*.[30] In a study of 38 patients, Rees *et al*. reported that those with *B7* had significantly higher serum creatinine concentrations at presentation, and a worse outlook in patients with *B7*, even though anti-GBM antibody titres were similar in both groups. Analysis of these data showed that these effects could not be explained merely by the strength of the known linkage disequilibrium between *B7* and *DR2* which suggests that *B7*-bearing haplotypes have independent effects on the severity of the disease.

Broadly similar data have been reported in minimal change nephrotic syndrome in children. Trompeter[57], in his original study, suggested that *HLA-B12* positive children with MCGN were more likely to have relapsed within three years of standard doses of cyclophosphamide therapy than those who had not inherited this antigen. On the other hand, O'Regan *et al.*[83] in a study of 54 Irish children suffering from MCGN demonstrated an increased frequency of both *B8* and *B12* class I antigens (relative risk = 3.5 and 1.3 respectively). They found that those with *HLA-B8* represented 72 per cent (13/18) of those in remission after 3 years, compared with 66 per cent (10/15) of those in relapse. Thus the strongest HLA associations are with class II genes, but extended haplotypes may influence the severity of disease or its response to therapy.

Hiki *et al*. recently confirmed the *B35* and *DR4* association with IgAN in the Japanese population but concluded that *DR4* plays no role in the long term prognosis of IgAN. The frequency of B35 was increased in the patients with stable disease (35.7 per cent) compared with those with progressive disease (20 per cent), but this difference was not statistically significant.[84]

Although there are no definite associations between HLA antigens and susceptibility to mesangio-capillary nephritis Welch *et al*. have suggested that, despite treatment with steroids, patients with the extended haplotype *HLA-B8*

DR3 are significantly more likely to develop renal failure than patients with other haplotypes.[85]

The data are much less certain for membranous nephropathy. Zucchelli has suggested that patients with the *HLA-DR3,B8* haplotype have a worse prognosis than others,[86] whereas Short *et al.*[87] reported that patients with *HLA-DR3,BfF1,B18* did worse than patients with other *DR3*-bearing haplotypes. These directly contradictary results suggest that HLA haplotypes have little influence on the severity of IMN, at least in comparison with other factors.

References

1. Wells, W.C. (1912). Observation on the dropsy which succeeds scarlet fever. *Transactions of the society for the improvement in medical and chirugical knowledge.* **13**, 167–86.
2. Moller, G. (1987). Antigenic requirements for activation of MHC restricted responses. *Immunol. Rev.* **98**, 1–187.
3. Bjorkman, P.J., Saper, M.A., Samaroui, B., Bennett, W.S., Strominger, J.L., and Wiley, D.C. (1987). Structure of the human class I histocompatibility antigen, HLA-Az. *Nature*, **329**, 506–12.
4. Bjorkman, P.J., Saper, M.A., Samraoui, B., Bennett, W.S., Strominger, J.L., and Wiley, D.C. (1987). The foreign antigen binding site and T cell recognition regions of class I histocompatibility antigens. *Nature,* **329**, 512–8.
5. Von Bohmer, H., Teh, H.S., and Kisieiow, P. (1989). The thymus selects the useful, neglects the useless and destroys the harmful. *Immunol. Today,* **10**, 57–61.
6. Nossal, G.J.V. (1989). Immunologic tolerance: collaboration between antigen and lymphokines. *Science*, **245**, 147–53.
7. Schwartz, R.H. (1986). Immune response (Ir) genes of the murine major histocompatibility complex. *Adv. Immunol.* **38**, 31–201.
8. Wraith, McDevitt, H.O., Steinman, L., and Acha-Orbea, H. (1989). T Cell recognition as the target for immune intervention in autoimmune disease. *Cell*, **57**, 709–15.
9. Acha-Orbea, H., Mitchell, D.J. Timmerman, L., and Wraith, D.C. (1988). Limited heterogeneity of T cell receptors from lymphocytes mediating autoimmune encephalomyelitis allows specific immune intervention. *Cell*, **54**, 263–73.
10. Urban, J.L., Kuman, V., Kono, D.H. *et al.* (1988). Restricted use of T cell receptor V genes in murine autoimmune encephalomyelitis raises possibilities for antibody therapy. *Cell*, **54**, 577–92.
11. Burns, F.R., Li, X., Shen, N. *et al.* (1989). Both rat and mouse T cell receptors specific for the encephalogenic determinants of myelin basic protein use similar Va and Vb chain genes. *J. Exp. Med.* **169**, 27–39.
12. Trowsdale, J. (1988). Molecular genetics of the MHC. *Immunology*, **1**, 21–3.
13. Campbell, R.D. and Bentley, D.R. (1985). Structure and genetics of C2 and factor B genes. *Immunol. Rev.* **87**, 19–37.
14. Dupont, B. (1987). *Histocompatibility testing*. Springer-Verlag New York.
15. Bjorkman, P.J., Saper, M.A., Samraoui, B., Bennett, W.S., Strominger, J.L., and Wiley, D.C. (1987). Structure of the human class I histocompatibility antigen, HLA-A2. *Nature*, **329**, 506–12.

16. Kaufman, J.F. and Strominger, J.L. (1982). HLA-DR light chain has a polymorphic N-terminal region and a conserved Ig-like C-terminal region. *Nature*, **297**, 694–7.
17. Brown, J.H., Jardetzky, T., Saper, M.A., Samaroui, B., Bjorkman, P.J., and Wiley, D.C. (1988). A hypothetical model of the foreign antigen binding site of class II histocompatibility molecules. *Nature*, **332**, 845–50.
18. Connealy, P.M. and Rivas, M.L. (1980). Linkage analysis in man. *Adv. Human Genetics*, **10**, 209–66.
19. Thompson, G. (1981). A review of theoretical aspects of HLA and disease. *Theoret. Pop. Biol.* **20**, 168–208.
20. Svejgaard, A., Platz, P., and Ryder, L.P. (1983). HLA and disease 1982—a survey. *Immunol. Rev.* **70**, 193–218.
21. Bengtsson, B.O. and Thomson, G. (1981). Measuring the strength of associations between HLA antigens and disease. *Tissue Antigens*, **18**, 315–22.
22. Lerner, R.A., Glassock, R.J., and Dixon, F.J. (1967). The role of anti-glomerular basement membrane antibody in the pathogenesis of human glomerulonephritis. *J. Exp. Med.* **126**, 989–1004.
23. Turner, N., Mason, P.J., Brown, R., Fox, M., Pobey, S., Rees, A.J., and Pusey, C.D. (1992). Molecular cloning of the human Goodpasture antigen demonstrates it to be in the alpha 3 chain of type IV collagen. *J. Clin. Invest.* **89**, 592–5.
24. Morrison, K.E., Mariyaince, M., Yang-Feng, T.L., and Reeders, S.T. (1991). Sequence and localization of a partial cDNA encoding the human $\alpha 3$ chain of type IV collagen. *Am. J. Human Genetics*, **49**, 545–54.
25. d'Apice, A.J., Kincaid Smith, P., Becker, G.H., Loughhead, M.G., Freeman, J.W., and Sands, J.M. (1978). Goodpasture's syndrome in identical twins. *Ann. Intern. Med.* **88**, 61–2.
26. Gossain, V.V., Gerstein, A.R., and Janes, A.W. (1972). Goodpasture's syndrome: a familial occurrence. *Am. Rev. Respir. Dis.* **105**, 621–4.
27. Simonsen, H., Brun, C., Thomsen, O.F., Larsen, S., and Ladefoged, J. (1982). Goodpasture's syndrome in twins. *Acta. Med. Scand.* **212**, 425–8.
28. Almkuist, R.D., Buckalew, V.M.Jr, Hirszel, P., Maher, J.F., James, P.M., and Wilson, C.B. (1981). Recurrence of anti-glomerular basement membrane antibody mediated glomerulonephritis in an isograft. *Clin. Immunol. Immunopathol.* **18**, 54–60.
29. Rees, A.J., Peters, D.K., Compston, D.A., and Batchelor, J.R. (1978). Strong association between HLA-DRW2 and antibody-mediated Goodpasture's syndrome. *Lancet,* **1**, 966–8.
30. Rees, A.J., Peters, D.K., Amos, N., Welsh, K.I., and Batchelor, J.R. (1984) The influence of HLA-linked genes on the severity of anti-GBM antibody-mediated nephritis. *Kidney Int.* **26**, 445–50.
31. Perl, S.I., Pussell, B.A., Charlesworth, J.A., Macdonald, G.J., and Wolnizer, M. (1981). Goodpasture's (anti-GBM) disease and HLA-DRw2 (letter). *N. Engl. J. Med.* **305**, 463–4.
32. Garovoy, M.R. (1982). Immunogenetic associations in nephrotic states. *Contemporary Issues in Nephrology*, **9**, 259–82.
33. Burns, A., Fisher, M., So, A., Pusey, C.D., and Rees, A.J. Analysis of HLA class II genes in patients with Goodpasture's syndrome (submitted).
34. Todd, J.A., Bell J.I., and McDevitt, H.O. (1987). HLA-DQb gene contributes to susceptibility and resistance to insulin-dependent diabetes mellitus. *Nature*, **329**, 599–604.
35. Sheehy, M.J., Scharf, S.J., Rowe, J.R. *et al.* (1989). A diabetes susceptible HLA

haplotype is best defined by a combination of HLA-DR and DQ alleles. *J. Clin. Invest.* **83**, 830–5.
36. Kerjaschki, S. and Farquhar, M.G. Immunocytochemical localisation of Heymann nephritis antigen (GP330) in glomerular epithelial cells from normal Lewis rats. *J. Exp. Med.* **157**, 667–86.
37. Sato, K., Oguchi, H., Hora, K. *et al.* (1987). Idiopathic membranous nephropathy in two brothers. *Nephron*, **46**, 174–8.
38. Short, C.D., Feehally, J., Gokal, R., and Mallick, N.P. (1984). Familial membranous nephropathy. *Br. Med. J. (Clin. Res.)* **289**, 1500.
39. Vangelista, A., Tazzari, R., and Bonomini, V. (1988). Idiopathic membranous nephropathy in 2 twin brothers (letter). *Nephron*, **50**, 79–80.
40. Papiha, S.S., Pareek, S.K., Rodger, R.S. *et al.* (1987). HLA-A,B,DR and Bf allotypes in patients with idiopathic membranous nephropathy (IMN). *Kidney Int.* **31**, 130–4.
41. Vaughan, R.W., Demaine, A.G., and Welsh, K.I. (1989). A DQA1 allele is strongly associated with idiopathic membranous nephropathy. *Tissue Antigens*, **34**, 261–9.
42. D'Amico, G. (1987). The commonest glomerulonephritis in the world: IgA nephropathy. *Q. J. Med.* **245**, 709–29.
43. Berger, J. (1984). IgA mesangial nephropathy. 1968–1983. *Contrib. Nephrol.* **40**, 4–6.
44. Lai, K.N., Lai, F.M., Chui, S.H. *et al.* (1987). Studies of lymphocyte subpopulations and immunoglobulin production in IgA nephropathy. *Clin. Nephrol.* **28**, 281–7.
45. Hale, G.M., McIntosh, S., Hiki, Y., Clarkson, A.R., and Woodruffe, A.J. (1986). Evidence for IgA specific B cell hyperactivity in patients with IgA nephropathy. *Kidney Int.* **29**, 718–24.
46. Ballardie, F.W., Brenchley, P.E.C., Williams, S., and O'Donoghue, D.J. Autoimmunity in IgA nephropathy. *Lancet*, **2**, 588–92.
47. Tolkoff Rubin, N.E., Cosmi, A.B., Fuller, T., Rubin, R.H., and Colvin, R.B. (1980). IgA nephropathy in HLA identical siblings. *Transplantation*, **29**, 505–6.
48. Sabatier, J.C., Genin, C., Assenat, H., Colon, S., Ducret, F., and Berthoux, F.C. (1979). Mesangial IgA glomerulonephritis in HLA identical brothers. *Clin. Nephrol.* **11**, 35–8.
49. Katz, A., Karanicolas, S., and Falk, J.A. (1980). Family study in IgA nephritis: possible role of HLA antigens. *Transplantation*, **29**, 505–6.
50. Montoliu, J., Darnell, A., Toras, A., Guadalupe, E., Valles, M., and Revert, L. (1980). Familial IgA nephropathy. Report of two cases and brief review of the literature. *Arch. Intern. Med.* **140**, 1374–5.
51. Kashiwabara, H., Shishido, H., Tomura, S., Tuchida, H., and Miyajima, T. (1982). Strong association between IgA nephropathy and HLA-DR4 antigen. *Kidney Int.* **22**, 377–82.
52. Julian, B.A., Quiggins, P.A., Thompson, J.S., Woodford, S.Y., Gleason, K., and Wyatt, R.J. (1985). Familial IgA nephropathy. Evidence of an inherited mechanism of disease. *N. Engl. J. Med.* **312**, 202–8.
53. Scolari, F., Savoldi, S., Scaini, P., Amoroso, A., Prati, E., and Maiorca, R. (1989). Familial occurrence of primary glomerulonephritis: experience in an Italian centre. *Nephrol. Dial. Transplant.* **4**, 441–2.
54. Moore, R.H., Hitman, G.A., Lucas, E. *et al.* (1989). Genetic susceptibility to autoimmunity in primary IgA nephropathy. *Nephrol. Dial. Transplant.* **4**, 428–9.
55. Li, P., Burns, A., So, A., Pusey, C.D., and Rees, A.J. (1992). HLA restrictions in IgA nephropathy. *Kidney Int.* (in press).

56. Thomson, P.D., Barratt, T.M., Stokes, C.R., and Turner, M.W. (1976). HLA antigens and atopic features in steriod-responsive childhood nephrotic syndrome. *Lancet*, **2**, 765–8.
57. Trompeter, R.S., Barrett, T.M., Kay, R., Turner, M.W., and Soothill, J.F. (1980). HLA atopy and cyclophosphamide in steroid responsive childhood nephrotic syndrome. *Kidney Int.* **17**, 113–7.
58. Komori, K., Nose, Y., Inouye, H. *et al.* (1983). Immunogenetical study in patients with chronic glomerulonephritis. *Tokai J. Exp. Clin. Med.* **8**, 135–48.
59. Dodge, W.F., Spargo, B.F., and Travis, L.B. (1967). Occurrence of acute glomerulonephritis in sibling contacts of children with sporadic acute glomerulonephritis. *Paediatrics*, **140**, 1028–30.
60. Rodriguez Iturbe, B., Rubio, L., and Garcia, R. (1981). Attack rates of post streptococcal glomerulonephritis in families. A prospective study. *Lancet*, **1**, 401–403.
61. Anthony, B.F., Kaplan, E.L., Wannamaker, L.W., Briese, F.W., and Chapman, S.S. (1969). Attack rates of acute nephritis after type 49 streptococcal infection of the skin and of the respiratory tract. *J. Clin. Invest.* **48**, 1697–704.
62. Layrisse, Z., Rodriguez Iturbe, B., Garcia Ramirez, R., Rodriguez, A., and Tiwari, J. (1983). Family studies of the HLA system in acute post-streptococcal glomerulonephritis. *Hum. Immunol.* **7**, 177–85.
63. Sasajuki, T., Hayase, R., Iwanoto, I., and Tsuchida, H. (1979). HLA and acute post-streptococcal glomerulonephritis. *N. Engl. J. Med.* **301**, 1184–5.
64. Naito, S., Kohara, M., and Arakawa, K. (1987). Association of class II antigens of HLA with primary glomerulopathies. *Nephron*, **45**, 111–4.
65. Falk, R.J. and Jennette, J.C. (1988). Anti-neutrophil cytoplasmic autoantibodies with specificity for myeloperoxidase in patients with systemic vasculitis and idiopathic necrotizing and crescentic glomerulonephritis. *N. Engl. J. Med.* **318**, 1651–7.
66. Muller, G.A., Gebhardt, M., Kompf, J., Baldwin, W.M., Ziegenhagen, D., and Bohle, A. (1984). Association between rapidly progressive glomerulonephritis and the properdin factor BfF and different HLA-D region products. *Kidney Int.* **25**, 115–8.
67. Elkon, K.B., Sutherland, D.C., Rees, A.J., Hughes, G.R., and Batchelor, J.R. (1983). HLA antigen frequencies in systemic vasculites: increase in HLA-DR2 in Wegener's granulomatosis. *Arthritis Rheum.* **26**, 98–101.
68. Meyer, O., Hauptmann, G., Trappeiner, G., Ochs, H.D., and Mascart-Lemone, F. (1985). Genetic deficiency of C4, C2, or C1q and lupus syndromes: associations with anti-Ro (SS-A) antibodies. *Clin. Exp. Immunol.* **62**, 678–84.
69. Glass, D., Raum, D., Gibson, D., Stillman, J.S., and Schur, P.H. (1976). Inherited deficiency of the second component of complement: rheumatic disease association. *J. Clin. Invest.* **58**, 853–61.
70. Fielder, A.H., Walport, M.J., and Bachelor, J.R. (1983). Family study of the MHC in patients with SLE: importance of null alleles in C4A and C4B in determining disease susceptibility. *Br. Med. J.* **286**, 425–8.
71. Christiansen, F.T., McCluskey, J., Dawkins, R.L., Kay, P.H., Uko, G., and Zilko, P.J. (1983). Complement allotyping in SLE: association with C4A null. *Aust. New Zealand J. Med.* **13**, 483–8.
72. Reveille, J.D., Arnett, F.C., Wilson, R.W., Bias, W.B., and McLean, R.H. (1985). Null alleles of the fourth component of complement and HLA haplotypes in familial systemic lupus erythematosis. *Immunogenetics*, **21**, 299–311.
73. Howard, P.F., Hochberg, M.C., Bias, W.B., Arnett, F.C., McLean, R.H. (1986). Relationship between C4 null genes, HLA-D region antigens, and genetic

susceptibility to systemic lupus erythematosis in Caucasion and Black Americans. *Am. J. Med.* **81**, 187–93.
74. Kemp, M.E., Atkinson, P.J., Skanes, V.M., Levine, R.P., and Chaplin, D.D. (1987). Deletion of C4A genes in patients with systemic lupus erythematosis. *Arthritis Rheum.* **30**, 1015–22.
75. Bachelor, J.R., Fielder, A.H.L., and Walport, M.J. (1987). Family study of the major histocompatibility complex in HLA-DR3 negative patients with systemic lupus erythematosis. *Clin. Exp. Immunol.* **70**, 364–71.
76. Hawkins, B.R., Wong, K.L., Chan, K.H., Dunckley, H., and Serjeantson, S.W. (1987). Strong association between the major histocompatibility complex and systemic lupus erythematosus in Southern Chinese. *J. Rheumatol.* **14**, 1128–31.
77. Woodrow, J.C. (1988). Immunogenetics of systemic lupus erythematosis. *J. Rheumatol.* **15**, 197–9.
78. Jacob, C.O. and McDevitt, H.O. (1988). Tumor necrosis factor-a in murine autoimmune lupus nephritis. *Nature*, **331**, 356–8.
79. Jacob, C.O., Fronek, Z., Lewis, G.D., Koo, M., Hansen, J.A., and McDevitt, H.O. (1990). Heritable major histocompatibility complex class II-associated differences in production of tumor necrosis factor a:Relevance to genetic predisposition to systemic lupus erythematosis: *Proc. Natl. Acad. Sci. USA*, **87**, 1233–7.
80. Bona, C.A. (1988). V genes encoding autoantibodies: molecular and phenotyping characteristics. *Ann. Rev. Immunol.* **6**, 327–58.
81. Petri, M., Bockenstedt, L., Colman, J. *et al.* (1988). Serial assessment of glomerular filtration rate in lupus nephropathy. *Kidney Int.* **34**, 832–9.
82. Inada, T. and Mims, C.A. (1984). Mouse Ia molecules are receptors for lactate dehydrogenase virus. *Nature*, **309**, 59–61.
83. O'Regan, O'Callaghan, Dundon, S., Reen, D.J. (1980). HLA antigens and steriod responsive nephrotic syndrome in childhood. *Tissue Antigens*, **16**, 147–51.
84. Hiki, Y., Kobayashi, Y., Ookubo, M., and Kashiwagi, N. (1990). The role of HLA-DR4 in long-term prognosis of IgA nephropathy. *Nephron*,
85. Welch, T.R., Beischel, L., Balakrishnan, K., Quinlan, M., and West, C.D. (1986). Major-histocompatibility-complex extended haplotypes in membranoproliferative glomerulonephritis. *N. Engl. J. Med.* **314**, 1476–81.
86. Zucchelli, P., Ponticelli, C., Cagnoli, L., Aroldi, A., and Tabacchi, P. (1987). Genetic factors in the outcome of idiopathic membranous nephropathy (letter). *Nephrol. Dial. Transplant.* **1**, 265–6.
87. Short, C.D., Dyer, P.A., Cairns, S.A. *et al.* (1983). A major histocompatibility system haplotype associated with poor prognosis in idiopathic membranous nephropathy. *Disease Markers*, **1**, 189–96.

Part 5 Glomerulonephritis

19 Immunopathology of glomerulonephritis

V. Cattell

Our understanding of the immunopathology of glomerulonephritis has evolved largely through studies of animal models. Although the aetiology of most human glomerulonephritides remains elusive, the value of experimental models has been in identifying in detail the pathological processes in glomerular injury (Hoedmaeker and Weening 1989). In this chapter the pathology of immune-induced injury will be briefly reviewed, and the morphological types of human glomerulonephritis described with current concepts of their pathogenesis.

Pathology of glomerular immune injury

Glomerular capillaries appear particularly vulnerable to immune injury as a result of their unique function and structure. The continuous production of a protein-free filtrate, generated by high intraluminal pressure, necessitates a complex capillary wall (fenestrated endothelium, size and charge-selective basement membrane, and a meshwork of epithelial cells), and a mechanism for clearing trapped residues (the mesangial cells and matrix). These features increase susceptibility to immune injury through trapping of circulating macromolecules, and also possibly by the variety of specific structural epitopes potentially capable of inducing antoantibodies.

It is generally accepted that most types of glomerulonephritis result from immune complex injury (Fig. 19.1). The possible antigens involved are more

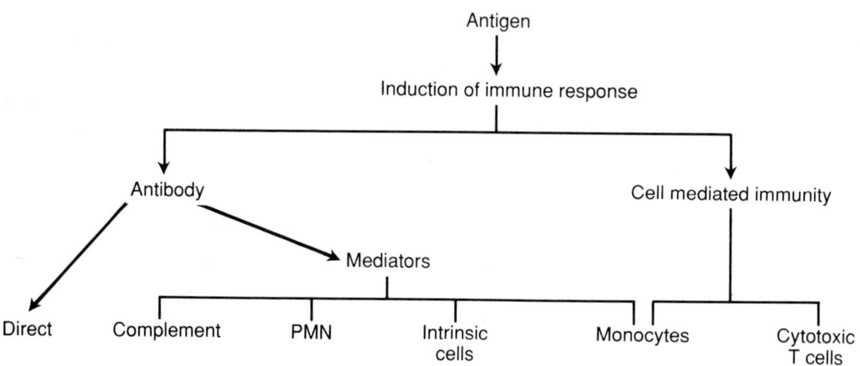

Fig. 19.1 Immune complex injury.

Table 19.1 Possible antigens involved in glomerulonephritis

(1) In trapped circulating immune complex
(2) *In situ*
  cell surface
    epithelial - gp330
    endothelial - cytokine induced
    mesangial - Thy 1

  leaking or secreted, or component of extracellular matrix
    Goodpasture
    laminin
    heparan sulphate
    ? phospholipid

  exogenous
    cationic
    infectious
    lectin

  immunoglobulin
    rheumatoid factors
    anti-idiotypic antibodies

complex and diverse than originally envisaged by early theories of antigen localizing in circulating immune complexes (Table 19.1). A variety of experimental models (Hoedmaeker and Weening 1989; Andres *et al.* 1988; Eddy and Michael 1989) show that glomerulonephritis is more easily induced by *in situ* reactions than by injection of preformed immune complexes. In *in situ* reactions the antigens are either structural components of the glomerulus, such as the epithelial cell antigen gp 330 of Heymann nephritis (Kerjaschki and Farquhar 1983), and Goodpasture antigen in antiglomerular basement membrane disease (Wieslander *et al.* 1987), or they are exogenous, and because of particular affinity become planted in the glomerulus. This can be demonstrated experimentally by renal perfusion with cationized antigens, which bind to the negative charges in the glomerular capillary walls (Fig. 19.2) (Oite *et al.* 1983). However, although *in situ* mechanisms are probably more frequent, in some instances injury may be the result of trapped circulating immune complexes, particularly in diseases in which subendothelial or mesangial deposits predominate.

Irrespective of the causes of complex formation, only certain types are pathogenic. Complexes which localize, persist and stimulate mediator systems are more likely to induce injury (Andres *et al.* 1988; Eddy and Michael 1989). The properties which lead to localization of persisting deposits are as follows.

(1) Ability to localize in glomerulus;

(2) Site of localization;

(a)

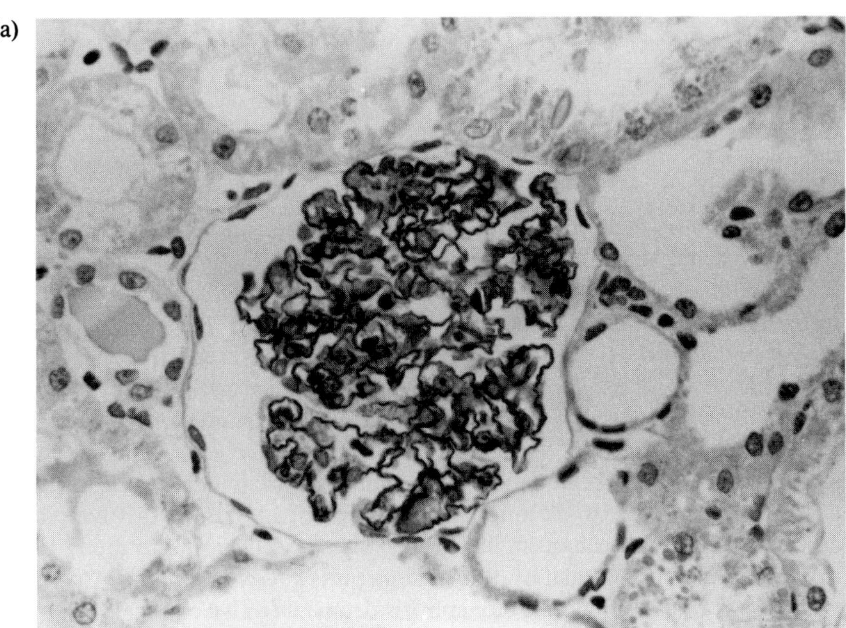

(b)

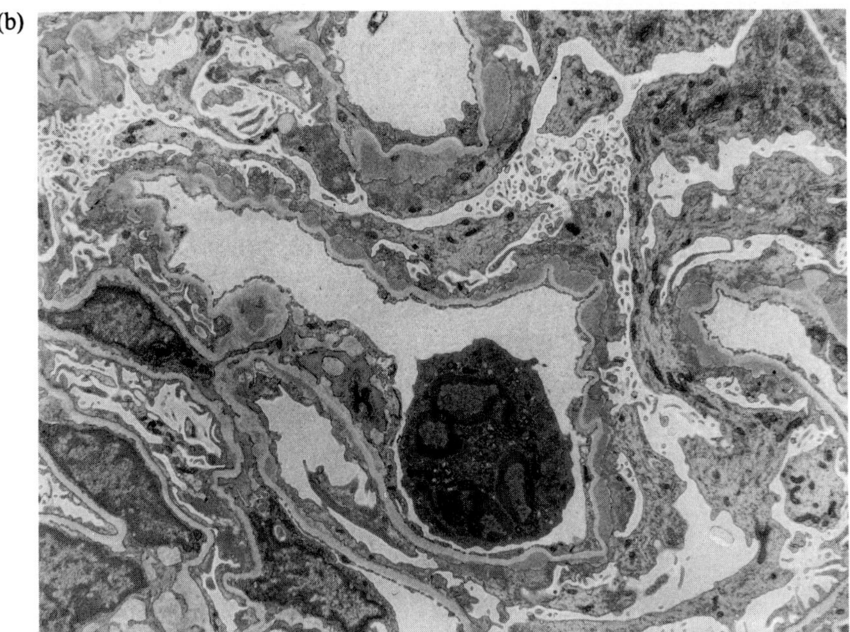

Fig. 19.2 Active immune complex glomerulonephritis induced by cationized human IgG. (a) Capillary wall localization of human IgG. (b) Multiple subepithelial electron-dense deposits.

(3) Undergo molecular rearrangement into deposits;

(4) Induce injury;

Properties include: amount, size, charge, cross-linking, interaction with acute phase proteins, binding to extracellular matrix, antibody-C fixing and inducer of mediators.

Some of these factors, such as amount, size and charge, have been known for many years. More recently the importance of interaction of complexes with matrix components, or with proteins, such as fibronectin which accumulate in tissues in acute phase responses, is being recognized.

The induction of the immune response is an area of research into glomerulonephritis in which rapid advances are being made. Most experimental models have been induced by large amounts of antigen, or antigen plus adjuvants, conditions of induction which are unlikely to apply to human glomerulonephritis. Human glomerulonephritis is more likely to be a result of abnormal induction of the immune response, or abnormal T-B lymphocyte co-operation. This mechanism has been shown to induce autoantibody formation in the mercuric model of glomerulonephritis (Goldman *et al.* 1988), suggesting that in human glomerulonephritis drugs, infective or other agents may lead to autoantibodies or cell-mediated immunity against endogenous antigens. This may occur against a background of genetic susceptibility (this is discussed further in Chapter 18).

The earliest immunopathological studies showed that antibody was deposited in glomeruli in glomerulonephritis. Different patterns of deposition were identified—granular patterns suggesting circulating immune complexes, and linear patterns indicating the fixation of antiglomerular basement membrane antibodies. It is now understood more clearly how antibody fixation can cause glomerular injury. Injury occurs either directly, through fixation of large quantities of antiglomerular basement membrane antibodies which induce proteinuria, presumably by altering the normal spatial arrangements within the basement membrane (Boyce and Holdsworth 1985), or through mediators.

The mediators activated by antibody fixation are complement, and products from infiltrating leucocytes (Table 19.2). In some instances injury depends on a single mediator, whereas in others a combination of mediators is necessary. The complement system mediates injury in at least two well-defined ways: (1) through recruitment of neutrophils via C3b and C5a (Cochrane *et al.* 1965), and (2) through lytic action of the membrane attack complex (C5–9) (Baker *et al.* 1989). Chemotactic fragments are thought to operate where complement fixes subendothelially, while attack complex injury is the result of subepithelial immune reaction. Where leucocyte products mediate injury, the reaction is essentially inflammatory.

Depletion or administration of specific antagonists shows that proteolytic enzymes, reactive oxygen species, cytokines, (see Chapter 20), platelet-

Table 19.2 Mediators of injury

Antibody
Complement
chemotaxis
membrane attack complex
Leucocytes
proteolytic enzymes
reactive oxygen species
cytokines
platelet-activating factor
procoagulants
? cytotoxicity

activating factor and procoagulant activity released by leucocytes are all capable of causing glomerular damage (Hoedmaeker and Weening 1989; Andres *et al.* 1988; Eddy and Michael 1989). Many of these substances are also produced by cultured mesangial cells so that *in vivo* mediators may be generated by intrinsic glomerular cells. In leucocyte-dependent models whether the infiltrate is predominately neutrophilic or monocytic may depend on the generation of different chemotaxins. In nephrotoxic nephritis heterologous phase injury is neutrophil-dependent (Cochrane *et al.* 1965), whereas in the antologous phase injury is monocyte-dependent (Schreiner *et al.* 1978). Complement chemotaxis attracts neutrophils, but the chemotactic factors for monocytes are not definitely established; monocyte chemotactic protein (MCP-1), a specific monocyte chemoattractant synthesized by cultured mesangial cells (Zoja *et al.* 1991), may be involved. MCP-1 also stimulates expression of adhesion molecules which are central to leucocyte migration and emigration.

A further difference between these leucocyte types is that, unlike neutrophils, monocytes only generate substantial amounts of mediators after transformation into activated macrophages; this activation occurs in glomerular inflammation (Cook *et al.* 1989), but the stimuli are unknown. More speculatively, macrophages may also cause injury by direct cytotoxicity. Recent evidence shows nitrite generation by nephritic glomeruli (Cattell *et al.* 1990; Cattell *et al.* 1991), probably synthesized by macrophages; the pathophysiological significance of this is not yet established. The presence of increased vasodilator activity during acute glomerular inflammation may have implications for glomerular haemodynamics. There may also be interaction with vasoactive eicosanoids which are known to be generated locally in experimental glomerular diseases (Lianos 1989).

One final mechanism requires consideration, i.e. cell mediated immunity. Recent evidence is more convincing than earlier attempts at injury induction by this means. Glomeruli contain resident class II antigen-expressing macrophages capable of antigen presentation (Schreiner *et al.* 1981). T lymphocytes in small numbers have been detected early in the course of experimental nephritis, preceding macrophage infiltration (Tipping *et al.* 1985), and in an avian model of glomerulonephritis in bursectomized chickens, a severe proliferative glomerulonephritis has been induced without deposition of antibodies (Bolton *et al.* 1984). These findings suggest a role for cell-mediated immunity.

Morphological types of human glomerulonephritis

With this background, the morphological types of human glomerulonephritis may now be examined, to see how relevant experimental studies are in understanding their pathogenesis (Table 19.3). Only primary glomerulonephritis where there is evidence suggesting an immune pathogenesis will be considered.

Table 19.3 Morphological classification of primary glomerulonephritis

Focal and segmental lesions
focal and segmental hyalinosis/sclerosis
Diffuse lesions
proliferative
endocapillary
mesangial
mesangiocapillary
crescentic
non-proliferative
membranous
sclerosing

Focal and segmental glomerulosclerosis

This is defined as a primary glomerulonephritis clinically associated with nephrotic syndrome which has IgM and C3 deposits in the sclerosing lesions. However, it is most improbable that these represent pathogenic immune complexes, as deposits are not found in non-sclerotic areas of glomeruli, and experimentally it can be shown that glomerular scars passively accumulate macromolecules (Fig. 19.3) (Melcion *et al.* 1984). Animal models of sclerosis suggest a haemodynamic origin. The likelihood of non-immunological mechanisms is supported by the very rapid recurrence of nephrotic levels of proteinuria in some patients with focal sclerosis who receive renal allografts; in these cases the earliest morphological change is loss of visceral epithelial cell foot processes without immune deposits, and sclerosis has been reported as early as 20 days after transplantation.

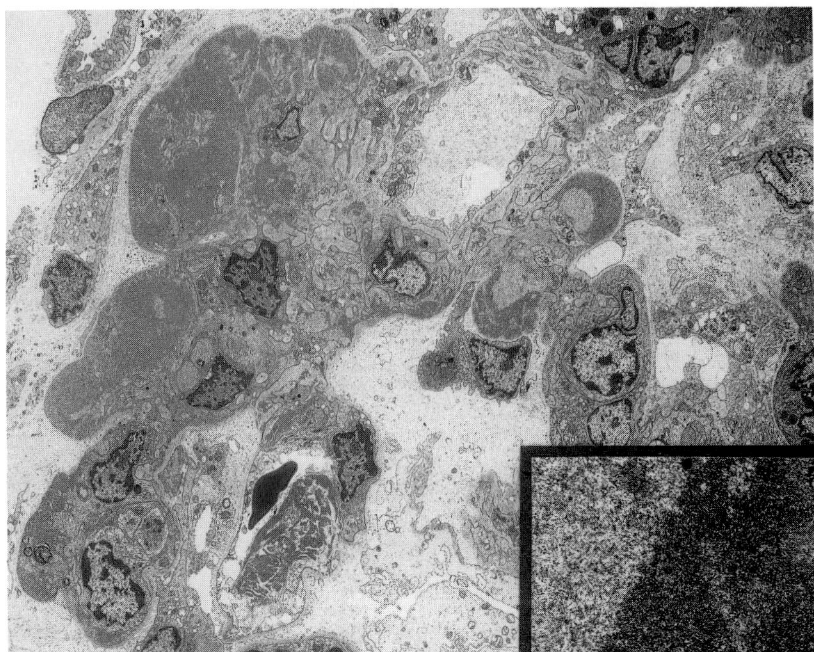

Fig. 19.3 Segmental glomerular sclerosis in the rat induced by amino-nucleoside and uninephrectomy—intravenously injected ferritin localized in sclerotic areas (insert). (From Melcion *et al.* 1984, with permission.)

Diffuse endocapillary proliferative glomerulonephritis

This is the archetypal example of acute immune complex glomerulonephritis. Although a post-streptococcal aetiology is now less common, most cases are post-infectious. The familiar appearance of global hypercellularity is identical to that of acute serum-sickness in rabbits induced by a single shot of foreign protein. The capillary lumens are full of leucocytes, predominantly macrophages, and these are the main mediators of injury (Holdsworth *et al.* 1981). By immunostaining and electron microscopy (Fig. 19.4) deposits are subepithelial, but the inflammatory cellular reaction suggests that, at least initially, complexes must form subendothelially. It is still not certain whether deposition of circulating complexes or an *in situ* reaction predominates, or whether the glomerular deposits contain infective antigen. Some recent data have been published showing subendothelial deposition of a streptococcal cytoplasmic antigen early in the course of post-streptococcal glomerulonephritis (Cronin *et al.* 1989).

Mesangial proliferative glomerulonephritis

In this type, the hypercellularity is due to increase in mesangial cells, and not leucocyte infiltration. The proliferation remains confined within an expanded

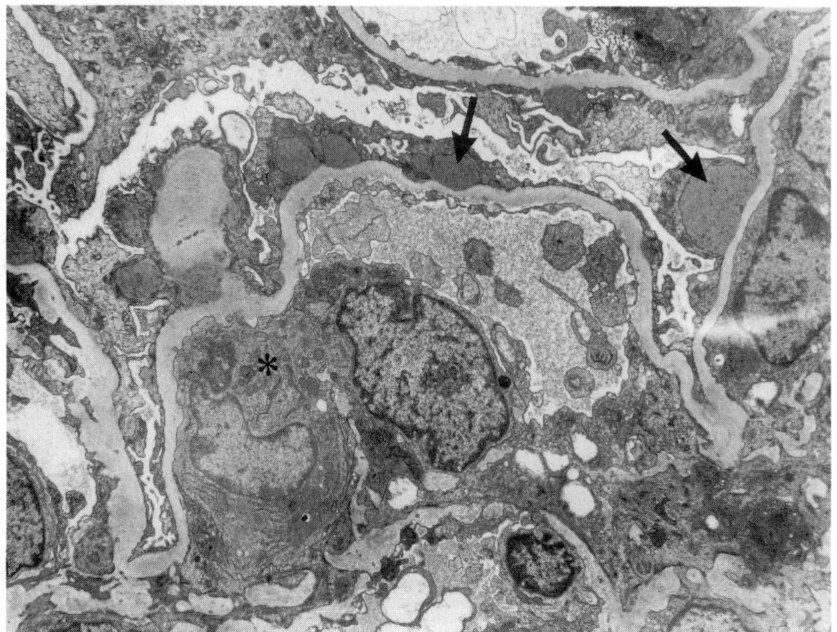

Fig. 19.4 Subepithelial electron dense deposits (arrowed) in diffuse endocapillary glomerulonephritis.

mesangium and the capillary walls and lumens are usually not involved. The immune deposits are found in the mesangium (Fig. 19.5) and are frequently IgA and C3 (IgA nephropathy). The cause of this relatively common type of human glomerulonephritis is unknown, but the presence of IgA, and experiments producing similar lesions by oral immunization suggest antigenic stimulation of gut-associated immunity. (IgA nephropathy is discussed more fully in Chapter 21). Although the aetiology of mesangial proliferative glomerulonephritis remains unknown, there have been recent advances in understanding mechanisms controlling mesangial cell proliferation. *In vivo* models of mesangial proliferation show a role for platelets (Johnson *et al.* 1990) and *in vitro* mesangial cells vigorously respond to a variety of growth factors (Silver *et al.* 1989).

Mesangiocapillary glomerulonephritis

This presents yet another morphological appearance. Here mesangial proliferation is more expansive than in mesangial proliferative glomerulonephritis, with interposition of mesangial cells into capillary walls between endothelium and basement membrane. This gives the capillaries a thickened and split appearance, which is enhanced by the presence of large subendothelial deposits (Fig. 19.6). There are also prominent mesangial deposits. As a primary

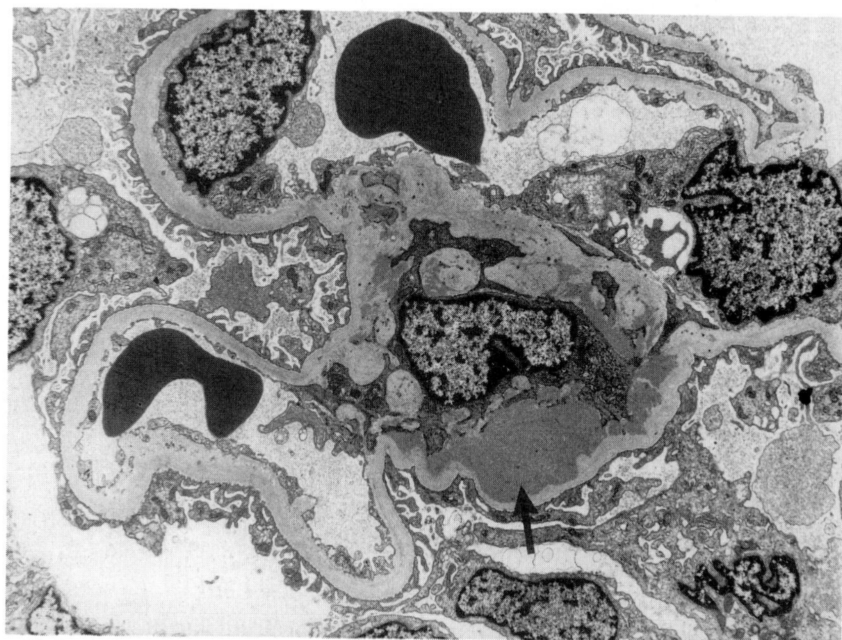

Fig. 19.5 IgA nephropathy—mesangial electron dense deposits (arrowed). A macrophage is infiltrating the capillary.

glomerulonephritis, the incidence of this type of glomerulonephritis is decreasing, but identical appearances are seen in some cases of systemic lupus erythematosus, cryoglobulinaemia and some chronic infections. There are no good animal models of primary mesangiocapillary glomerulonephritis experimentally induced by immune complexes; however, similar appearances are seen in murine systemic lupus erythematosus, and in dogs it is the commonest form of spontaneous glomerulonephritis (Fig. 19.7) (MacDougall *et al.* 1986).

Crescentic glomerulonephritis

This is defined as a glomerulonephritis in which more than 50 per cent of glomeruli have crescents; the underlying glomerulus most frequently shows a focal necrotizing glomerulonephritis. Many of these cases are associated with vasculitis (discussed in Chapters 22 and 23); others are of idiopathic immune complex origin or due to anti-glomerular basement membrane (anti-GBM) disease.

Experimental anti-GBM disease, usually initiated by intravenous injection of heterologous anti-GBM sera has provided invaluable insight into the pathogenesis of acute crescentic immune injury. The injury can occur through several mechanisms; as reviewed in the first part of this chapter—the direct fixation

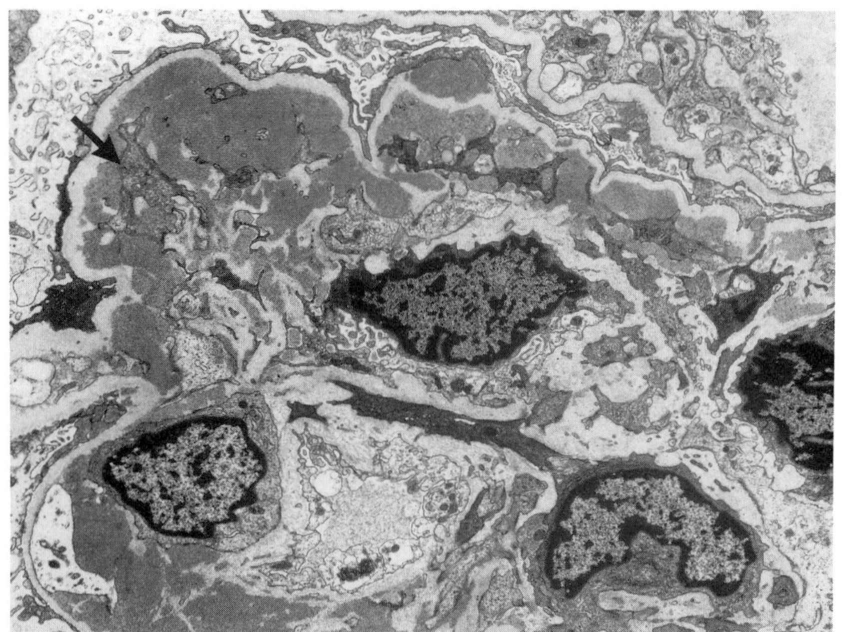

Fig. 19.6 Mesangiocapillary glomerulonephritis—capillary walls are thickened by large subendothelial electron dense deposits and mesangial cell interposition (arrowed). There is also a solitary subepithelial deposit.

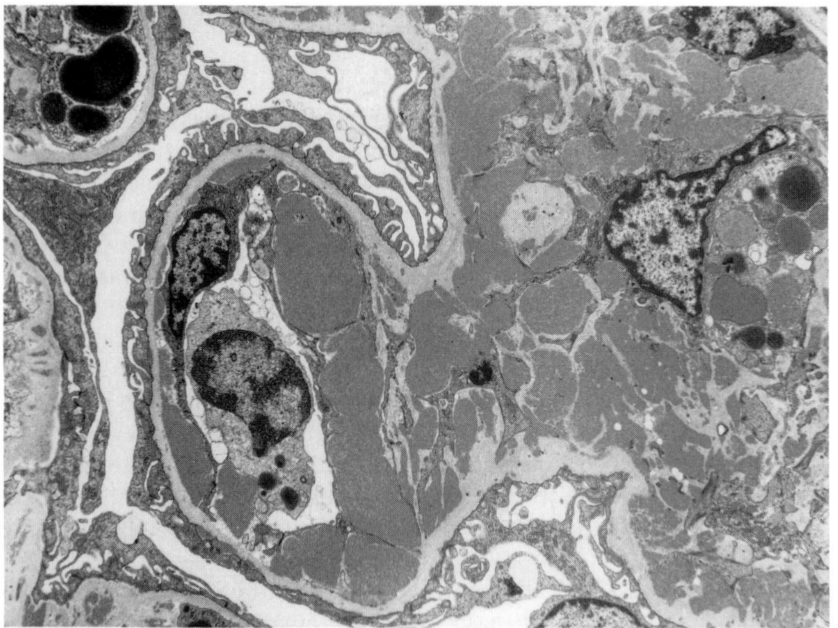

Fig. 19.7 Mesangiocapillary glomerulonephritis in the dog. There are large subendothelial and mesangial electron dense deposits, as in the human disease.

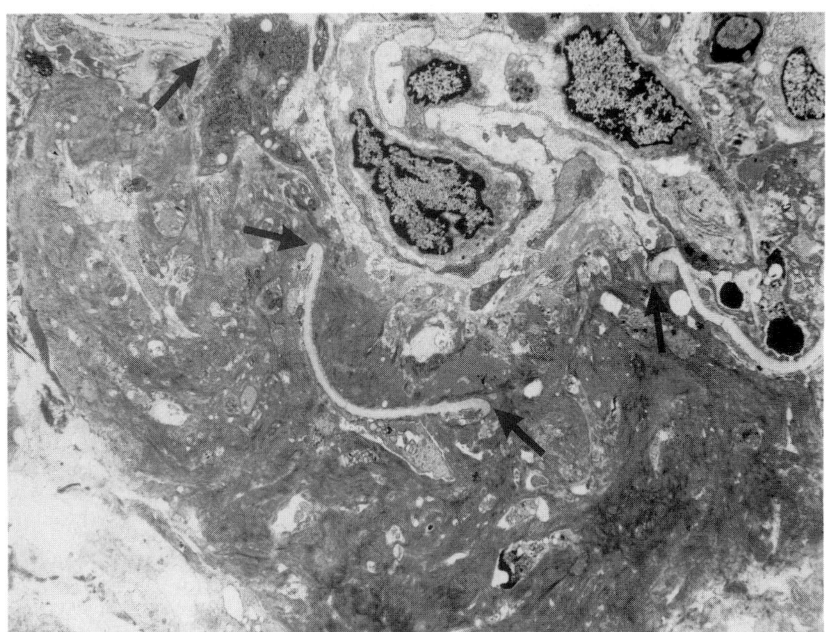

Fig. 19.8 Human crescentic glomerulonephritis—crescent formation. There are two breaks in the glomerular basement membrane (arrowed) with loss of endothelium and fibrin masses in the loop and Bowman's space.

of antibody, complement, or leucocytes. Crescents form when focal breaks occur in the basement membrane (Fig. 19.8) and fibrin leaks into Bowman's space. This stimulates migration of infiltrating monocytes from capillaries into Bowman's space, which together with proliferating epithelial cells (mainly of parietal origin) form cellular crescents (Cattell and Jamieson 1978). In human glomerulonephritis glomerular antigen in this *in situ* reaction is the non-collagenous globular domain of type IV collagen. The factors leading to autoantibody against this epitope are currently under investigation and may involve an initiating injury which exposes this epitope or induces breakdown in normal mechanisms of immune tolerance, as discussed earlier.

Membranous glomerulonephritis

This is a chronic non-proliferative type of glomerulonephritis in which diffuse capillary wall thickening is due to multiple subepithelial immune deposits (Fig. 19.9). With unremitting disease, the progressive deposition of complexes stimulates production of basement membrane components (principally laminin) which accumulate between the complexes first as spikes (stages II) and then completely enclosing deposits causing further thickening of the capillary wall (stage III). The reasons for this stimulation of matrix are

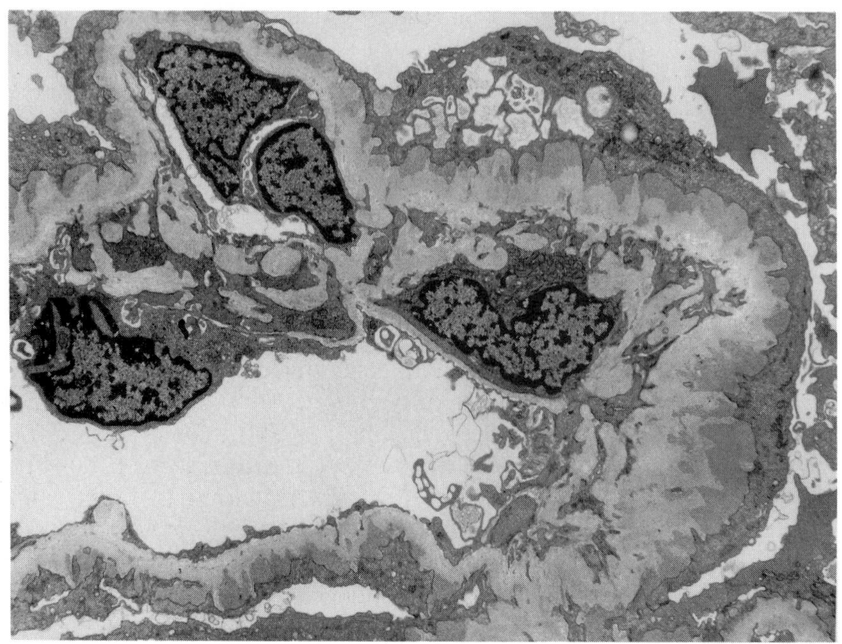

Fig. 19.9 Membranous glomulonephritis (stage II). Multiple subepithelial electron dense deposits are present with intervening spikes of basement membrane-like material.

unknown. There are several animal models of this type of immune complex deposition, including repeated injection of cationized antigens (Rahman *et al.* 1987) and Heymann nephritis (Hoedmaeker and Weening 1989; Andres *et al.* 1988; Eddy and Michael 1989) which have advanced our understanding of how these deposits form and cause injury.

In Heymann nephritis in rats the complexes form by an *in situ* reaction. The principal antigen is a glycoprotein (gp 330) (Kerjaschki and Farquhar 1983) expressed on glomerular epithelial cells and tubular cell brush borders. When antibody against renal tubular epithelial antigen is either passively injected (Van Damme *et al.* 1978) or induced by active immunization with extracts of rat kidney or purified renal tubular antigens (Heymann *et al.* 1959), deposits form subepithelially beneath the foot processes. Complement activation with insertion of the membrane attack complex (C5–9) into the epithelial cell membrane causes proteinuria (Baker *et al.* 1989). Leucocytes are not thought to be involved.

The cause of human membranous glomerulonephritis is almost certainly a subepithelial *in situ* reaction. The appearances of Heymann nephritis are so similar that it is most likely that at least in some human cases a structural epithelial antigen is involved, though not gn 330 which appears to be species specific, and research is now directed towards its identification. In other cases, a planted exogenous antigen, possibly cationic, may be responsible. The development of membranous glomerulonephritis in some patients receiving heavy metals or drugs suggests, that as in the model of mercuric chloride–induced glomerulonephritis (Goldman *et al.* 1988), these agents, which have never been identified within glomerular deposits, lead to a breakdown in immune tolerance with induction of autoantibodies.

There is no good evidence that immune mechanisms cause progressive glomerular sclerosis, therefore this very important aspect of glomerulonephritis will not be considered here.

Conclusion

Thus, recent research into glomerulonephritis supports and extends the early concepts of immunological glomerular injury. Mediators defined by the use of experimental models can account for most of the pathological changes found in human glomerulonephritis. Exciting areas of current and future investigation concern identification of the antigens involved in human glomerulonephritis, the way in which immunity develops against them, and immunological or pharmacological manipulations to prevent or resolve disease.

References

Andres, G., Yuzalva, Y., and Cavalot, F. (1988). Recent progress in renal immunopathology. *Human Pathology*, **19**, 1132–43.

Baker, P.J., Ochi, R.F., Schulze, M. *et al.* (1989). Depletion of C6 prevents development of protein uria in experimental membranous nephropathy in rats. *Amer. J. Pathol.* **135**, 185–94.

Boyce, N.W. and Holdworth, S.R. (1985). Anti-glomerular basement membrane antibody-induced experimental glomerulonephritis. Evidence for dose-dependent, direct antibody and complement induced, cell independent injury. *J. Immunol.* **135**, 3918–21.

Bolton, W.K., Tucker, F.L., and Sturgill, B.C. (1984). New avian model of experimental glomerulonephritis consistent with mediation by cellular immunity. Nonhumorally mediated glomerulonephritis in chicken. *J. Clin. Invest.* **73**, 1263–76.

Cattell, V. and Jamieson, S.W. (1978). The origin of glomerular crescents in experimental nephrotoxic nephritis in the rabbit. *Lab. Invest.* **39**, 584–90.

Cattell, V., Cook, T., and Moncada, S. (1990). Glomeruli synthesize nitrate in experimental nephrotoxic nephritis. *Kid. Int.* **38**, 1056–60.

Cattell, V., Largen, P., de Heer, E., and Cook, T. (1991). Glomeruli synthesize nitrite in active Heymann nephritis; the source is infiltrating macrophages. *Kid Int*, **40**, 847–51.

Cochrane, C.G., Unanue, E.R., and Dixon, F.J. (1965). A role of polymorphonuclear leukocytes and complement in nephrotoxic nephritis. *J. Exp. Med.* **122**, 99–116.

Cook, H.T., Smith, J., Salmon, J.A., and Cattell, V. (1989). Functional characteristics of macrophages in glomerulonephritis in the rat. 02 generation, MHC Class II expression, and synthesis. *Amer. J. Pathol.* **134**, 431–7.

Cronin, W., Deol, H., Azadegan, A., and Lange, K. (1989). Endostreptosin: isolation of the probable immunogen of acute post-streptococcal glomerulonephritis (PSGN). *Clin. Exp. Immunol.* **76**, 198–203.

Eddy, A.A. and Michael, A.F. (1989). Immunopathogenetic mechanism of glomerular injury. In *Renal pathology* (ed. Tisher and Brenner). J.B. Lippincott.

Goldman, M., Baran, D., and Druet, P. (1988). Polyclonal activation and experimental nephropathies. Editorial review. *Kid. Int.* **34**, 141–50.

Heymann, W., Hackel, D.B., Harwood, S.G.G. *et al.* (1959). Production of the nephrotoxic syndrome in rats by Freund's adjuvants and rat kidney suspensions. *Proc. Soc. Exp. Biol. Med.* **100**, 660–4.

Hoedmaeker, P.J. and Weening, J.J. (1989). Relevance of experimental models for human nephropathology. *Kid. Int.* **35**, 1015–25.

Holdsworth, S.R., Neale, T.J., Wilson, C.B. (1981). Abrogation of macrophage-dependent injury in experimental glomerulonephritis in the rabbit. *J. Clin. Invest.* **68**, 686–98.

Johnson, R.J., Garcia, R.L., Pritzl, P., and Alpers, C.E. (1990). Platelets mediate glomerular cell proliferation in immune complex nephritis induced by anti-mesangial cell antibodies in the rat. *Amer. J. Pathol.* **136**, 369–74.

Kerjaschki, D. and Farquhar, M.G. (1983). Immunocytochemical localisation of the Heymann antigen (gp 330) in glomerular epithelial cells of normal Lewis rat. *J. Exp. Med.* **157**, 667–86.

Lianos, E. (1989). Eicosanoids and the modification of glomerular immune injury. *Kid. Int.* **35**, 985–92.

MacDougall, D.F., Cook, T., Steward, A.P., Cattell, V. (1986). Canine chronic renal disease: prevalence and types of glomerulonephritis in the dog. *Kid. Int.* **29**, 1144–5.

Melcion, C., Smith, J., and Cattell, V. (1984). Ferritin deposition in the glomerular deposits of focal glomerular sclerosis in the rat. *J. Pathol.* **144**, 45–55.

Oite, T., Shimitzu, F., Kihara, I. *et al.* (1983). An active model of immune complex glomerulonephritis in the rat employing cationised antigen. *An. J. Pathol.* **112**, 185–94.

Rahman, M.A., Emancipator, S.N., and Dunn, M.J. (1987). Immune comples effects on glomerular eicosanoid production and renal haemodynamics. *Kid. Int.* **31**, 1317–26.

Schreiner, G.F., Cotran, R.S, and Pardo, V. (1978). A mononuclear cell component in experimental immunological glomerulonephritis. *J. Exp. Med.* **147**, 369–84.

Schreiner, G.F., Kiely, J.M., Cotran, R.S. *et al.* (1981). Characterization of resident glomerular cells in the rat expressing Ia determinants and manifesting genetically restricted interactions with lymphocytes. *J. Clin. Invest.* **68**, 920–31.

Silver, B.J., Jaffer, F.E., and Abboud, H.E. (1989). Platelet derived growth factor synthesis in mesangial cells: induction by multiple peptide mitogens. *Proc. Nat. Acad. Sci. USA*, **86**, 1056–61.

Tipping, P.G., Neale, T.J., and Holdsworth, S.R. (1985). T lymphocytes participation in antibody-induced experimental glomerulonephritis. *Kid. Int.* **27**, 530–7.

Van Damme, B.J.G., Fleuren, G.J., Barker, W.W. *et al.* (1978). Experimental glomerulonephritis in the rat induced by antibodies directed against tubular antigens. IV. Fixed glomerular antigens in the pathogenesis of heterologous immune complex glomerulonephritis. *Lab. Invest.* **38**, 502–10.

Wieslander, J., Kataja, M., and Hudson, B.G. (1987). Characterisation of the human Goodpasture antigen. *Clin. Exp. Immol.* **69**, 332–40.

Zoja, C., Wang, J.M., Bettoni, S., Sironi, M., Renzi, D. *et al.* (1991). Inteleukin – 1β and tumour necrosis factor α induce gene expression and production of leukocyte chemotactic factors, colony-stimulating factors and interleukin-7 in human mesangial cells. *Amer. J. Pathol*, **138**, 991–1003.

20 Cytokines and glomerulonephritis

J.D. Williams and M. Davies

The deposition of immune reactants within the renal glomerulus is usually followed by an inflammatory response which may lead to an alteration in structure and function. This process frequently involves the influx of leucocytes and platelets, and the subsequent changes are likely to result from an interaction between these infiltrating cells and those cells intrinsic to the glomerulus. The potential effect of this interaction, as well as the interaction of leucocytes with the glomerular matrices, has been highlighted by tissue culture studies, by a detailed analysis of appropriate animal models, and by the application of immunohistochemistry to renal biopsy material. It is likely that the whole process of inflammation is accompanied by, and probably controlled by, the synthesis and release of a plethora of soluble mediators, derived either from the infiltrating or intrinsic cells.

This chapter will detail the recent evidence of the involvement of cytokines in such a process. Cytokines are a group of low molecular weight glycosylated polypeptides, known variously as lymphokines, monokines and growth factors (Le and Vilcek 1987; Dinarello 1988*a, b*; Beutler and Cerami 1988; Sherry and Cerami 1988; Balkwill and Burke 1989). In general, cytokines possess the ability to stimulate or augment cell proliferation, initiate protein synthesis, and modify the production of several proinflammatory molecules, such as eicosanoids, platelet activating factor, and procoagulant activity. In addition, this distinct group of molecules act in both an autocrine and paracrine manner, are active in the 10^{-12} M range, and interact with target cells via specific cell surface receptors.

To date, the cytokine family consists of eight well-characterized interleukins (IL-1–8) (Hamblin 1988; Balkwill and Burke 1989), tumour necrosis factor (TNF) (Le and Vilcek 1987; Beutler and Cerami 1988; Sherry and Cerami 1988), and interferons α, β and γ (Johnson 1985). In addition, a number of factors which are usually referred to as growth factors or mitogens must be considered. These include platelet-derived growth factor (PDGF) (Ross *et al.* 1986; Hannink and Donoghue 1989), epidermal growth factor (EGF) (Carpenter and Cohen 1979), transforming growth factor (TGF) (Roberts and Sporn 1985), fibroblast growth factor (FGF) (Gospodarowicz *et al.* 1987), insulin-like growth factor-1 (somatomedin IGF-1) (D'Ercole 1987), and endothelin (Yanagiasawa and Masaki 1989).

Those features of cytokines salient to the kidney are illustrated in Tables 20.1–20.3.

Table 20.1 General properties of cytokines

Group of glycosylated polypeptides variously called: lymphokines, monokines, interferons, interleukins and growth factors
Low molecular weight (<80 kDa)
Involved in immunity, inflammation, repair
Produced transiently and locally acting in a paracrine/autocrine manner
Act at picomolar concentrations
Interact with specific cell surface receptors
Multiple overlapping cell regulatory actions
Interact in cytokine network.
(1) Inducing each other
(2) Transmodulating cytokine cell surface receptors
(3) Synergistic, additive or antagonistic interactions

Glomerular cell–cytokine interaction

The potential pathophysiological role of cytokines in glomerulonephritis has been illustrated by the *in vitro* culture of the different cells that constitute the glomerulus. In the main, these studies have centered on the mesangial cell and its interaction with mononuclear phagocytes. The glomerular epithelial cell has to date received limited attention, and present knowledge of the interactions involving the endothelial cell is largely based on data gained from cells of non-renal sources, with the exception of recent studies (Ballermann 1989; MacKay *et al.* 1989).

Initial studies of the interaction of the macrophage and defined glomerular cells described a protein factor in the condition medium of rat peritoneal phagocytes, that triggered the proliferation of rat mesangial cells (Lovett *et al.* 1983). Further analysis of the phagocyte condition medium demonstrated that it contained a number of factors including interleukin 1 (IL-1), tumour necrosis factor (TNFα) and platelet-derived growth factor (PDGF). The activated mesangial cells, as well as proliferating, also responded by synthesising and releasing a number of inflammatory mediators, including cyclooxygenase products (Topley *et al.* 1989), type IV collagenase (Martin *et al.* 1986), and proteoglycans (Davies *et al.* 1989; Border *et al.* 1990). In addition, the demonstration of the release of oxygen-derived free radicals in response to IL-1 and TNF implies that NADPH oxidase synthesis was also enhanced (Radeke *et al.* 1990). These observations not only show the potential of the mesangial cell to respond to a cytokine stimulus by proliferation, but also to produce molecules relevant to the pathogenesis of glomerulonephritis.

Mesangial cell culture medium has also been reported to contain biological activity synonymous with a number of cytokines, including IL-1 and TNF (Lovett *et al.* 1986; Baud *et al.* 1989). Confirmation of these molecules as products of the mesangial cells depend on their identification, both at the

Table 20.2 Some physical characteristics and biological properties of cytokines

Cytokine	Structure	Precursor	Receptor	Cell origin	Biological properties
IL-1α	159 aa Mr 17 500	271 aa Mr 31 000	Mr 80 000	Macrophage/MNC Endothelial cells T and B cells Fibroblasts Mesangial cells	Induction of PGE_2 synthesis. Mitogen for mesenchymal cells. ICAM-1 expression. Collagenase synthesis Fever. Growth and differentiation of T and B cells.
IL-1β	153 aa Mr 17 000	269 aa Mr 31 000			
IL-6	184 aa Mr 26 000	212 aa		Macrophage/MNC Endothelial cells Fibroblasts Mesangial cells	B cell differentiation. Acute phase protein synthesis. Plasmacytoma growth. T cell activation and differentiation.
TNF (Cachectin)	157 Mr 17 000	233	Mr 75 000 and 95 000	Macrophage/MNC T cells NK cells Mesangial cells	Tumourcidal activity. Induction of bone resorption, PCA and ICAM-1 expression. Mitogen for mesenchymal cells.
Endothelin	21 aa Mr 25 000	221 aa		Endothelial cells Epithelial cells	Increase vascular resistance. Decreasing glomerular filtration rate, RBF, glomerular ultrafiltation coefficient. Increase Na^+ reabsorption. Contract smooth muscle cells. Mitogen of rat mesangial cell.

Table 20.3 Some physical characteristics and biological properties of cytokines

Growth factor	Structure	Receptor	Cell origin	Biological properties
PDGF	Heterodimer of 2 disulphide-linked chains A = chain Mr 17 000 B = chain Mr 14 000	Tyrosine kinase	Platelets Endothelial cell Macrophage/MNC SMC/mesangial cell	Mitogen for mesenchymal cells. Competence factor.
TGF-α	50 aa	Competes for same receptors as EGF	Macrophage/MNC	Mitogen for mesenchymal cells.
TGF-β	Mr 25 000	Mr 280 000	Macrophage/MNC Platelets	Mitogen or growth inhibitor. Induces matrix synthesis. Inhibits protease degradation of ECM.
IGF-1	Mr 7 650	Mr 540 000	Fibroblasts	Mitogen for mesenchymal cells. Progression factor.
EGF	Mr 6 000	Mr 170 000	Macrophage/MNC	Mitogen for mesenchymal cells. Progression factor.
IFN-γ	Mr 25 000 (dimer) Glycoprotein		T-helper cells that make IL-2	Induce class II MHC molecules. Activate macrophage. Activate endothelial cells. B cell maturation.

messenger RNA levels and that of the secreted product (Baud *et al.* 1989). Such evidence has recently been presented for IL-6 (Ruef *et al.* 1990). Rat mesangial cells have been shown to express IL-6 mRNA by Northern blot analysis, and to secrete a molecule that has been characterized both biochemically and biologically as IL-6. Furthermore, it was clear that, as well a possible paracrine role within the glomerulus, mesangial IL-6 may also have an autocrine function in relation to mesangial cell proliferation.

The control of mesangial cell replication and proliferation is clearly a complex issue. These cells exhibit a proliferative response to a number of cytokines and growth factors, including TGFα and β, PDGF, IGF, and EGF. In a normal adult kidney, the intrinsic glomerular cells have a very low replicative rate. In contrast, mesangial cells in culture have a relatively fast turnover. If however, they are held in G0 (by culture in serum-free medium), the addition of serum to the culture medium induces the cells to leave the resting phase, traverse G1, and enter the S phase (Pledger *et al.* 1978). Such an effect of serum is attributed to its constituent growth factors. These molecules function at precise stages of the cell cycle. Broadly speaking, mitogens are placed into one of two groups: (1) competence factors (for example PDGF, FGF, and thrombin) affect early G0/G1 events. (2) progression factors (for example EGF, IGF) function at a later stage of G1. The G2 phase has no requirement for specific growth factors.

This means that cells exposed first to a competence factor, and then to a progression factor, are capable of completing the cell cycle. Thus, the addition of PDGF to fibroblasts in G0 will trigger DNA synthesis, but the cells require a progression factor, for example EGF to achieve full mitogenesis (MacKay *et al.* 1989). These conditions do, however, vary from cell to cell. In the case of human mesangial cells, PDGF appears to function as a complete growth factor, and the addition of a competence factor does not seem to be necessary (Jaffer *et al.* 1989; Floege *et al.* 1990). This unusual response is not restricted to PDGF alone, since EGF, FGF, TGFα and TNFα also affect mesangial cells in a similar manner.

As has been reported for IL-1, exogenous PDGF appears to generate the synthesis of a PDGF like factor by mesangial cells. This molecule seems to be identical with that purified from human platelets. Furthermore, antibodies to PDGF are capable of partially neutralizing the mitogenic response of mesangial cells to other growth factors, such as EGF. This suggests that EGF induces the synthesis of PDGF by mesangial cells, which then acts on the cells in an autocrine manner.

It should also be borne in mind that cytokines may down-regulate the mitogenic affect of growth factors. Thus the role of IL-1 and TNF is somewhat conflicting. Lovett *et al.* (1986) have demonstrated that IL-1 can serve as a progression factor, but Topley *et al.* (1990) have suggested that it may delay the proliferation response of mesangial cells to PDGF. Similarly, TNF has been reported to enhance and suppress mitogenesis, depending on the conditions under which it is exposed to the cells (Silver *et al.* 1988; Jaffer *et al.* 1989; Martin *et al.* 1989; Davies *et al.* 1989; Topley *et al.* 1990).

The principal action of IL-1 and TNF on mesangial cells, however, is to up-regulate cellular metabolism and increase the expression of genes coding for biologically active molecules. TNF can indeed stimulate catabolic events in glomerular mesangial cells, but its major role in glomerular disease is most likely to relate to its ability to up-regulate neutrophil activity (Petersen *et al.* 1990; Steadman *et al.* 1990) and induce changes in endothelial membranes that favour coagulation and neutrophil adhesion.

While much information has been gained from tissue culture studies on the factors that modify mesangial cell function, comparatively little is known about factors that modulate glomerular epithelial and endothelial cells. Glomerular epithelial cells have been reported to proliferate in response to conditioned medium from macrophages (Baud *et al.* 1985) and to respond to IL-1 with enhanced type IV collagen synthesis (Hänsch *et al.* 1988). In contrast, TGF-β inhibited the proliferation of glomerular epithelial (and endothelial) cells (MacKay *et al.* 1989) induced by EGF, IGF-1, and PDGF, but did not augment collagen production. TGF-β did, however, increase fibronectin production in glomerular epithelial cells.

To date, few studies have investigated in any detail cytokine-mediated activation of glomerular endothelial cells (Ballermann 1989; MacKay *et al.* 1989), although it has been suggested that human glomerular endothelial cells require PDGF to proliferate (Striker *et al.* 1984). In one study (MacKay *et al.* 1989) TGF-β inhibited endothelial proliferation, but in another study (Ballermann 1989) FGF-α in the presence of heparin and FGF-β augmented the growth

Table 20.4 Effect of IL-1/TNF on endothelial cells

Procoagulant effects
Increased tissue factor
Decreased thrombomodulin
Decreased tPA
Increased PA inhibitor
Leukocyte adhesion
Induction of ELAM-1
Stimulation of ICAM-1
Stimulation of growth
Secretion of PDGF and CSF(S)
Induction of angiogenesis
Other effects
Increased PGI_2 synthesis
Increased IL-1 secretion
Stimulation of class I MHC antigens (TNF)
Morphological changes

rate of subconfluent cells. In addition, it has been reported that supernatants from mesangial cells in culture can influence the growth of endothelial cells. Since medium conditioned by confluent glomerular mesangial cells enhances proliferation of subconfluent endothelial cells it is apparent that these glomerular cells, as with mesangial cells, are affected by both autocrine and paracrine mechanisms. A further important feature of the interaction of cytokines with endothelial cells, however, is their potential to influence the balance between the anticoagulant/pro-coagulant status of the vessel lining. The result of IL-1 and TNF/endothelial cell interaction is to increase the chances of intravascular coagulation (Table 20.4).

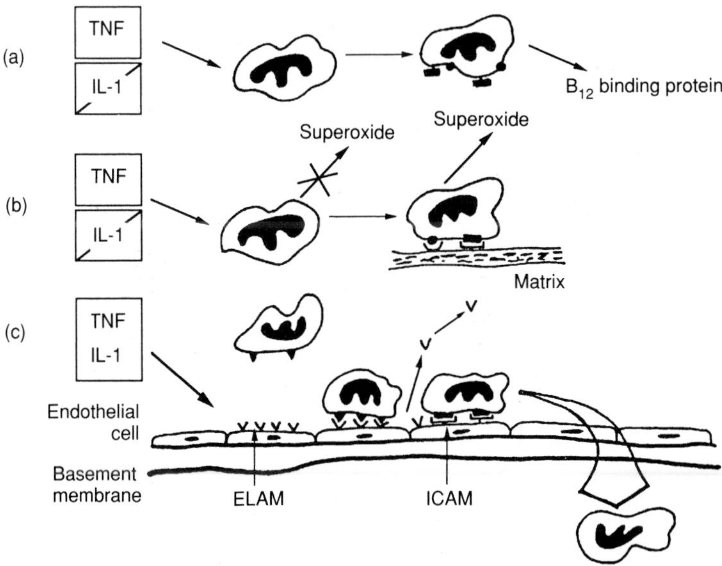

Fig. 20.1 Interaction of neutrophils and cytokines (a) TNFα, but not IL-1, triggers the release of secondary granules by PMN. This results in the increased expression of CD11/CD18 family of heterodimers on the surface of the PMN, as well as the release of highly charged cationic binding proteins, which may increase cell/cell interaction. (b) TNFα, but not IL-1, primes the PMN, such that when the cell binds to the matrix through its surface integrin receptors, then reactive oxygen products are released by the phagocyte. (c) Both TNF and IL-1 increase the expression of surface binding molecules or integrins on the endothelial cell. One proposed mechanism for PMN binding is illustrated. ELAM receptors on the endothelial cell wall cause the primary adherence by interacting with their complementary molecules on the PMN. Separate CD11/CD18 molecules are then expressed by the PMN, which themselves bind to ICAM molecules on the endothelial cell. This allows the initial binding proteins, (ELAM and its counterpart) to be shed. The secondary binding is of much less avidity than the primary binding and allows the PMN to move along the endothelial surface and then through it by diapedesis.

In addition to pro-coagulation effects, the activation of endothelial cells by cytokines, in particular IL-1 and TNF, is to promote leucocyte adhesion. This is achieved by the induction of specific cell surface adhesion molecules on endothelial cells, which interact with their counterparts on circulating inflammatory cells and facilitate the movement of PMN to sites of imflammation.

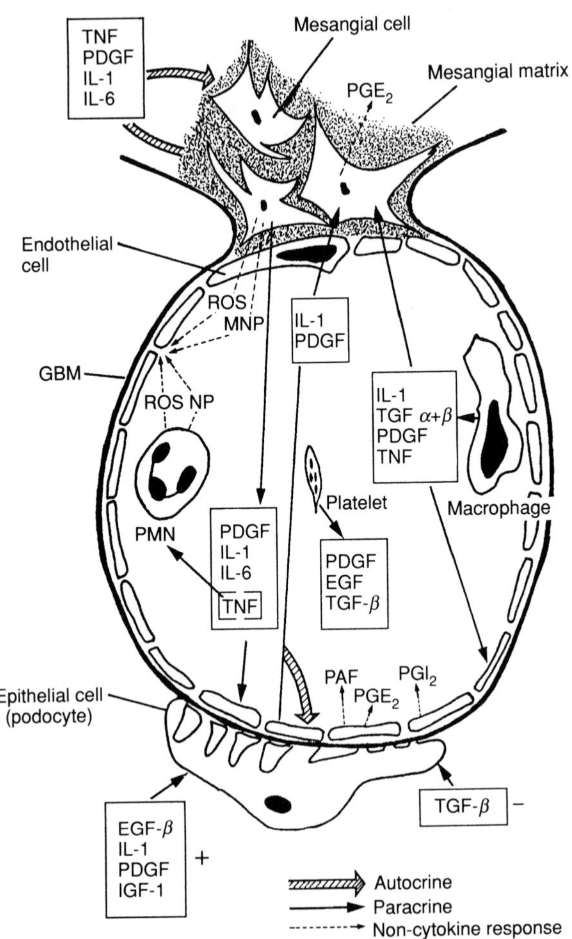

Fig. 20.2 Possible cell–cell interactions among intrinsic glomerular cells and infiltrating leucocytes mediated through cytokines. The diagram also indicates that response of the endothelium/mesangium in terms of proinflammatory products other than cytokines. The data on glomerular epithelial cells is sparse but an indication of the proliferative response (+ or −) to cytokines is shown. PMN, polymorphonuclear cell; MØ, macrophage; ROS, reactive oxygen species; MNP, metallo neutral proteinases; NP, neutral proteinases; PAF, platelet activating factor; PGE_2, prostaglandin E_2; PGI_2, prostacyclin; PCP, pro-coagulant products; ACP, anti-coagulant products.

The role of cytokines in the activation of PMN, their priming of PMN for the release of inflammatory mediators, as well as their role in facilitating the movement of PMN on, and through the capillary wall, illustrates the complexity of the mechanism of the inflammatory response (Fig. 20.1).

Thus, glomerular cells (both intrinsic and extrinsic) once activated, may produce cytokines that have autocrine activity or act in a paracrine manner to stimulate the other cells of the glomerulus (Fig. 20.2). Most, but not all, of the cell/cell interactions are up-regulatory and result in enhanced cell proliferation and/or the secretion of pro-inflammatory products. In contrast, TGF-β (MacKay *et al.* 1989) and in one report IL-1 (Topley *et al.* 1990) have a down-regulatory action. This suggests a role for cytokines in both positive and negative feedback mechanisms in glomerular cells. In addition, secretory products from glomerular epithelial cells (probably a heparan sulphate), can modulate the proliferative response of mesangial cells (Castellot *et al.* 1985). In other cell systems, cytokine-induced synthesis of proteins appears to be regulated by PGE_2 and cAMP (Lehmann *et al.* 1988; Raz *et al.* 1988; Kunkel *et al.* 1988). Furthermore, in mouse macrophages the consumption of arginine and the production of ornithine, a metabolic pathway important for the induction of T cell-mediated immune response *in vivo*, is regulated by an autoregulatory loop which involves the PGE_2-dependent elevation of cAMP (Lehmann *et al.* 1988).

Cytokines and experimental nephritis

There is now increasing evidence that the above observations in tissue culture studies may well have an *in vivo* counterpart. This gap between cell culture and the disease state has to some extent been bridged by experiments, for example which have demonstrated that the glomeruli isolated from animals with immune complex nephritis have raised IL-1 gene expression (Werber *et al.* 1987).

Similarly, experiments focusing on a role for extrinsic cells have also shown that macrophages isolated from glomeruli of nephritic animals were clearly in an activated state and secreted increased amounts of proinflammatory products, such as IL-1 and TNF. This has been recently confirmed in an experimental model of nephrotoxic nephritis, in which the synthesis of IL-1 was dependent on the infiltration of macrophages into the glomerulus (Tipping *et al.* 1991). In addition, there is a small but increasing amount of evidence from animals experiments that cytokines themselves can directly cause renal damage.

In one study in rabbits, the direct perfusion of TNF (8000 U/kg/h) over 5 h induced a rise in serum creatinine (Bertani *et al.* 1989). The principal morphological changes observed were endothelial cell damage and the accumulation of neutrophils, together with some evidence of fibrin deposition. There was no significant proteinuria and 24 h after the cessation of the infusion, glomerular morphology had returned to normal. In a separate model of glomerular disease induced by anti-glomerular basement membrane antibody (Tomosgui *et al.* 1989) LPS, IL-1 and TNF all enhanced the antibody-mediated injury, as demonstrated by a significant increase in albuminuria and the deposition of thrombi within capillaries. Anti-rat TNF antibodies significantly reduced the cytokine-induced proteinuria in the animals treated with LPS, suggesting that TNF was involved in LPS-induced enhancement of glomerular injury of nephrotoxic nephritis in rats. The above animal model may explain the clinical observation that microbial infection in patients with established anti-GBM nephritis is associated with a deterioration in renal function even in the absence of circulating anti-GBM antibodies.

More recently, the potential importance of IL-6 in renal disease has been illustrated (Suematsu *et al.* 1989). Transgenic mice carrying the human IL-6 genome, fused with a human immunoglobulin heavy chain enhancement, not only developed myeloma, but also mesangial proliferative glomerulonephritis. These findings were extended by the observation that patients with mesango-proliferative glomerulonephritis secreted significantly raised levels of urinary IL-6 (Horii *et al.* 1989). In a separate study (Richards *et al.* 1990) there was a selective increase in urinary IL-6 in patients with mesangio-proliferative and focal segmental glomerulosclerosis, but not in individuals with membranous nephropathy or minimal change disease. These experiments suggest that cytokines may be involved not only in the acute events surrounding glomerulo-nephritis, but also in regulating the growth status of the intrinsic glomerular cells, as well as the synthesis and production of extracellular matrix (Foellmer *et al.* 1987). The potential role, for example, of TGFβ in matrix production has been recently emphasized by the observation that this growth factor promotes the production of the proteoglycans biglycan and decorin by rat mesangial cells (Border *et al.* 1990). Furthermore, in an animal model of mesangio-proliferative GN, the administration of an antibody to TGFβ suppressed the increased production of matrix and attenuated the histological manifestation of the disease.

Cytokines in human disease

Despite this accumulating evidence, a defined role for cytokines in the pathogenesis of human disease remains speculative. The recent reports of increased IL-6 levels in certain types of glomerular disease is intriguing, but at present remains a phenomenon. There are a few immunomorphological studies demonstrating increased TNF and integrin expression in diseased human kidneys and there is a suggestion that in vasculitic conditions, there is a correlation between tumour necrosis factor and the expression of ICAM-1 (Waldherr *et al.* 1990). In addition, the screening of a group of young nephrotics, revealed a five-fold increase in circulating TNFα levels, but no change in IL-1, IL-2, or the interferons, α and γ (Suranyi *et al.* 1990).

The exacerbation of immunologically mediated renal disease during episodes of superadded infection is well established (Rees *et al.* 1977; Praga *et al.* 1985). This may be partly explained by the observation that circulating TNF levels are elevated in human volunteers injected with endotoxin (Michie *et al.* 1988). A direct cause and effect association, however, is not established, and will require studies into either specific inhibitors or receptor antagonists to the cytokines in question. Nevertheless, the availability of specific antibodies should allow a better understanding of the role of cytokines in glomerular disease in the near future.

References

Balkwill, F.R., and Burke, F. (1989). The cytokine network. *Immunology Today*, **10**, 299–304.

Ballermann, B.J. (1989). Regulation of bovine glomerular endothelial cell growth in vitro. *Am. J. Physiol.* **256**, C182–9.

Baud, L., Sraer, J., Perez, J., Nivez, M.P., and Ardaillou, R. (1985). Leukotriene C4 binds to human glomerular epithelial cells and promotes their proliferation in vitro. *J. Clin. Invest.* **76**, 374–77.

Baud, L., Oudinet, J.-P., Bens, M., Noe, L., Peraldi, M.-M., Rondeau, E., Etienne, J., and Ardaillou, R. (1989). Production of tumour necrosis factor by rat mesangial cells in response to bacterial lipopolysaccharide. *Kidney Int.* **35**, 1111–8.

Bertani, T., Abbate, M., Zoja, C., Corna, D., Perico, N., Ghezzi, P., and Remuzzi, G. (1989). Tumour necrosis factor induces glomerular damage in the rabbit. *Am. J. Pathol.* **134**, 419–30.

Beutler, B. and Cerami, A. (1988). Tumour necrosis, cachexia, shock, and inflammation: a common mediator. *Ann. Rev. Biochemistry*, **57**, 505–18.

Border, W.A., Okuda, S., Languino, L.R., and Ruoslahti, E. (1990). Transforming growth factor-β regulates production of proteoglycans by mesangial cells. *Kidney Int.* **37**, 689–95.

Carpenter, G. and Cohen, S. (1979). Epidermal growth factor. *Ann. Rev. Biochem.* **48**, 193–216.

Castellot, J.J., Hoover, R.L., Harper, P.A., and Karnovsky, M.J. (1985). Heparin and glomerular epithelial cell-secreted heparin like species inhibit mesangial cell proliferation. *Amer. J. Pathol.* **120**, 427–35.

D'Ercole, A.J. (1987). Somatomedins/insulin-like growth factors and foetal growth. *J. Dev. Physiol.* **9**, 481–95.

Davies, M., Shewring, L., Thomas, G., and Jenner, L. (1989). Stimulation of proteoglycan synthesis in rat mesangial cells in response to tumour necrosis factor. *Clin. Sci.* **76**, (20), 25(A).

Dinarello, C.A. (1988*a*). Biology of interleukin-1. *F.A.S.E.B. J.* **2**, 108–15.

Dinarello, C.A. (1988*b*). Interleukin-1 and its biologically related cytokines. *Ann. Rev. Immunology*, **44**, 153–205.

Floege, J., Topley, N., Wessel, K., Kaever, V., Radeke, H., Hoppe, J., Kishimoto, T., and Resch, K. (1990). Monokines and platelet-derived growth factor modulate prostanoid production in growth arrested human mesangial cells. *Kidney Int.* **37**. 859–70.

Foellmer, H.G., Perfetto, M., Kashgarian, M., and Sterzel, R.B. (1987). Matrix constituents promote adhesion and proliferation in glomerular mesangial cells in culture. *Kidney Int.* **31**, 318(A).

Gospodarowicz, D., Neufeld, G., and Schweiggerer, L. (1987). Fibroblast growth factor: structural and biological properties. *J. Cell. Physiol.* **5**, 15–26.

Hamblin, A.S. (1988). Lymphokines and interleukins. *Immunology*, **1**, 39–41.

Hannink, M. and Donoghue, D.J. (1989). Structure and function of platelet-derived growth factor (PDGF) and related proteins. *Biochim. Biophys. Acta*, **989**, 1–10.

Hänsch, G.-M., Torbohm, I., Kempis, J., and Rother, K. (1988). Modulation of collagen synthesis in glomerular epithelial cells by interleukin-1 and the supernatants of mesangial cells. *Kidney Int.* **33**, 317(A).

Horii, Y., Muraguchi, A., Iwano, M., Matsuda, T., Hirayama, T., Yamade, H., Fugi, Y., Dohi, K., Ishikawa, H., Ohmoto, Y., Yoshizaki, K., Hirano, T., and Kishimoto, T. (1989). Involvement of IL-6 in mesangial proliferative glomerulonephritis. *J. Immunol.* **143**, 3949–55.

Jaffer, F., Saunders, C., Shultz, P., Throckmorton, D, Weinshell, E., and Abboud, H.E. (1989). Regulation of mesangial cell growth by polypeptide mitogens. Inhibitory role of transforming growth factor beta. *Am. J. Path.* **135**, 261–9.

Johnson, H.M. (1985). Mechanisms of interferon gamma production and assessment of immunoregulatory properties. *Lymphokines*, **11**, 33.

Kunkel, S.L., Spengler, M., May, M.A., Spengler, R., Larrick, J. and Remick, D. (1988). Prostaglandin E_2 regulates macrophage-derived tumour necrosis factor gene expression. *J. Biol. Chem.* **263**, 5380–4.

Le, J. and Vilcek, J. (1987). Tumour necrosis factor and interleukin-1: cytokines with multiple overlapping biological activities. *Lab. Invest.* **56**, 234–48.

Lehmann, V., Benninghoff, B., and Dröge, W. (1988). Tumour necrosis factor-induced activation of peritoneal macrophages is regulated by prostaglanin E_2 and cAMP. *J. Immunol.* **141**, 587–91.

Lovett, D.H., Ryan, J.L., and Sterzel, R.B. (1983). Stimulation of rat mesangial cell proliferation by macrophage interleukin-1. *J. Immunol.* **131**, 2830–6.

Lovett, D.H., Szamel, M., Ryan, J.L., Sterzel, R.B., Gemsa, D., and Resch, K. (1986). Interleukin-1 and the glomerular mesangium. I. Purification and characterization of a mesangial cell-derived autogrowth factor. *J. Immunol.* **136**, 3700–5.

MacKay, K., Striker, L.J., Stauffer, J.W., Dol, T., Agodao, L.Y., and Striker, G.E. (1989). Transforming growth factor-β. Murine glomerular receptors and responses of isolated glomerular cells. *J. Clin. Invest.* **83**, 1160–7.

Martin, J., Lovett, D.H., Gemsa, D., Sterzel, R.B., and Davies M. (1986). Enhancement of glomerular mesangial cell neutral proteinase secretion by macrophages: role of interleukin-1. *J. Immunol.* **137**, 525–9.

Martin, M., Schwinzer, R., Schellekens, H., and Resch, K. (1989). Glomerular mesangial cells in local inflammation: Induction of the expression of MHC Class II antigens by IFN-γ. *J. Immunol.* **142**, 1887–94.

Michie, H.R., Manogue, K.R., Springs, D.R., Revhaug, A., O'Dwyer, S., Dinarello, G.A., Cerami, A., Wolff, S.M., and Wilmore, D.W. (1988). Detection of circulating tumour necrosis factor after endotoxin administration. *New Engl. J. Med.* **318**, 1481–6.

Petersen, M., Steadman, R., Hallet, M.B., Matthews, N., and Williams, J.D. (1990). Zymosan induced leukotriene B4 generation by human neutrophils is augmented by rhTNFα but not chemotactic peptide. *Immunology*, **70**, 75–81.

Pledger, W.J., Stiles, C.D., Antoniades, H.N., and Scher, C.D. (1978). An ordered sequence of events is required before BA1B c3T3 cells become committed to DNA synthesis. *Proc. Natl. Acad. Sci. USA*, **75**, 2839–43.

Praga, M., Gutierrez-Mille, V., Navas, J.J., Ruilope, L.M., Morrales, J.M., Alcazar, J.M., Bello, I., and Rodicio, J.L. (1985). Acute worsening of renal function during episodes of macroscopic hematuria in IgA nephropathy. *Kidney Int.* **28**, 69–74.

Radeke, H.H., Meier, B., Topley, N., Floege, J., Habermehl, G.G., and Resch, K. (1990). Interleukin-1α and tumour necrosis factor-α induce oxygen radical production in mesangial cells. *Kidney Int.* **37**, 767–75.

Raz, A., Wyche, A., Siegel, N., and Needleman, P. (1988). Regulation of fibroblast cyclooxygenase synthesis by interleukin-1. *J. Biol. Chem.* **263**, 3022–8.

Rees, A.J., Lockwood, C.M., and Peters, D.K. (1977). Enhanced allergic tissue injury in Goodpasture's syndrome by intercurrent bacterial infection. *Br. Med. J.* **2**, 723.

Richards, N.T., Gorden, C., Richardson, K., Emery, P., Howie, A.J., Adu, D., and Michael, J. (1990). Urinary IL-6, a marker for mesangial proliferation? *J.A.S.N.* **1**, 556.

Roberts, A.B., and Sporn, M.B. (1985). Transforming growth factors. *Cancer Surv.* **4**, 683–705.

Ross, R., Raines, W., and Bowen-Pope, D.R. (1986). The biology of platelet derived growth factor. *Cell*, **46**, 155–69.

Ruef, C., Budde, K., Lacy, J., Northemann, W., Baumann, M., Sterzel, R.B., and Coleman, D.L. (1990). Interleukin-6 is an autocrine growth factor for mesangial cells. *Kidney Int.* **38**, 249–57.

Sherry, B., and Cerami, A. (1988). Cachectin/tumour necrosis factor exerts endocrine, paracrine, autocrine control of inflammatory response. *J. Cell. Biol.* **107**, 1269–77.

Silver, B.J., Jaffer, F.E., and Abboud, H.E. (1988). Platelet-derived growth factor synthesis in mesangial cells: Induction by multiple peptide mitogens. *Proc. Natl. Acad. Sci. USA*, **86**, 1056–60.

Steadman, R., Petersen, M., Topley, N., Williams, D., Matthews, N., Spur, B., and Williams, J.D. (1990). Differential augmentation by recombinant human tumour necrosis factor α of neutrophil responses to particulate zymosan and glucan. *J. Immunol.* **144**, 2712–8.

Striker, G.E., Soderland, D., Bowen-Pope, F., Gown, A.M., Schmer, G., Johnson, A., Luchtel, D., Ross, R., and Striker, L.J. (1984). Isolation, characterization and propagation in vitro of human glomerular endothelial cells. *J. Exp. Med.* **160**, 323–8.

Suematsu, S., Matsuda, T., Aozasa, K., Akira, S., Nakano, N., Ohno, S., Miyazaki, J.I., Yamamura, K.I., Hirano, T., and Kishimoto, T. (1989). IgG1 plasmacytosis in interleukin-6 transgenic mice. *Proc. Natl. Acad. Sci. USA*, **86**, 7547–51.

Suranyi, M.G., Quiza, C., Gausch, A., Newton, L., Myers, B.D., and Hall, B.M. (1990). Cytokine levels in patients with the nephrotic syndrome. *Kidney Int.* **37**, 445(A).

Tipping, P.G., Low, M.G., and Holdsworth, S.R. (1991). Glomerular interleukin-1 production is dependent on macrophage infiltration in anti-GBM glomerulonephritis. *Kidney Int.* **39**, 103–10.

Tomosgui, N., Cashman, S.J., Hay, H., Pusey, C.D., Evans, D.J., Shaw, A., and Rees, A.J. (1989). Modulation of antibody mediated glomerular injury in vivo by bacterial lipopolysaccharide, tumour necrosis factor, and IL-1. *J. Immunol.* **142**, 3083–90.

Topley, N., Floegge, J., Wessel, K., Hass, R., Radeke, H.H., Kaever, V., and Resch, K. (1989). Prostaglandin E_2 production is synergistically increased in cultured human glomerular mesangial cells by combination of IL-1 and tumour necrosis factor. *J. Immunol.* **143**, 1989–95.

Topley, N., Floege, J., Hoppe, J., Kishimoto, T., and Resch, K. (1990). Lymphokines modulate rather than induce proliferation in cultured human mesangial cells. *Kidney Int.* **37**, 204A.

Waldherr, R., Eberlein-Gonska, M., Noronha, I.L., Andrassy, K., and Ritz, E. (1990). TNFα and ICAM-1 expression in renal disease. *J.A.S.N.* **1**, 544.

Werber, H.I., Emancipator, S.N., Tykocinski, M.L., and Sedor, J.R. (1987). The interleukin-1 gene is expressed by rat glomerular mesangial cells and is augmented in immune complex glomerulonephritis. *J. Immunol.* **138**, 3207–12.

Yanagisawa, M., and Masaki, T. (1989). Molecular biology and biochemistry of endothelin. *Trends in Pharmacol. Sci.* **10**, 374–8.

21 Immunoglobulin A-Related nephropathies; pathogenesis and treatment

F.W. Ballardie

Insight into the immunopathogenesis of immunoglobulin A (IgA)-related nephropathies has provided our most direct evidence of induction of nephritis by infection and has improved our understanding of the diverse interactions of the immune system with the environment and the kidneys. The nephropathies have become the most frequently detected renal diseases in developed countries (D'Amico 1987). Policies for renal biopsy are likely to account for some of the variation in reported incidences in the UK, Europe, and possibly the USA (Ballardie *et al.* 1987*a*; Feehally *et al.* 1989; Julian *et al.* 1988). Once considered benign, an inexorable progression of disease often occurs in the largest subgroup, with low grade urinary abnormalities (Emancipator *et al.* 1985), and studies suggest the nephropathies may be among the commonest causes of end stage renal failure (Simon *et al.* 1988), although this requires further evaluation (Wing *et al.* 1990).

The cardinal symptom in the early descriptions, of recurrent macroscopic haematuria of early onset in upper respiratory tract infections, is a feature in a minority of affected patients only. The prerequisite for defining IgA nephropathy is accepted as immunoglobulin A deposited in glomerular mesangium as dominant or co-dominant immunoglobulin, in the absence of serological markers of systemic lupus erythematosus. This diagnostic criterion and the designation IgA nephropathy, is now applicable to a diverse spectrum of renal disease (Welch *et al.* 1988; Brouhard 1988), encompassing most clinical forms of glomerulonephritis (Table 21.1). The close temporal relationship between infection with environmental pathogens at mucosal surfaces and clinical evidence of nephritis in some disease types is a central part of explanations of the immunopathogenesis.

Considerable recent advances in our understanding of the extent of perturbations in the immune system includes autoimmunity and dysregulation in the cytokine network. The immunopathogenetic factors responsible for the diversity of disease expression have not been properly established. Glomerular immunoglobulin A itself does not appear a sufficient explanation, and experimental models of IgA nephropathy have, to date, not reproduced their intended counterpart in humans with macroscopic haematuria.

Table 21.1 Clinical presentation of IgA-related nephropathies

Primary:		Secondary
Nephritic	Nephrotic syndrome	Others
Rapidly progressive nephritis Haematuria with no significant proteinuria: recurrent macroscopic protracted microscopic Haematuria and proteinuria Nephritis and vasculitis: Henoch–Schonlein purpura	- Steroid responsive - Steroid unresponsive	Nephropathy + gastrointestinal disease hepatic cirrhosis, porta-caval shunt, Crohn's, coeliac disease, complicating inflammatory arthropathies

Conceptually, analysis of immune system dysfunction in IgA-related nephropathies has focused on two areas of regulatory disorder: processes of local mucosal immunity and immune complex production as a primary cause of nephritis, and increasing evidence that in some forms of IgA-related nephropathy systemic immune responses originate from bone marrow derived cells. Disease in humans has features in common with experimental models of nephritis, in which polyclonal B cell activation occurs. Familial occurrence of the nephropathies and of immune system abnormalities also argues for an inherited mechanism of disease, but most genetic factors studied are weak associations and vary between populations. Patients exhibit variable hypersensitivity to environmental antigens. The existence of immune system autoreactivity with the presence in sera of autoantibodies which recognize self-antigens on vascular endothelial cells, immunoglobulin constant regions, and mesangial cells, is in favour of an autoimmune pathogenesis.

These emerging concepts in IgA-related nephropathies have provided new avenues for clarification of their cause. Key developments in our understanding of the immunopathogenesis of the IgA nephropathies and Henoch–Schonlein purpura, and implications for design of therapies for this important renal disease in humans will be examined in this chapter.

IgA-related nephropathies as a mucosal serum sickness

It is widely believed that IgA-related nephropathies arise from a mucosal form of serum sickness, (Lamm 1987; Bene and Faure 1987). Pathogen-derived antigen from a viral or bacterial infection of the upper respiratory or gastrointestinal tract is introduced via the mucosal route (Nicholls *et al.* 1984). This elicits a secretory IgA response from mucosal-associated lymphoid tissue (MALT), forming an immune complex which deposits in the glomerular mesangium, causing injury. Increased permeability to low molecular weight tracers and carbohydrates has been shown in patients whose serum levels of IgA-containing complexes are directly related to permeability defects (Davin *et al.* 1988; Jenkins *et al.* 1988). Enhanced mucosal permeability is unlikely to be a primary cause of raised immune complex levels, since similar abnormalities are found in unrelated bowel injuries (Forget *et al.* 1984).

Abnormalities in mucosal surface immune mechanisms in IgA-related nephropathies

Various disturbances of plasma cell numbers have been described at the sites of mucosal antigen challenge. Whilst there is no quantitative difference in plasma cells in the small intestine between patients and controls (Westberg *et al.* 1983), tonsillar abnormalities have been found consistently. In patients with IgA-related nephropathy (Bene *et al.* 1983; Egido *et al.* 1984*a*) and Henoch–Schonlein purpura (Bene *et al.* 1986), expansion occurs in the J-chain positive, dimeric IgA, secreting cell population compared with IgG positive cells. Similar findings are reported in episclera (Bene *et al.* 1984). Both IgA subclasses, IgA1 and 2, are found in pharyngeal secretions (Tomino *et al.* 1983) of patients, and IgA-containing lambda light chains dominate in tonsillar as well as glomerular deposits (Lai *et al.* 1988*a*).

However, evidence for a mucosal surface origin of glomerular IgA deposits does not receive support from immunochemical characterisation of IgA subclass. Glomerular deposits, defined by monoclonal reagents, are predominantly IgA1 subclass co-localizing with J-chain determinants in the primary nephropathies, whereas secretory component is detectable in hepatic disease-related forms (Andre *et al.* 1980; Lomax-Smith *et al.* 1983; Tomino *et al.* 1981) (Table 21.2). The absence of secretory component in primary disease places doubt on the significance of mucosal surface derived immune complexes. In IgA-related nephropathies secondary to hepatic and some other forms of gastrointestinal disease, these links are more firmly established.

Table 21.2 Immunochemical characterization of mesangial IgA deposits in IgA-related nephropathies using monoclonal reagents

IgA-related Nephropathy	IgA1		IgA2	J-chain	Secretory component	Light chains: κ and λ
Nephritic and nephrotic	+	$\gg$	±	+	−	$\kappa \ll \lambda$
Hepatic cirrhosis	+		+	+	+	Analysis not available

Hypersensitivity to environmental antigens

The selective induction of specific secretory IgA antibodies is desirable phylogenetically for prevention of infection contracted through contact with pathogens at mucosal surfaces. Antigen is eliminated, potentially with minimal induction of local inflammation, since IgA, unless it is in the form of aggregates, has little capability of complement fixation through the alternate pathway. After oral immunization antigen-specific immune responses involving IgA and also IgG and IgM are detectable in peripheral blood lymphocytes in normal subjects. Continued ingestion of foreign antigen results in systemic unresponsiveness, although local mucosal surface antigen-specific IgA responses persist (Challacombe and Tomas 1980). Most antigens displayed by environmental pathogens require T-B cell co-operation for priming of B lymphocytes to synthesize dimeric IgA antibodies, but a minority are capable of priming B lymphocytes in a T-independent manner. Populations of B lymphocytes migrate to blood via lymph, and thence to mucous or exocrine glands, where there is completion of maturation to IgA-dimer secreting plasma cells. The mechanisms of antigen transport and processing, sites of synthesis of specific antibodies and mechanisms governing the formation of polymeric or monomeric antibodies are unknown in patients with IgA-related nephropathies.

In normal subjects, antibodies with specificity for many microbial and food antigens are found (Mestecky *et al.* 1987); these are predominantly IgA1 subclass, in polymeric and monomeric forms. After prolonged exposure to certain viral antigens, however, serum IgA antibodies shift from polymeric to monomeric (Mestecky and McGhee 1987) suggesting there is a change in the site of production of antibodies, from mucosa, where IgA2 B cells predominate, to bone marrow, where there are mainly IgA1 positive B cells (Van den Wall Bake *et al.* 1988a).

Patients with primary IgA-related nephropathies have altered kappa/lambda ratios of both IgA and IgG in serum (Lai *et al.* 1988a), and enhanced IgA-lambda light chain synthesis following mitogen stimulation of lymphocytes. Highly selective lambda light chain deposition of immunoglobulins is found in glomerular mesangium (Lai *et al.* 1985). This suggests there may be a selective population expansion of IgA1 and IgG secreting B cells in these patients.

Potential mechanisms of glomerular injury

There are three mechanisms whereby serum antibodies to exogenous immunogens can result in glomerular injury. These are, first, by formation of an immune complex, capable of depositing in the glomerular mesangium, secondly, by cross-reactivity of antibody with endogenous determinants within the glomerulus, and thirdly by the so-called 'planted antigen' mechanism, whereby an extrinsic antigen is trapped within the glomerulus, binding circulating antibody *in situ*.

Each of these mechanisms receives some support from work in experimental models of IgA-related nephropathy and disease in humans. In healthy adults, undegraded dietary antigens and associated IgG class antibodies are frequently detected in sera. Increased responsiveness of IgA isotype antibodies to environmental antigens in sera of patients with primary forms of IgA-related nephropathy is *prima facie* evidence for a fundamental regulatory defect in the immune system. IgA antibodies in patients' sera are found frequently with specificity for many common food and bacterial antigens: for example, determinants present on ovalbumin (Sancho *et al.* 1983) are capable of binding IgA. There is also evidence implicating hypersensitivity of IgA class antibodies, particularly to gliadin. Serum IgA binding to crude gliadin and to a gliadin fraction occur with variable frequency in IgA-related nephropathy (Nagy *et al.* 1988; Fornasieri *et al.* 1987), although these abnormalities are absent in Henoch–Schonlein purpura (O'Donoghue *et al.* 1992).

IgA antibodies with activity to reticulin and endomyosin – markers of coeliac disease and dermatitis herpetiformis – are absent in IgA-related nephropathy (Rostoker *et al.* 1988), and IgA-containing complexes in sera are dependent on dietary gluten intake (Coppo *et al.* 1990). There is no association, however, between IgA nephropathy and coeliac disease (Kumar *et al.* 1988), nor dependence of activity of nephritis on these IgA-immune complexes. In contrast to these clinical observations, a gluten–mesangial cell complex is demonstrable *in vitro* (Amore *et al.* 1988), suggesting that glomerular deposition of IgA might be a consequence of *in situ* formation of a complex, following systemic immunity to gluten.

The majority of analyses have not included examination of the specificity of immunoglobulins eluted from glomeruli of patients with IgA-related nephropathy. Conceptually, it is difficult to formulate a mechanism by which chronic exposure to antigens and ensuing enhanced immunity can result in a dramatic, infection-associated, recurrent macroscopic haematuria of subgroups of patients with IgA-related nephropathy, without invoking other factors exaggerating glomerular injury, such as processes of amplification. The lack of longitudinal studies of patients' immunity to food antigens and association with activity of nephritis reinforces this view. However, in patients with protracted low-grade urinary abnormalities diet-induced renal injury cannot be excluded.

Immune aggregates

The physicochemical constituents of glomerular deposits are not fully established, but there is evidence that IgA is present in polymeric form, usually associated with IgG. There is little evidence that deposits are true immune complexes, incorporating either exogenous or endogenous antigens. The nephritogenicity of macromolecular aggregates in sera remains an unresolved issue. Attempts to detect IgA-containing aggregates in sera in the IgA-related nephropathies were initially unsuccessful, since complement-based assays often failed to capture these species from sera. At least ten distinct characteristics have been exploited for their detection (Table 21.3). A number of constituents have been described, including polymeric and monomeric IgA (Sanchio *et al.* 1983), fibronectin (Nagy 1987*a*), IgM and IgG (Woodruffe *et al.* 1980; Egido *et al.* 1984*b*). Initial reservations concerning their signifi-

Table 21.3 Characteristics of immunoglobulin-containing aggregates in sera and glomeruli of patients with IgA-related nephropathies

Circulating			Mesangial deposits		
Physicochemical characteristic		Assay	Size		
Solubility	Cryoglobulin		>21S	17–21S (<1%)	
	Cryofibrinogen				
	Polyethylene glycol precipitate				
IgA	Multimeric	Raji cell,	7S,	5–9S (50%)	1000 kDa
	Polymeric	inhibitions;	11S	10S	
	Subclass 1	latex particle,	to	to	to
		anti-IgA;	21S	17S (45%)	<160 kDa
		conglutinin binding;			
	Antiglobulin activity	Rheumatoid factor.			
IgG, IgM	Multimers	C1q binding. Raji cell. Conglutin binding. Fc receptor-related platelet agglutination. Inhibition of complement dependent lymphocyte resetting.			
Complement binding					
Anti-C3	Aggregate C3	Anti-C3			
Anti-C3	Aggregate C3d	Anti-C3d			
Complexed antigen	Anti-RSA antibodies	Raji			
Fibronectin binding	IgA-fibronectin/collagen complex	ELISA			
Erythrocyte complexed	Aggregate affinity to erythrocyte CR1 receptor	Red blood cell-IgA binding			

cance, a consequence of detection of IgA-containing complexes in normal subjects have been resolved by IgA subclass analysis. Macromolecular IgA1 is found exclusively in patients with primary IgA-related nephropathies, with increased total IgA1/2 ratios in serum (Valentijn *et al.* 1984) – further evidence for a selected B cell population proliferation.

The majority of circulating aggregates are of intermediate size on density grading centrifugation (11S to 21S) (Valentijn *et al.* 1983), and instances of detection of foreign antigen suggest that at least some are immune complexes (Yap *et al.* 1987). The properties of IgA-containing aggregates (Table 21.3) give little insight into their potential to deposit and cause glomerular injury. Renal biopsy eluants have shown a more restricted range of features of deposited IgA aggregates, to date, principally characterized by size, IgA subclass, and charge. It is clear that anionic IgA multimers deposit preferentially in glomeruli (Monteiro *et al.* 1985). These are predominantly of subclass IgA1. Polymers of IgA are considered nephritogenic, since nephritis is reported with detection of IgA polymers in sera more commonly with multimeric complexes (Hernando *et al.* 1986). Polymeric IgA-producing lymphocytes are present in blood (Lozano *et al.* 1987), and may be of intestinal origin (Kutteh *et al.* 1982). These associations, however, are equally consistent with other immune species, secreted at the time of IgA polymer synthesis, contributing to glomerular injury.

Kinetics and clearances of aggregates

Clearances of the immune aggregates in humans are poorly understood. In a murine model there is renal deposition from sera of only a small fraction of dimeric IgA (Rifai and Mannik 1983), the majority being cleared by hepatic phagocytic activity. Size, charge, and avidity are likely to alter considerably the tendency for mesangial deposition. There is a brief half-life in the glomerulus of polymeric IgA complexes in murine systems; although no analysis of glomerular turnover in humans has been possible, clearance of IgA within weeks of deposition has occurred in a transplanted kidney (Silva *et al.* 1982). Delayed clearances of polymeric IgA-containing complexes after food challenge in patients with IgA-related nephropathy suggests a defect in hepatic clearance of these species (Sancho *et al.* 1983); in normal subjects and patients with coeliac disease, IgA polymeric antibodies to gluten have a shorter serum half-life than monomeric antibodies. Although clearance from the blood both of monomeric and of polymeric IgA-containing complexes have similar biphasic components in murine models, renal deposits contain only polymers (Rifai and Mannik 1983; Rifai and Millard 1985). This important regulatory dysfunction of polymer versus monomer formation and clearance is poorly understood in humans.

The capability of primary deposits to accrete further monomeric immunoglobulin or other protein in the glomerulus, or to trap circulating proteins through nonspecific means has recently been demonstrated (Chen *et al.* 1988).

Furthermore, stabilization of mesangial IgA deposits by patients' sera *in vitro* is less effective than by normal human sera (Tomino *et al.* 1987). Complement may play a role, since C3b, a product of alternate pathway activation, enhances antigen–antibody aggregate solubility (Miller and Nussenzweig 1975) by insertion of C3b in the lattice, limiting its size and altering its solubility characteristics. This phenomenon may contribute to impaired clearances of mesangial IgA deposition of aggregates, although the underlying mechanisms are not clear.

Immune reactants and cytokines in glomerular injury

The mechanisms by which immune reactants generate the great diversity of forms of glomerular injury in the IgA-related nephropathies in humans are controversial and poorly understood. Direct evidence for the roles of mediator systems is lacking, implying that the various immunoglobulin species present in sera and eluted from glomeruli are capable of causing injury. There is new evidence that components of the cytokine network have effects in the B cell system, as well as directly on glomerular cells. Interleukin 2 secretion and receptor expression is increased in peripheral blood T cells (Lai *et al.* 1988b; Schena *et al.* 1989). Monocytes, but not T cells, synthesize greatly increased levels of interleukin 6 during haematuria, and there is evidence for both paracrine and autocrine functions of this cytokine (Ballardie *et al.* 1990, 1992).

Experimental IgA-related nephropathies

Experimental models have focused principally on the means by which the characteristics, kinetics and clearances of the different forms of IgA can be altered to achieve glomerular deposition: it is clear that species containing IgA alone are weakly phlogistic. Glomerular injury is mediated by immune species in other experimental nephropathies, notably Masugi nephritis: antibody binding alone to nephritogenic antigen of the glomerular basement membrane; the membrane attack complex of complement; antibody dependent polymorph-, and macrophage dependent injury, are all important at different phases of disease. Further amplification of these responses with lipopolysaccharides, and components of the cytokine network in an additive or synergistic manner is also established.

IgA-complex induced tissue injury in the alveolar epithelium of rats is complement-dependent and occurs in the presence of activated macrophages and oxygen radicals, but is neutrophil independent (Emancipator and Lamm 1986). Complexes containing other immunoglobulin isotypes, notably IgG, are capable of localizing neutrophils similarly, presumably via more effective chemotaxis from complement cleavage products and more effective Fc-dependent mechanisms than IgA, although Fc-receptors for both immunoglobulin isotypes are present on neutrophils and mononuclear cells. Glomerular inflammatory cell infiltrate in humans and in experimental models has been described frequently as nil or minimal. However, during overt haematuria in IgA nephritis renal biopsy appearances show increased inflammatory cells with extracapillary glomerular cell proliferation and crescents, with fibrin staining (Nicholls *et al.* 1984), and interstitial, but not glomerular, phagocyte infiltration—also present during disease quiescence (Lupo *et al.* 1987).

In early attempts to generate glomerular deposits, preformed complexes of specific IgA with DNP were injected into Balb/c mice. This resulted in diffuse glomerular IgA deposition with mesangial expansion, but urinary abnormalities were minor and there was no apparent renal dysfunction (Rifai *et al.* 1979). Multimeric fractions of the IgA complexes were essential for glomerular deposition, variations in antigen–antibody ratio had no effect on nephritogenicity, and active immunization forms, with intra-peritoneal, anionic dextran-derived polysaccharide antigen resulted in an IgA-IgM mesangial proliferative nephropathy. More extensive glomerular morphological change was apparent with C3 deposition, when a different dextran-derived antigen carried a cationic charge (Isaacs and Miller 1982). Oral immunization of mice with bovine gamma globulin, bovine serum albumin, or ferritin (Emancipator *et al.* 1983) also induced serum IgA antibody responses and glomerular IgA deposition, but without detection of antigen or antibody in the glomerulus (Imai *et al.* 1985). Persistent virus infection is also capable of inducing glomerular IgA and other deposits: lymphocytic choriomeningitis (IgA and IgG) in mice; aleutian mink disease—a progressive multisystem parvovirus-induced disease (IgM, IgG and IgA); and oncorna virus-induced disease.

Complement

Current evidence favours complement activation by the alternate pathway in IgA-related nephropathy in humans. Studies have been cross-sectional, and it is not clear whether glomerular deposition of immune reactants other than IgA at different phases of activity of intermittent nephritis is also capable of inducing activation via the classical pathway. In passive murine models of IgA-related nephropathy, microscopic haematuria can be induced when IgA is complexed to anionic dextran–antigen, independently capable of inducing localization of complement (Isaacs and Miller 1982). This minor glomerular injury may be prevented by complement depletion.

Multimeric IgA complexed with a variety of antigens, including pneumococcal polysaccharide can induce glomerular complement deposition (Rifai *et al.* 1987), although the potential nephrogenicity of the complexes is uncertain. In humans, the detection of glomerular properdin and infrequent detection of C1q and C4 (Chen *et al.* 1988) suggests alternate pathway activation and the presence of C9 neoantigen, C3c and C5 suggests terminal complement activation and participation of the membrane attack complex (MAC). Immunoelectron microscopy shows co-localization of IgA, IgG, C3, C5 and C9, but no relationship of glomerular protein leak to intensity of MAC immunostaining

(Miyamoto *et al.* 1988). Mesangial cells in culture can be stimulated to produce interleukin I and prostaglandins by the MAC (Lovett *et al.* 1987), but the significance of these findings *in vivo* is unknown. Intraglomerular IgA is detected frequently with components C3–9 and properdin, but the early classical pathway components C1q and C4 are present only when IgG is also deposited; codeposition of IgM does not increase the classical pathway components further (Tomino 1980).

Immune-mediated injury and urine sediment

There is little information relating immune, morphological, and urinary abnormalities in patients with IgA-related nephropathies. The mechanisms of haematuria are largely unknown. Glomerular bleeding implies a mechanical disruption to the glomerular structure substantially greater than that when only proteinuria is present. The sites of injury leading to glomerular bleeding in IgA-related nephropathies have not been identified, but in a modified Masugi nephritis model in rabbits, gaps in the glomerular basement membrane were proposed to be the source (Makino *et al.* 1988). The propensity of subgroups of the mesangial IgA nephropathies to exhibit this phenonomen does suggest a mesangial or capillary loop site of injury.

Biopsy coincident with nephritic episodes of macroscopic haematuria frequently shows crescent formation, and patients with microscopic haematuria show increased urinary erythrocyte excretion following respiratory tract infection, in contrast to other glomerular disease (Miura *et al.* 1984). IgA, IgG, and IgM are present in relapses, whereas IgA only is more often the sole immunoglobulin during quiescence. Urinary red cell morphology is variable and associated with glomerular injury, increased mesangial and extra-capillary proliferation leading to crenated and contracted erythrocytes (Kaneko *et al.* 1987). Urinary platelet excretion is similarly related to glomerular abnormalities. (Tomino *et al.* 1988) and the degree of proteinuria is reflected in the severity of scarring and mesangial proliferative change, independent of renal function (Neelakantappa *et al.* 1988).

Urinary IgA and IgG characteristics do not reflect those in serum of patients with IgA-related nephropathy and Henoch–Schonlein purpura—monomeric not multimeric IgA is found (Galla *et al.* 1985). Urinary low molecular weight protein content relates to tubulo-interstitial injury (Nagy *et al.* 1987b). These findings suggest there are associated changes in immune-mediated injury, glomerular morphology, and bleeding with consequent erythrocyte deformity. However, protein loss is likely to be dependent on more complex factors.

Autoimmunity

Autoimmunity in the pathogenesis of IgA-related nephropathies has advanced our understanding of immune system perturbations, in nephritis. Glycoprotein and other structures within the glomerulus are a source of many potential antigens, but their characterization as autoantigens, and proof of their nephritogenic role in humans has proved to be technically difficult. Autoantibodies may be nephritogenic by interacting with discrete glomerular structures, initiating or compounding pre-existing injury, and, depending on their characteristics, targeting mediators of inflammation, such as complement or inflammatory cells.

Mechanisms

There are two plausible mechanisms by which autoreactive B cell clones are activated: first, by specific triggering—selective clonal expansion in response to exogenous antigen challenge, with secretion of an autoantibody cross-reactive with endogenous determinants within the glomerulus; secondly by nonspecific means—polyclonal B cell activation (PCBA) in response to exogenous mitogen, the B cell repertoire secreting autoreactive species of several immunoglobulin isotypes and subclasses. Most evidence favours the latter mechanism in IgA-related nephropathies.

The mechanism of induction of B cell activation in the IgA-related nephropathies is unclear. Tolerance to self antigens is generally maintained by T cells, in some instances by T suppressor cells, and in others by clonal deletion. In humans PCBA can be triggered nonspecifically by microbial products, such as lipopolysaccharides, peptidoglycans and purified protein derivative, with production of several autoantibodies from within the B cell repertoire, including rheumatoid factor and anti-DNA antibodies. PCBA can also be triggered by persistent viral and parasitic infection. Within the spectrum of PCBA, there is preferential immunoglobulin isotype production; for example, lipopolysaccharide stimulation results in dominant IgM, IgG and IgE, (for review, see Goldman *et al.* 1988).

Autoantibodies

Autoimmunity was first suggested in IgA-related nephropathies when an IgM class antinuclear factor was described, which was absent in patients with systemic lupus erythematosus, and which was cold reactive to nuclei in liver and renal tissue (Nomoto and Sakai 1979). IgA eluted from renal biopsies recombined with autologous tissue, but not with normal human tissue, both in patients with IgA-related nephropathies and with Henoch–Schonlein purpura (Tomino *et al.* 1982). Affinity of IgA eluants for tonsillar cell nuclei has also been shown (Tomino *et al.* 1985). IgA from sera complexes with fibronectin and collagen to form aggregates. However it is uncertain whether this represents true antibody activity through the F(ab′)2 site of IgA or nonspecific binding to Fc (Cederholm *et al.* 1988).

Autoimmunity directed at glomerular antigens, antimesangial cell autoantibodies in sera, of IgG isotype with specific F(ab′)2 binding (IgG-MESCA), were first described by our own group (Ballardie *et al.* 1987*b*, 1988*a*,*b*). The autoantigen has components of 48–55 kDa and is apparently localized to glomerular mesangium and capillary loops. It has recently been shown that the target antigen is a human mesangial cell constituent, initial studies suggesting localization to the cell membrane: mesangial cell autoantigens are the only autoantigens detectable in whole normal glomeruli (O'Donoghue *et al.* 1989, 1991). Antimesangial cell antibodies were detected in Henoch–Schonlein purpura and in primary IgA-related nephropathies (O'Donoghue *et al.* 1988 *a,b*, 1991), but not in patients with hepatic disease. Antivascular endothelial cell antibodies have also been found, with binding in most instances to endothelial cell determinants other than HLA class 1 antigens (Yap *et al.* 1988). It is not clear that this antigen system is represented in human glomeruli.

IgA autoantibodies with antiglobulin activity—IgA rheumatoid factors—have been detected in sera by several groups. These autoantibodies have similar characteristics in both primary IgA-related nephropathies and Henoch–Schonlein purpura: they are polymeric, of IgA1 subclass (Czerkinsky *et al.* 1986), they appear to recognize determinants on the constant regions of human IgG, and they contain IgG and IgA rheumatoid factor (Saulsbury 1987).

The presence of IgA rheumatoid factors provides insight into immune system perturbations in IgA-related nephropathies. Rheumatoid factors may be part of a primal immune system: the relationship between monoclonal rheumatoid factor and the polyclonal rheumatoid factors of the IgA-related nephropathies is unclear, but they may represent secretion products of polyclonally stimulated and partially differentiated B cell populations.

Autoantibody nephrotoxicity

The human glomerulus contains a multiplicity of potentially antigenic structures, but those which may function as autoantigens are limited, dependent on the genetic constitution of the individual, the synthesis of autoantibody, and accessibility and display of the relevant epitopes. Hyper-reactivity of the immune system, and sensitivity to environmental antigens in IgA-related nephropathies create difficulties in interpreting the increasing array of apparent autoantigenic substances cited: a number of immunoassays have used substrates derived from other mammalian species. Those using only substrates extracted from the human glomerulus are more likely to measure species implicated in autoimmune processes. Further analysis of cross-reactivity and polyspecificity of serum and biopsy eluant antibody activity in humans and in experimental models is needed. There is so far no proven direct nephrotoxicity of autoantibodies to environmental antigens.

IgG-MESCA in IgA-related nephropathies and Henoch–Schonlein purpura were intermittently detected in sera of patients who had IgG present in most renal biopsies, and were found with a high frequency during macroscopic haematuria (O'Donoghue *et al.* 1989). IgA antivascular endothelial antibody levels correlate with the degree of proteinuria (Yap *et al.* 1988). Longitudinal data on clinical–pathological correlations between immune reactants in sera and disease in humans have provided some circumstantial evidence for the role of autoantibodies in glomerular injury. In experimental models, Wistar rats receiving murine, complement fixing, IgG1 class monoclonal antibody specific to the Thy1 antigen, present on mesangial cells and other body tissues, developed a crescentic nephritis with mesangial cell lysis, proliferation and heavy proteinuria (Bagchus *et al.* 1986). A similar toxicity has been shown with anti-thymocyte serum in Lewis rats (Wilson *et al.* 1988). The nephritogenicity of these species is critically dependent on antibody characteristics, including ability to fix complement: other monoclonals recognizing Thy1 were unable to induce glomerular injury (Bagchus *et al.* 1986).

The contributions and interdependence of the immune reactants in glomerular injury remains uncertain. Occasionally patients may enter complete clinical remission and have no urinary abnormality yet have significant, but reduced, IgA deposits present on renal biopsy (Costa *et al.* 1987); during episodes of acute glomerular inflammation IgG and IgM are often present (Nicholls *et al.* 1984).

Patients may secrete IgG antimesangial cell antibodies either together with, or distinct from, IgA rheumatoid factor (Ballardie *et al.* 1988b). Patients with episodic macroscopic haematuria more frequently secrete these two autoantibodies concurrently, whereas there is dissociation in those with microscopic haematuria. Progressive loss of renal function is associated with certain autoimmune profiles—notably, presence of IgG-MESCA, IgA-cardiolipin, and IgG-laminin binding (O'Donoghue *et al.* 1990). Thus, IgG autoantibodies may contribute to glomerular injury in both the acute and chronic phases of disease activity, and IgG and IgA autoantibodies may produce distinct and additive glomerular injury, potentially via *in situ* complex formation.

Viruses and autoantigens

Intercurrent and persistent viral infection is capable of inducing or modifying nephritis through distinct mechanisms. Each is exemplified in experimental human nephritis. Mechanisms include polyclonal B cell activation, perturbations of T cell mediated immunity including T cell dependent regulatory circuits, immune complex formation from viral antigen, or *in situ* complex formation with virus-encoded glomerular neoantigens or by modulation or up-regulated expression of an intrinsic glomerular antigen.

In IgA-related nephropathy, there are several abnormalities associated with viral infection. Haematuria with intercurrent virus infection is well described. Patients have preferential virus-specific IgA1 antibody synthesis in response to vaccination with influenza virus (Van den Wall Bake *et al.* 1989). In an endemic area persistent hepatitis B antigenaemia is more common in patients with IgA disease than in the general population; virus antigen and DNA are present in glomeruli (Lai *et al.* 1988*c*). Cytomegalovirus has also been recently implicated (Gregory *et al.* 1988), although this has not been confirmed (Waldo *et al.* 1989). Rheumatoid factor autoantibodies might cross-react with cytomegalovirus antigens, since there is considerable amino acid homology between monoclonal (IgM) rheumatoid factor and monoclonal cytomegalovirus antibody. Cytomegalovirus DNA, if inserted into host genome, could mimic polymorphisms present in the DQB region, as homology of codings for at least six amino acids is described (Newkirk *et al.* 1987) with potential consequences for immunoregulation. Glomerular cytomegalovirus antigen and DNA can be present, however, without glomerulopathy, or with minor IgG deposits only, during active infection.

Immunoregulation: genetics and cytokines

There is a need in primary IgA nephropathies for a better understanding of the immunobiology of stimulatory factors, including cytokines, which result in selective responses of IgA1, IgA2 and IgG to antigen challenge. There is a remarkable absence of constant HLA associations in these diseases, although they are definable as clinically homogeneous subgroups. This necessitates further examination of DQβ and non-HLA genetic systems.

In IgA-related nephropathies, activation of the IgA-secreting system occurs during haematuria, but studies of cell function and phenotype have provided conflicting results. Receptor up-regulation for Fc on T and B cells, and for interleukin 2 on T cells is found in patients (Lai *et al.* 1988*b*). Fc expression is inducible *in vitro* by co-culture with IgA, and reduced expression in patients with selective IgA deficiency may be corrected by conditioned, cell-free medium (Adachi *et al.* 1983); in contrast, patients with IgA-related nephropathy have maximal Fc density expressed. These findings suggest that there is receptor expression secondary to a primary defect in cytokine production. Variable and inconsistent associations of T cell populations with IgA-related nephropathy have been described; these may be explicable in future clonal analyses. Patterns of IgG production parallel those of IgA, particularly in patients with marked glomerular injury (Cosio *et al.* 1982). Pokeweed mitogen stimulated polymeric IgA synthesis is increased and found more frequently in patients with macroscopic haematuria (Lozano *et al.* 1987), but the dependence of other potential disease markers or antigen-specific cell systems during active disease remains uncertain.

Genetic factors which may influence expression of these disturbances in the immune system in IgA-related nephropathies have recently been reviewed (Egido *et al.* 1987). Familial occurrence of disease without association with HLA antigens supports the existence of non-HLA linkages. Polymorphisms within the immunoglobulin heavy chain switch region have been associated with primary IgA nephropathy with recurrent macroscopic haematuria (Demaine *et al.* 1988): significant decreases in cu-related phenotype and increases in C 1 were also present. Transcription from polymorphisms in these loci, of importance in immunoglobulin isotype switching, is likely to influence expression of gene products in IgA-related nephropathies.

Secondary disease

Hepatic dysfunction in cirrhosis leads to an increase in total serum IgA, together with low grade proliferative and glomerulosclerotic lesions, with IgA deposition in 33–90 per cent of cases and IgG and IgM in 4–10 per cent (Berger *et al.* 1977; Sinniah 1984). Clinical evidence of nephropathy, with significant proteinuria or microscopic haematuria, is present in a minority only, typically 5–10 per cent of those with immune deposits (Bene *et al.* 1988). Loss of renal function is generally a consequence of renal haemodynamic factors and not glomerulopathy. The distinction from the primary IgA nephropathies is further supported by immunochemical differences of glomerular deposits (Table 21.2) and absence of evidence of activity of the systemic immune system (Ballardie 1988*b*). Much information on our understanding of clearance kinetics of IgA and aggregates has originated from studies of experimental and human hepatic dysfunction. The liver is the major site of catabolism of both polymeric and monomeric IgA in the mouse (Moldoveanu *et al.* 1988).

Parenchymal liver disease in humans, but not biliary obstruction, results in a preferential rise in serum polymeric IgA, accompanied by a reduction in fractional catabolic rate (Edlacroix *et al.* 1982). In a murine system, cholestasis also promotes deposition of circulating IgA complexes, with specific antigen detectable in glomeruli and in cirrhotic rats, there is a rise in IgA- and IgM-containing aggregates with glomerular deposition which may include IgG (Iida *et al.* 1985). Foreign antigen complexed with oligomeric IgA is cleared via hepatocyte receptors which have affinity for both Fc and secretory component, whereas Kupffer cells express Fc only (Sancho *et al.* 1986). Polymeric IgA and polymeric IgA antigen complexes are transported by the same mechanism (Phillips *et al.* 1988).

Glomerular morphological abnormalities or proteinuria do not occur in these experimental systems, unless there is additional nonspecific blockage of the hepatic reticuloendothelial system, which results in greater intensity of staining of immunoglobulin deposits (Ogata *et al.* 1988). These models appear to represent the minor glomerular disease found in the secondary IgA nephropathies in humans.

Henoch–Schonlein nephritis

The diagnosis of Henoch–Schonlein purpura (HSP), a systemic disease with leucocytoclastic vasculitis proven on cutaneous biopsy, or implicit from bowel and joint symptoms generally requires detection of IgA in an organ of injury, for confirmation (Knight and Cameron 1987). When IgA is found in glomeruli, together with urinary abnormalities the term Henoch–Schonlein nephritis is best applied. The concept of HSP as a vasculitis superimposed on the nephritis of a primary IgA nephropathy has important implications in our understanding of the pathogenesis and interdependence of autoimmunity-induced vascular and glomerular injury. In adults, HSP is rare in contrast to its frequency in paediatric practice, but it is now recognized to carry a higher long-term risk of renal dysfunction (Fogazzi *et al.* 1988) than considered previously (Roth *et al.* 1985). Mortality may be significant, although bowel vasculitis is reported as steroid responsive or self-limiting in some cases.

Serological abnormalities found in patients with HSP closely reflect those of primary IgA nephropathies already described, and are reviewed by Meadow and Scott (1985). Circulating IgG autoantibodies recognizing glomerular antigenic components of the same molecular weight as IgA nephropathies are detectable in HSP and the autoantigen in HSP has recently been localized to the mesangial cell, as in IgA-related nephropathy (O'Donoghue *et al.* 1989, 1991).

These comparisons of immune system perturbations do not explain the disparity of organ injuries between HSP and IgA-related nephropathy. HSP with microscopic haematuria is described in a patient with a history of recurrent macroscopic haematuria due to IgA-related nephropathy (Ballardie *et al.* 1987*b*). Each disease precipitated by the same antecedent infection, has been described in twins (Meadow and Scott 1985), and *formes fruste* of HSP exist, with IgA nephritis-related uveitis (Yamabe *et al.* 1988). The pathogenetic mechanism of these diseases may be clarified by recent findings of autoimmunity. IgG anti-mesangial cell antibodies are found in sera during activity of Henoch–Schonlein nephritis, but are absent when organs other than the kidney are affected in Henoch–Schonlein purpura (O'Donoghue *et al.* 1988*a*, *b*). The vasculitis of HSP is accompanied by presence of IgA and IgG autoantibodies recognizing anti-neutrophil cytoplasmic antigens (Van den Wall Bake *et al.* 1988*b*). Autoantibody recognition of antigens in the mesangium and vasculature may hold the key to expression of organ injuries.

Treatments

Therapeutic intervention in the IgA-related nephropathies has been designed to arrest decline in renal function, or to modify disease expression by induction of remission in nephrotic syndrome. Treatment has also been used to suppress severe forms of refractory vasculitis which may occur in adults and in a minority of children with Henoch–Schonlein purpura. Patients whose disease merits consideration for treatment have been in these categories, or those at 'high risk' of progressive renal failure, with hypertension, significant proteinuria, or with protracted features, classified as grade IV or V, on renal biopsy (Table 21.4). Circumstantial evidence in some series suggests that co-deposition of IgG and/or IgM is associated with more severe disease (Sinniah 1985).

In evaluating acceptability of therapeutic regimens a clear perspective of relative risk and benefit is essential. Low toxicity is a prerequisite for those with slowly progressive disease who receive long-term therapy. Actuarial kidney survival in affected patients has varied between 100 and 80 per cent at 10 years (Sinniah 1985; Kincaid-Smith and Nicholls 1983), but figures at 20 years show renal survival is in the order of 70 per cent (Beukhof *et al.* 1983), with less than half maintaining renal function within normal limits. This slow deterioration in renal function in most patients has contributed to the paucity of adequate prospective trails, and no therapy is yet established as beneficial.

Treatments evaluated have attempted to modify components of the diverse interactions of the immune system with the environment and the kidney. Many aspects have been incorporated in study design (Table 21.5). Outcome parameters have included measures of the immune profile of patients' sera, and clinical effects, such as proteinuria or haematuria. The lack of direct information relating glomerular injury to immune abnormalities has complicated interpretation of the latter. Further evaluation is required to enable the important distinction of immune-mediated glomerular injury from non-immune processes, particularly hypertension and other processes leading to progressive scarring.

In a controlled study tonsillectomy resulted in a reduction in episodic macroscopic haematuria and in proteinuria. Immune system markers, including serum total IgA, polymeric IgA and PEG-precipitable immune complexes were also reduced (Walker *et al.* 1990). In an uncontrolled study examining

Table 21.4 WHO morphological criteria for classification of glomerular lesions in IgA-related nephropathies

Class	Mesangium	Glomerular sclerosis	Tubulo-interstitial	Arteriole sclerosis
I	Normal*	–	–	–
II (30%)	Widening, ↑ cellularity <3 cells/glomerulus	–	–	–
III (30%)	Widening, ↑ cellularity >3 cells/glomerulus	<50% focal, segmental	±	+
IV	Diffuse proliferative	>50% focal, segmental diffuse sclerosing, segmental crescents	+, oedema	+
V	Diffuse proliferative		+, fibrosis	++

* Electron dense deposits (Sinniah 1985).

the effects of a gluten-free diet for up to 4 years in 29 patients, at least 50 per cent reductions from pretreatment values of IgA complexes, and IgA with affinity for gliadin, ovalbumin and bovine serum albumin were observed in patients. Proteinuria was reduced, but microscopic haematuria and progression of disease were unaffected (Coppo *et al.* 1990).

Cyclophosphamide reduces GALT IgA responses to cholera toxin in murine systems, but only when administered in high doses prior to antigen challenge. It has been used in conventional dosage together with antiplatelet agents and warfarin in IgA-related nephropathies. Decline in renal function over a period of up to three years was less than in controls, with reduction in proteinuria (Woo *et al.* 1987). Repeated renal biopsies showed reduction in progressive scarring (Woo *et al.* 1988). Arrest of decline in function with this regimen has not been confirmed in a recent short-term study (Walker *et al.* 1990), using cyclophosphamide (1–2 mg/kg body weight). Corticosteroids, administered for up to 19 months, followed by non-steroidal anti-inflammatory or anti-platelet drugs, reduced persistent proteinuria and decline in renal function in patients with creatinine clearance greater than 70 ml/min who received treatment (Kobayashi *et al.* 1986). Haemodynamic effects of treatment in these uncontrolled studies cannot be discounted. Remission can be induced in patients with severe nephrotic syndrome with steroids, but without benefit on longer term effects on renal function (Kobayashi *et al.* 1988; Lai *et al.* 1986). Sodium cromoglycate reduced proteinuria when compared with controls, although its effect on disease progression was not evaluated (Sato 1990). Urokinase, administered repeatedly and parenterally has been claimed to reduce proteinuria and progressive scarring on repeat biopsy, and to reduce decline in function (Miura 1990). Antiplatelet agents used in isolation (Chan *et al.* 1987), non-steroidals (Lagrue *et al.* 1981), eicosapentanoic acid (Bennett *et al.* 1989), danazol (Tomino *et al.* 1984), dapsone (Deteix *et al.* 1984), and phenytoin (Clarkson *et al.* 1980; Egido *et al.* 1984*c*) are of unproven or no benefit (Table 21.5).

Five children with rapidly progressive nephritis and IgA-related nephropathy proved resistant to pulse corticosteroids (Simon *et al.* 1988). However, there are anecdotal reports of recovery of renal function with the use of plasma exchange, cyclophosphamide and prednisolone in a patient with 30 per cent of glomeruli affected by crescents (Coppo *et al.* 1985). Our experience in this rare form of IgA-related nephropathy is less encouraging: two adults with 80 and 95 per cent crescentic change failed to improve after similar therapy (Ballardie, unpublished data). Plasma exchange without immunosuppresive drugs had no effect on patients with slowly progressive IgA-related nephropathy, but improved in two with rapidly progressive Henoch–Schonlein nephritis, one of whom received prednisolone (Hene and Kater 1983). Plasma exchange has also been systematically evaluated in an uncontrolled prospective study of patients with progressive IgA-related nephropathy (Nicholls *et al.* 1990), in which repeated exchange over months was found to slow the rate of

Table 21.5 Treatment forms evaluated for IgA-related nephropathy

Potential foreign antigen load reduction:
chronic infection eradication
dietary manipulation
Modification of immune response:
removal of immunocompetent tissue - tonsillectomy
immunosuppressive therapy
cyclophosphamide
cyclosporin A
sodium cromoglycate
dapsone
plasma exchange
phenytoin
Modification of glomerular inflammatory response:
steroids
NSAIDs
eicosapentanoic acid
Rheology
antiplatelet agents
anticoagulation
danazol
urokinase

decline in renal function. There remain doubts over the cost-effectiveness or acceptability of this form of intervention.

Severe Henoch–Schonlein nephritis is reported to be steroid responsive (Rose *et al.* 1981; Austin and Barlow 1983). Vasculitis affecting bowel HSP can be dramatically improved with corticosteroids alone (Roth *et al.* 1985). Anti-mesangial cell autoantibodies in individual patients with persistent activity of HSP are rapidly suppressed during treatment with corticosteroids alone, with remission of nephritis and vasculitis: relapse with further evidence of autoimmunity followed withdrawal of therapy (Ballardie 1988*b*), but proof of longer term benefit in patients remains in doubt (Lai *et al.* 1986).

Future therapies, and the design and application of treatment regimens, are likely to be radically influenced by our rapidly expanding knowledge of the origins of dysregulation of the immune system in IgA-related nephropathies. Attention is likely to focus on new developments implicating the cytokine network, as it has in other diseases, such as systemic lupus, which have an autoimmune component. Discovery of safe, long-term treatment for most patients with progressive disease will require a greater understanding of the processes of immune injury in the IgA-related nephropathies.

References

Adachi, M., Yodi, J., and Masuda, T. (1983). Altered expression of lymphocyte Fc receptor in selective IgA deficiency and IgA nephropathy. *J. Immunol.* **131**, (3), 1246–51.

D'Amico, G. (1987). The commonest glomerulonephritis in the world: IgA nephropathy. *Q. J. Med.* **64**, (245), 709–27.

Amore, A., Emancipator, S., and Coppo, R. (1988). Specific binding of gliadin to rat mesangial cells in culture. *Proc. EDTA-ERA*, XXV Congress, 47.

Andre, C., Berthoux, F.C., and Andre, F. (1980) Prevalence of IgA2 deposits in IgA nephropathies. *New Eng. J. Med.* **303**, (23), 1343–6.

Austin, H.A., and Barlow, J.E. (1983). Henoch Schonlein nephritis: prognostic features and the challenge of therapy. *Am. J. Kid. Dis.* **2**, 512–20.

Bagchus, W.M., Hoedemaeker, P.H.J., and Rozing, J. (1986). Glomerulonephritis induced by monoclonal anti-Thy 1. 1 antibodies. A sequential histological and ultrastructural study in the rat. *Lab Invest.* **55**, (6), 680–7.

Ballardie, F.W., O'Donoghue, D.J., and Feehally, J. (1987*a*). Increasing frequency of adult IgA nephropathy in the UK? *Lancet*, **ii**, 1205.

Ballardie, F.W., Williams, S., and Brenchley, P.E.C. (1987*b*). IgG antibodies to glomerular antigens in IgA nephropathy: detection and clinical significance. *Nephrol. Dial. Transplant.* **2**, 422.

Ballardie, F.W., Brenchley, P.E.C., and Williams, S. (1988*a*). Autoimmunity in IgA nephropathy. *Lancet*, **ii**, 598–2.

Ballardie, F.W., Brenchley, P.E.C., and O'Donoghue, D.J. (1988*b*). Autoantibodies in IgA nephropathy: differential secretion of immunoglobulin isotypes during haematuria. *Nephrol. Dial. Transplant.* **3**, (4), 524–5.

Ballardie, F.W., Baker, P., Green, M., and Vose, B.M. (1990). Cytokine IL-6 and regulation of autoimmunity in IgA nephropathy. *Proc. XI Int. Cong. Nephrology, Tokyo.*

Ballardie, F.W., Campbell, D., Booth, C.G., Baker, P. and Vox, B.M. (1992). Human and murine control of IgA production. In: *Autoimmunity in nephritis*, pp 35–47). Harwood Acad., Reading.

Bene, M.C., and Faure, G.C. (1987). Mucosal immunity and IgA nephropathies. *Seminars in nephrology*, Vol. II, 4, (ed. N.A. Kurtzman) pp. 297–300. Grune Stratton.

Bene, M.C., Faure, G., and Hurault de Ligny, B. (1983). Quantitative immunohistomorphometry of the tonsillar plasma cells evidences an inversion of the immunoglobulin A versus immunoglobulin G secreting cell balance. *J. Clin. Invest.*, **71**, 1342–7.

Bene, M.C., Hurault de Ligny, B., and Sirbat, D. (1984). IgA nephrology: dimeric IgA-secreting cells are present in episcleral infiltrate. *Am. J. Clin. Path.*, **82**, 608–11.

Bene, M.C., Hurault de Ligny, B., and Faure, G. (1986). Histoimmunological discrepancies in primary IgA nephropathy and anaphylactoid purpura sustain relationships between mucosa and kidney. *Nephron*, **43**, 214–16.

Bene, M.C., de Korwin, J.D., and de Ligny, B.H. (1988). IgA nephropathy and alcoholic liver cirrhosis. A prospective necropsy study. *Am. J. Clin. Pathol.* **89**, 769–73.

Bennett, W.M., Walker, R.G., and Kincaid-Smith, P. (1989). Treatment of IgA nephropathy with eicosapentanoic acid (EPA): a two-year prospective trial. *Clin. Nephrol.* **3**, 128–31.

Berger, J., Yaneva, H., and Nabarra, B. (1977). Glomerular changes in patients with cirrhosis of the liver. *Adv. Nephrol.* **7**, 3–14.

Beukhof, J.R., Anema, J., and Halie, L.M. (1983). Prognosis of adult, primary IgA nephropathy. *Kidney Int.* **24**, 408–15.

Brouhard, B.H. (1988). The spectrum of IgA nephropathy. *Am. J. Dis. Child.* **142**, 709–10.

Cederholm, B., Wieslander, J., and Bygren, P. (1988). Circulating complexes containing IgA and fibronectin in patients with primary IgA nephropathy. *Proc. Natl. Acad. Sci. USA.* **85**, 4865–8.

Challacombe, S.J., and Tomasi, T.B. (1980). Systemic intolerance and secretory immunity after oral immunisation. *J. Exp. Med.* **153**, 1459–72.

Chan, M.K., Kwan, S.Y.L., and Chan, K.W. (1987). Controlled trial of antiplatelet agents in mesangial IgA glomerulonephritis. *Am. J. Kid. Dis.* **9**, (5), 417–21.

Chen, A., Wong, S.S., and Rifai, A. (1988). Glomerular immune deposits in experimental IgA nephropathy. A continuum of circulating and in situ formed immune complexes. *Am. J. Path.* **130**, (1), 216–22.

Clarkson, A.R., Seymore, A.E., and Woodroffe, A.J. (1980). Controlled trial of phenytoin therapy in IgA nephropathy. *Clin. Nephrol.* **13**, 215–18.

Coppo, R., Basolo, B., and Giachino, O. (1985). Plasmapheresis in a patient with rapidly progressive idiopathic IgA nephropathy: removal of IgA-containing circulating immune complexes and clinical recovery. *Nephron*, **40**, 488–90.

Coppo, R., Roccatello, D., and Amore, A. (1990). Effects of a gluten-free diet in primary IgA nephropathy. *Clin Nephrol.*, **33**, (2), 72–86.

Cosio, F.G., Lam, S., and Folami, A.O. (1982). Immune regulation of immunoglobulin production in IgA-nephropathy. *Clin. Immunol. Immunopathol.* **23**, 430–6.

Costa, R.S., Droz, D., and Noel, L.H. (1987). Long-standing spontaneous clinical remission and glomerular improvement in primary IgA nephropathy (Berger's disease). *Am. J. Nephrol.* **7**, 440–4.

Czerkinsky, C., Koopman, W.J., and Jackson, S. (1986). Circulating immune complexes and immunoglobulin. A rheumatoid factor in patients with mesangial immunoglobulin A nephropathies. *J. Clin. Invest.* **77**, 1931–8.

Davin, J.C., Forget, P., and Mahieu, P.R. (1988). Increased intestinal permeability to (51 Cr) EDTA is correlated with IgA immune complex-plasma levels in children with IgA-associated nephropathies. *Acta. Paediatr. Scand.* **77**, 118–24.

Demaine, A.G., Rambausek, M., and Knight, J.F. (1988). Relation of mesangial IgA glomerulonephritis to polymorphism of immunoglobulin heavy chain switch region. *J. Clin. Invest.* **81**, 611–4.

Deteix, P., Colon, S., and Leitienne, P.H. (1984). Prospective controlled therapeutic trial with diaminodiphenylsulfone-dapsone (DDS) in primitive IgA nephropathy (IgAN). *Kidney Int.* **26**, 493.

Edlacroix, D.L., Elkon, K.B., and Geubel, A.P. (1982). Changes in size, subclass, and metabolic properties of serum immunoglubulin A in liver diseases and in other diseases with high serum immunoglobulin A. *J. Clin. Invest.* **71**, 358–67.

Egido, J., Blasco, R., and Lozano, L. (1984a). Immunological abnormalities in the tonsils of patients with IgA nephropathy: inversion in the ratio of IgA: IgG bearing lymphocytes and increased polymeric IgA synthesis. *Clin. Exp. Immunol.* **57**, 101–6.

Egido, J., Sancho, J., and Hernando P. (1984*b*). Presence of specific IgA immune complexes in IgA nephropathy. *Contr. Nephrol.* **40**, 80–6.

Egido, J., Rivera, F., and Sancho, J. (1984*c*). Phenytoin in IgA nephropathy: a long-term controlled trial. *Nephron*, **38**, 30–9.

Egido, J., Julian, B.A., and Wyatt, R.J. (1987). Genetic factors in primary IgA nephropathy. *Nephrol. Dial. Transplant.* **2**, 134–42.

Emancipator, S.N., and Lamm, M.E. (1986). Pathways of tissue injury initiated by humoral immune mechanisms. *Lab. Invest.* **54**, (5), 475–8.

Emancipator, S.N., Gallo, G.R., and Lamm, M.E. (1983). Experimental IgA nephropathy induced by oral immunization. *J. Exp. Med.* **157**, 572–82.

Emancipator, S.N., Gallo, G.R., and Lamm, M.E. (1985). IgA nephropathy: perspectives on pathogenesis and classification. *Clin. Nephrol.* **24**, (4), 161–79.

Feehally, J. (1988). Immune mechanisms in glomerular IgA deposition. *Nephrol. Dial. Transplant.* **3**, (4), 361–78.

Feehally, J., O'Donoghue, D.J., and Ballardie, F.W. (1989). Current nephrological practice in the investigation of haematuria: relationship to incidence of IgA nephropathy. *J. Royal Coll. Phys.*, **23**, 228–31.

Fogazzi, G.B., Pasqualim, S., and Moriggi, M. (1988). Long-term outcome of Schonlein–Henoch nephritis in the adult. *Clin. Nephrol.* **198**, 60–6.

Forget, P., Sodoyez-Goffauz, F, and Zappitelli, A. (1984). Permeability of the small intestine to (51 Cr) EDTA in children with acute gastroenteritis or aczema. *J. Pediatr. Gastroenterol. Nutr.* **4**, 393–6.

Fornasieri, A., Sinico, R.A., and Maldifassi, P. (1987). IgA-antigliadin antibodies in IgA mesangial nephropathy (Berger's disease). *Br. Med. J.* **295**, 78–80.

Forrest, B.D. (1988). Identification of an intestinal immune response using peripheral blood lymphocytes. *Lancet*, **i**, 81–3.

Galla, J.H., Spotswood, M.F., and Harrison, L.A. (1985). Urinary IgA in IgA nephropathy and Henoch–Schoenlein purpura. *J. Clin. Immunol.* **5**, (5), 298–306.

Goldman, M., Baran, D., and Druet, P. (1988). Polyclonal activation and experimental nephropathies. *Kidney Int.* **34** 141–50.

Gregory, M.C., Hammond, M.E., and Brewer, E.D. (1988). Renal deposition of cytomegalovirus antigen in immunoglobulin A nephropathy. *Lancet*, **i**, 11–14.

Hene, R.J., and Kater, L. (1983). Plasmapheresis in nephritis associated with Henoch–Schonlein purpura and in primary IgA nephropathy. *Plasma Ther. Transtus. Technol.* **4**, 165–73.

Hernando, P., Egido, J., and de Nicolas, R. (1986). Clinical significance of polymeric and monomeric IgA complexes in patients with IgA nephropathy. *Am. J. Kid Dis.* **8**, (6), 410–16.

Iida, H., Izumino, K., and Matsumoto, M. (1985). Glomerular deposition of IgA in experimental hepatic cirrhosis. *Acta. Pathol. Jpn.* **35**, (3), 561–7.

Imai, H., Nakamoto, Y., and Asukara, K. (1985). Spontaneous glomerular IgA deposition in ddY mice: an animal model of IgA nephritis. *Kidney Int.* **27**, 756–61.

Isaacs, K.L., and Miller, F. (1982). Role of antigen size and charge in immune complex glomerulonephritis. I. Active induction of disease with dextran and its derivatives. *Lab. Invest.* **47**, (2), 198–205.

Jenkins, D.A.S., Bell, G.M., and Ferguson, A. (1988). Intestinal permeability in IgA nephropathy. *Nephron*, **50**, 390.

Julian, B.A., Waldo, F.B., and Rifai, A. (1988). IgA nephropathy, the most common glomerulonephritis worldwide. A neglected disease in the United States? *Am J. Med.* **84**, 129–32.

Kaneko, Y., Tomino, Y., and Ikeda, T. (1987). Comparative studies among morphological changes of urinary sediments and histolpathological injuries in patients with IgA nephropathy. *Tokai. J. Exp. Clin. Med.* **12**, (2), 103–8.

Kincaid-Smith, P. and Nicholls, K. (1983). Mesangial IgA nephropathy. *Am. J. Kid. Disease*, **3**, (2), 90–102.

Knight, J.F. and Cameron, J.S. (1987). Henoch Schonlein nephritis in adults: a report from the UK MRC glomerulonephritis registry. *Nephrol. Dial. Transplant.* **2**, 415.

Kobayashi, Y., Fujii, K., and Hiki, Y. (1986). Steroid therapy in IgA nephropathy: a prospective pilot study in moderate proteinuric cases. *Quart. J. Med. New Series*, **61**, (234), 935–43.

Kobayashi, Y., Fujii, K., and Hiki, Y. (1988). Steroid therapy in IgA nephropathy: a retrospective study in heavy proteinuric cases. *Nephron*, **48**, 12–17.

Kumar, V., Sieniawska M., and Beutner, E.H. (1988). Are immunological markers of gluten-sensitive enteropathy detectabel in IgA nephropathy? *Lancet*, **ii**, 1307.

Kutteh, W.H., Prince, S.J., and Mestecky, J. (1982). Tissue origins of human polymeric and monomeric IgA. *J. Immunol.* **128**, (2), 990–5.

Lagrue, G., Sadreux, T., and Laurent, J. (1981). Is there a treatment of mesangial IgA glomerulonephritis? *Clin. Nephrol.* **16**, 161–6.

Lai, K.-N., Chan, K.W., and Lai, F.M.-M. (1985). The immunochemical characterization of the light chains in the mesangial IgA deposits in IgA nephropathy. *Am. J. Clin. Pathol.* **85**, (5), 548–51.

Lai, K.N., Lai, E.M., and Ho, C.P. (1986). Corticosteroid therapy in IgA nephropathy with nephrotic syndrome: a long-term controlled trial. *Clin. Nephrol.* **26**, (4), 174–80.

Lai, K.-N., Chui, S.-H., and Lai, F.M.-M. (1988*a*). Predominant synthesis of IgA with lambda light chain in IgA nephropathy. *Kidney Int.* **33**, 584–9.

Lai, K.N., Leung, J.C.K., and Mac-Moune Lai, F. (1988*b*). In vitro study of expression of interleukin-2 receptors in T-lymphocytes from patients with IgA nephropathy. *Clin Nephrol.* **30**, (6), 330–4.

Lai, K.N., Lai, F.M., and Tam, J.S. (1988*c*). Strong association between IgA nephropathy and hepatitis B surface antigenaemia in endemic areas. *Clin Nephrol.*, **29**, (5), 229–35.

Lamm, M.E. (1987). The mucosal immune system and IgA nephropathy. *Seminars in nephrology*, Vol 2, 4, (ed. N.A. Kurtzman), pp. 280–82. Grune Stratton.

Lomax-Smith, J.D., Zabrowarny, L.A., and Howarth, G.S. (1983). The immunochemical characterization of mesangial IgA deposits. *Am. J. Pathol.* **113**, (3), 359–6.

Lovett, D.H., Haensch, G.M., and Goppelt, M. (1987). Activation of glomerular mesangial cells by the terminal membrane attack complex of complement. *J. Immunol.* **138**, 2473–80.

Lozano, L., Garcia-Hoyo, R., and Egido, J. (1987). IgA nephropathy: association of a history of macroscopic haematuria episodes with increased production of polymeric IgA. *Nephron*, **45**, 98–103.

Lupo, A., Rugiu, C., and Cagnoli, L. (1987). Acute changes in renal function in IgA nephropathy. *Seminars in Nephrology* Vol. 2, 4, ed. (N.A. Kurtzman), pp. 359–62. Grune Stratton.

Makino, H., Nishimura, S., and Takaoka, M. (1988). Mechanism of hematuria. *Nephron*, **50**, 143–50.

Meadow, S.R. and Scott, D.G. (1985). Berger disease: Henoch–Schonlein syndrome without the rash. *J. Pediatr.* **106**, 27–32.

Mestecky, J. and McGhee, J.R. (1987). Immunoglobulin A (IgA): molecular and cellular interactions involved in IgA biosynthesis and immune response. *Advances in immunology*, **40**, pp. 153–245, Academic Press.

Mestecky, J., Czerkinsky, C., and Russell, M.W. (1987). Induction and molecular properties of scretory and serum IgA antibodies specific for environmental antigens. *Ann. Allergy*, **59**, 54–9.

Miller, G.W., and Nussenzweig, V. (1975). A new complement function: solublisation of antigen-antibody aggregate. *Proc. Natl. Acad. Sci. USA*, **72**, 418–22.

Miura, M., Tomino, Y., and Suga, T. (1984). Increase in proteinuria and/or microhematuria following upper respiratory tract infections in patients with IgA nephropathy. *Tokai. J. Exp. Clin. Med.* **9**, (2), 139–45.

Miyamoto, H., Yoshioka, K., and Takemura, T. (1988). Immunohistochemical study of the membrane attack complex in IgA nephropathy. *Virchows Archiv. A. Pathol. Anat.* **413**, 77–86.

Moldoveanu, Z., Epps, J.M., and Thorpe, S.R. (1988). The sites of catabolism of murinie monomeric IgA. *J. Immunol.* **141**, (1), 208–13.

Monteiro, R.C., Halbwachs-Mecarelli, L., and Roque-Barreira, M.C. (1985). Charge and size of mesangial IgA in IgA nephropathy. *Kidney. Int.* **28**, 666–71.

Nagy, J., Ambrus, M., and Paal, M. (1987*a*). Cryloglobulinaemia and cryofibrinogenaemia in IgA nephropathy: a follow-up study. *Nephron*, **46**, 337–42.

Nagy, J., Miltenyi, M., and Dobos, M. (1987*b*). Tubular proteinuria in IgA glomerulonephritis. *Clin. Nephrol.* **27**, (2), 76–8.

Nagy, J., Scott, H., and Brandtzaeg, P. (1988). Antibodies to dietary antigens in IgA nephropathy. *Clin. Nephrol.* **29**, (5), 274–9.

Neelakantappa, K., Gallo, G.R., and Baldwin, D.S. (1988). Proteinuria in IgA nephropathy. *Kidney Int.* **33**, 716–21.

Newkirk, M.M., Ostberg, L., and Wasserman, R.W. (1987). Human rheumatoid factors of the Wa idiotypic family appear to use highly homologous variable region genes as a human anti-cytomegalovirus antibody. *Fed. Proc.* **46**, 916.

Nicholls, K.M., Fairley, K.F., and Dowling, J.P. (1984). The clinical course of mesangial IgA associated nephropathy in adults. *Q. J. Med.* **210**, 227–50.

Nicholls, K.M., Becker, G., Walker, R., Wright, C., and Kincaid-Smith, P. (1990). Plasma exchange in progressive IgA nephropathy. *J. Clin. Apheresis*, 128–32.

Nomoto, Y., and Sakai, H. (1979). Cold-reacting antinuclear factor in sera from patients with IgA nephropathy. *J. Lab. Clin. Med.* **94**, 76–87.

O'Donoghue, D.J., Brenchley, P.E.C., and Ballardie, F.W. (1988*a*). Autoimmunity to glomerular antigens in adult Henoch–Schonlein nephritis and IgA nephropathy. *Clinical Science*, **75**, 52.

O'Donoghue, D.J., Brenchley, P.E.C., and Ballardie, F.W. (1988*b*). Autoimmunity to glomerular antigens in adult Henoch–Schonlein nephritis. *Nephrol. Dial. Transplant.* **3**, (4), 512.

O'Donoghue, D.J., Darvill, A., and Brenchley, P.E.C. (1989). Mesangial cell autoantigens in IgA nephropathy and Henoch–Schonlein purpura. *Kidney Int.* **35**, (1), 372.

O'Donoghue, D.J., Darvill, A., and Ballardie, F.W. (1990). Antibodies to glomerular and other antigens in IgA nephropathy: influence on disease state and progression. *Nephrol. Dial. Transplant.* **5**, 298.

O'Donoghue, D.J., Darvill, A., and Ballardie, F.W. (1991). Mesangial cell autoantigens in IgA nephropathy and Henoch Schonlein purpura. *J. Clin. Invest.* **88**, 1522–30.

O'Donoghue, D.J., Dewkes, F.J., Postlethwaite, R.B., and Ballardie, F.W. (1992). Autoimmunity to glomerular antigens in Henoch–Schonlein nephritis. *Clin. Sci.* (in press).

Ogata, I., Fujiwara, K., and Nishi, T. (1988). Contribution of hepatic reticuloendothelial system to glomerular IgA deposition in rat liver injury. *Am. J. Pathol.* **131**, (3), 411–17.

Phillips, J.O., Komiyama, K., and Epps, J.M. (1988). Role of hepatocytes in the uptake of IgA and IgA-containing immune complexes in mice. *Molecular Immunol.* **25**, (9), 873–9.

Rifai, A., and Mannik, M. (1983). Clearance kinetics and fate of mouse IgA immune complexes prepared with monomeric or dimeric IgA. *Immunol.* **130**, (4), 1826–32.

Rifai, A., and Millard, K. (1985). Glomerular deposition of immune complexes prepared with monomeric or polymeric IgA. *Clin. Exp. Immunol.* **60**, 363–8.

Rifai, A., Small, P.A., and Teague, P.O. (1979). Experimental IgA nephropathy. *J. Exp. Med.* **150**, 1161–73.

Rifai, A., Chen, A., and Imai, H. (1987). Complement activation in experimental IgA nephropathy: an antigen-mediated process. *Kidney Int.* **32**, 838–44.

Rose, G.M., Cole, B.R., and Robson, A.M. (1981). The treatment of severe glomerulopathies in children using high dose intravenous methylprednisolone pulses. *Am. J. Kid. Dis.* **1**, (3), 148–56.

Rostoker, G., Laurent, J., and Andre, C. (1988). High levels of IgA antigliadin antibodies in patients who have IgA mesangial glomerulonephritis but not coelic disease. *Lancet*, **i**, 356–7.

Roth, D.A., Wilz, D.R., and Theil, G.B. (1985). Schonlein–Henoch syndrome in adults. *Quart. J. Med. New Series*, **55**, (217), 1450–52.

Sanchio, J., Gido, J., and Gonzalis, E. (1983). Simplified method for determining polymeric Iga-containing immune complexes. *J. Immunol. Method.* **60**, 305–17.

Sancho, J., Egido, J., and Rivera, F. (1983). Immune complexes in IgA nephropathy: presence of antibodies against diet antigens and delayed clearance of specific polymeric IgA immune complexes. *Clin. Exp. Immunol.* **54**, 194–202.

Sancho, J., Gonzalez, E., and Egido, J. (1986). The importance of the Fc receptors for IgA in the recognition of IgA by mouse liver cells: its comparison with carbohydrate and secretory component receptors. *Immunol.* **57**, 37–42.

Saulsbury, F.T. (1987). The role of IgA rheumatoid factor in the formation of IgA-containing immune complexes in Henoch–Schonlein purpura. *J. Clin. Lab. Immunol.* **23**, 123–7.

Schena, F.P., Masterolitti, G., Jirillo, E., Munno, I., Pellegrino, N., Fracasso, A.R., and Aventaggiato, L. (1989). Increased production of interleukin 2, and IL-2 receptor in Iga nephropathy. *Kidney Int.* **35**, (3), 875–9.

Silva, F.G., Chander, P., and Pirani, C. (1982). Disappearance of glomerular mesangial IgA deposits after renal allograft transplantation. *Transplantation*, **33**, (2), 214–16.

Simon, P., Bramee, M.S., and Ang, K.S., (1988). Idiopathic IgA mesangial nephropathy is the main cause of end stage renal failure in a French area. *Kidney Int.* **34**, (4), 566.

Sinniah, R. (1984). Heterogenous IgA glomerulonephropathy in liver cirrhosis. *Histopathology*, **8**, 947–62.

Sinniah, R. (1985). IgA mesangial nephropathy: Berger's disease. *Am. J. Nephrol.* **5**, 73–83.

Tomino, Y. (1980). Complement system in IgA nephropathy. *Toka. J. Exp. Clin. Med.* **5**, (1), 15–22.

Tomino, Y., Endoh, M., and Nomoto, Y. (1981). Immunoglobulin A1 in IgA nephropathy. *New Eng. J. Med.* **305**, 1159–60.

Tomino, Y., Endoh, M., and Nomot, Y. (1982). Specificity of eluted antibody from renal tissues of patients with IgA nephropathy. *Am. J. Kid. Dis.* **1**, (5), 276–80.

Tomino, Y., Endoh, M., and Kaneshige, H. (1983). Increase of IgA in pharyngeal washings from patients with IgA nephropathy. *Am. J. Med. Sci.* **286**, (2), 15–21.

Tomino, Y., Sakai, H., and Miura, M. (1984). Effect of danazol on solubilization of immune deposits in patients with IgA nephropathy. *Am. J. Kid. Dis.* **9**, (2), 135–40.

Tomino, Y., Sakai, H., and Miura, M. (1985). Specific binding of circulating IgA antibodies in patients with IgA nephropathy. *Am. J. Kidney Dis.* **4**, (3), 149–53.

Tomino, Y., Sakai, H., and Woodroffe, A.J. (1987). Studies on glomerular immune solubilization by complement in patients with IgA nephropathy. *Acta. Pathol. Jpn.* **37**, (ii), 1763–7.

Tomino, Y., Ma, Y., and Sakai, H. (1988). Detection of platelets in urinary sediments from patients with 'advanced' stages of immunoglobulin A nephropathy. *J. Clin. Lab. Anal.* **2**, 241–4.

Valentijn, R.M., Kauffmann, R.H., and de la Riviere, G.B. (1983). Presence of circulating macromolecular IgA in patients with hematuria due to primary IgA nephropathy. *Am. J. Med.* **74**, 375–81.

Valentijn, R.M., Radl, J., and Haaijman, J.J. (1984). Circulating and mesangial secretory component-binding IgA-1 in primary IgA nephropathy. *Kidney Int.* **26**, 760–6.

Waldo, F.B., Britt, W.J., and Tomana, M. (1989). Nonspecific mesangial staining with antibodies against cytomegalovirus in immunoglobulin A nephropathy. *Lancet*, **i**, 129–31.

Walker, R.G., Yu, S.H., Owen, J.E., and Kincaid-Smith, P. (1990). The treatment of mesangial IgA nephropathy with cyclosphosphamide, dipyridamole and warfarin: a two-year prospective trial. *Clin. Nephrol.* **34**, (3), 103–7.

Van den Wall Bake, A.W.L., Daha, M.R., and Radl, J. (1988*a*). The bone marrow as production site of the IgA deposited in the kidneys of patients with IgA nephropathy. *Clin. Exp. Immunol.* **72**, 321–5.

Van den Wall Bake, A.W.L., Lobatta, S., and Jonges, L. (1988*b*). IgA antibodies directed against cytoplasmic antigens of polymorphonuclear leucocytes in patients with Henoch–Schonlein purpura. *Adv. Exp. Biol.* 1593–8.

Van den Wall Bake, A.W.L., Beyer, W.E., and Jeanette, H. (1989). Immune response to influenza virus in primary IgA nephropathy. *7th Int. Cong. Imm.* (Abs).

Welch, T.R., McAdams, A.J., and Berry, A. (1988). Rapidly progressive IgA nephropathy. *Am. J. Dis. Child.* **142**, 789–93.

Westberg, N.G., Baklien, K., and Schmekel, B. (1983). Quantitation of immunoglobulin-producing cells in small intestinal mucosa of patients with IgA nephropathy. *Clin. Immunol. Immunopathol.* **26**, 442–5.

Wilson, C.B., Yamamoto, T., and Moullier, P. (1988). Selective glomerular mesangial cell immune injury—antimesangial cell antibodies. *Nephrology*, **1**, pp. 509–22, (ed. A.M. Davidson), Baillière, Tindall, London.

Wing, A., J., Brunner, F.P., and Brynger, H. (1990). IgA nephropathy in patients treated for end stage renal failure in Europe. *Kidney Int.* (In press).

Woo, K.T., Edmondson, R.P.S., and Yap, H.K. (1987). Effects of triple therapy on the progression of mesangial proliferative glomerulonephritis. *Clin. Nephrol.* **27**, 56–64.

Woo, K.T., Chaing, G.S.C., and Yap, H.K. (1988). Controlled therapeutic trial of IgA nephritis with follow-up renal biopsies. *Am. Acad. Med.* **17**, (2), 226–31.

Woodruffe, A.J., Gormly, A.A., and McKenzie, P.E. (1980). Immunologic studies in IgA nephropathy. *Kidney Int.* **18**, 366–74.

Yamabe, H., Ozawa, K., and Fukushi, K. (1988). IgA nephropathy and Henoch–Schonlein purpura nephritis with anterior uveitis. *Nephron*, **50**, 368–50.

Yap, H.K., Sakai, R.S., and Woo, K.T. (1987). Detection of bovine serum albumin in the circulating IgA immune complexes of patients with IgA nephropathy. *Clin. Immunol. Immunopath.* **43**, 395–402.

Yap, H.K., Sakai, R.S., and Bahn, L. (1988). Anti-vascular endothelial cell antibodies in patients with IgA nephropathy: frequency and clinical significance. *Clin. Immunol. Immunopathol.* **49**, 450–62.

Part 6 Vasculitis

22 Cell biology of vasculitis

C.O.S. Savage

The individual cellular components which make up a vasculitic lesion have been recognized for many years on the basis of their morphology. It is known, for example, that neutrophils, macrophages and lymphocytes infiltrate and surround the vessel wall. Sometimes one cell type appears to dominate, giving rise to the concepts of neutrophil-rich (leucocytoclastic vasculitis) and lymphocyte plus macrophage-rich lesions (mononuclear vasculitis).[1] This has prompted questions as to what the significance of these differences might be, and whether they represent sequential stages in an inflammatory reaction, or different degrees or modes of injury. Yet other lesions were found to contain large numbers of eosinophils whose presence is characteristic of Churg–Strauss vasculitis.[1] Attempts to relate the type of inflammatory infiltrate to the disease outcome recognized that subtle differences in the cellular infiltrate might have pathogenic consequences.

Cells of the blood vessel wall itself also have the potential to contribute to the development of vascular injury. Initially, the vascular wall was regarded mainly as a target for damage. The presence of fibrinoid necrosis was a characteristic feature of this, and detailed morphological studies showed necrosis of endothelial cells.[2] More recently, it has been recognized that vascular cells, particularly endothelial cells, may participate actively in the development of inflammatory responses, by altering their expression of cell surface molecules which have pro-inflammatory and pro-coagulant properties, and by secreting soluble cytokines which help direct immune responses.[3]

The precise sequence of events that triggers the peri-vascular inflammatory response and leads to tissue injury in the systemic vasculitides is in the early stages of investigation only (see Savage for review).[4] Therefore, by necessity, leucocyte–endothelial cell interactions will be considered in a broad context, with reference to how such mechanisms may relate to vasculitis. The relevance of autoantibodies, both anti-neutrophil cytoplasm antibodies (ANCA) and anti-endothelial cell antibodies (AECA), to cellular interactions will also be discussed. Cytokines play pivotal roles in all these interactions, being secreted by, and having effects on, both vascular endothelial cells and leucocytes. Cytokines known to be secreted by endothelial cells include interleukin (IL)-1, IL-6, IL-8, and various colony-stimulating factors.

The role of the endothelial cell in the inflammatory response

The neutrophil-predominant and mononuclear-predominant forms of vasculitis are probably broadly equivalent to the immunological concepts of inflammation, i.e. acute (neutrophil mediated) and delayed-type hypersensitivity (T cell mediated). Both involve leucocyte recruitment and leucocyte activation at the site of the antigenic stimulus. Recruitment requires delivery of leucocytes to the site, followed by their adhesion to the endothelial cell surface, and finally their transmigration into extravascular tissues. Both adhesive events and leucocyte activation are of particular relevance to vasculitis. Adhesion involves the interaction of endothelial and leucocyte surface molecules.[5] The cytokine-induced expression of specific endothelial cell surface adhesion molecules helps to localize the site of inflammation in the periphery and to determine the cell types recruited. Leucocytes that have been recruited to the inflammatory site require full activation to mediate the effector phase of the response. There is evidence that the endothelial cell can play a major role in lymphocyte activation which has particular relevance to the immune inflammatory response.

Leucocyte—endothelial cell adhesive interactions

The endothelial cell lining of normal blood vessels presents a surface that has low adherence properties for circulating leucocytes. However, *in vitro* studies suggest that the endothelial cell surface changes under the influence of cytokines and other inflammatory mediators, to allow expression of molecules which promote leucocyte adherence.[5,6] This process has been referred to as 'endothelial activation'.[6] Cytokines, such as tumour necrosis factor (TNF), and IL-1 can orchestrate a series of changes which lead to an increased adherence of neutrophils, followed by increased adhesion for lymphocytes. Some of the endothelial and leucocyte molecules which are involved in these processes have been identified (Fig 22.1). Endothelial adhesion molecules may be classified as 'selectins' (which include ELAM-1 and GMP-140) and immunoglobulin-like molecules (which include ICAM-1, ICAM-2 and VCAM-1/INCAM-110). Treatment of cultured human umbilical vein endothelial cells with IL-1 results in *de novo* synthesis and expression of ELAM-1 which peaks at 4–6 h and then rapidly declines; *in vitro*, this correlates with increased neutrophil adhesion, although the neutrophil molecule which recognizes ELAM-1 is not known.[5,6] Interleukin-1 also causes increased expression from basal levels of ICAM-1 which plateaus at 24 h and remains elevated as long as IL-1 remains in the culture system.[5,6] ICAM-1 is a ligand for both CD11a/CD 18 (LFA-1) and CD11b/CD18 (MAC-1). LFA-1 is expressed on all white blood cells, while MAC-1 is found on neutrophils and monocytes but only some lymphocytes. Antibody blocking studies have implicated ICAM-1 in neutrophil and B cell adhesion to endothelium. It is likely that ICAM-1 and LFA-1 also mediate T cell and monocyte adhesion to endothelium

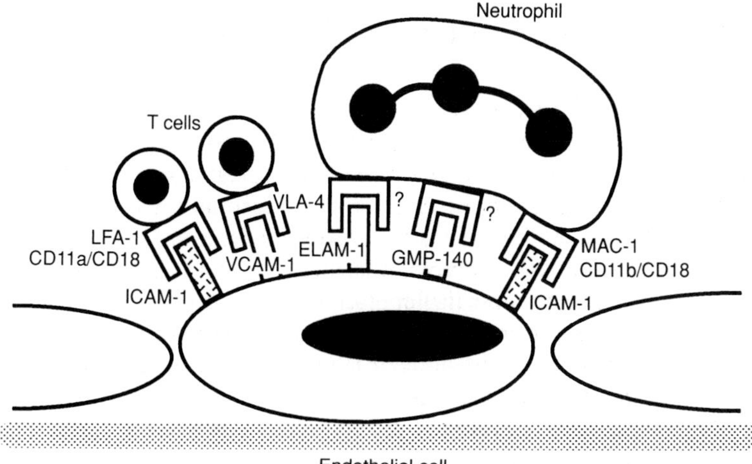

Fig. 22.1 Endothelial cell and leucocyte adhesion molecules. Endothelial cell—T cell adhesion molecules include the ligand pairs ICAM-1 and LFA-1, and VCAM-1 (INCAM-110) and VLA-4. Endothelial adhesion molecules for neutrophils include ELAM-1 and GMP-140 (the neutrophil ligands for these molecules have not been identified yet), as well as ICAM-1 which may be recognized by both neutrophil MAC-1 and LFA-1. A second ligand for LFA-1, called ICAM-2, has been identified. Macrophages (not shown) probably adhere to endothelial ICAM-1 (via LFA-1) and VCAM-1.

since T cells and myeloid cells bind to purified ICAM-1.[7] However, antibody blocking studies have been equivocal, so other ligands may be involved too. *In vivo* studies involving injection of TNF or IL-1 intradermally demonstrate *de novo* expression of ELAM-1 and increased expression of ICAM-1 on perivascular tissues, which correlate with the development of neutrophil and mononuclear cell infiltration respectively.[8] The effects of TNF on ELAM-1 and ICAM-1 mirror those of IL-1 although it acts through a different receptor.

Another recently identified and IL-1 inducible molecule which is expressed by endothelial cells is VCAM-1,[9] which is probably the same molecule as INCAM-110.[10] This molecule is an adhesion ligand for lymphocytes and monocytes whose expression is maximal at 4–6 h and is sustained for at least 48 h.

Finally, GMP-140 is a molecule with structural similarities to ELAM-1, which is contained within the membrane surrounding endothelial cell Weibel–Palade bodies. Inflammatory mediators, such as thrombin and histamine, cause a rapid (within 10 min) redistribution of GMP-140 to the cell surface, where it acts as an adhesive molecule for neutrophils.[11]

Adhesion is a preliminary step to leucocyte transmigration into tissues at the site of localized endothelial cell activation. More generalized endothelial activation, or prolonged expression of adhesion molecules, could lead to widespread or chronic inflammation.

The role of the endothelial cell in immune inflammation

Cytokine-inducible endothelial capacities may play important roles in the development of immune inflammation (delayed-type hypersensitivity reactions), by facilitating T cell recognition of antigen and T cell activation, by facilitating adhesion and diapedesis of various leucocyte populations, and by development of leakiness to macromolecules.

An essential cell in immune inflammation is the $CD4^+$ helper T cell. T cells of this phenotype recognize and engage their cognate antigen via their specific T cell receptor only when the antigen is presented to them by being held in association with an MHC class II molecule on the surface of an antigen-presenting cell. Following antigen recognition, T cells have the potential to become fully activated, providing that additional co-stimulatory signals are received. These co-stimulatory signals may be supplied by the antigen-presenting cell. On full activation, the T cell secretes cytokines, including IL-2 (which acts as an autocrine growth factor for T cells thereby amplifying the immune response), and TNF, lymphotoxin, IL-4 and interferon-γ, (which help to recruit other mononuclear cells, such as cytotoxic T cells and mononuclear phagocytes, which effect the immune response).

If endothelial cells had the capacity to function as antigen presenting cells, they would be in a powerful position to direct specific immune reactions in the periphery;[12] such reactions might be fundamental to the development of vasculitis. It has not been shown definitively that endothelial cells have this capacity, but two features suggest that they are well suited to presenting antigen to circulating T cells in the peripheral vasculature, thereby localizing the immune response to the site of antigen presentation. Firstly, endothelial cells may express class II molecules.[13] These are not constitutively expressed by endothelial cells in humans, but expression can be induced by interferon-γ (IFN-γ). Secondly, endothelial cells can provide co-stimulatory signals to T cells which could secure an ongoing immune response.[14] Co-stimulatory signals are provided through direct contact with the T cell membrane (via endothelial cell LFA-3 and T cell CD2) rather than via soluble cytokines.

Pathogenic implications of leucocyte–endothelial cell to cell interactions

Immunocytochemical staining of tissues from patients with a variety of inflammatory disorders (including vasculitis), demonstrate that endothelial activation occurs *in vivo*.[15] Extrapolation from *in vitro* studies suggests that the ELAM-1 expression indicates the presence of an acute lesion, while increased expression of ICAM-1 or MHC class II molecules might accompany an older lesion.

Adherence of leucocytes to the cytokine-activated endothelial cell surface is believed to be fundamental to the development of an inflammatory response with subsequent migration of leucocytes to the extravascular tissues. It is presumed that the extravascular inflammatory site is also the main place for leucocyte effector functions with elimination of the source of foreign antigen.

However, in pathological circumstances, this sequence of events may not occur and the activated endothelial surface may permit or promote vascular injury. Possible examples of this include: (1) The increased expression of molecules, such as MHC antigens or ICAM-1, which could increase the potential of endothelial cells to mount anti-self responses. Further, prolonged or very generalized expression might encourage chronic or widespread inflammation, as noted earlier. (2) The activated endothelial cell surface may pathologically alter the ability of the effector phase of the immune response to recognize endothelium. For example, there is increased binding of AECA to activated endothelial cells in Wegener's granulomatosis, microscopic polyarteritis and Kawasaki disease (see below). (3) Activation of leucocytes following adherence to the endothelial cell surface (rather than in tissues) may result in endothelial cell damage, particularly under circumstances where the endothelial self-protective mechanisms are damaged. For example, cytomegalovirus and herpes simplex infection of endothelial cells *in vitro* cause increased adherence of neutrophils which in turn demonstrate increased cytotoxicity towards the endothelial cells.[16] There have been repeated suggestions that viral infection of endothelial cells may play a role in vasculitis, although as yet, there is no definitive evidence of this occurring in humans.

Potential roles of vasculitis autoantibodies in development of injury to vascular cells

Wegener's granulomatosis and microscopic polyarteritis

The first major group of antibodies to be identified in Wegener's granulomatosis and microscopic polyarteritis were ANCA which, by virtue of their reactivity with normal neutrophil constituents, were designated autoantibodies.[17] These antibodies react with neutrophil enzymes (including myeloperoxidase and proteinase III) contained within the primary granules.[18,19] A second group of antibodies which are detectable in the serum of patients with these vasculitides are AECA, which react with cultured endothelial cells isolated from human umbilical veins.[20,21] The targets recognized by these autoantibodies have not yet been identified, but appear to differ from targets recognized by ANCA. The targets recognized by AECA appear to be up-regulated on the endothelial cell surface by TNF, IL-1 and IFN-γ (unpublished observations).

What are the likely effects of autoantibodies on cellular components in vasculitis? (1) The presence of B cell derived autoantibodies are indicative of immune activation in the broadest sense, with the likely presence of antigen-specific CD4$^+$ T cells which secrete soluble cytokines to provide help to B cells and, potentially, to antigen-specific cytotoxic T cells. (2) Antibodies may interact with an intermediary target cell to alter its function and increase its cytotoxic potential for bystander vascular cells. For example, ANCA mediate granule release and oxygen radical release from tumour necrosis factor (TNF)-primed neutrophils.[22] (3) Antibodies might interact with the released cell products, such as the neutrophil enzymes proteinase III and myeloperoxidase, to prolong or inhibit their function. These effects of ANCA on neutrophil functions and on neutrophil-secreted enzymes, could have direct implications for endothelial cell injury if they occurred in close proximity to the endothelial cell, in the so-called 'protected' microenvironment which develops between these cells following adherence of neutrophils to the endothelial cell surface. (4) Antibodies may bind also to endothelial cells. Following binding they might interfere with endothelial cell functions or might direct effector elements of the immune response to the endothelial cell by promoting complement-mediated cytotoxicity, antibody-dependent cytotoxicity, or neutrophil adherence and activation. There is some evidence for antibody-dependent cellular cytotoxicity in occasional patients with Wegener's granulomatosis and microscopic polyarteritis (unpublished observations), but no evidence for complement dependent cytotoxicity.

Kawasaki syndrome

Kawasaki syndrome is a childhood vasculitis characterized by generalized immune activation. AECA also develop and mediate complement dependent cytotoxicity towards cytokine-inducible determinants on the surface of endothelial cells. At least two sets of antigens are recognized, some induced by TNF or IL-1, and others induced by IFN-γ.[23,24] In this disorder, AECA lyse only those cultured endothelial cells which have been pretreated with cytokines.

Other effects of cytokines on endothelium of relevance to the development of vasculitis

Cytokines do not appear to be direct toxins for normal human endothelial cells, but they may promote vascular damage in other ways besides those related to endothelial activation. For example, cytokines may alter the endothelial cell surface so that it favours, rather than hinders, coagulation.[15] The basal state of endothelial cells is anticoagulant due to the expression of heparan sulphate (which catalyzes anti-thrombin III) and due to the expression of thrombomodulin and the secretion of protein S (which catalyse the activated protein C pathway). Endothelial cells also synthesize prostacyclin, which inhibits platelet aggregation. Cytokines, such as IL-1 and TNF, induce endothelial cells to express tissue factor so that they acquire the capacity to bind factor VIIa and to initiate the extrinsic clotting pathway. At the same time, the expression of thrombomodulin by endothelial cells is inhibited. Although cytokines may have other effects relevant to haemostasis (such as increase of prostacyclin production), *in vitro* and *in vivo* studies suggest that cytokines such as IL-1 or TNF mediate a shift to a procoagulant state, mainly via effects on endothelial cells. This is supported by infusion of IL-1 into rabbits, which causes activation of coagulation and deposition of fibrin on the endothelial cell lining. In vasculitis a procoagulant state, even if restricted to certain areas of the microvascular system, would be likely to contribute to tissue damage.

Summary

The inflammatory response that develops in vasculitis appears to be associated with a cytokine-activated endothelial cell state. This may be essential to the development of the inflammatory reaction as well as permitting or promoting pathophysiological interactions between circulating leucocytes and vascular cells. Autoantibodies may contribute to these interactions by altering the function of inflammatory cells or their products, by directing the effector mechanisms of the immune response, or by other as yet unidentified mechanisms.

References

1. Churg, J. and Churg, A. (1989). Idiopathic and secondary vasculitis: a review. *Modern Pathol.* **2**, 144–60.
2. Jones, R.E. Jr (ed). (1985). Identification of small vessel vasculitis by conventional microscopy. *Am. J. Dermatopathol.* **7**, 181.
3. Pober, J.S. and Cotran, R.S. (1990). Cytokines and endothelial cell biology. *Physiol. Review*, **70**, 427–51.
4. Savage, C.O.S. (1991) Pathogenesis of systemic vasculitis. In *Systemic vasculitis.* (ed. J. Churg and A. Churg). Igaku-Shoin, New York.
5. Albelda, S.M. and Buck, C.A. (1990) Integrins and other cell adhesion molecules. *F.A.S.E.B.J.* **4**, 2868–80.
6. Pober, J.S. (1988). Cytokine-mediated activation of vascular endothelium. *Am. J. Pathol.* **133**, 426–33.
7. Makgoba, M.W., Sanders M.E., Ginther Luce, G.E., Dustin, M.L., Springer, T.A., Clark, E.A., Mannoni, P., Shaw, S. (1988). ICAM-1 a ligand for LFA-1-dependent adhesion of B, T and myeloid cells. *Nature*, **331**, 86–8.
8. Munro, J.M., Pober, J.S., and Cotran, R.S. (1990). Tumor necrosis factor and interferon-γ induce distinct patterns of endothelial activation and leukocyte accumulation in skin of *Papio anubis*. *Am. J. Pathol.* **135**, 121–33.
9. Osborn, L., Hession, C., Tizard, R., Vassallo, C., Luhowskyj, S., Chi-Rosso, G., and Lobb, R. (1989). Direct expression cloning of vascular cell adhesion molecule 1, a cytokine-induced endothelial protein that binds to lymphocytes. *Cell*, **59**, 1203–11.
10. Rice, G.E., and Bevilacqua, M.P. (1989). An inducible endothelial cell surface glycoprotein mediates melanoma adhesion. *Science*, **246**, 1303–6.
11. Geng, J.-G., Bevilacqua, M.P., Moore, K.L., and McIntyre, T.M., Prescott, S.M., Kim, J.M., Bliss, G.A., Zimmerman, G.A., and McEver, R.P. (1990). Rapid

neutrophil adhesion to activated endothelium mediated by GMP-140. *Nature*, **343**, 757–60.

12. Pober, J.S., Doukas, J., Hughes, C.C.W., Savage, C.O.S., Munro, J.M., and Cotran R.S. (1990). The roles of vascular endothelium in immune reactions. *Human Immunol.* **28**, 258–62.
13. Pober, J.S., Collins, T., Gimbrone, M.A., Jr., Cotran, P.R.S., Gitlin, J.D., Fiers W., Clayberger, C., Krensky, A.M., Burakoff, S.J., and Reiss, C.S. (1983). Lymphocytes recognise human vascular endothelial and dermal fibroblast la antigens induced by recombinant immune interferon. *Nature*, **305**, 726–9.
14. Hughes, C.C.W., Savage, C.O.S., and Pober, J.S. (1990). Endothelial cells augment T cell IL-2 production by a contact-dependent mechanism involving CD2: LFA-3 interaction. *J. Exp. Med.* **171**, 1453–68.
15. Cotran, R.S., and Pober, J.S. (1989). Effects of cytokines on vascular endothelium: their role in vascular and immune injury. *Kidney Int.* **35**, 969–75.
16. Visser, M.R, Jacob, H.S., Goodman, J.L., McCarthy, J.B., Furcht, L.T., and Vercellotti, G.M. (1989). Granulocyte-mediated injury to herpes simplex virus-infected human endothelium. *Lab. Invest.* **60**, 296–304.
17. Van der Woude, F.J., Rasmussen, N., Lobatto, S., Wiik, A., Permin, H., Van ES, L.A., Van der Giessen, M., Van der Hem, G.K., The, T.H. (1985). Autoantibodies against neutrophils and monocytes; tool for diagnosis and marker of disease activity in Wegener's granulomatosis. *Lancet*, **i**, 425–9.
18. Falk, R.J. and Jennette, J.C. (1988). Anti-neutrophil cytoplasm autoantibodies with specificity for myeloperoxidase in patients with systemic vasculitis and idiopathic necrotising crescentic glomerulonephritis. *New. Engl. J. Med.* **318**, 1651–7.
19. Niles, J.L., McCluskey, R.T., Ahmad, M.F., and Arnaout, M.A. (1989). Wegener's granulomatosis autoantigen is a novel neutrophil serine proteinase. *Blood*, **74**, 1888–93.
20. Ferraro, G., Meroni, P.L., Tincani, A., Sinico, A., Barcellini, W., Radice, A., Gregorini, G., Froldi, M., Borghi, M.O., and Balestrieri, G. (1990). Anti-endothelial cell antibodies in patients with Wegener's granulomatosis and micropolyarteritis. *Clin. Exp. Immunol.* **79**, 47–53.
21. Pottinger, B., Savage, C.O.S., and Pearson, J.D. (1989). Endothelial cell damage in Wegener's granulomatosis and microscopic polyarteritis: presence of auto-antibodies recognizing endothelium and their relation to anti-neutrophil cytoplasmic antibodies. *Br. J. Haematol.* **73**, 430.
22. Falk, R.J., Terrell, R.S., Charles, L.A., and Jennette, J.C. (1990). Anti-neutrophil cytoplasmic autoantibodies induce neutrophils to degranulate and produce oxygen radicals *in vitro*. *Proc. Natl. Acad. Sci. USA*, **87**, 4115–19.
23. Leung, D.Y.M., Geha, R.S., Newburger, J.W., Burns, J.S., Fiers, W., Lapierre, L.A., and Pober, J.S. (1986). Two monokines, interleukin 1 and tumor necrosis factor, render cultured vascular endothelial cells susceptible to lysis by antibodies circulating during Kawasaki syndrome. *J. Exp. Med.* **164**, 1958–72.
24. Leung, D.Y.M., Collins, T., Lapierre, L.A., Geha, R.S., and Pober, J.S. (1986). IgM antibodies in the acute phase of Kawasaki syndrome lyse cultured endothelial cells stimulated by gamma interferon. *J. Clin. Invest.* **77**, 1428–35.
25. Naworth, P.P., Handley, D.A., Esmon, C.T., and Stern, D.M. (1986). Interleukin 1 induces endothelial cell procoagulant while suppressing cell-surface anticoagulant activity. *Proc. Natl. Acad. Sci. USA*, **83**, 3460–4.

23 Autoimmunity in systemic vasculitis

G. Gaskin and C.D. Pusey

Classification of systemic vasculitis

The term primary systemic vasculitis encompasses a spectrum of conditions characterized by inflammation and necrosis of blood vessels throughout the body (Fauci *et al.* 1978; Gaskin and Pusey 1991). The syndromes may be classified according to the size of vessel involved and by the presence or absence of granulomata (Table 23.1). This scheme identifies Wegener's granulomatosis (characterized by granulomatous inflammation of the upper and lower respiratory tract) and microscopic polyarteritis as affecting small vessels, and showing prominent renal involvement. Isolated focal necrotizing glomerulonephritis (FNGN) without antibodies against the glomerular basement membrane may be considered to represent small vessel vasculitis confined to the kidney, and usually presents clinically with rapidly progressive renal failure (Couser 1988). Churg–Strauss syndrome, which comprises granulomatous vasculitis with asthma and eosinophilia, may involve medium-sized or small vessels and sometimes leads to FNGN. Polyarteritis nodosa involves medium-sized muscular arteries, and may lead to renal arterial aneurysms and occlusions, but only rarely causes FNGN (presumably due to an 'overlap' syndrome.) Takayasu's arteritis and giant cell arteritis have very rarely been associated with renal disease and will not be discussed further. This chapter will concentrate on the small vessel vasculitides, which are of particular interest to the nephrologist because they may present as rapidly progressive glomerulonephritis.

Table 23.1 Classification of primary systemic vasculitis

Size of blood vessels involved	With granulomata	Without granulomata
Small	Wegener's granulomatosis	Microscopic polyarteritis (including isolated FNGN)
Medium	Churg–Strauss syndrome	Polyarteritis nodosa
Large	Giant cell arteritis and Takayasu's arteritis (both *may* show granulomatous arteritis)	

Non-specific immunological findings

Early evidence for an immune aetiology underlying small vessel vasculitis came from the finding of non-specific immunological abnormalities in patients' sera, including hyperglobulinaemia and the presence of rheumatoid factors and immune complexes. These findings were variable, and reported in between one third and two thirds of cases (Ronco *et al.* 1983; Serra *et al.* 1984). Immune deposits in the glomeruli were, however, scanty or absent. The phenotype of inflammatory cells present in renal biopsies from these patients, which includes both CD4 and CD8-positive T lymphocytes, suggests immunologically-mediated attack (Bolton *et al.* 1987; Nolasco *et al.* 1987). The favourable response to non-specific immunosuppressive therapy also supports an immune aetiology (Pusey *et al.* 1990, see Chapter 24); the typical response to treatment with a combination of corticosteroids, cyclophosphamide and plasma exchange is illustrated in Fig. 23.1.

Active systemic vasculitis is also accompanied by a vigorous acute phase response, with normochromic anaemia, leucocytosis and thrombocytosis. C-reactive protein is increased, albumin low (even in the absence of heavy proteinuria) and alkaline phosphatase is typically raised (Gaskin and Pusey

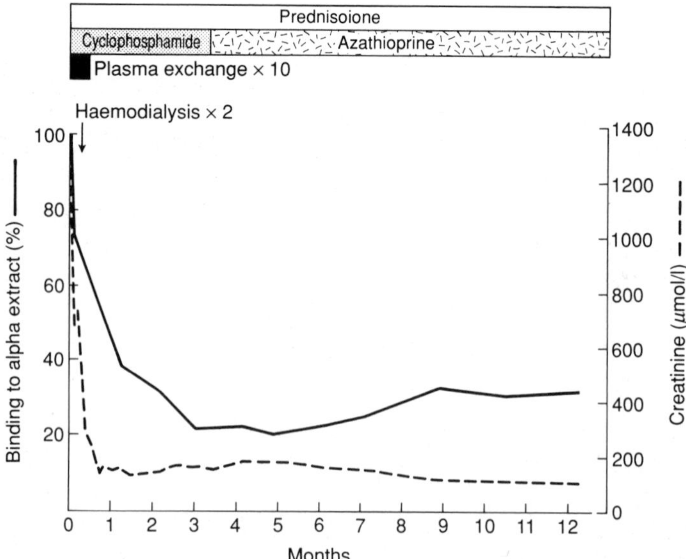

Fig. 23.1 The effect of treatment in a patient with Wegener's granulomatosis who presented with dialysis-dependent renal failure, illustrating: (1) renal recovery to near-normal serum creatinine values; (2) fall in ANCA titre, as measured by an ELISA detecting antibodies binding to an extract of neutrophil primary granules (normal range is less than 16 per cent).

1991). Circulating Von Willebrand factor levels are also high (Savage *et al.* 1991), although it is uncertain whether this is part of the acute phase response, or whether it results from release by damaged vascular endothelium.

Further support for an immune aetiology for vasculitis comes from the association with particular MHC genes. Small studies using serological tissue typing techniques have suggested an association with *HLA DR2* (Elkon *et al.* 1983; Müller *et al.* 1984); more recently, molecular biological methods have demonstrated a positive association with *DQw7* and a negative association with *DR3* (Spencer *et al.*, 1992). The role of the MHC in the induction of T cell-dependent immune responses is discussed further in Chapter 18.

Anti-neutrophil cytoplasmic antibodies (ANCA)

Recently, the discovery of specific autoantibodies has attracted considerable attention. In 1982, Davies *et al.* reported finding antibodies binding to normal human neutrophils in patients with febrile illnesses associated with focal necrotizing glomerulonephritis—then attributed to a viral infection (Davies *et al.* 1982). In 1985, Van der Woude *et al.* analysed sera from patients with Wegener's granulomatosis using an indirect immunofluorescence assay (Van der Woude *et al.* 1985), and detected antibodies which stained the cytoplasm

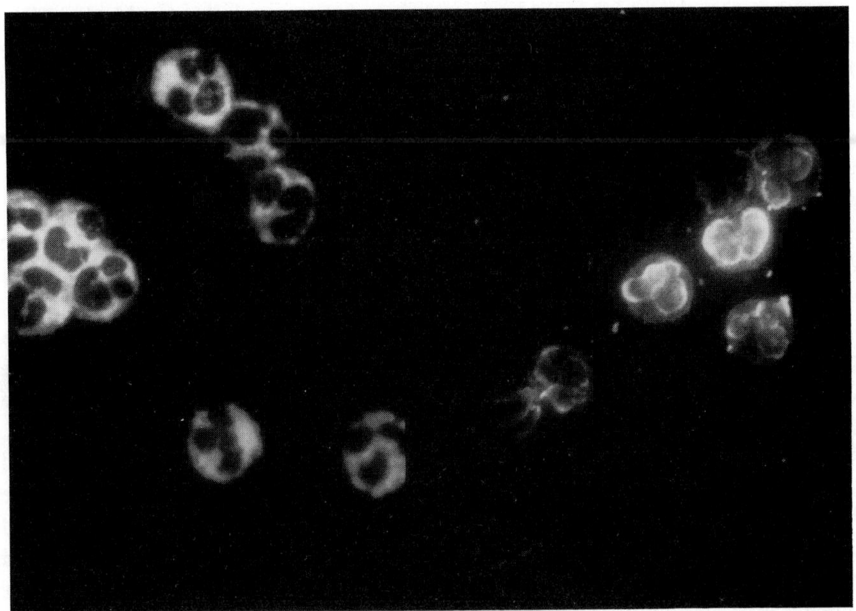

Fig. 23.2 The appearance of C-ANCA (left) and P-ANCA (right) by indirect immunofluorescence on ethanol-fixed normal human neutrophils.

of neutrophils and monocytes in a granular pattern—anti-neutrophil cytoplasmic antibodies (ANCA). This finding was soon reproduced by other groups, and it was found that the assay had a high specificity and sensitivity in active generalized disease. It also became apparent that sera from patients with microscopic polyarteritis could produce similar results (Savage *et al.* 1987). An indirect immunofluorescence assay was standardized at the First International ANCA Workshop in 1988, and is now widely available (Wiik 1989).

A second immunofluorescence pattern, with staining concentrated round the neutrophil nucleus, was described in 1988 (Falk and Jennette 1988). This was termed perinuclear ANCA, abbreviated to P-ANCA, in contrast to cytoplasmic, or C-ANCA. The C-ANCA pattern has a high specificity for primary systemic vasculitis, but the P-ANCA pattern may be seen in rheumatoid and related diseases—in which it was described as granulocyte-specific ANA (Wiik *et al.* 1974) and may be reproduced by anti-nuclear factors in systemic lupus erythematosus. The typical appearances of C-ANCA and P-ANCA by indirect immunofluorescence on ethanol-fixed neutrophils are shown in Fig. 23.2.

ANCA specificity

In the last two years the specificities of C-ANCA and P-ANCA have largely been defined. Most C-ANCA appear to recognize a 29 kDa serine proteinase present in the azurophil (primary) granules of the neutrophil (Goldschmeding *et al.* 1989; Ludemann *et al.* 1990; Niles *et al.* 1990). A cDNA for this enzyme has now been isolated (Campanelli *et al.* 1991); it has been designated proteinase 3, shows considerable sequence homology to neutrophil elastase, and is probably identical to myeloblastin, a factor involved in regulation of myeloid proliferation and differentiation (Jenne *et al.* 1990). Not all C-ANCA recognize proteinase 3, however, and some antibodies bind to cationic antimicrobial protein (CAP 57), a 57 kDa neutrophil granule protein also known as bactericidal permeability increasing factor (Falk 1990). Rarely, antibodies to myeloperoxidase may be found in sera giving a C-ANCA pattern.

When sera from patients with primary systemic vasculitis giving a P-ANCA pattern are analysed, specificity is usually for the enzyme myeloperoxidase (Falk and Jennette 1988). Myeloperoxidase is a constituent of the same granule population as proteinase 3, and thus the perinuclear pattern is artefactual. Ethanol fixation of the neutrophil appears to permeabilize the granules, and allows redistribution of myeloperoxidase to the perinuclear region—hence binding on formalin-fixed neutrophils shows a cytoplasmic pattern (Falk and Jennette 1988). In contrast, P-ANCA patterns in rheumatoid arthritis and systemic lupus erythematosus are not generally associated with antibodies against myeloperoxidase, and on formalin-fixed neutrophils the pattern either remains perinuclear or is abolished. The granule enzymes elastase and lactoferrin may rarely be a target of ANCA (Goldschmeding *et al.* 1989; Dolman *et al*

1990). Antibodies to lactoferrin have also been reported in patients with systemic lupus erythematosus.

Disease associations with ANCA

Particular clinical syndromes appear to be associated with distinct antibody specificities (see Table 23.2). In Wegener's granulomatosis, the immunofluorescence pattern is almost always cytoplasmic, with solid phase assay reactivity against proteinase 3 in most, but probably not all, cases. In microscopic polyarteritis and isolated FNGN, both C-ANCA and P-ANCA are seen. The ratio between the two patterns varies in different series, but P-ANCA with anti-myeloperoxidase antibodies are more common in certain studies of isolated FNGN (Falk and Jennette 1988; Cohen Tervaert *et al.* 1990*a*). In Churg–Strauss syndrome, small series generally report the presence of antibodies against myeloperoxidase (Cohen Tevaert *et al.* 1990*a*; Gaskin *et al.* 1991). There is presently conflicting data as to whether ANCA are generally a feature of polyarteritis nodosa; we have seen three patients with biopsy and angiographic proof of muscular artery involvement, but without glomerulonephritis, who had P-ANCA and antibodies against myeloperoxidase. Results of our study of ANCA by immunofluorescence in patients with systemic vasculitis are shown in Fig. 23.3. ANCA have also been reported in the childhood vasculitis Kawasaki syndrome; the immunofluorescence pattern is atypical and the precise target has not yet been defined (Savage *et al.* 1989).

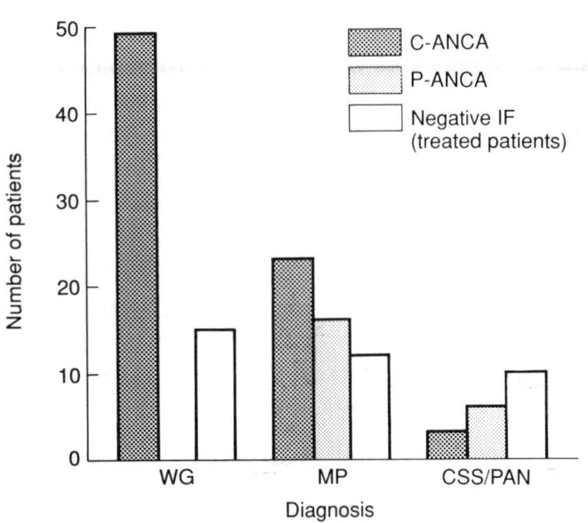

Fig. 23.3 The relationship between ANCA immunofluorescence pattern and clinical diagnosis in a cross-sectional study of patients treated at Hammersmith Hospital, London, UK. WG, Wegener's granulomatosis; MP, microscopic polyarteritis and isolated FNGN; CSS/PAN, Churg–Strauss syndrome and polyarteritis nodosa.

ANCA are generally of the IgG class and are predominantly IgG_1 (and probably IgG_4) subclass (Van der Woude *et al.* 1989). Jayne *et al.* (1989) described a subgroup of patients with IgM ANCA only, characterized clinically by necrotizing glomerulonephritis with alveolar haemorrhage. Other groups, however, have reported similar patients with IgG antibodies to myeloperoxidase. IgA ANCA have occasionally been detected in adults with Henoch–Schonlein purpura, but this finding has not been widely confirmed (Van der Woude *et al.* 1989).

Vasculitis and FNGN induced by certain drugs may be accompanied by the presence of ANCA. Antibodies to myeloperoxidase and elastase are reported in patients with FNGN associated with hydralazine therapy (Nässberger *et al.* 1990). Additionally, we have treated a patient with FNGN and alveolar haemorrhage following penicillamine therapy who had antibodies to myeloperoxidase.

Pathological effects of ANCA

A pathogenetic role for ANCA is suggested by the relationship between the presence and titre of antibody and the activity of disease. Van der Woude *et al.* (1985) reported a high sensitivity in active disease, and Nölle and Specks reported a correlation between ANCA titre by immunofluorescence and disease activity (Nölle *et al* 1989; Specks *et al.* 1989). Recently, we have analysed immunofluorescence results in patients with systemic vasculitis under follow-up over a period of 50 months (Gaskin *et al.* 1991). Patients who relapsed during the study invariably had detectable ANCA at the time of relapse, and in some the relapse was preceded by reappearance of ANCA. Patients with consistently negative ANCA did not relapse. Cohen Tervaert *et al.* studied patients with Wegener's granulomatosis and C-ANCA and found that a four-fold rise in immunofluorescence titre predicted clinical relapse (Cohen Tervaert *et al.* 1989). A subsequent clinical trial by the same group randomized patients showing a fourfold rise in titre, either to receive pre-emptive treatment, or

Table 23.2 ANCA specificity and disease characteristics

	C-ANCA	P-ANCA
IF appearance	Granular, cytoplasmic	Perinuclear on ethanol-fixed neutrophils Cytoplasmic on formalin-fixed neutrophils
Targets	Proteinase 3 (CAP 57)	Myeloperoxidase (Elastase)
Disease associations	Wegener's granulomatosis Microscopic polyarteritis (including isolated FNGN)	Microscopic polyarteritis (including isolated FNGN) Churg–Strauss syndrome Polyarteritis nodosa

to delay treatment until clinical evidence of relapse (Cohen Tervaet *et al.* 1990*b*). The control group showed a high incidence of relapse, while no relapses occurred in the treated group. They concluded that rising ANCA titre was a sufficiently good predictor of relapse to dictate the need for immunosuppressive therapy.

Nonetheless, in all of the published series, ANCA could be present without overt disease, suggesting that even if these antibodies do have a pathogenetic role other factors must be involved. It has been suggested that only certain immunoglobulin subclasses might lead to tissue injury, and that ANCA in the absence of disease might be of a non-damaging subclass. There is evidence for a shift in IgG subclass distribution during the course of disease (Van der Woude *et al.* 1989, Jayne and Lockwood 1990).

More direct evidence for a pathogenetic role for ANCA comes from experiments on activation of neutrophils. The underlying hypothesis is that ANCA bind to their antigens on the neutrophil surface; activation of the neutrophil then leads to release of mediators, which could cause tissue (particularly endothelial) damage. Incubation of human autoantibodies, (both C-ANCA and P-ANCA), or of monoclonal antibodies against the same targets, with normal neutrophils primed with TNF leads to a respiratory burst and also to degranulation (Falk *et al.* 1990). $F(ab)_2$ fragments of ANCA were able to reproduce these effects, but control antibodies (prepared in the same way) were not. Preliminary data also suggest that co-incubation of ANCA or monoclonal antibodies with TNF-primed neutrophils and TNF-treated endothelial cells can lead to up-regulation of endothelial injury (Ewart *et al.* 1992). Ludemann *et al.* have shown an inhibitory effect of C-ANCA on elastinolytic activity (Ludemann *et al.* 1990), and since proteinase 3 appears to be identical to myeloblastin, they suggest a pathogenetic role through effects on myeloid differentiation and proliferation (Jenne *et al.* 1990).

Other autoantibodies in systemic vasculitis

In patients with Wegener's granulomatosis, Abbott *et al.* identified binding to glomerular epithelial and endothelial cells in culture (Abbott *et al.* 1989). Several groups have identified binding of patients' sera to cultured human umbilical vein endothelial cells by radioimmunoassay or enzyme-linked immunoassay—so-called anti-endothelial cell antibodies (AECA) (Brasile *et al.* 1989; Ferraro *et al.* 1990; Savage *et al.* 1991). These occur in only a proportion of patients, and are not specific for primary systemic vasculitis; similar results have been found in scleroderma, systemic lupus erythematosus and other connective tissue diseases. Cross-absorption studies do not support cross-reactivity between such anti-endothelial antibodies and ANCA, and their pathogenetic significance is unclear.

Occasionally, ANCA may co-exist with antibodies to the glomerular basement membrane (GBM) (Jayne *et al.* 1990; O'Donaghue *et al.* 1990). Estimates vary, but approximately 10 per cent of patients with anti-GBM antibodies also have ANCA. Patients with both antibodies often have extrarenal clinical manifestations. It is not yet clear whether this subgroup of patients shows a different response to therapy than that seen in patients with either ANCA or anti-GBM antibodies alone.

Conclusions

In summary, there is mounting evidence that small vessel systemic vasculitis is autoimmune in aetiology. There is a close association with specific autoantibodies (ANCA) which appear capable of activating neutrophils, and which may have a role in injuring vascular endothelium. However, it remains likely that T cell responses are important in the initiation of autoimmunity, and in the mediation of tissue injury. The role of the MHC, which is central to the induction of T-cell dependent immune responses, has been discussed in Chapter 18; recent work demonstrates both positive and negative associations between HLA class II genes and systemic vasculitis. Greater understanding of autoimmune mechanisms should allow the development of more specific therapy for systemic vasculitis in the future.

References

Abbott, F., Jones, S., Lockwood, C.M., and Rees, A.J. (1989). Autoantibodies to glomerular antigens in patients with Wegener's granulomatosis. *Nephrology Dialysis Transplantation*, **4**, 1–8.

Bolton, W.K., Innes, D., Sturgill, B.C., and Kaiser, D.L. (1987). T cells and macrophages in rapidly progressive glomerulonephritis: clinicopathological correlates. *Kidney International*, **32**, 869–76.

Brasile, L., Kremer, J.M., Clarke, J.L., and Cerilli, J. (1989). Identification of an autoantibody to vascular endothelial cell-specific antigens in patients with systemic vasculitis. *American Journal of Medicine*, **87**, 74–86.

Campanelli, D. *et al.* (1990). Cloning of cDNA for proteinase 3: a serine protease, antibiotic, and autoantigen from human neutrophils. *Journal of Experimental Medicine*, **172**, 1709–15.

Cohen Tervaert, J.W. *et al.* (1989). Association between active Wegener's granulomatosis and anticytoplasmic antibodies. *Archives of Internal Medicine*, **149**, 2461–5.

Cohen Tervaert, J.W. *et al.* (1990*a*). Autoantibodies against myeloid lysosomal enzymes in crescentic glomerulonephritis. *Kidney International*, **37**, 799–806.

Cohen Tervaert, J.W. *et al.* (1990*b*). Prevention of relapses in Wegener's granulomatosis by treatment based on antineutrophil cytoplasmic antibody titre. *Lancet*, **336**, 709–11.

Couser, W.G. (1988). Rapidly progressive glomerulonephritis: classification, pathogenetic mechanisms, and therapy. *American Journal of Kidney Diseases*, **11**, 449–64.

Davies, D.J., Moran, J.E., Niall, J.F., and Ryan, G.B. (1982). Segmental necrotising glomerulonephritis with antineutrophil antibody: possible arbovirus aetiology? *British Medical Journal*, **285**, 606.

Dolman, K.M., Goldschmeding, R., Sonnenberg, A., and von dem Borne, A.E.G.Kr. (1990). ANCA related antigens. *Acta Pathologica, Microbiologica et Immunologica Scandanavica*, **98**, (19), 28.

Elkon, K.B., Sutherland, D.C., Rees, A.J., Hughes, G.R.V., and Batchelor, J.R. (1983). HLA antigen frequencies in systemic vasculitis: increase in HLA-DR2 in Wegener's granulomatosis. *Arthritis and Rheumatism*, **26**, (1), 102–5.

Ewert, B.H. Jennette, J.C., and Falk, R.J. (1992). Anti-myeloperoxidase antibodies stimulate neutrophils to damage human endothelial cells. *Kidney International*, **41**, 375–83.

Falk, R.J. (1990). ANCA-associated renal disease. *Kidney International*, **38**, 998–1010.

Falk, R.J. and Jennette, J.C. (1988). Anti-neutrophil cytoplasmic antibodies with specificity for myeloperoxidase in patients with systemic vasculitis and idiopathic necrotising and crescentic nephritis. *New England Journal of Medicine*, **318**, 1651–7.

Falk, R.J., Terrell, R.S., Charles, L.A., and Jennette, J.C. (1990). Anti-neutrophil cytoplasmic autoantibodies induce neutrophils to degranulate and produce oxygen radicals *in vitro*. *Proceedings of the National Academy of Sciences USA*, **87**, 4115–9.

Fauci, A.S., Haynes, B.F., and Katz, P. (1978). The spectrum of vasculitis. Clinical, pathologic, immunologic, and therapeutic considerations. *Annals of Internal Medicine*, **89**, (1), 660–76.

Ferraro G. *et al.* (1990). Anti-endothelial cell antibodies in patients with Wegener's granulomatosis and micropolyarteritis. *Clinical and Experimental Immunology*, **79**, 47–53.

Gaskin, G. and Pusey, C.D. (1992). Systemic vasculitis. *Oxford textbook of clinical nephrology*. Oxford University Press, 612–36.

Gaskin, G., Ryan, J.J., Rees, A.J., and Pusey, C.D. (1990). Anti-myeloperoxidase antibodies in vasculitis: relationship to ANCA and clinical diagnosis. *Acta Pathologica Microbiologica et Immunologica Scandanavica*. **98**, (19), 33.

Gaskin, G. *et al.* (1991). Anti-neutrophil cytoplasmic antibodies in long-term follow-up of seventy patients with systemic vasculitis. *Nephrology Dialysis Transplantation*. **6**, 689–94.

Goldschmeding, R. *et al* (1989). Wegener's granulomatosis autoantibodies identify a novel diisopropylfluorophosphate-binding protein in the lysosomes of normal human neutrophils. *Journal of Clinical Investigation*, **84**, 1577–87.

Jayne, D.R.W. and Lockwood, C.M. (1990). The isotype distribution of ANCA in systemic vasculitis. *Acta Pathologica Microbiologica et Immunologica Scandanavica*, **98**, (19), 45.

Jayne, D.R.W., Jones, S.J., Severn, A., Shaunak, A., Murphy, J., and Lockwood, C.M. (1989). Severe pulmonary haemorrhage and systemic vasculitis in association with anti-neutrophil cytoplasm antibodies of IgM subclass only. *Clinical Nephrology*, **32**, 101–6.

Jayne, D.R.W., Marshall, P.D., Jones, S.J., and Lockwood, C.M. (1990). Autoantibodies to GBM and neutrophil cytoplasm in rapidly progressive glomerulonephritis. *Kidney International*, **37**, 965–70.

Jenne, D.E., Tschopp, J., Lüdemann, J., Utecht, B., and Gross, W.L. (1990). Wegener's autoantigen decoded. *Nature*, **346**, 520.

Lüdemann, J., Utecht, B., and Gross, W.L. (1990). Anti-neutrophil cytoplasm antibodies in Wegener's granulomatosis recognize an elastinoolytic enzyme. *Journal of Experimental Medicine*, **171**, 357–62.

Müller, G.A., Gebhardt, M., Kömpf, J., Baldwin, W.M., Ziegenhagen, D., and Bohle, A.(1984). Association between rapidly progressive glomerulonephritis and the properdin factor BfF and different HLA-D region products. *Kidney International*, **25**, 115–8.

Nässberger, L., Sjoholm, A.G., Jonnson, H., Sturfelt., G., and Akesson, A. (1990). Autoantibodies against neutrophil cytoplasm components in systemic lupus erythematosus and in hydralazine-induced lupus. *Clinical and Experimental Immunology*, **81**, 3803.

Niles, J.L., McCluskey, R.T., Ahmad, M.F., and Arnaout, M.A. (1989). Wegener's granulomatosis autoantigen is a novel neutrophil serine proteinase. *Blood*, **74**, (6), 1888–93.

Nolasco, F.E., Cameron, J.S., Hartley, B., Coelho, A., Hildreth, G., and Reuben, R. (1987). Intraglomerular T cell and monocytes in nephritis. Study with monoclonal antibodies. *Kidney International*, **31**, 1160.

Nölle, B., Specks, U., Lüdemann, J., Rohrbach, M.S., DeRemee, R.A., and Gross, W.L. (1989). Anticytoplasmic autoantibodies: their immunodiagnostic value in Wegener's granulomatosis. *Annals of Internal Medicine*, **111**, 28–40.

O'Donaghue, D.J., Short, C.D., Brenchley, P.E.C., Lawler, W., and Ballardie, F.W. (1989). Sequential development of systemic vasculitis with anti-neutrophil cytoplasmic antibodies complicating anti-glomerular basement membrane disease. *Clinical Nephrology*, **32**, 251–5.

Pusey, C.D., Gaskin, G., and Rees, A.J. (1990). Treatment of primary systemic vasculitis. *Acta Pathologica Microbiologica et Immunologica Scandanavica*, **98**, (19), 48–50.

Ronco, P. *et al.* (1983). Immunopathological studies of polyarteritis and Wegener's granulomatosis: a report of 43 patients with 51 renal biopsies. *Quarterly Journal of Medicine*, **52**, 212–23.

Savage, C.O.S., Winearls, C.G., Jones, S., Marshall, P., and Lockwood, C.M. (1987). Prospective study of radioimmunoassay for antibodies against neutrophil cytoplasm in diagnosis of systemic vasculitis. *Lancet*, **i**, 1389–93.

Savage, C.O.S., Tizard, J., Jayne, D.R.W., Lockwood, C.M., and Dillon, M.J. (1989). Anti-neutrophil cytoplasm antibodies in Kawasaki syndrome. *Archives of Diseases of Childhood*, **64**, 462.

Savage, C.O.S., Pottinger, B., Gaskin, G., Lockwood, C.M., Pusey, C.D., and Pearson, J.D. (1991). Vascular damage in Wegener's granulomatosis and microscopic polyarteritis; presence of anti-endothelial antibodies and their relation to anti-neutrophil cytoplasm antibodies. *Clinical and Experimental Immunology*, **85**, 14–19.

Serra, A., *et al.* (1984). Vasculitis affecting the kidney: presentation, histopathology and long-term outcome. *Quarterly Journal of Medicine*, **53**, 181–207.

Specks, U., Wheatley, C.L., McDonald, T.J., Rohrbach, M.S., and DeRemee, R.A. (1989). Anticytoplasmic autoantibodies in the diagnosis and follow-up of Wegener's granulomatosis. *Mayo Clinic Proceedings*, **64**, 28–36.

Spencer, S.J.W., Burns, A., Gaskin, G., Pusey, C.D., and Rees, A.J. (1992). HLA Class II specificities in vasculitis with antibodies to neutrophil cytoplasmic antigens. *Kidney International*, **41**, 1059–63.

Van der Woude, F.J., *et al.* (1985). Autoantibodies against neutrophils and monocytes: tool for diagnosis and marker of disease activity in Wegener's granulomatosis. *Lancet*, **i**, 425–9.

Van der Woude, F.J., Daha, M.R., and Van Es, L.A. (1989). The current status of neutrophil cytoplasm antibodies. *Clinical and Experimental Immunology*, **78**, 143–8.

Wiik, A. (1989). Delineation of a standard procedure for indirect immunofluorescence detection of anti-neutrophil cytoplasmic antibodies (ANCA). *Acta Pathologica Microbiologica et Immunologica Scandanavica*, **97**, (6), 12–13.

Wiik, A., Jensen, E., and Friis, J. (1974). Granulocyte-specific antinuclear factors in synovial fluids and sera from patients with rheumatoid arthritis. *Annals of Rheumatic Diseases*, **33**, 515–22.

24 Treatment of systemic vasculitis

D.R.W. Jayne and C.M. Lockwood

The primary vasculitides are a group of disorders which have in common blood vessel inflammation of such severity that necrosis can be seen on histological examination. Because their aetiology and pathogenesis remain unclear, classification has hitherto relied upon the eponymous definition of clinicopathological syndromes, such as Wegener's granulomatosis and Takayasu's arteritis (Table 24.1) (Fahey *et al.* 1954). However, this nosological approach is inadequate, because patients may present at a stage with insufficient clinical features of a disease to satisfy the criteria for a particular syndrome, or with features which could qualify them for more than one form of primary vasculitis. Nevertheless, there is some justification in grouping these disorders under the general term of systemic vasculitis, because the different clinical syndromes do show certain similarities in their serological markers, response to treatment and natural history (Carrington and Liebow 1966).

Recently, the finding that circulating antibodies to the neutrophil cytoplasm (ANCA) are strongly associated with the development of certain primary vasculitides has allowed the introduction of the first serological tests for these disorders, and has implicated autoimmune mechanisms in their aetiology (Van der Woude *et al.* 1985). Now, growing evidence shows that ANCA are not only commonly found in small vessel systemic vasculitides, such as

Table 24.1 Histological classification of primary systemic vasculitis based on the size of vessel involved and the presence of granulomata. (WG, Wegener's granulomatosis; MPA, microscopic polyarteritis; HSP, Henoch–Schonlein purpura; KS, Kawasaki disease; PAN, polyarteritis nodosa; CS, Churg–Strauss angiitis; TA, Takayasu's arteritis; GCA/PR, giant cell arteritis/polymyalgia rheumatica.)

Vessel size	Granulomata absent	Granulomata present
Small	Microscopic polyarteritis Henoch–Schonlein purpura Renal-limited vasculitis	Wegener's granulomatosis
Medium	Polyarteritis nodosa Kawasaki disease	Churg–Strauss angiitis
Large	Giant cell arteritis	Takayasu's arteritis

Wegener's granulomatosis and microscopic polyarteritis, but may also be found in some patients with larger vessel disease and their differing specificities may allow differentiation between the syndromes (Table 24.2) (Van der Woude *et al.* 1989). ANCA have also confirmed the overlaps between vasculitis and other autoimmune diseases, for example, anti-glomerular basement membrane disease and rheumatoid arthritis (Jayne *et al.* 1990).

For the nephrologist, the typical renal lesion of vasculitis is a focal necrotizing glomerulonephritis, but milder disease with focal proliferation may be seen (Heptinstall 1990). Primary vasculitis is the most frequent cause of rapidly progressive glomerulonephritis, the change from the indolent to this accelerated phase being marked histologically by the development of cellular crescents around the glomerulus. Whether ANCA play a pathogenetic role, in a manner analogous to the renal injury brought about by anti-glomerular basement membrane antibodies, is the subject of considerable interest. This chapter aims to show how greater understanding of the relationship of ANCA to the development of vasculitis nephritis has helped shape strategies for its treatment.

Background

Before the introduction of cortisone, patients with a diagnosis of polyarteritis nodosa or Wegener's granulomatosis had a five year survival of less than 20 per cent. Steroids alone suppressed disease, but were required in high dose and were not curative; they were subsequently combined with various cytotoxic agents, which produced major improvements in disease control and survival (Bouroncle *et al.* 1967; Capizzi and Bertino 1971; McIlvanie 1966). A survey of 64 patients with polyarteritis nodosa, who received either steroids and cytotoxics, steroids alone or no treatment, showed five year survivals of 80, 53 and 12 per cent, respectively (Leib *et al.* 1979). The current use of prednisolone and cyclophosphamide is largely based on retrospective, uncontrolled studies involving fewer than 20 patients, and proper prospective controlled studies have not been carried out (Fauci *et al.* 1979).

The treatment of anti-glomerular basement membrane disease has been held as a paradigm for treatment of other forms of crescentic nephritis, because a rational approach to therapy has evolved based on a knowledge of pathogenesis (Wilson and Dixon 1973). Circulating pathogenetic anti-glomerular basement membrane antibodies are removed by plasma exchange, and cytotoxic drugs combined with steroids are used to suppress further antibody synthesis and reduce inflammation. Furthermore, the titre of circulating anti-glomerular basement membrane antibodies can be monitored by sensitive radioimmunoassay, thus allowing therapy to be altered to treatment response (Lockwood *et al.* 1976). When this protocol was applied to crescentic nephritis associated with vasculitis it was also successful (Lockwood *et al.* 1977), and later, the discovery of ANCA enabled similar monitoring of disease and tailoring of therapy.

Table 24.2 A serological classification based on ANCA positivity and the size of vessel involved for the same diagnostic categories as Table 24.1. ANCA ? column corresponds to conditions where preliminary reports are, as yet, unconfirmed.

Vessel size	ANCA positive	ANCA ?	ANCA negative
Small	Microscopic polyarteritis Wegener's granulomatosis Renal-limited vasculitis	Henoch-Schonlein purpura	
Medium	Churg–Strauss angiitis Kawasaki disease	Polyarteritis nodosa	
Large		Takayasu's arteritis	Giant cell arteritis Polymyalgia rheumatica

Current therapy

There is now a broad consensus that treatment strategy should involve the use of higher dose immunosuppression to induce a remission, with lower dosage being continued during remission to prevent relapse. The dose and duration of therapy vary between units; our patients receive cyclophosphamide 3 mg/kg/d for six weeks, or until remission is achieved, in combination with prednisolone 60 mg/d, reducing stepwise to 10 mg/d over six weeks. Doses are reduced in patients over 55 years of age in whom the risks of immunosuppression are higher. Azathioprine 2 mg/kg/d, with lower cumulative toxicity, is substituted for cyclophosphamide and continued with prednisolone 10 mg/d for at least a year. If leucopaenia occurs and the total white count falls below 3.0×10^9/l then the cytotoxic drugs are withdrawn temporarily.

Remission is achieved in 80–90 per cent of cases with this combination, but relapse is common. Relapse rates depend on the immunosuppressive regimen and range from 10–30 per cent per year; they are particularly common after reductions in therapy and are treated by re-introducing the initial high dose therapy.

Protocols have been developed for patients with either severe systemic disease or renal involvement, which may not be appropriate for those with mild or localized disease, or for the elderly, who tolerate immunosuppression poorly. These groups may benefit from alternative approaches described below.

Is current therapy adequate?

There is debate about the role of additional therapy in systemic vasculitis with renal involvement. In a prospective controlled study comparing plasma exchange plus cyclophosphamide and prednisolone with drugs alone, significant benefit was seen in the group receiving plasma exchange if they were dialysis dependent at presentation (Pusey *et al.*, in press). The fact that additional benefit was not seen in the groups with less severe renal involvement was attributed to the success achieved with drugs alone. In view of this result, it is perhaps not surprising that a second study failed to show a significant benefit from plasma exchange when patients were not stratified on the severity of their renal disease (Glückner *et al.* 1988).

Used as sole therapy, plasma exchange can induce remission in selected patients, without a rebound in ANCA or disease activity (Jayne and Lockwood 1990). This may be useful when there are compelling reasons to avoid steroids and cytotoxic drugs, as plasma exchange alone has few adverse effects. Should the pathogenetic role of circulating reactants in vasculitis be confirmed, then the theoretical argument for plasma exchange would strengthen.

Intravenous methyl prednisolone has also been advocated as an additional agent for rapidly progressive glomerulonephritis, allowing up to 70 per cent of patients who are dialysis-dependent to recover independent renal function

(Bolton and Sturgill 1989). Methyl prednisolone is cheaper and more convenient to administer than plasma exchange, but the evidence for its use is not as strong as that for plasma exchange, and a direct comparison is now required.

Other situations in which additional therapy has been required are acute pulmonary haemorrhage and disease refractory to conventional therapy, but prospective data on the use of plasma exchange or methylprednisolone for these indications is lacking. In the case of pulmonary haemorrhage, an interesting observation has been the association of IgM ANCA with this complication. If IgM ANCA plays a role in injury, plasma exchange would certainly appear to be an appropriate treatment, as this antibody is confined to the vascular compartment (Jayne *et al.* 1989).

Adverse effects of therapy

The major complications of immunosuppressive therapy are due to infection, which has a mortality greater than 10 per cent, while morbidity associated with less severe infection approaches 100 per cent (Cohen *et al.* 1982; Bradley *et al.* 1989). The infective risk is increased in elderly patients and is also associated with total steroid dosage, severity of presenting disease and leucopaenia. Myelosuppression is frequent and reversible, but requires a reduction in therapy, with the danger of incomplete disease control. In the longer term, as well as the cumulative toxicity of prednisolone, cyclophosphamide can cause pulmonary fibrosis, haemorrhagic cystitis and infertility and leads to an increased risk of neoplasia (Bradley *et al.* 1989). In systemic lupus erythematosus with nephritis, intravenous cyclophosphamide pulses are less toxic than a daily oral dose, but the efficacy of this route in vasculitis has not been confirmed (Schroeder *et al.* 1987). Prophylaxis against pneumocystis carinii pneumonia with low dose trimethoprin/sulfamethoxazole may be worthwhile.

Monitoring of disease activity

Management of patients with vasculitis requires a careful balance between the dangers of therapy and those of relapse. The non-specific indicators of disease activity, erythrocyte sedimentation rate and C-reactive protein (CRP), are influenced by factors apart from the vasculitic process, such as infection, anaemia and hypoproteinaemia. ANCA are specific diagnostic markers for vasculitis and their titre is associated with disease activity (Fig. 24.1) (Cohen Tervaert *et al.* 1989). Monitored sequentially, ANCA are superior to CRP in the prediction and diagnosis of relapse (Fig. 24.2) (Jayne *et al.* 1990*b*). Management may be improved by basing treatment on the ANCA result, rather than waiting for a clinical relapse; a group so treated had fewer relapses and a lower total exposure to immunosuppressive therapy (Cohen Tervaert *et al.* 1990).

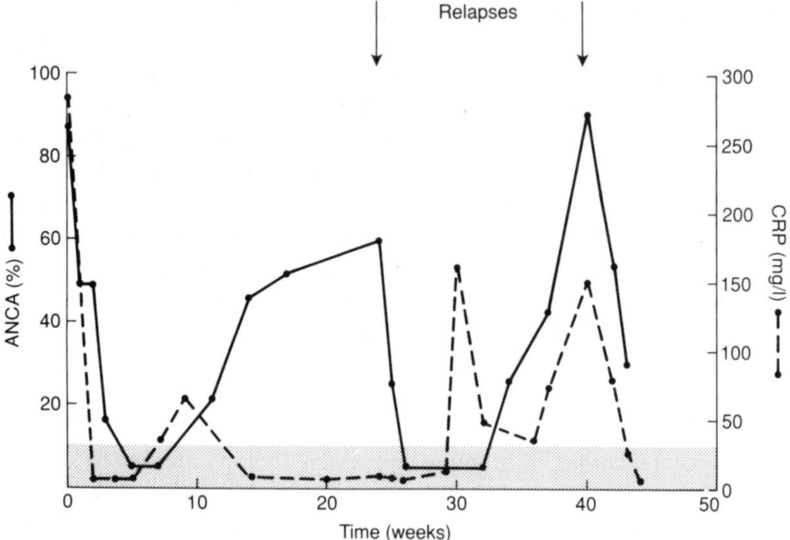

Fig. 24.1 Sequential ANCA percentage binding and C-reactive protein (CRP) plotted against time for a patient with systemic vasculitis. Following a good clinical response to initial treatment, ANCA and CRP fell to normal. Subsequent clinical relapses, indicated by the arrows, were preceded by rises in ANCA but did not correlate as well with CRP. (The shaded area represents the ANCA normal range = < 16 per cent).

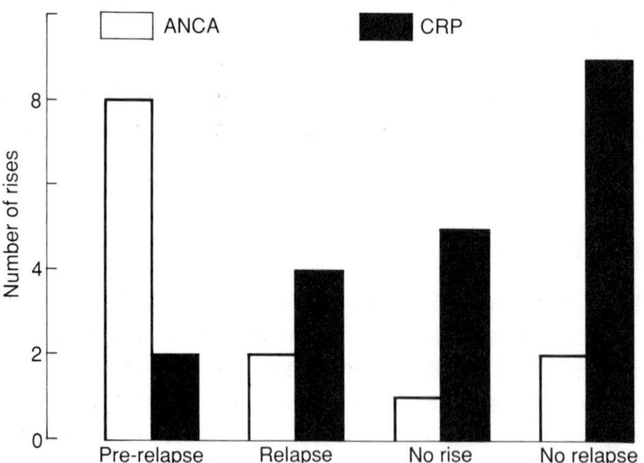

Fig. 24.2 Episodes of rise in ANCA and CRP in 20 patients with systemic vasculitis followed prospectively over one year. In 9 patients 11 clinical relapses occurred; they were predicted by rises in ANCA in 8, a mean of 4.2 weeks earlier, and by CRP in only 2. By the time of relapse, ANCA had risen in a further 2 and CRP in 5. ANCA rose in 2 and CRP in 7, without subsequent relapse during the follow-up period. One clinical relapse occurred in the presence of a negative ANCA.

White cell scanning using ^{111} indium-labelled cells is a sensitive indicator of upper respiratory tract disease in Wegener's granulomatosis and may also reveal asymptomatic disease and involvement of other organs (Llewelyn *et al.* 1990). Scans may be repeated during remission to assess disease resolution.

Optimal management of individual patients requires regular clinical assessment and measurement of multiple indices. Perhaps with improved monitoring, it will be possible to give lower doses of immunosuppression for a shorter time during remission, with a rapid increase in treatment in those predicted to be at high risk of relapse.

Alternatives to current therapy

The antibiotic combination trimethoprim/sulfamethoxazole has been reported to control disease in Wegener's granulomatosis, and the number of patients treated exceeds 60 (Deremee 1989). The mechanism of action is unknown, but may not be due to the antibiotic potential, since other antibiotics are not so effective. No controlled studies or comparisons with conventional immunosuppression are available, and there has been a high incidence of allergy. This approach may be most beneficial in patients, otherwise untreated, whose disease is localized to the upper respiratory tract.

Monoclonal antibody therapy has been given in a single case of systemic vasculitis: a humanized rat monoclonal antibody, Campath 1H, which recognizes a determinant on T-cells, B-cells and monocytes was used in combination with a rat anti-CD4 monoclonal antibody (Mathieson *et al.* 1990). This approach was based on the assumption of a significant T-cell contribution to pathogenesis and the experimental evidence that tolerance to an antigen can be induced if it is presented at the same time as anti-CD4 (Waldmann 1989). The patient, who was resistant to conventional therapy, has had a lasting remission from his disease.

Circulating anti-idiotypic antibodies have been found in post-recovery sera which bind to idiotypes on ANCA in acute sera, pointing to a role for network regulation of ANCA (Rossi *et al.*, in press). These anti-idiotypic antibodies recognize public idiotypes on ANCA and inhibit its binding to autoantigen and modulate the biological activity of ANCA *in vitro*. Similar anti-idiotypic antibodies are also present in pooled immunoglobulin (IVIg). The latter has been used as effective therapy for Kawasaki disease, an ANCA positive childhood vasculitis, as well as for other diseases mediated by pathogenetic autoantibodies. A therapeutic benefit of intravenous pooled intravenous immunoglobulin was seen in six out of seven patients with active systemic vasculitis.

There are also reports of remission following cyclosporin A in three patients with severe vasculitis. Two had failed to respond to conventional therapy and two recovered independent renal function, having been dialysis dependent (Borleffs *et al.* 1987; Gremmel *et al.* 1988). Immunoabsorption with staphylococcal protein A offers more selective removal of immunoglobulin than plasma exchange, and may be an effective alternative to plasma exchange (Palmer *et al.* 1988). Other forms of immunoabsorption using tryptophan, phenylalanine or monoclonal anti-idiotype antibody columns, aimed at more specific removal of autoantibodies, have been tried in systemic lupus erythematosus and may be of use in vasculitis.

Conclusions

The important factors for optimal management of systemic vasculitis are early diagnosis and treatment, and effective monitoring of disease activity to minimize exposure to immunosuppressive drugs and avoid relapse. With these aims in mind, the discovery of specific associations between ANCA and disease activity has offered the possibility of widely available, rapid assays to facilitate diagnosis and participate with other agents in monitoring activity. The trend to increasing degrees of immunosuppression is now being balanced by recognition of the serious adverse effects of treatment and by definition of particular subgroups which will benefit, for example, from the use of plasma exchange in the presence of severe renal disease.

The many problems associated with conventional therapy have stimulated the search for alternative agents, and several are now available whose role is currently being determined. In certain other autoimmune diseases and in experimental models, underlying regulatory abnormalities, such as restrictions of MHC and T-cell receptor gene usage, have been identified, which have led to the development of peptides and monoclonal antibodies capable of specifically blocking the autoreactive response (Vandenbark *et al.* 1989). However, only weak MHC associations have been reported in vasculitis to date, and investigations into T-cell responses in vasculitis are at an early stage (Elkon *et al.* 1983). Nevertheless, the future of immunotherapy in vasculitis relies on a better understanding of pathogenesis and elucidation of the mechanisms by which the physiological control of autoreactivity can be restored.

Acknowledgements

D.R.W.J. has a Clinical Research Fellowship from Gonville and Caius College Cambridge, and C.M.L. is a Wellcome Senior Lecturer.

References

Bolton, W.K. and Sturgill, B.C. (1989). Methylprednisolone therapy for acute crescentic rapidly progressive glomerulonephritis. *Am. J. Nephrol.* **9**, 368.

Borleffs, J.C.C., Derksen, R.H.W.M., and Hené, R.J. (1987). Treatment of Wegener's granulomatosis with cyclosporin. *Ann. Rheum. Dis.* **46**, 175.

Bouroncle, B.A., Smith, E.J., and Cuppage, F.E. (1967). Treatment of Wegener's granulomatosis with Imuran. *Am. J. Med.* **42**, 314.

Bradley, J.D., Brandt, K.D., and Katz, B.P. (1989). Infectious complications of cyclophosphamide treatment for vasculitis. *Arthritis Rheum.* **32**, 45.

Capizzi, R.L. and Bertino, J.R. (1971). Methotrexate therapy of Wegener's granulomatosis. *Ann. Intern. Med.* **74**, 74.

Carrington, C.B. and Liebow, A.A. (1966). Limited forms of angiitis and granulomatosis of Wegener's type. *Am. J. Med.* **41**, 497.

Cohen Tervaert, J.W., Van Der Woude, F., Fauci, A.S., Ambrus, J.L., Velosa, J., Keane, W.F., Meijer, S., Van Der Giessen, M., The, T.H., Van Der Hem, G.K., and Kallenberg, C.G.M. (1989). Association between active Wegener's granulomatosis and anticytoplasmic antibodies. *Arch. Intern. Med.* **149**, 2461.

Cohen Tervaert, J.W., Huitema, M.G., Hene, R.J., Sluiter, W.J., The, T.H., Van Der Hem, G.K., and Kallenberg, C.G.M. (1990). Prevention of relapses in Wegener's granulomatosis by treatment based on antineutrophil cytoplasm antibody titre. *Lancet*, **ii**, 709.

Cohen, J., Pinching, A.J., Rees, A.J., and Peters, D.K. (1982). Infection and immunosuppression: a study of the infective complications of 75 patients with immunologically mediated disease. *Q. J. Med.* **51**, 1.

Deremee, R.A. (1989). The treatment of Wegener's granulomatosis with Trimethoprim/Sulfamethoxazole: illusion or vision?. *Arthritis Rheum.* **31**, 1068.

Elkon, K.B., Sutherland, D.C., and Rees, A.J. (1983). HLA antigen frequencies in systemic vasculitis: increase in HLA DR4 in Wegener's granulomatosis. *Arthritis Rheum.* **26**, 102.

Fahey, J.L., Leonard, E., Churg, J., and Godman, G. (1954). Wegener's granulomatosis. *Am. J. Med.* **46**, 168.

Fauci, A.S., Katz, P., Haynes, B.F., and Wolff, S.M. (1979). Cyclophosphamide therapy of severe systemic necrotizing vasculitis. *N. Engl. J. Med.* **301**, 235.

Glückner, W.M., Sieberth, H.G., Wichmann, H.E., Backes, E., Bambauer, R., Boesken, W.H., Bohle, A., Daul, A., Graben, N., and Keller, F. (1988). Plasma exchange and immunosuppression in rapidly progressive glomerulonephritis: a controlled, multi-center study. *Clin. Nephrol.* **29**, 1.

Gremmel, F., Druml, W., Schmidt, P., and Graninger, W. (1988). Cyclosporin in Wegener granulomatosis. *Ann. Intern. Med.* **108**, 491.

Heptinstall, R.H. (1990). Schonlein-Henoch syndrome; Lung haemorrhage and glomerulonephritis, or Goodpasture's syndrome. In *Pathology of the kidney*. pp. 741–92. Little, Brown, Boston.

Jayne, D.R.W. and Lockwood, C.M. (1990). New approaches to therapy in systemic vasculitis. In *Recent advances in systemic vasculitis*, (ed. G.S.Panayi) *J. Roy. Soc. Med.* **83**, 407.

Jayne, D.R.W., Jones, S.J., Severn, A., Shaunak, S., Murphy, J., and Lockwood, C.M. (1989). Severe pulmonary hemorrhage and systemic vasculitis in association with circulating anti-neutrophil cytoplasm antibodies of IgM class only. *Clin. Nephrol.* **32**, 101.

Jayne, D.R.W., Marshall, P.D., Jones, S.J., and Lockwood, C.M. (1990*a*). Autoantibodies to GBM and neutrophil cytoplasm in rapidly progressive glomerulonephritis. *Kidney Int.* **37**, 965.

Jayne, D.R.W., Heaton, A., Evans, D.B., and Lockwood, C.M. (1990*b*). Sequential anti-neutrophil cytoplasm antibody titres in the management of systemic vasculitis. *Nephrol. Dial. Transplant.* **4**, 309.

Leib, E.S., Restivo, C., and Paulus, H.E. (1979). Immunosuppressive and corticosteroid therapy of polyarteritis nodosa. *Am. J. Med.* **67**, 941.

Llewelyn, M.B., Wraight, P., and Lockwood, C.M. (1990). Indium leucocyte imaging in systemic vasculitis. *J. Am. Soc. Nephrol.* **1**, (4), 563.

Lockwood, C.M., Rees, A.J., Pearson, T.A., Evans, D.J., and Peters, D.K. (1976). Immunosuppression and plasma exchange in the treatment of Goodpasture's syndrome. *Lancet*, **i**, 711.

Lockwood, C.M., Pinching, A.J., Sweny, P., and Peters, D.K. (1977). Plasma exchange and immunosuppression in the treatment of fulminating immune-complex nephritis. *Lancet*, **i**, 63.

Mathieson, P.W., Cobbold, S.P., Hale, G., Clark, M.J., Oliveira, D.B.G., Lockwood, C.M., and Waldmann, H. (1990). Monoclonal antibody therapy in systemic vasculitis. *N. Engl. J. Med.* **323**, 250.

McIlvanie, S.K. (1966). Wegener's granulomatosis: successful treatment with chlorambucil. *J.A.M.A.* **197**, 139.

Palmer, A., Severn, A., Gjorstrup, P., Parsons, V., Welsh, K., and Taube, D. (1988). Successful treatment of rapidly progressive glomerulonephritis (RPGN) with extracorporeal immunoabsorption (IA). *Nephrol. Dial. Transplant.* **2**, 891.

Pusey, C.D., Rees, A.J., Evans, D.J., Peters, D.K., and Lockwood, C.M. A randomised controlled trial of plasma exchange in rapidly progressive glomerulonephritis without anti-GBM antibodies. *Kidney Int.* (In press).

Rossi, F., Jayne, D.R.W., Lockwood, C.M., and Kazatchkine, M.D. Anti-idiotypes to anti-neutrophil cytoplasm antibodies are present in pooled immunoglobulin and post-recovery sera. *Clin. Exp. Immunol.* (In press).

Schroeder, J.O., Euler, H.H., and Löffler, H. (1987). Synchronization of plasmapheresis and pulse cyclophosphamide in severe systemic lupus erythematosus. *Ann. Intern. Med.* **107**, 344.

Van Der Woude, F.J., Rasmussen, N., Lobatto, S., Wiik, A., Permin, H., Van Eś, L.A., Van Der Giessen, M., Van Der Hem, G.K., and The, T.H. (1985). Autoantibodies against neutrophils and monocytes; tool for diagnosis and marker of disease activity in Wegener's granulomatosis. *Lancet*, **I**, 425.

Van Der Woude, F.J., Daha, M.R., and Van Eś, L.A. (1989). The current status of neutrophil cytoplasmic antibodies. *Clin. Exp. Immunol.* **78**, 143.

Vandenbark, A.A., Hashim, G., and Offner, H. (1989). Immunization with a synthetic T-cell receptor V-region peptide protects against experimental autoimmune encephalomyelitis. *Nature*, **341**, 541.

Waldmann, H. (1989). Manipulation of T-cell responses with monoclonal antibodies. *Ann. Rev. Immunol.* **7**, 407.

Wilson, C.B. and Dixon, F.J. (1973). Anti-glomerular basement membrane antibody-induced glomerulonephritis. *Kidney Int.* **3**, 74.

Part 7 Acute Renal Failure

25 Pathophysiology of acute tubular necrosis

P.J. Ratcliffe

Acute renal failure may arise from a variety of pathologies, including glomerulonephritis, interstitial nephritis, and arterial, venous and urinary obstruction. In hospital, however, the majority arise from a different entity, which is most commonly called acute tubular necrosis. The term acute tubular necrosis refers to a type of acute renal failure which occurs in seriously ill patients, usually when there has been a severe haemodynamic disturbance or exposure of the kidney to toxic substances. Since the pathogenesis is imperfectly understood, the justification for regarding acute tubular necrosis as a single entity is clinical; the syndrome has recognizable clinical features, loss of excretory function with urinary abnormalities suggestive of tubular dysfunction, it complicates specific clinical settings, and has a definable prognosis—recovery within days or weeks of removal of the initiating cause. Confusion is generated by the name itself, since it poorly reflects the clinical pathology; necrosis of tubular cells is not usually extensive, and may not be present at all. Injury to the tubules is a consistent feature, but this may not be discernible morphologically.

Much of what is understood about the cause and pathophysiology of acute tubular necrosis has been deduced from experimental studies in animal models, and it is important to be aware of the uncertainties in this method. In most of these models a form of acute renal failure resembling acute tubular necrosis can reliably be produced by complete renal ischaemia, or by exposure to nephrotoxins, such as heavy metal salts. It is presumed that factors identified as associated with acute renal failure in clinical practice cause renal damage by similar mechanisms, although at the current level of knowledge they might more accurately be termed risk factors than causes. Thus, haemodynamic risk factors are presumed to cause renal damage by ischaemia. Factors associated with pre-renal failure are also associated with acute tubular necrosis, and it is probable, but unproven, that this reflects greater reduction in renal perfusion, causing ischaemic damage. Direct nephrotoxicity is often presumed when no haemodynamic disturbance seems likely. For some factors, such as heavy metals, organic solvents, and polyene antibiotics, a consistent association with renal damage is observed in humans and can be reproduced experimentally, so that the status of nephrotoxin is clear. For many more commonly implicated risk factors the clinical association is less consistent and more difficult to reproduce experimentally.

This chapter will consider principally the pathogenesis of haemodynamically mediated acute tubular necrosis in which renal ischaemia is believed to be the injurious agent. Two questions will be considered: first, how does injury arise, and secondly how does injury compromise function?

The mechanism of ischaemic injury

When organ dysfunction follows circulatory shock, the kidney is often compromised most severely. Nevertheless, renal susceptibility to ischaemia during hypoperfusion appears paradoxical, since under normal physiological conditions renal blood flow is high, so that oxygen delivery greatly exceeds consumption, and a sophisticated control system for defence of renal perfusion exists (Brezis *et al.* 1984*a*).

There are three possible explanations for this apparent paradox. First, it could be that ischaemia is not the mode of injury; secondly, oxygen delivery might be inhomogenous, so that despite adequate overall oxygen delivery certain susceptible areas are at risk of ischaemic damage; and thirdly, during shock the kidney might be at risk of a disproportionately severe reduction in total renal blood flow.

The classical microdissection work of Oliver *et al.* (1951) distinguished ischaemic from non-ischaemic injury, on the basis that ischaemically injured cells were not confined to any one nephron segment but occurred sporadically throughout the nephron and were associated with rupture of the tubular basement membrane, termed tubullorhexis. However, this distinction has not been widely confirmed and there is currently no specific clinical or histopathological hallmark to distinguish damaging renal cellular ischaemia.

The following discussion of mechanisms by which haemodynamic compromise might lead to ischaemic cell injury is therefore given in the knowledge there is no absolute proof that ischaemia is the mode of renal injury in haemodynamically compromised patients. Nevertheless, in many cases, no factor other than ischaemia appears likely to account for injury and the presumption that renal ischaemia occurs is by far the most likely hypothesis.

Inhomogenous renal oxygenation

The possibility that inhomogenous renal oxygenation might leave areas at risk of injury from only a small reduction of perfusion has been argued strongly by Epstein and colleagues (Brezis *et al.* 1984*a*). The existence of profoundly hypoxic regions within the tissue of normal kidneys was first postulated in explanation for the low pO_2 of intrapelvic urine and has been confirmed by direct measurement using microelectrodes (Leichtweiss *et al.* 1969). In the renal cortical tissue, pO_2 varies widely between sites of close proximity, with many values falling well below those of the renal venous blood. In the renal medulla, tissue pO_2 is even lower (Leichtweiss *et al.* 1969). Theoretically, inefficient delivery of oxygen could arise from arterio-venous shunting of whole blood, separation of blood from plasma (plasma skimming) with the more rapid transit of the oxygen-carrying red cells, or from direct diffusion of oxygen from arterial to venous vessels. However, strong support for shunt diffusion as the major contribution to inhomogenous renal oxygenation is provided by the appearance of oxygen ahead of labelled red cells in the renal venous blood after bolus renal arterial injection, the occurrence of inhomogenous oxygen delivery in blood-free perfused kidney, and the absence of evidence for direct arterio-venous shunting in kidney, or for plasma skimming.

Evidence of inhomogenous renal oxygen delivery arising from shunt diffusion is therefore convincing, but the extent to which it contributes to renal damage, particularly in the clinical setting, is still unclear. Convincing evidence that regional hypoxia may lead to zonal damage in advance of severely limited oxygen delivery has been obtained in the blood-free isolated perfused rat kidney (Brezis *et al.* 1984*a*), but this model differs in several important ways from hypoperfusion *in vivo*. First, in the blood-free perfused kidney, oxygen is carried at high pO_2 in free solution, conditions which enhance shunt diffusion. Because of the oxygen binding characteristics of haemoglobin, physiological oxygen delivery via oxyhaemoglobin is likely to produce quite a different pattern of intrarenal oxygenation. The addition of erythrocytes to the perfusate alters the site of injury in isolated perfused kidneys (Endre *et al.* 1989) with deep proximal tubules, rather than the medullary thick ascending limbs becoming most susceptable to injury.

Secondly, in the isolated perfused kidney glomerular filtration and tubular transport continue despite hypoxia and these continuing high rates of transport work are important in the pathogenesis of the hypoxic injury which develops in the medullary thick ascending limbs of the loops of Henle (Brezis *et al.* 1984*b*). When renal perfusion pressure is reduced by haemorrhage, glomerular filtration, which will dictate the energy demand of reabsorption, may be reduced early so that this severe imbalance does not occur in the same way (Ratcliffe *et al.* 1989). Nevertheless, in a different model where unilateral nephrectomy, sodium depletion, indomethacin and radiographic contrast media were used together to produce renal injury, isolated injury to medullary thick ascending limbs has been reported *in vivo* (Heyman *et al.* 1988).

One clinical point deserves emphasis in this argument. If only a small reduction in total renal oxygen delivery were required to damage a susceptible zone, then one might expect renal damage to be a predictable consequence of reduced oxygen delivery by hypoxia, anaemia or hypoperfusion. Clearly, this is not the case. Indeed, the progression from pre-renal failure to established acute renal failure is notoriously unpredictable and acute tubular necrosis complicates certain clinical settings, such as sepsis, much more frequently than others, such as gastrointestinal haemorrhage. Other factors, possibly involving the breakdown of protective circulatory control mechanisms, are presumably involved.

Reduction in overall renal perfusion

The possibility that, in shock, the kidney suffers a disproportionately severe reduction in blood supply was suggested by Trueta *et al.* (1947) on the basis of direct and radiographic observations of vessel calibre. Marked, but rather unpredictable, constriction of the renal vessels, but not mesenteric vessels, could be provoked in rabbits by hind limb trauma. The cortical and medullary circulations appeared to behave differently, and it was proposed that blood flow was diverted from cortical to juxtamedullary glomeruli, whose effluent perfused the medullary capillaries. Subsequently, there have been many attempts to measure regional renal blood flow in acute renal failure. Methodological doubts still surround these studies (Aukland 1980), but there is agreement that in established acute tubular necrosis total renal blood flow is reduced only to 30–50 per cent of normal (Hollenberg *et al.* 1968) and in some circumstances can be increased to a normal level without resumption of renal function. Thus, it has been argued that reduced renal blood flow *per se* is not responsible for persistent renal failure in established acute tubular necrosis.

Unfortunately, some confusion has surrounded this conclusion and it should be made clear that the events responsible for initiation of ischaemic damage may be quite different from those involved in maintenance of renal failure. It remains possible that a very severe, but possibly quite transient, reduction in renal blood flow is responsible for the induction of ischaemic damage, a possibility supported by angiographic studies in humans. Such an event would presumably involve breakdown of the complex control systems outlined previously.

Unfortunately, at present there is no simple means of detecting the onset of renal ischaemia. Therefore, although it seems likely that renal ischaemia is responsible for the occurrence of acute tubular necrosis in many situations this likelihood, along with the exact reasons for its occurrence, cannot be proven. Techniques which permit recurrent measurement of renal blood flow, such as Doppler ultrasonography and thermodilution measurement of flow in the renal veins may soon answer the central question as to whether or not a profound reduction in total blood flow is necessary to initiate injury.

Septic shock

The reason for the increased risk of renal damage in sepsis is unknown, although it seems clear that vascular injury compounds the effects of reduced perfusion pressure in producing organ injury. Renal failure appears to occur predominantly in two clinical settings: sudden dramatic shock from bacterial septicaemia and the progressive failure of multiple organs in patients on the intensive care unit. In the latter, increasing difficulty in supporting the circulation and failure of lungs, kidney and the gastrointestinal system are reminiscent of the pattern of illness in bacterial septicaemia, but progression is less rapid and organisms are rarely cultured from blood.

Recently, some aspects of the pathogenesis of septic shock have been clarified. It has long been known that the features of bacterial septicaemia can be reproduced by the injection of endotoxin, a lipopolysaccharide component of many Gram-negative and some Gram-positive organisms (Thomas 1954), and it is possible that progressive multi-organ failure is mediated by bacterial endotoxin, leaking from the bowel. Immunotherapy directed against lipopolysaccharide has been reported to improve survival from septic shock but has not yet been used widely. Endotoxin itself is not directly toxic to most tissue and it appears that endogenous cytokines rather than lipopolysaccharide itself mediate endotoxic injury. Interest has surrounded the demonstration that many of the features of septic shock can be reproduced by the injection of one such substance, tumour necrosis factor (TNF).

The production by recombinant DNA technology of pure TNF was followed by the demonstration that intravenous injection of pathophysiologically appropriate quantities reproduced the effects of injection of endotoxin, and that monoclonal antibodies directed against TNF are protective against injection of live *E. coli* (Tracey *et al.* 1987). The organ pathology observed, pulmonary inflammation and haemorrhage, ischaemic and haemorrhagic lesions of the gut, and acute tubular necrosis clearly resembles that of human endotoxic shock, but the reasons for organ susceptibility are unknown. TNF acts directly on endothelial cells. It can alter their haemostatic properties, leading to disseminated intravascular coagulation, and can alter antigenic expression, leading to adherence of blood cells. It also releases other cytokines; platelet activating factor, leukotrienes and interleukin-1 have been implicated in the pathogenesis of endotoxic shock and probably form part of a cascade following release of TNF (Beutler and Cerami 1987). Both primary and secondary injury to other organs, such as myocardial depression and massive release of catecholamines, may contribute to renal injury.

How does injury comprise function?

Continuing controversy surrounds the mechanism by which the major manifestation of diseases principally affecting the renal tubules is an apparent failure of glomerular filtration.

The principal mechanisms proposed to explain the excretory failure in acute tubular necrosis are a persistent alteration of renal blood flow, an alteration of glomerular capillary ultrafiltration coefficient, obstruction of the renal tubules and back-leakage of filtration from damaged tubules. All these mechanisms interact, and it is likely that several factors contribute to the excretory failure.

Tubular mechanisms: back-leakage of filtrate and obstruction

The possibility that in a damaged kidney glomerular filtration might persist, but be disguised by the backleakage of filtrate was suggested by Richards (1929), who observed the passage of dye into the proximal tubule of frog kidneys which had been damaged by mercuric chloride and passed no urine. Oliver *et al.* (1951), noting disruption of the tubular epithelium (tubullorhexis), considered back-leakage of filtrate to be likely in human acute renal failure. Persisting glomerular filtration in human oliguric acute tubular necrosis has also been suggested by appearances during intravenous urography, in which a dense nephrogram appears early and persists; appearances interpreted by Fry and Cattell (1972) as indicating filtration of contrast with leakage into the parenchyma.

Using micropuncture techniques, back-leakage of filtrate can be directly demonstrated by loss of filtration markers following micro-injection into the renal tubules. Direct observation of the leakage of dyes, such as lissamine green (Arendshorst *et al.* 1975), microscopic evidence of leakage of macromolecules, such as horseradish peroxidase (mw 40 000), and recovery of microinjected radiolabelled inulin from the contralateral kidney or non-recovery from the ipsilateral kidney (Donohoe *et al.* 1978), all provide evidence that back-leakage occurs. The extent of recovery of injected marker provides a guide to the severity of leakage, but since filtered inulin might behave differently from microinjected material, these techniques cannot determine the precise extent of persisting filtration and back-leakage.

Back-leakage of filtrate has been demonstrated in some, but not all, models of both ischaemic and nephrotoxic acute tubular necrosis, and is associated with severe structural injury in the renal tubule. In ischaemia, correlation with the severity of structural damage is illustrated by the findings of Donohoe *et al.* (1978), who studied rat kidneys after relief of total ischaemia ranging from 15–60 minutes in duration. Contralateral excretion of microinjected inulin (the minimum estimate of back-leakage) was 35 per cent after 60 minutes' ischaemia, when structural damage was severe, but only 11 per cent after 25 minutes, and negligible after 15 minutes, when structural damage was less striking.

Thus, in animal models, back-leakage of filtrate requires severe tubular damage, though not necessarily the disruption of the tubular basement membrane suggested by Oliver *et al.* (1951).

Back-leakage cannot account for reduced inulin clearance in milder models of acute tubular necrosis, and does not account for the reduced inulin clearance during the later stages of recovery from severe tubular injury (Finn and Chevalier 1979). In human acute tubular necrosis, tubular damage is believed to be mild, but Myers and colleagues (Myers and Moran 1986) have provided evidence of significant back-leakage of inulin. Measurements of the fractional urinary clearance of dextran molecules of graded size revealed that fractional clearance (relative to inulin) was increased in patients with protracted acute tubular necrosis, and that clearances of those molecules slightly larger than inulin actually exceeded inulin clearance. This phenomenon cannot arise from glomerular haemodynamics and is most easily explained by back-leakage of filtrate with differential rates of leakage dependent on molecular size, thus strongly supporting the observations of Fry and Cattell.

Tubular obstruction

The observation of pigmented casts in the tubules of kidney of patients dying from crush syndrome lead Bywaters and Beall (1941) to propose that tubular obstruction contributed to renal failure. Simplistically, obstruction of renal tubular lumina by precipitated protein or cellular debris arising from damaged tubular epithelium should be visible morphologically, but since a tubule may be obstructed along a variable length the number of obstructing lesions in a given section will be a poor guide to the number of tubules obstructed. Tubular obstruction may also occur as a consequence of cell swelling with competition for space, or from other hydrodynamic abnormalities which might be difficult or impossible to detect in fixed tissue.

In addition, luminal debris may accumulate when flow is slow in unobstructed tubules. Thus, although morphological appearances may on occasion be highly suggestive of obstruction they can ultimately neither confirm nor refute its existence.

Intratubular pressures can be measured directly by micropuncture, but the interpretation is complex. The pressure attained proximal to a tubular obstruction will clearly be influenced by back-leakage of intratubular urine and by the net glomerular ultrafiltration pressure. Obstruction to urinary flow is followed after a period of time by alteration in glomerular haemodynamics which reduce this filtration force. Therefore, whilst the demonstration of high intratubular pressure at normal flow rates will confirm the presence of obstruction, normal or even low pressures do not entirely exclude it.

In no model can uniformly raised intratubular pressures be demonstrated, but in models involving total ischaemia a consistent pattern is seen hours after relief of ischaemia in which the majority of intratubular pressures are markedly raised, but the scatter is very wide with some intratubular pressures lying below the normal value (Finn and Chevalier 1979). The raised pressures are not sustained but decline over a period of days, most probably due to adaptive changes in glomerular haemodynamics.

Experimental evidence supporting this possibility was obtained by Arendshorst *et al.* (1975), who volume expanded animals and noted a return of intratubular pressures to high levels. However, in their study of recovery from renal ischaemia, Finn and Chevalier (1979) noted that after 1–2 weeks intratubular pressures fall to normal even in volume-expanded animals, this change being associated with the disappearance of intratubular debris.

Significant renal failure continued, despite the evidence of relief of tubular obstruction.

Haemodynamic factors

Failure to account completely for renal failure by the mechanisms outlined previously led to the proposal that a persisting alteration in renal perfusion could reduce the glomerular filtration rate, and has led some authors to use the term vasomotor nephropathy (Oken 1984).

Renal blood flow is reduced in human acute tubular necrosis, and in most animal models, but in all cases the reduction in glomerular filtration rate is much greater, so that filtration fraction is greatly reduced. Renal blood flow remains low after relief of total ischaemia in animal models but then increases and may reach normal levels either spontaneously or after pharmacological manoeuvres. Renal blood flow is also reduced initially in most models of nephrotoxic acute renal failure, but this is not a universal occurrence and renal blood flow usually recovers spontaneously to near normal levels. These increases in blood flow are not paralleled by an increased glomerular filtration rate.

Clearly, the reduced perfusion alone cannot account for reduced filtration. Furthermore, a major redistribution of blood flow through a glomerular bypass cannot be sustained either anatomically or from attempted measurements of regional renal blood flow. However, two haemodynamic alterations could very severely reduce filtration; severe reduction in glomerular capillary ultrafiltration pressure by afferent vasoconstriction and efferent vasodilation, or reduction in the glomerular ultrafiltration coefficient, Kf. The majority of measurements in experimental animal models of acute renal failure have demonstrated reduction in Kf (Williams *et al.* 1981; Baylis *et al.* 1977).

Kf cannot be measured in humans, but the failure of glomerular filtration rate to respond to renal vasodilation would be consistent with a reduction in Kf. On the other hand, Oken has argued from the radiographic appearance in humans that increased preglomerular resistance could reduce glomerular capillary hydraulic pressure (P_{GC}) to very low levels, and that no other mechanism is theoretically necessary to account for the very low filtration fraction. Radiographic appearances would suggest that the site of the increased resistance was the interlobar artery, rather than the afferent arteriole, and direct observation of the interlobular artery has demonstrated that a number of substances act at that site (Steinhausen *et al.* 1988). There is, of course, no reason why preglomerular vasoconstriction and reduced Kf should not act together to reduce the glomerular filtration rate.

How might such haemodynamic changes arise and why do they persist? There are many vasoactive influences active at the time of haemodynamic compromise, such as catecholamines, renal nerves, adenosine, renin and the angiotensin system, which are capable of altering Kf or arteriolar resistances. However, the duration of renal failure greatly outlasts any overt evidence of activation, and pharmacological intervention in the maintenance phase is of little or no benefit.

Continuing debate has surrounded the possibility that the haemodynamic alteration is secondary to the tubular damage. Poorly understood haemodynamic responses to tubular obstruction have been described. A second possibility is of persistent activation of the tubulo-glomerular feedback mechanism.

Thurau and Boylan (1976) argued that the reduced transport capacity of the damaged proximal nephron would lead to activation of tubulo-glomerular feedback and reduction in glomerular filtration. Several points, however, seem to be at odds with this hypothesis. Although tubular transport is clearly impaired, at the reduced intratubular flow rates which occur in acute tubular necrosis, it is not clear that the physiological signal (raised NaCl concentration at the macula densa) actually occurs. Particularly, frusemide which is a powerful inhibitor of tubulo-glomerular feedback has little or no effect on the glomerular filtration rate in the maintenance phase of acute tubular necrosis in animal models and humans. However, neither of these objections could be sustained if it were proposed that the signal arose from pathological feedback control rather than physiological control, i.e. that direct damage to the transporting epithelium at or near the macula densa was responsible for pathological activation of the signal.

Conclusions

To summarize the pathophysiological mechanisms accounting for loss of excretory function in damaged kidneys is clearly difficult. First, it is apparent from the experimental work in animals that no single one of the mechanisms proposed can account for renal failure. Different mechanisms operate in different models, at different times. Species and dosage add further complexity. Some generalizations are possible. When tubular injury is severe in the early maintenance phase after ischaemia or high doses of toxins, tubular obstruction and/or leakiness can usually account for renal failure, but almost certainly cannot account for renal failure in milder models or for the renal failure in later stages of recovery from severe injury.

In humans it is usually held that tubular injury is milder than in animal models, even allowing for the much greater difficulty in histological assessment of human material. It might therefore be argued that these factors are unlikely to be important in humans, but there is one important proviso; the hydrodynamic conditions for obstruction or leakage in relation to the amount of cell damage might be different in humans. This is not at all unlikely, since the human nephron is obviously very much longer than the rat nephron, and in fact very convincing evidence that back-leakage of filtrate occurs in humans has been provided by Myers (Myers and Moran 1986).

Nevertheless, an important role for altered glomerular haemodynamics is proven in animal models and appears very likely to play a major role in the human syndrome, particularly in the milder non-oliguric cases, but none of the known mediators of renal circulatory control currently seems able to account for the time course, severity, or resistance to pharmacological intervention observed.

References

Arendshorst, W.J., Finn, W.F., and Gottschalk, C.W. (1975). Pathogenesis of acute renal failure following temporary renal ischaemia in the rat. *Circulation Research*, **37**, 558–68.

Aukland, K. (1980). Methods for measuring renal blood low: total flow and regional distribution. *Annual Review of Physiology*, **42**, 543–55.

Baylis, C., Rennke, H.R., and Brenner, B.M. (1977). Mechanisms of the defect in glomerular ultrafiltration associated with gentamicin administration. *Kidney International*, **12**, 344–53.

Beutler, B. and Cerami, A. (1987). Cachectin: more than a tumor necrosis factor. *New England Journal of Medicine*, **316**, 379–85.

Brezis, M., Rosen, S., Silva, P., and Epstein, F.H. (1984*a*). Renal ischaemia: a new perspective. *Kidney International*, **26**, 375–83.

Brezis, M., Rosen, S., Silva, P., and Epstein, F.H. (1984*b*). Transport activity modifies thick ascending limb damage in the isolated perfused kidney. *Kidney International*, **25**, 65–72.

Bywaters, E.G.L. and Beall, D. (1941). Crush injuries with impairment of renal function. *British Medical Journal*, **1**, 427–32.

Donohoe, J.F., Venkatachalam, M.A., Bernard, D.B., and Levinsky, N.G. (1978). Tubular leakage and obstruction after renal ischaemia: Structural-functional correlations. *Kidney International*, **13**, 208–22.

Endre, Z.H., Ratcliffe, P.J., Tange, J.D., Ferguson, D.J.P., Radda, G.K., and Ledingham, J.G.G. (1989). Erythrocytes alter the pattern of renal hypoxic injury: predominance of proximal tubular injury with moderate hypoxia. *Clinical Science*, **76**, 19–29.

Finn, W.F. and Chevalier, R.L. (1979). Recovery from postischemic acute renal failure in the rat. *Kidney International*, **16**, 113–23.

Fry, I.K. and Cattell, W.R. (1972). The nephrographic pattern during excretion urography. *British Medical Bulletin*, **28**, 227–32.

Heyman, S.N., Brezis, M., Reubinoff, C.A., Greenfield, Z., Lechene, C., Epstein, F.H., and Rosen, S. (1988). Acute renal failure with selective medullary injury in the rat. *Journal of Clinical Investigation*, **82**, 401–12.

Hollenberg, N.K., Epstein, M., Rosen, S.M., Basch, R.I., Oken, D.E., and Merrill, J.P. (1968). Acute oliguric renal failure in man: evidence for preferential renal cortical ischemia. *Medicine*, **47**, 455–74.

Leichtweiss, H-P., Lubbers, D.W., Weiss, Ch., Baumgartl, H., and Reschke, W. (1969). The oxygen supply of the rat kidney: measurements of intrarenal pO2. *Pflugers Archives*, **309**, 328–49.

Myers, B.D. and Moran, S. (1986). Hemodynamically mediated acute renal failure. *New England Journal of Medicine*, **314**, 97–105.

Oken, D.E. (1984). Hemodynamic basis for human acute renal failure (vasomotor nephropathy). *American Journal of Medicine*, **76**, 702–10.

Oliver, J., MacDowell, M., and Tracy, A. (1951). The pathogenesis of acute renal failure associated with traumatic and toxic injury. Renal ischaemia, nephrotoxic damage and the ischemuric episode. *Journal of Clinical Investigation*, **30**, 1307–51.

Richards, A.N. (1929). Direct observations of change in function of the renal tubule caused by certain poisons. *Transactions of the Association of American Physicians*, **44**, 64–7.

Steinhausen, M., Holz, F.G., and Parekh, L. (1988). Regulation of pre and post glomerular resistances visualized in the split hydronephrotic kidney. *Proceedings of the Xth International Congress of Nephrology* (Vol. I), pp. 37–45, Baillière Tindall, London.

Thomas, L. (1954). The physiological disturbances produced by endotoxins. *Annual Review of Physiology*, **16**, 467–90.

Thurau, K. and Boylan, J.W. (1976). Acute renal success. The unexpected logic of oliguria in acute renal failure. *American Journal of Medicine*, **61**, 308–15.

Tracey, K.J. *et al.* (1987). Anti-cachectin/TNF monoclonal antibodies prevent septic shock during lethal bacteraemia. *Nature*, **330**, 662–4.

Trueta, J., Barclay, A.E., Daniel, P.M., Franklin, K.J., and Prichard, M.M.L. (1947). *Studies of the renal circulation*. Blackwell Scientific, Oxford.

Williams, R.H., Thomas, C.E., Navar, G., and Evans, A.P. (1981). Hemodynamic and single nephron function during the maintenance phase of ischemic acute renal failure in the dog. *Kidney International*, **19**, 503–15.

26 Urinary tract infection and acute renal failure

W.R. Cattell

Standard textbooks or review articles rarely mention non-obstructive urinary tract infection (UTI) among the causes of acute renal failure. This may be because it is indeed rare, or, as suggested by Eknoyan (1988), it is overlooked.

A review of the literature is confusing. While there are scattered references to acute renal failure due to non-obstructive pyelonephritis, there are in addition many more references to urinary tract sepsis as a major causal factor for acute renal failure. It is unclear whether these refer to septicaemia due to UTI, obstructive pyonephrosis or non-obstructive pyelonephritis. Acute renal failure associated with UTI in pregnant women is well documented and will not be discussed further (Krane 1988).

Given this confusion, this chapter reviews the possible and reported mechanisms for the development of acute renal failure in association with UTI. Diagnosis, management and prognosis are discussed. The possible mechanisms for acute renal failure are shown diagramatically in Fig. 26.1.

UTI and obstructive uropathy

This is the most clearly recognized cause for acute renal failure associated with UTI and is not uncommon. There are two possible mechanisms. First, ascending infection superimposed on pre-existing subacute obstruction can, because of the inflammatory reaction, result in complete obstruction and an obstructed pyonephrosis. Where there is a single functioning kidney, this results in acute oliguric renal failure. A second mechanism may, however,

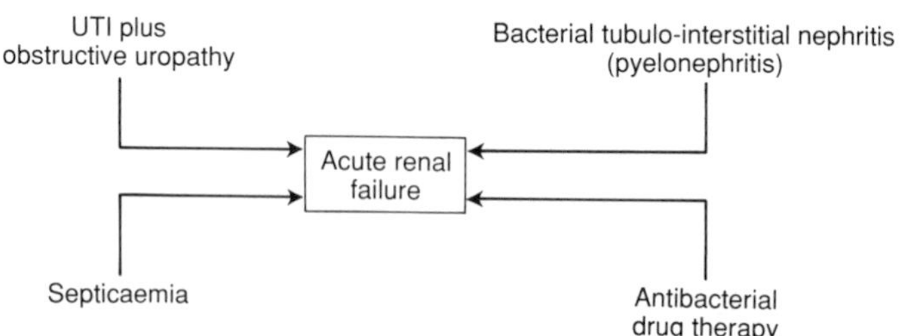

Fig. 26.1 Mechanisms of acute renal failure associated with urinary infection.

operate when a unilateral pyonephrosis is complicated by bacteraemia. Septicaemia is not uncommon in this situation and may be associated with 'septic shock'. This, in turn, may result in acute tubular necrosis or vaso-motor nephropathy in the contralateral non-obstructed kidney, resulting in oliguric renal failure.

The computerized tomography scan in a young man who presented desperately ill with acute oliguric renal failure is shown in Fig. 26.2. The scan shows an obstructed pyonephrosis due to infection and stone disease. A perinephric abscess is also present. The contralateral kidney was slightly enlarged but otherwise normal. It did not have the computerized tomography appearance of severe bacterial tubulo-interstitial nephritis (Cattell *et al.* 1989). The abscess and the pyonephrosis were dealt with by free drainage and aggressive antibacterial treatment, but the patient remained oliguric and dialysis-dependent for 10 days. Diuresis then occurred, with rapid improvement in kidney function.

Without renal biopsy evidence, it may be argued that there is no proof of acute tubular necrosis. Biopsy of the better kidney in these circumstances is not justified. The pattern of recovery is highly suggestive of a vaso-motor nephropathy.

Patients with an obstructive pyonephrosis are usually febrile, ill and complain of loin pain and tenderness. There may or may not be a preceding history

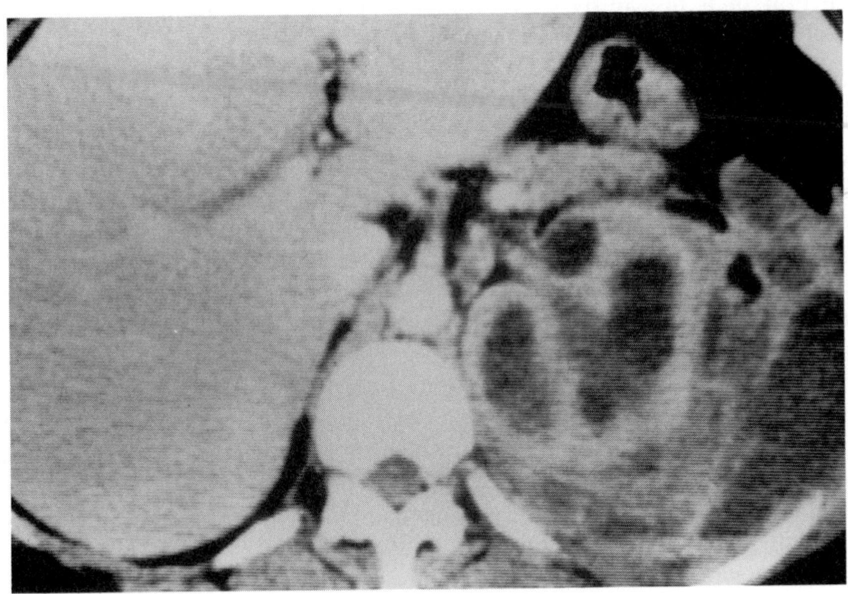

Fig. 26.2 Abdominal computerized tomography scan showing obstructed pyelonephrosis of left kidney, with perinephric abscess.

of frequency and dysuria. Whenever suspected, the condition must be treated as a surgical emergency, since delay in diagnosis and appropriate treatment increases the likelihood of irreversible kidney damage and septicaemia.

Diagnosis is based on the demonstration of bacteriuria plus upper tract dilatation. The advent of ultrasonography has simplified both diagnosis and treatment. Imaging can be carried out quickly and non-invasively. If the typical features of an obstructed infected system are found (Cattell *et al.* 1989) percutaneous drainage should be established immediately. Urine (or more often pus) must be sent for culture and antibiotic sensitivity testing. Blood cultures should also be obtained. Drainage is best covered by an initial loading dose of intravenous or intramuscular gentamicin.

Infection plus obstruction results in severe diffuse bacterial tubulo-interstitial nephritis. Even with the establishment of free drainage, eradication of infection commonly requires prolonged (6 weeks') antibacterial therapy, with an antibiotic which achieves effective tissue concentrations. The choice of antibiotic should be based on bacterial sensitivity testing. If aminoglycosides are used, and especially if there is renal impairment, trough and peak blood levels must be measured. Management of the associated acute renal failure with respect to fluid balance, diet, and dialysis, is as for acute renal failure in general. Careful follow-up with post-treatment urine cultures is essential to ensure eradication of infection. Surgery may be required for removal of stones or treatment of pelvi-ureteric junction obstruction. Full recovery of function in a previously obstructed and infected kidney is unusual, and papillary necrosis is a not uncommon residuum.

Septicaemia

Septicaemia (Fig. 26.1) is often quoted as a cause of acute tubular necrosis, and septicaemia secondary to UTI is not uncommon. Most often associated with obstructive uropathy, it may also occur with non-obstructive pyelonephritis or bacterial prostatitis.

Rayner *et al.* (1990) have recently reported a study of 209 patients with community acquired bacteraemia. In 58 patients (24 per cent) bacteraemia was secondary to UTI. Renal impairment was recorded in 31 per cent of these 58, but only 9 (15 per cent) required dialysis. No details are given as to whether the UTI was associated with stones or obstruction. Similarly, whilst for inclusion all had bacteraemia, no details are given as to whether septic shock had been observed. Bacteraemia secondary to hospital-acquired UTI is more common (Siroky 1976) and bacteraemia complicating UTI increases with age, especially among men (Gleckman *et al.* 1982).

The mechanisms for acute renal failure in association with septicaemia are probably multifactorial (Fig. 26.3). Severe hypotension is a common feature of septicaemia, and may result in severe renal under-perfusion and ischaemic renal damage—acute tubular necrosis or vaso-motor nephropathy. Wardle

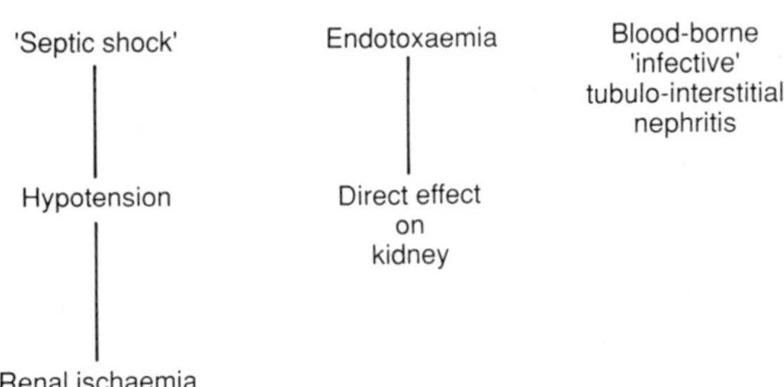

Fig. 26.3 Mechanisms of acute renal failure associated with septicaemia.

(1982) has argued persuasively that endotoxaemia itself may result in acute vaso-motor nephropathy and may be the dominant mechanism. Finally, Mayaud *et al.* (1977) have described acute renal failure attributable to severe blood-borne bacterial interstitial nephritis in patients with UTI. It is not clear from their paper, how the distinction was made between ascending bacterial tubulo-interstitial nephritis and blood-borne infection.

Pursuing the arguments for and against septicaemic shock versus endotoxaemia, it is interesting to note that Rasmussen and Ibels (1982), analysing the risk factors for the development of acute renal failure in hospitalized patients, were unable to identify sepsis as a risk factor if this was dissociated from hypotension. These workers concluded that hypotension was the dominant factor. Conversely, the following case report rather favours the hypothesis of Wardle. A 15-year-old Afro-Caribbean girl was admitted to hospital with a 5 day history of dysuria, loin pain, fever and severe lassitude. On admission it was recorded that she was dehydrated, but her blood pressure was not remarkably low for her age, at 100/70 mmHg. She had marked right loin tenderness. She was febrile (38°C) and had a marked polymorpho-nuclear leucocytosis (white blood cell count 17.7×10^9/l). Blood urea was 17.3 mmol/l, serum creatinine 400 μmol/l and serum potassium 4.7 mmol/l. Urine microscopy showed marked pyuria and moderate haematuria. Urine culture yielded a significant growth of a coliform ($> 10^5$/ml).

She was treated with intravenous amoxycillin and intravenous fluids, but, because of oliguria, increasing uraemia and hyperkalaemia, she was transferred 3 days later. At the time of transfer she was ill, febrile (38°C), fluid overloaded with a blood pressure of 110/80 mmHg and oliguric. Blood urea was 29.2 mmol/l, serum creatinine 700 μmol/l and serum potassium 6.2 mmol/l. She was sickle cell negative and blood cultures were negative. Urine microscopy showed modest pyuria but many fragmented red cells. Culture was negative. Renal ultrasound showed normal-sized, non-obstructed kidneys with increased echogenicity and prominent pyramids.

Because of the absence of any clear cut evidence of hypotension, the persisting oliguria despite anti-bacterial treatment and the presence of fragmented red cells in the urine, a double diagnosis of UTI and glomerulonephritis was entertained and renal biopsy performed. This showed no evidence of glomerular or vascular disease (Fig. 26.4). There was moderate but unequivocal evidence of tubular necrosis, with a moderate interstitial inflammatory response.

On admission, she was given a single dose of intramuscular gentamicin (80 mg) and treatment with intravenous amoxycillin was continued, as the original coliform recovered from the urine was fully sensitive. She required haemodialysis to control hyperkalaemia on two occasions. Seventy-two hours after her renal biopsy she had an increasing diuresis and renal function rapidly improved. Treatment with amoxycillin was continued for 6 weeks in all. Six months after her admission to hospital she was completely well. EDTA clearance was 127 ml/min per 1.73 m^2 and her intravenous urogram was normal.

This young woman therefore had UTI without evidence of severe septic shock or of severe bacterial tubulo-interstitial nephritis. The case favours the Wardle (1982) hypothesis for acute vaso-motor nephropathy due to Gram-negative endotoxaemia.

The mechanisms postulated by Wardle are as follows (Fig. 26.5). Gram-negative septicaemia results in severe endotoxaemia. Endotoxin, and in particular lipid A, is a potent vasoconstrictor. It can also precipitate disseminated intravascular coagulation with the potential for microthrombi in the renal micro-circulation. Damage to endothelial cells results in loss of endothelial prostacyclin and its intra-renal vasodilator effect. Finally, endotoxaemia may precipitate the complement cascade. Although in the case quoted there was no evidence of disseminated intravascular coagulation or of capillary thrombi, vasocontrictor effects of endotoxin could well have precipitated her vaso-motor nephropathy.

The conclusion from this well-documented case is that absence of significant septic shock need not exclude acute vaso-motor nephropathy as a result of severe Gram-negative infection. The other important observation is the complete recovery of function, which indicates both a good prognosis and the absence of any pre-existing renal disease. Diagnosis requires renal biopsy, and treatment requires intensive anti-bacterial therapy plus conventional management for acute renal failure.

Anti-bacterial drug therapy

Acute renal failure in association with UTI always requires careful review of drug treatment (Fig. 26.1). In the analysis of risk factors in the development of acute renal failure already quoted (Rasmussen and Ibels 1982) a major factor was toxic levels of gentamicin in the blood. Aminoglycosides are effective antibacterial drugs in the treatment of UTI, but they can be nephrotoxic. Nephrotoxicity may be increased by the simultaneous prescription of thiazide

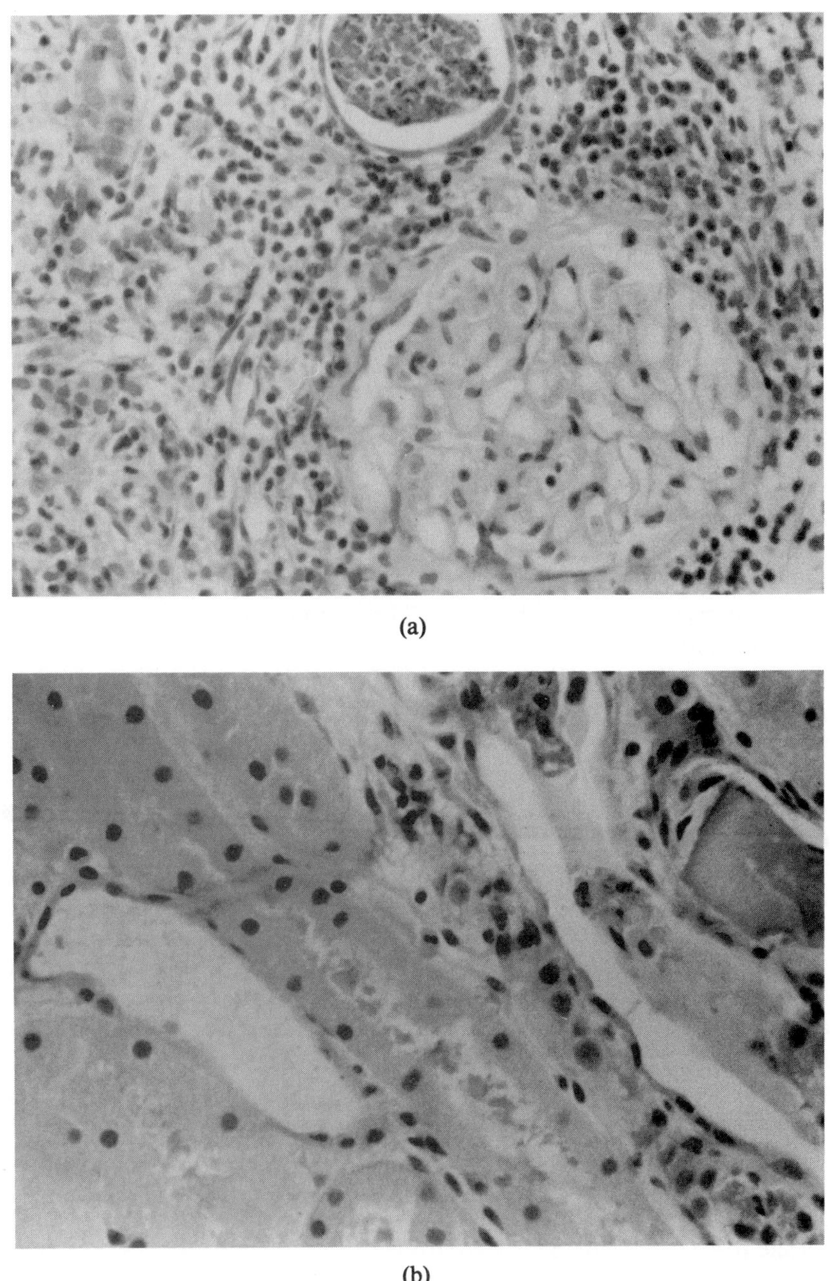

(a)

(b)

Fig. 26.4 Renal biopsy appearances in a case of acute renal failure associated with urinary tract infection, showing (a) normal glomerular and (b) vascular appearances, with tubular necrosis and interstitial inflammatory response.

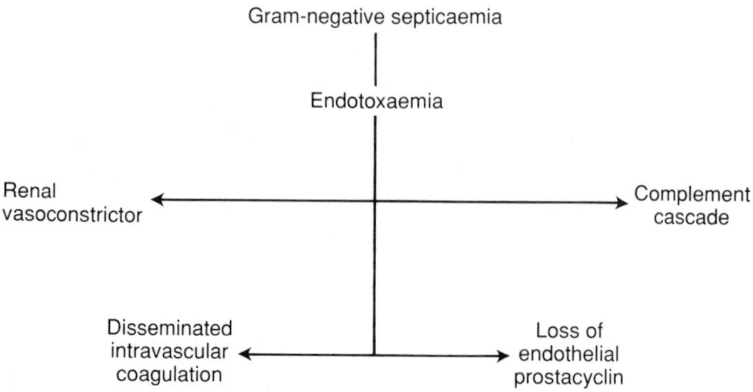

Fig. 26.5 Postulated mechanism of acute renal failure in Gram-negative endotoxaemia.

diuretics. It is not known how frequently nephrotoxicity occurs with conventional dose schedules in the absence of pre-existing renal impairment. It is wise, however, in patients with severe UTI to monitor blood levels if gentamicin is administered.

Apart from direct nephrotoxicity, antibacterial treatment may, rarely, be associated with a hypersensitivity tubulo-interstitial nephritis. This is most common with the penicillin family of drugs but has been reported following many antibiotics. Hypersensitivity to sulphonamides is less common now that they are used less frequently, but is still a risk with the combined antibacterial co-trimoxazole. The hypersensitivity reaction may develop within days of commencing treatment or may be delayed for up to 6 weeks. Usually associated with a skin rash, this may be minimal or absent. The renal failure is commonly non-oliguric.

Diagnosis depends on constant alertness to this possible cause. Urine culture is usually negative, depending on the appropriateness of the original treatment. Eosinophiluria is reputed to occur, but is in fact rare. Clinical suspicion demands renal biopsy. However, a distinction between infective tubulo-interstitial nephritis and drug hypersensitivity tubulo-interstitial nephritis may not be clear cut. In the former there is diffuse infiltration with polymorphonuclear cells with possible micro-abscess formation. In the latter, the cellular infiltrate is mainly of plasma cells and lymphocytes with the possible association of eosinophils.

While there is no satisfactory controlled trial, it is our practice not only to withdraw the offending drug, but also to prescribe steroid therapy. In the rare event of continuing bacteriuria or clinical evidence of infection, this must be combined with the use of a different antibacterial, to which the organism is sensitive.

Infective tubulo-interstitial nephritis (unobstructed pyelonephritis)

While these other mechanisms for acute renal failure in association with UTI are recognized and accepted, a major uncertainty is the frequency with which acute renal failure complicates non-obstructive pyelonephritis. There is further uncertainty as to whether this can occur in patients with previously normal kidneys.

Because of poor documentation, it is extremely difficult from review of the literature to define the frequency of acute renal failure in association with non-obstructive pyelonephritis. In 1979, Baker *et al.* reported 5 cases. Three of the 5 had consumed analgesics. These cases had been collected over a period of 7 years in a department treating approximately 40–50 patients with acute renal failure annually. This gives a rough incidence of 2–3 per cent of patients seen with acute renal failure.

In 1978, Richet and Mayaud reported 30 cases of acute renal failure with biopsy-proven acute tubulo-interstitial nephritis. On the basis of the histological finding of an acute polymorpho-nuclear cell infiltrate with micro-abscesses, they considered that 21 of the 30 had an infective tubulo-interstitial nephritis. In all 21 there was clinical evidence of sepsis and positive blood cultures were obtained in 16. In 11 of the 21 cases they stated that 'the source of infection was urological', but no further details are given. It is therefore not possible to identify how many of these patients had non-obstructive pyelonephritis, nor what the incidence of acute renal failure due to this cause was in their population of patients.

Recently, Lorentz *et al.* (1990) reported acute renal failure due to pyelonephritis in a 3-year-old child with a single kidney. However, this kidney had previously been shown to have vesico-ureteric reflux. Further poorly documented cases are reported with acute renal failure due to pyelonephritis in single kidneys.

In 1982, Bailey and Maling reported three women with acute renal failure associated with non-obstructive pyelonephritis. All were chronic alcoholics, a condition recorded as having a high prevalence of papillary necrosis (Edmondson *et al.* 1966). In 1985, Cattell *et al.* reported reversible renal failure in four diabetic patients with non-obstructive pyelonephritis, none of whom had diabetic glomerulopathy. There are also several scattered references to acute renal failure associated with non-obstructive pyelonephritis in patients taking non-steroidal anti-inflammatory drugs (NSAIDs) (Atkinson *et al.* 1986).

The largest review of acute renal failure covering a period of more than 20 years, is still being carried out by a group in Leeds, UK. As yet unpublished data (J. Turney, personal communication) show that out of a total of 1506 cases presenting with acute renal failure, and excluding pregnant women, only 20 patients with non-obstructive pyelonephritis could be identified. This is an incidence of 1.3 per cent. There were 3 diabetic patients among this group and 3 with evidence of previous tuberculosis. As previously found, it has proved difficult in this series to exclude other possible predisposing factors, such as analgesic abuse.

From the published reports it thus seems clear that non-obstructive pyelonephritis can indeed result in acute renal failure. However, it appears to be uncommon. There is little evidence that it ever occurs in patients with previously normal kidneys; rather, it appears to occur where pyelonephritis is superimposed on some pre-existing condition. Patients at particular risk are those with diabetes mellitus, those who abuse alcohol and/or analgesics and those taking NSAIDs (Table 26.1). To this list may be added patients with sickle cell disease or trait.

Table 26.1 Risk factors for acute renal failure in unobstructed acute pyelonephritis

Diabetes mellitus
Non-steroidal anti-inflammatory drug usage
Chronic alcohol abuse
(Analgesic abuse)

This is well illustrated by the following case. A 53-year-old white man was admitted to hospital with 4 months' history of increasing left loin pain. There was a 1 day history of dysuria and occasional hesitancy. He had no vomiting. He had had no significant illness in the past. He admitted to drinking at least 10 pints of beer daily for many years and had taken 2 paracetamol tablets every 4 hours for 4 years. On admission he was flushed, ill and febrile (38°C). His blood pressure was 130/70 mmHg. There was left loin tenderness and hepatomegaly. He also had gynaecomastia. His haemoglobin was 9.2 g/dl, mean corpuscular volume 110 fl, white blood cell count 37.5×10^9/l, urea 14.3 mmol/l, serum creatinine 407 μmol/l. Liver enzymes were elevated and his prothrombin ratio (INR) was 1.6. Both blood and urine culture yielded a coliform. Ultrasound examination showed that both kidneys were unobstructed, but that the left kidney was slightly enlarged.

He was treated with intravenous cefuroxime, but because of increasing uraemia was transferred four days later. On transfer he was afebrile but still ill and complaining of left loin pain. He was euvolaemic and non-oliguric. Blood pressure was 120/80 mmHg, blood urea was 24.9 mmol/l, rising to 35 mmol/l. Serum creatinine was 580 μmol/l rising to 1370 μmol/l. Blood and urine cultures were repeatedly negative. Because of his failure to improve on antibacterial therapy, renal biopsy of the left (larger) kidney was performed. This showed normal glomeruli (Fig. 26.6) with intense interstitial oedema and infiltration with polymorpho-nuclear cells, plasma cells and lymphocytes.

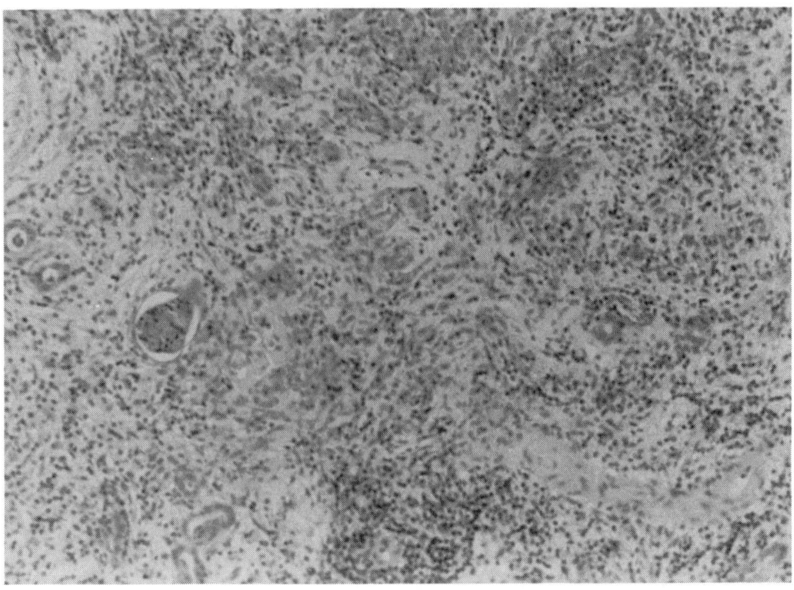

Fig. 26.6 Renal biopsy appearances in a case of acute pyelonephritis superimposed on chronic tubulointerstitial disease. The glomeruli are normal, and there is marked interstitial oedema and inflammatory infiltrate.

A section of the medullary zone showed necrosis consistent with papillary necrosis (Fig. 26.7). A diagnosis of infective tubulo-interstitial nephritis was made. His treatment was changed from cefuroxime to ciprofloxacin and he made a slow improvement requiring haemodialysis on three occasions. Serum creatinine steadily fell to 350 μmol/l and stabilized at this level. Because of the failure in complete recovery, renal biopsy was repeated 8 months after discharge. This showed extensive fibrosis (Fig. 26.8). In summary, this chronic alcoholic and analgesic abuser developed acute non-obstructive pyelonephritis, almost certainly on a background of chronic tubulo-interstitial nephritis.

Unlike patients with acute renal failure due to septicaemic shock or endotoxic vaso-motor nephropathy, the prognosis for recovery of full function in this group is not good. Intravenous urography following improvement in function not uncommonly shows evidence of papillary necrosis. Healing of the acute tubulo-interstitial nephritis is associated with severe interstitial fibrosis and tubular loss. This pattern of recovery has also been described by Richet and Mayaud (1978). Prolonged antibacterial treatment is required to eradicate infection, and patients must be carefully followed up. All patients should be given advice on prophylactic measures against further infection, and warned to seek medical attention should they develop any symptoms suggestive of recurrence of infection.

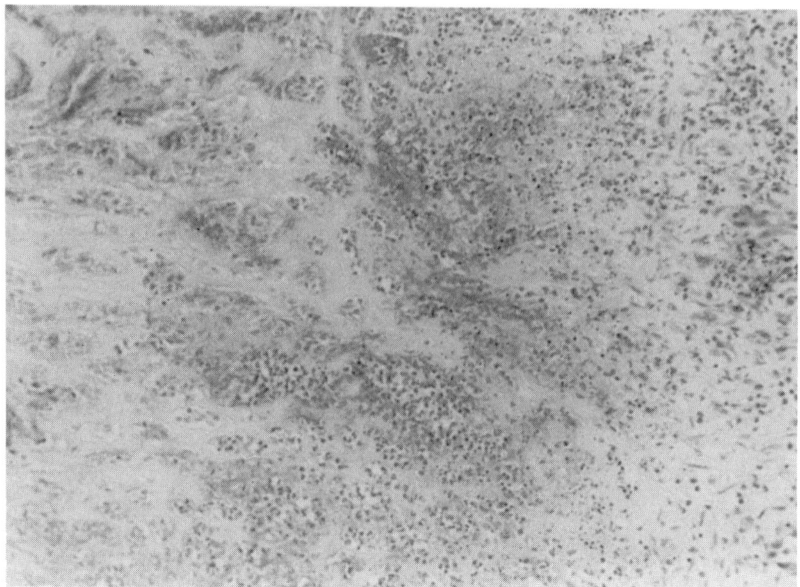

Fig. 26.7 Medullary zone of renal biopsy shown in Fig. 26.6, demonstrating papillary necrosis.

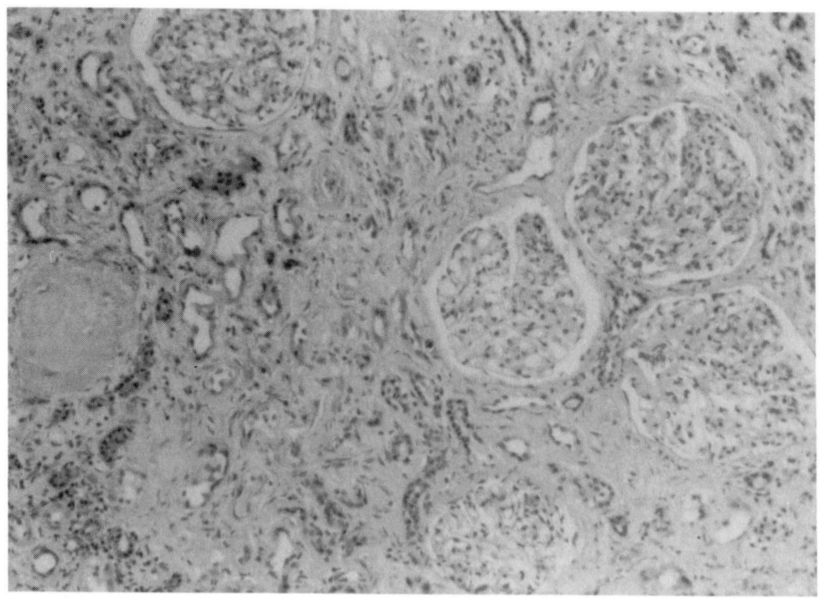

Fig. 26.8 Repeat renal biopsy after 8 months, showing extensive fibrosis.

Finally, the incidence, causation and natural history of acute renal failure in association with UTI remains poorly documented. A plea must therefore be made for careful assessment of all patients in whom this is suspected. In patients with normal sized non-obstructed kidneys, renal biopsy should be undertaken with the specific purpose of distinguishing between vaso-motor nephropathy and acute infective tubulo-interstitial nephritis, since the prognosis in this latter group appears to be less good, the need for aggressive antibacterial treatment greater, and long-term follow up mandatory.

References

Atkinson, L.K., Goodship, T.H.J., and Ward, M.K., (1986). Acute renal failure associated with acute pyelonephritis and consumption of non-steroidal anti-inflammatory drugs. *British Medical Journal*, **292**, 97–8.

Bailey, R.R. and Maling, T.M., (1982). Acute renal failure due to non-obstructive acute pyelonephritis. In *Acute renal failure*. (ed. H.E. Eliahou) pp. 145–148. John Libby, London.

Baker, L.R.I., Cattell, W.R., Fry, I.K., and Mallinson, W.J.W. (1979). Acute renal failure due to bacterial pyelonephritis. *Quarterly Journal of Medicine*, **48**, 603–12.

Cattell, W.R., Greenwood, R.N., and Baker, L.R.I. (1985) Reversible renal failure due to interstitial infection of the kidney. In *recent advances in chemotherapy*. (ed. J. Ishigami) pp. 225–8. University of Tokyo Press, Tokyo.

Cattell, W.R., Webb, J.A.W., and Hilson, A.J.W. (1989). *Clinical renal imaging*. John Wiley, Chichester.

Eknoyan, G. (1988). Acute renal failure associated with tubulo interstitial nephropathies. In *Acute renal failure*. (2nd edn) (ed. B.M. Brenner and J.M. Lazarus) pp. 491–534. Churchill Livingstone, Edinburgh.

Gleckman, R., Blagg, N., Gibert, D., Hall, A., Crowley, M., Pritchard, A., and Warren, W. (1982). Acute pyelonephritis in the elderly. *Southern Medical Journal*, **75**, 551–6.

Krane, N.K. (1988). Acute renal failure in pregnancy. *Archives of Internal Medicine*, **148**, 2347–57.

McMurray, S.B., Luft, F.C., Maxwell, D.R., Hamburger, R.J., Futty, D., Szwed, J.J., Lavelle, K.J., and Kleit, S.A. (1978). Prevailing patterns and predictor variables in patients with acute tubulo necrosis. *Archives of Internal Medicine*, **138**, 950–5.

Mayaud, C., Morel-Maroger, L., Sraer, J.D., Kanfer, A., Marsac, J., and Richet, G. (1977). L'insufficiance renale aigue des nephrites interstitielles en l'absence d'obstacle urologique. *Proceedings Second World Congress on Intensive Care. Intensive Care Medicine,* **3**, (Abstract).

Rasmussen, H.H. and Ibels, L.S. (1982). Acute renal failure, *American Journal of Medicine*, **73**, 211–18.

Rayner, B.L., Willcox, P.A., and Pascoe, M.D. (1990). Acute renal failure in community acquired bacteraemia. *Nephron*, **54**, 32–5.

Richet, G. and Mayaud, C. (1978). The course of acute renal failure in pyelonephritis and other types of interstitial nephritis. *Nephron*, **22**, 124–7.

Siroky, M.B., Moylan, R.A., Austen, G., and Olsson, C.A. (1976). Metastatic infection secondary to genitourinary tract sepsis. *American Journal of Medicine*, **61**, 351–5.

Wardle, N. (1982). Acute renal failure in the 1980s; the importance of septic shock and endotoxaemia. *Nephron*, **30**, 193–200.

27 Outcome in acute renal failure

J.H. Turney

Replacement of renal function by haemodialysis has more than halved the mortality of severe acute renal failure. Series of oliguric acute renal failure, predominantly post-traumatic or obstetric, from the 1940s and 1950s variously reported a mortality of 45–100 per cent. From the mid-1950s, there were reports of apparently similar, but presumably selected, cases treated by haemodialysis, with mortalities ranging from 0–70 per cent, the outcome being determined by the circumstances in which the acute renal failure arose. However, since the advent of haemodialysis, the overall mortality of unselected cases of acute renal failure has remained at about 50 per cent or worse. Despite considerable technical advances, the outcome of acute uraemia remains depressingly poor (Abreo *et al.* 1986; Beaman *et al.* 1987).

Since 1956, more than 1500 patients with severe acute renal failure have been treated at Leeds General Infirmary, UK (Turney *et al.* 1990). The size of this series allows direct comparison of the clinical features of acute renal failure over a long period of time, during which haemodialysis has been available, and therefore allows analysis of trends in cause and outcome. Such trends may provide indications for present-day management of acute renal failure and its associated conditions, and may influence the allocation of the necessary resources. In the Leeds series, there were numerically similar groups of patients in the decades 1960–69 and 1980–89 and in this chapter the clinical features and outcome for patients in these two periods are compared.

Patient case mix and survival

There were similar numbers and survival of medical and surgical cases in the two decades (Table 27.1). However, the median age of the patients in the 1980s was about 10 years greater than in the 1960s (Fig. 27.1). As outcome is adversely affected by increasing age (Fig. 27.2), the crude survival data obscures an improvement in outcome in the more numerous older patients encountered recently (Rodgers *et al.* 1990). The disparity between the two decades is increased when all cases are compared; the age of the 1980s patients is 15 years greater, but the overall survival is less. This is because of the large number of obstetric cases in the 1960s (Fig. 27.3) which had an excellent prognosis.

Comparison of those clinical categories in which there was an appreciable difference in outcome (Table 27.2, Fig. 27.4) indicates some of the changes in casemix that have occurred. There has been a marked increase in age in

Table 27.1 Acute renal failure: numbers, ages, outcome of major diagnostic categories

	1960–1969				1980–1989			
	Number of patients	Deaths	One-year Survival (%)	Median age (years)	Number of patients	Deaths	One-year Survival (%)	Median age (years)
Surgery	273	149	45.4	54.7	221	128	41.4	64.2
Medicine	201	101	49.6	43.3	279	140	49.4	59.4
All	579	283	51.1	46.3	506	269	46.3	61.7

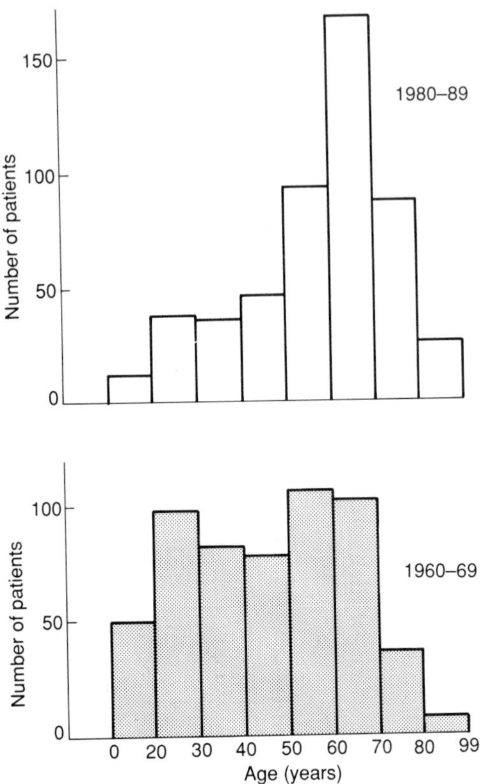

Fig. 27.1 Acute renal failure: numbers and ages in 1960–9 and 1980–9.

all groups except trauma and obstetrics cases, both of which have become uncommon in the 1980s. The decline in acute renal failure associated with obstetric mishaps has resulted both from developments in maternal care and from the disappearance of illegal abortions (Turney *et al*. 1989). Post-traumatic acute renal failure is now uncommon because of early effective resuscitation and surgery (Guly and Turney 1990).

These two categories underline the importance of prevention rather than treatment of acute renal failure. These young patients with 'uncomplicated' acute renal failure were disproportionately represented in many of the early series of treated acute renal failure, and this largely explains the often spectacular survival figures. The prognosis for 'uncomplicated' acute renal failure has been good since the advent of haemodialysis. It is in medical and surgical patients that progress in terms of outcome appears to have been limited (Wheeler *et al*. 1986).

Although general surgical cases were less frequent in the 1980s, the outcome

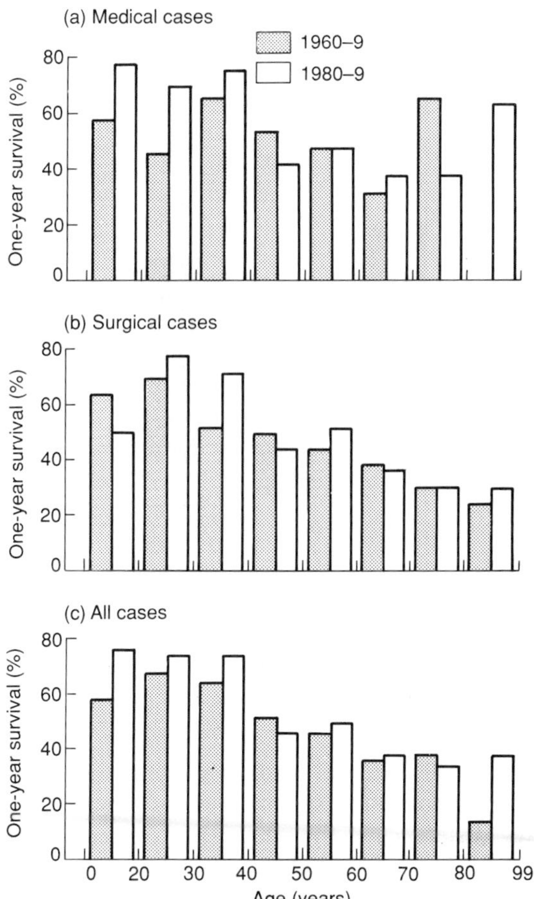

Fig. 27.2 Acute renal failure: percentage actuarial one-year survival in medical, surgical, and all cases according to age, 1960–9 and 1980–9.

remained the same (42.3 per cent survival in 1960–69; 43.0 per cent in 1980–89). The lack of improvement may be partly related to increased age (58.6 years in 1960–69; 64.9 years in 1980–89) and to the larger proportion of patients in whom sepsis appeared to be the major precipitating cause of acute renal failure (22 per cent in 1960–69; 38 per cent in 1980–89). However, survival in the septic surgical patients did not alter, being 30.2 per cent in 1960–69 and 30.8 per cent in 1980–89. It would appear that the outcome of acute renal failure could be enhanced by attention to the management of these patients, particularly those with postoperative abdominal sepsis. If the outcome in general surgical patients has remained constant, the results of acute renal failure complicating general medical conditions has noticeably declined

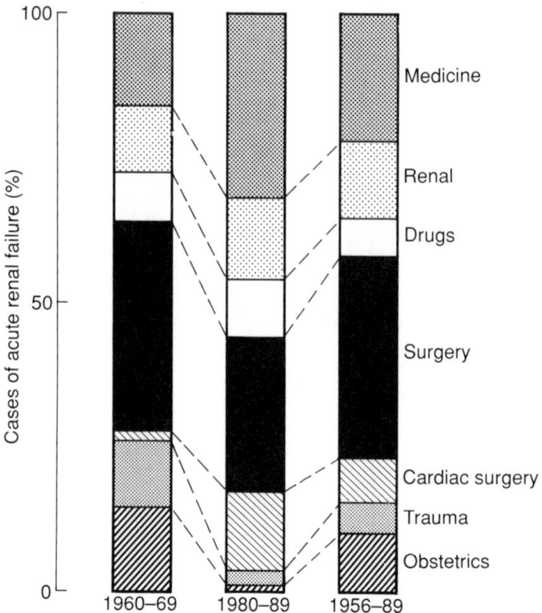

Fig. 27.3 Acute renal failure: percentage distribution of causes.

(Table 27.2). In the 1980s, general medical patients were more common and were older, but the proportion of septic cases has remained constant at 40–45 per cent. The key difference is the occurrence of acute renal failure in the context of cardiovascular disease, haematological malignancy, and with multi-organ failure complicating other medical (or surgical) conditions.

Outcome in renal parenchymal disease

The one group in which there has been progress in management, and of which there are comparable numbers, are those patients presenting as an acute uraemic emergency as a result of intrinsic renal parenchymal disease. The age distribution and causes of rapidly progressive glomerulonephritis have altered; the vasculitides are now by far the commonest diagnosis, with other crescentic nephritides probably declining, and post-infectious glomerulonephritis virtually absent. The most important factor in the improvement in outcome of patients with renal parenchymal disease has been the institution of immunosuppressive therapy. Despite the fact that these patients usually have a multisystem disorder and receive potentially hazardous treatment, intrinsic renal disease now has the best prognosis for survival (but not necessarily for renal recovery) of all categories of acute renal failure (Beaman *et al.* 1987).

Table 27.2 Acute renal failure: comparison of clinical categories with a significant difference between 1960–69 and 1980–89

	1960–1969				1980–1989			
	Number	Deaths	One-year Survival (%)	Median age (years)	Number	Deaths	One-year Survival (%)	Median age (years)
Trauma	66	30	54.5	40	13	4	69.2	35
Cardiovascular	12	7	41.7	50	73	48	33.8	64.7
Obstetrics	81	14	82.7	28.1	6	2	66.7	27.5
General medicine	90	42	53.3	51.7	176	110	36.6	61.9
Renal parenchymal disease	64	50	21.3	31.7	63	20	68.3	56.8
Sepsis	109	58	46.8	51.7	142	94	33.3	63.1

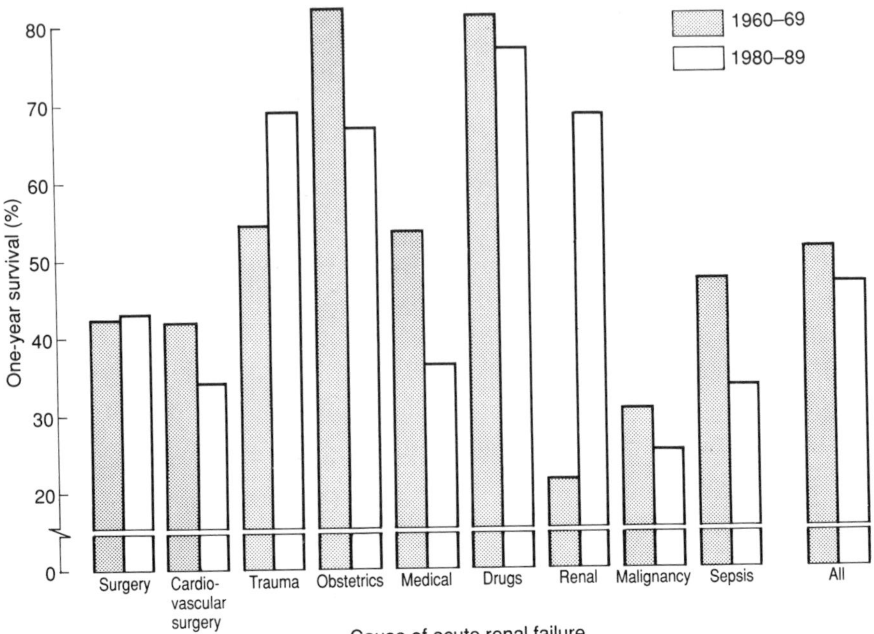

Fig. 27.4 Acute renal failure: percentage actuarial one-year survival according to cause, 1960–69 and 1980–89.

The improvement in outcome for patients with renal parenchymal disease clearly illustrates the crucial importance of the underlying disorder in determining the outcome of acute renal failure. Unfortunately, acute renal failure arising from self-limiting conditions in young patients, for example post-trauma or obstetrics, is increasingly uncommon. The outcome of acute renal failure is determined by the clinical setting in which it occurs, the prognosis being largely that of the precipitating cause (Abreo *et al.* 1986; Hov *et al.* 1983; Shusterman *et al.* 1987). It therefore follows that developments in the management of uraemia *per se*, for example alternative modes of renal replacement therapy, will at best have only a marginal effect on overall outcome.

Trends in clinical course and cause of death

The underlying condition was considered to be the major cause of death in by far the largest group of patients and this proportion has increased (Table 27.3). Of the other causes of death, well over half the cardiovascular and septic deaths were related to the underlying disease, and did not arise *de novo* during the course of acute renal failure. There is perhaps now a greater tendency to withdraw active treatment in apparently hopeless cases.

Table 27.3 Acute renal failure: comparison of causes of death 1960–9 and 1980–9. (NB, Total greater than 100 per cent because of combinations of causes of death)

	1960–69		1980–89	
	Number of deaths	%	Number of deaths	%
Haemorrhage	16	6.3	5	2.3
Infection	94	37	93	43.5
Neurological	12	4.7	6	2.8
Cardiovascular	54	21.3	75	35
Treatment withdrawn	14	5.5	33	15.4
Non-recovery Renal function	45	17.7	5	2.3
Underlying disorder	137	53.9	146	68.2
Other	5	2	1	0.5
Total	254		214	

Non-recovery of renal function is no longer a significant cause of death, as these patients are now established on the regular haemodialysis programme, where they fare as well as those with other causes of end stage renal failure. Haemorrhage, predominantly gastrointestinal, is less frequent because of active prophylactic measures, and perhaps because haemodialysis is commenced at a lower level of uraemia.

In a study of traumatic acute renal failure (Guly and Turney 1990), it was shown that although the length of time between injury and the institution of haemodialysis remained the same, more recent patients are dialysed at a lower level of blood biochemistry. This suggests firstly that haemodialysis is now more intensive and secondly that resuscitative measures and nutrition are more effective in controlling uraemia than formerly.

The benefit of intensive haemodialysis in controlling the side-effects of uraemia was demonstrated more than 30 years ago. Not only has the cause of death changed over the study period, but also the length of survival before eventual death has increased significantly. In the 1980s, 13.2 per cent of patients died more than 30 days after the onset of acute renal failure, compared with 8.9 per cent in the 1960s; the median survival of these late deaths was 52 days in 1960–69 and 68 days in 1980–89. These differences, which are statistically significant, suggest that modern intensive therapy often merely postpones inevitable death, usually from the underlying disorder, in a significant proportion of patients. This prolongation of the illness in recent years is again illustrated by post-traumatic acute renal failure, in which the median time to discharge or transfer has increased from 33 to 98 days, in a period in which the length of stay of patients in general hospitals has dramatically shortened. This also suggests that recent patients, even in directly comparable groups, present a more complicated clinical picture. The effect on health service resources is self-evident.

Acute renal failure largely or solely due to sepsis comprises a large proportion of the total and has a generally poor prognosis which appears to have deteriorated in recent years (Table 27.2). Sepsis was the cause of 18.8 per cent of cases in 1960–69 and 28.1 per cent in 1980–89. In the 1960s, 18.3 per cent of cases of sepsis were due to illegal abortions, but this only partially accounts for the lower ages and higher survival. In contrast, sepsis arising in the context of a medical condition has increased from 36.7 per cent to 51.4 per cent of the total of this group of patients. Of the septic patients, only 2 per cent received mechanical ventilation in the 1960s, compared with 41 per cent (27.5 per cent survival) in the 1980s. This difference may result either from notably increased severity of illness in later years, or from the fact that patients now survive, as a result of intensive therapy, for long enough to develop and be treated for acute renal failure, whereas formerly they rapidly succumbed from overwhelming sepsis.

Prediction of outcome

The increased severity of the condition of recent patients is clearly shown by the APACHE II severity of illness index (Table 27.4), determined from the worst recorded observations on the day of presentation to the renal unit. In all cases, the scores were high, but were significantly greater in the 1980s. Not only are the recent patients more severely ill, but the APACHE scoring system also shows that their outcome has in fact improved, in that the scores

Table 27.4 Acute renal failure due to sepsis: APACHE II severity of illness scores in medical and surgical patients, 1960–69 and 1980–89

	1960–69			1980–89			
	Number	Median	Range	Number	Median	Range	p
Surgery	50	32	22–41	37	36	26–49	<0.0001
Medicine	37	32	23–45	56	35	25–46	=0.0078
Survivors*	32	27.5	22–40	23	33	25–40	=0.0042
Deaths**	55	33	23–45	70	36	26–49	=0.0006
ALL	87	32	22–45	93	35	25–49	<0.0001

* 1960–9 survivors versus deaths, $p < 0.0001$.
** 1980–9 survivors versus deaths, $p = 0.0002$.
Surgery versus medicine, not significant.
Deaths 1960–9 versus survivors 1980–9, $p = 0.3699$.

of the survivors in the 1980s are the same as those of the non-survivors from the 1960s. The increased survival of the more severely ill is largely, if not exclusively, due to developments in general supportive care. However, whilst the treatment may be getting better, the patients are getting worse, and to such an extent that any effect of therapeutic advances is masked and the overall outcome in septic acute renal failure is deteriorating.

The APACHE scores also illustrate the inadequancy of all current methods of predicting outcome in acute renal failure. These patients tend to cluster in the middle-range of the potential APACHE score, thus reducing its discriminatory value and rendering the method imprecise for individual, as opposed to group, prognosis. However, no recovery has been achieved in patients with acute renal failure with an initial APACHE score greater than 40. At present there is little better than simple summation of failed organ systems, or the trained clinical impression, in predicting individual outcome. Nevertheless, the APACHE system is valuable for comparisons between series, audit, and prediction of resource requirements.

Conclusions

In summary, the apparent lack of improvement in outcome of acute renal failure since the advent of haemodialysis can be explained, at least in part, by the changes in casemix, age, and underlying disorders of the patient population. Crude survival figures mask real improvements in the management of elderly patients who develop acute renal failure during the course of a complicated medical or surgical illness. As uraemia *per se* is no longer a cause of death in acute renal failure, any future advances are more likely to derive from the general management of these patients, rather than from developments in haemodialysis techniques (Butkus 1983). The analysis of the Leeds series of acute renal failure patients also has ethical, economic, and clinical implications, which will be heightened by future demographic trends and by developments in surgery and medicine. These developments will present nephrologists with an increasing burden of elderly patients in the future, who will survive long enough to develop and require treatment for acute renal failure, in all its complexity.

References

Abreo, K., Moorthy, V., and Osborne, M. (1986). Changing patterns and outcome of acute renal failure requiring haemodialysis. *Arch. Intern. Med.* **146**, 1338–41.

Beaman, M., Turney, J.H., Rodger, R.S.C., McGonigle, R.S.J., Adu, D., and Michael, J. (1987). Changing pattern of acute renal failure. *Quart. J. Med.* **62**, 15–23.

Butkus, D.E. (1983). Persistent high mortality in acute renal failure. Are we asking the right questions? *Arch. Intern. Med.* **143**, 209–11.

Guly, U.M. and Turney, J.H. (1990). Post-traumatic acute renal failure, 1956–1988. *Clin. Nephrol.* **34**, 79–83.

Hou, S.H., Bushinsky, D.A., Wish, J.B., Cohen, J.J., and Harrington, J.T. (1983). Hospital-acquired renal insufficiency: a prospective study. *Amer. J. Med.* **74**, 243–8.

Rodgers, H., Staniland, J.R., Lipkin, G.W., and Turney, J.H. (1990). Acute renal failure: a study of elderly patients. *Age and Ageing*, **19**, 36–42.

Shusterman, N., Strom, B.L., Murray, T.G., Morrison, G., West, S.L., and Maislin, G. (1987). Risk factors and outcome of hospital-acquired acute renal failure. *Amer. J. Med.* **83**, 65–71.

Turney, J.H., Ellis, C.M., and Parsons, F.M. (1989). Obstetric acute renal failure: 1956–1987. *Br. J. Obstet. Gynaecol.* **96**, 679–87.

Turney, J.H., Marshal, D.H., Brownjohn, A.M., Ellis, C.M., and Parsons, F.M. (1990). The evolution of acute renal failure, 1956–1988. *Quart. J. Med.* **74**, 83–104.

Wheeler, D.C., Feehally, J., and Walls, J. (1986). High risk acute renal failure. *Quart. J. Med.* **61**, 977–84.

28 Acute and chronic renal failure in the tropics

A.O. Ogunlesi and R.E. Phillips

Renal failure accounts for significant morbidity and mortality in tropical regions of the world. It is not the tropical climate *per se* which imposes additional stress on the kidneys of indigenous peoples but the fact that the majority of poor developing countries are situated between the tropics of Cancer and Capricorn (Fig. 28.1). The pattern of renal disease reflects the prevailing health status of these communities: endemic parasitic diseases, malnutrition and poor hygiene are the most important factors predisposing to disease. In essence, the problems of medical care in tropical developing countries are more closely related to, and dependent upon, socioeconomic factors than on climatic or geographical factors (Hutt and Wing 1971).

The true prevalence of renal failure in the tropics is not known, and is difficult to determine. Most estimates are based on case studies and hospital-acquired data. Population-based surveys with any relevance to kidney disease have concentrated on indices such as urinalysis, urine microscopy, and blood

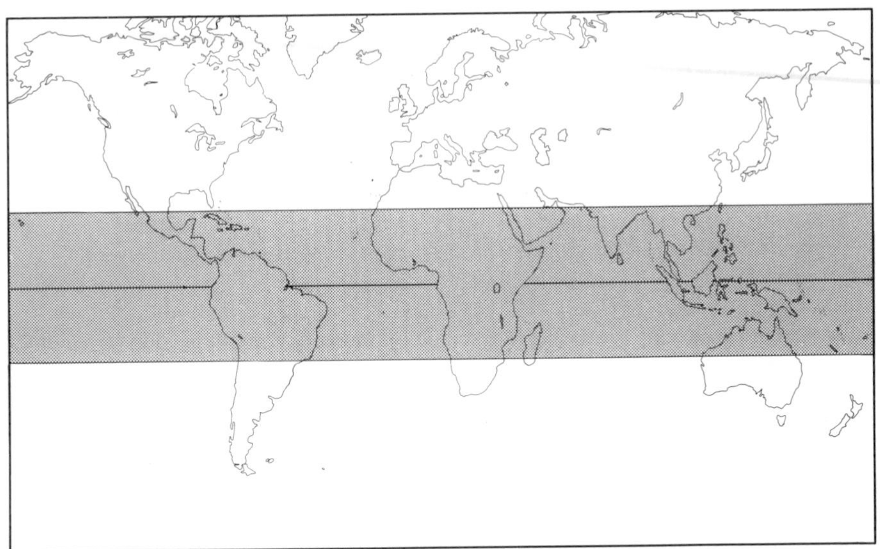

Fig. 28.1 Global map showing the tropical regions.

pressure estimation, rather than on overall renal function. The physiological renal reserve allows individuals with impaired renal function to remain asymptomatic, and so makes accurate estimates of renal failure difficult in the tropics, where medical help is generally sought only as a last resort.

Available data show that the incidence of renal failure in the tropics varies widely. The most important variable determining the incidence of acute renal failure is the precipitating factor (Table 28.1). Snake-bite is an important cause of acute renal failure in Burma (Thein-Than *et al.* 1991), but not in Ghana, where typhoid and Gram-negative septicaemia were the precipitating factors in over 50 per cent of cases with acute renal failure in one series (Adu *et al.* 1976).

Chronic renal failure accounted for 1–5 per cent of all medical admissions into a teaching hospital in tropical Africa (Adetuyibi *et al.* 1977), while 3 per cent of all autopsies in the same hospital showed evidence of chronic glomerulonephritis (Edington and Mainwaring 1966). The clinical features of chronic renal failure are essentially the same in all these areas but the important aetiological factor(s) vary from region to region; calculus nephropathy is the second most important cause of chronic renal failure in the Sudan, accounting for over 12 per cent of cases (Abboud *et al.* 1989), whereas chronic glomerulonephritis and accelerated hypertension account for more than two thirds of cases of chronic renal failure in Nigeria (Oyediran and Akinkugbe 1970; Akinsola *et al.* 1989).

Table 28.1 Factors precipitating acute renal failure in the tropics

(1) Haemodynamic/pre-renal
- Severe gastroenteritis (cholera, typhoid)
- Acute severe haemorrhage (road traffic accidents, antepartum and postpartum haemorrhage, oesophageal varices)
- Massive intravascular haemolysis with haemoglobinuria (G6PD-deficiency, acute falciparum malaria, incompatible blood transfusion)
- Snake-bite envenoming (Russell's viper)
- Septicaemia (Gram-negative, post-abortal, instrumentation—'bougie fever')

(2) Renal/toxic
- Pre-eclampsia
- Direct nephrotoxicity (native herbal remedies, mercury-containing skin-lightening creams)
- Snake-bite envenoming
- Acute post-streptococcal glomerulonephritis
- Rhabdomyolysis with myoglobinuria (pyomyositis, rabies, snake-bite, crush injury)

(3) Post-renal
- Obstructive uropathy (carcinoma of the cervix, prostatic hypertrophy, renal calculi)
- Urethral stricture with acute fulminating pyelonephritis or necrotizing papillitis

There is great ethnic and climatic diversity in the tropics. In tropical Africa, apart from the different peoples (Fig. 28.2), the environment also varies greatly with vegetation from mangrove swamp to arid desert. This influences the prevailing parasitic diseases and the pattern of renal failure. Earlier attempts at defining tropical nephropathies probably did not examine fully the host–parasite relationships in the tropics; thus, the association between *Plasmodium malariae* (quartan malaria) and the development of the nephrotic syndrome (QMNS) in children from malarious areas, though strong (Gilles and Hendrickse 1963; Kibukamasoke *et al.* 1967), cannot be accepted as causal. More recently, the role of certain genetic influences, especially the major histocompatibility complex (MHC) genes in determining aspects of host–parasite

Fig. 28.2 Distribution of the major peoples of Africa.

relationships in parasitic disease has been highlighted (Hill *et al.* 1991). It is likely that genetic factors have some bearing on the responses of the different ethnic groups in the tropics to common parasites and possibly the pattern of renal damage especially when this is immunologically mediated.

Renal failure, as seen in West Africa, typifies that seen in most tropical regions, although there are important exceptions. Against this background, this chapter reviews renal failure as seen in tropical West Africa, drawing attention to features that distinguish it from that commonly seen in more developed countries. Other renal failure syndromes native to the tropics, but uncommon in West Africa are also reviewed.

Aetiology and clinical features

Acute renal failure

The majority of patients developing acute renal failure in the tropics, regardless of precipitating cause, have acute tubular necrosis because most of the precipitating factors are haemodynamic in nature. Effective reversal of the precipitating factor(s) and adequate management of the renal failure usually assures the good prognosis characteristic of acute tubular necrosis.

The sudden occurrence of oliguria/anuria and dark urine in a previously healthy individual is a dramatic event, and this ensures early presentation in hospital—usually within 72 h of onset. The patient complains of oliguria (or anuria), dark-coloured urine (as a result of more concentrated urine, rather than haemglobinuria) and symptoms of generalized oedema. These symptoms are frequently accompanied with features of a rapidly progressive uraemic state; hyperkalaemia, azotaemia and hypertension. The precipitating factor is evident from the history and more than one factor is usually present. Local customs may be helpful in diagnosis: for example in many areas of West Africa, it is not unusual for the relatives of the patient to bring along the dead snake in suspected snake-bite envenoming. This practice is to be encouraged, for it facilitates the identification of snake species and administration of the correct antivenom (Warrell *et al.* 1977).

The most important evaluation the admitting physician has to make when presented with a patient with renal failure in the tropics is to determine the reversibility of the renal failure. Simply put, which patients have genuine acute renal failure and which have acute-on-chronic renal failure? Although not an easy task this is mandatory where resources are scarce. Uraemic pigmentation is difficult to detect in dark-skinned West Africans and anaemia has too many other causes to be of any value (Adu *et al.* 1976). Features of long-standing hypertension should be sought: retinopathy, ventricular enlargement and thickening of peripheral vessels. Renal fat shadows are difficult to visualize on plain abdominal X-rays in Nigerians, presumably because of scanty renal fat (Lagundoye 1975). Where the necessary facilities exist, ultrasonography of the kidneys has proved to be a rapid and sensitive screening method. More than 75 per cent of cases of chronic renal failure in Nigeria result from chronic glomerulonephritis and accelerated hypertension, with bilaterally shrunken granular kidneys on renal ultrasound (Akinsola *et al.* 1989). A normal-sized kidney with good cortico-medullary differentiation virtually excludes irreversible renal failure (Ogunlesi *et al.* 1991*a*). This emphasis on 'reversible renal failure' also highlights an important clinical and ethical problem in the tropics: acute tubular necrosis has a good prognosis, with short hospital stay and a definitive end-point, whereas acute-on-chronic renal failure will require more capital outlay and, eventually, renal replacement therapy. Even then, the long-term prognosis is poor.

Urine microscopy is a useful indicator of tubular necrosis, where the presence of tubular casts is diagnostic. Urine osmolarity and urinary sodium excretion should be measured where feasible, as 24-h urine sodium excretion has been found to be lower in healthy rural people, compared with healthy urban people in parts of tropical Africa, possibly signifying varied sodium homoeostasis (Poulter *et al.* 1990). The urine/plasma sodium ratio is a useful guide to management of fluid balance. Very high plasma urea and creatinine levels may be a reflection of the hypercatabolic (sepsis, fever) acute renal failure frequently present in these patients (Adu *et al.* 1976; Ogunlesi *et al.* 1991*a*).

Chronic renal failure

The diversity of aetiological factors precipitating acute renal failure is not so marked with chronic renal failure. Most reports have identified chronic glomerulonephritis as the most important cause of chronic renal failure in the tropics (Akinsola *et al.* 1989; Sobh *et al.* 1988; Abboud *et al.* 1989). The prevailing circumstances in the tropics, with endemic parasites and rudimentary early detection programmes for renal disease probably favour the high prevalence of chronic glomerulonephritis. The incidence of chronic renal failure is 1–5 per cent of all medical admissions in Ibadan, Nigeria (Adetuyibi *et al.* 1977). Most other centres report figures similar to this (Paton 1974).

A striking feature of chronic renal failure in tropical West Africa is the relative youth of the patients; they are often in their second or third decade, with bilaterally shrunken kidneys and in end stage renal failure (Akinsola *et al.* 1989). Anaemia is invariable, and hypertension frequently severe. Retinopathy may not correlate with the severity of the blood pressure (Akinkugbe 1967). Proteinuria is profuse, signifying the underlying glomerular dysfunction. Many patients are seen for the first time with a syndrome of rapidly progressive uraemia, oliguria and massive pulmonary oedema from left ventricular failure. The high frequency of infections of the urinary tract in these patients suggests that this abrupt deterioration may be related to infection (Akinsola *et al.* 1989; Akinkugbe O.O., personal communication). Sadly, many of these patients die because of the high cost and limited facilities for renal replacement therapy, coupled with late presentation at hospital.

In more developed countries, chronic glomerulonephritis, polycystic kidney disease, essential hypertension, and diabetic nephropathy are the major causes of end stage renal disease (Klahr *et al.* 1988). The prevalence of hypertension is approximately 8 per cent of adults in Nigeria (Akinkugbe and Ojo 1969; Ogunlesi *et al.* 1991b) while type II (NIDDM) diabetes occurs in about 2 per cent of adults in Nigeria (Ohwovoriole *et al.* 1988). The contribution of these diseases to end stage renal disease in Nigerians is not as high as would have been expected from their prevalence. The lower life expectancy in the tropics and the burden of parasitic diseases may be some of the factors masking their importance. As preventive measures and vaccination reduce the toll of parasitic disease, there are signs that hypertension and type II diabetes may become principal factors in the aetiology of chronic renal failure in the tropics (Beevers and Price 1991; Barnett 1991).

Renal bone disease is uncommon in patients with chronic renal failure in the tropics (Hutt and Wing 1971). Because osteoporosis is also uncommon, dietary factors with production of a more alkaline urine, abundant ultraviolet radiation from sunlight (hence vitamin D), and a shorter life expectancy after diagnosis of chronic renal failure, compared with those in more developed countries, have been suggested as factors which may explain the difference. No studies have examined the genetic factors which may determine this interesting observation. Indirect evidence suggests that there is no difference in endocrine responses of uraemic Nigerians compared with those of other non-tropical ethnic groups (Ogunlesi *et al.* 1990).

Management and prognosis

Acute renal failure

The principles of management are broadly the same when dealing with acute renal failure in the tropics as elsewhere. As soon as the diagnosis is made, the patient is best managed in a renal intensive-care unit with facilities for haemodialysis (rare in most parts of tropical Africa) to improve the chances of survival (Akinkugbe and Abiose, 1967; Adu *et al.* 1976).

Successful management depends on the judicious use of antibiotics and fluid therapy. It is worth considering prophylactic antibiotic therapy where blood cultures cannot be obtained routinely, because of the high incidence of sepsis-induced acute renal failure and the immune suppression accompanying it (Adu *et al.* 1976; Ogunlesi *et al.* 1991*a*). Doses of antibiotics, especially aminoglycosides should be adjusted carefully to suit the renal function. In practice, parenteral administration of a combination of ampicillin, gentamicin and metronidazole effectively covers the range of organisms encountered. Chloramphenicol remains the drug of choice for treatment of typhoid, even when these patients develop acute renal failure. Monitoring blood levels of drugs is the ideal, but is hardly feasible in most developing countries. As an alternative, individual doses should be reduced and administration intervals attenuated. Antimalarial drugs should be given when indicated, as the widespread emergence of chloroquine-resistant falciparum malaria in West Africa (Greenwood 1986) means that persistent pyrexia in some patients may be due to treatment failure with chloroquine. The use of primaquine is contraindicated in glucose-6-phosphate dehydrogenase (G6PD) deficient patients.

The most important emergency in acute renal failure is hyperkalaemia. Urgent steps should be taken to reduce the potential threat to life. An electrocardiogram is mandatory on admission to look for tenting of the T waves, which is characteristic of hyperkalaemia. Where practicable, electrocardiogram monitoring should be continuous at plasma potassium levels above 6 mmol/l, and intravenous frusemide should be given as it promotes kaliuresis and reduces ventricular pre-load, relieving hyperkalaemia and acute pulmonary oedema. Intravenous calcium gluconate (10 ml of 10 per cent solution) should be administered slowly to protect against asystole; insulin/glucose and sodium bicarbonate (provided the patient is not sodium overloaded) should also be given to sequester potassium intracellularly. Dopamine has been used in acute malarial renal failure to promote diuresis in the oliguric phase (Lumlertgul *et al.* 1988). This experience is limited to a few centres and cannot be recommended for all cases of acute renal failure in the tropics.

All of these measures are only temporary. Dialysis is required in all but a few cases and should be the mainstay of treatment (Adu *et al.* 1976). Peritoneal dialysis is cheap, effective, and does not require highly skilled operators. Various adaptations have made its use effective in the tropics (Akinkugbe and Abiose 1967; Ojogwu 1983; Onwubalili 1989), but problems of catheter tunnel infection, blockage and peritonitis are common. Scrupulous asepsis and improved nursing technique reduces these problems. Haemodialysis (with or without haemofiltration) is very effective, but remains limited to a few centres, requiring skilled personnel and capital outlay. (Adu *et al.* 1976; Odutola *et al.* 1989). Exchange blood transfusion may help in cases of severe falciparum malaria with very heavy parasitaemia and/or massive intravascular haemolysis (Kramer *et al.* 1983; Looareesuwan *et al.* 1990).

Deficiency of erythrocyte G6PD and infection (usually septicaemia) is associated with acute renal failure in the tropics (Owusu *et al.* 1972; Adu *et al.* 1976). The clinical features of G6PD-deficiency are similar to classical blackwater fever, but it is best not classified as such, so that the specific aetiology can be highlighted. The high prevalence of G6PD-deficiency in many parts of West Africa (Adu *et al.* 1976) makes it obligatory to exclude this diagnosis in any patient with fever, jaundice and acute renal failure. G6PD assay is unreliable during the acute phase of haemolysis, because reticulocytes and young red blood cells have sufficient enzyme activity to give a normal result on assay. (Carson and Frischer 1966). Management of these patients is conservative, with avoidance of antioxidant drugs and management of the acute renal failure along standard lines.

Fluid therapy deserves special mention because as a group, haemodynamic factors are the most important precipitants of acute renal failure in the tropics. An accurate fluid chart is helpful but can be unreliable. When input and output are poorly documented, this may be more dangerous than having no records at all. Signs of hypovolaemia should be sought (loss of skin turgor, dry tongue low blood pressure, etc) and the circulating volume restored quickly. The choice of fluid (0.9 per cent normal saline or 0.45 per cent normal saline in 5 per cent dextrose) is usually dictated by availability rather than any other factor, but normal saline has been advocated for patients with hyponatraemia accompanying severe falciparum malaria (Warrell *et al.* 1991). To restore blood volume in cases with acute haemorrhage, preferably fresh whole blood should be given. This minimizes transfusion of haemolysed red blood cells and their by-products, which are present in substantial amounts in stored blood. Necessary precautions should be taken when handling blood and blood products, as both the humans immunodeficiency virus (HIV) and hepatitis B virus (HBV) are prevalent in West Africa (Anderson *et al.* 1991).

When deciding the amount of fluid replacement required, an allowance of 1000–1500 ml/24 h should be made for insensible water loss, because of the high ambient temperature in tropical regions and the large number of patients with hypercatabolic acute renal failure (Adu *et al.* 1976). In the diuretic phase, urine output can exceed 5–10 l/24 h, so adequate oral and intravenous intake must be provided, otherwise severe fluid depletion can occur (Ogunlesi *et al.* 1991*a*).

Although not necessary for management of acute renal failure, a renal biopsy may be indicated in cases in which the oliguric phase is unduly prolonged (>6–8 weeks), or in which recovery is still incomplete (heavy proteinuria after 12–18 months). Renal biopsy should only be performed in centres where a nephrologist trained in the procedure and a histopathologist experienced in examining the tissue obtained are on hand. The outcome of acute tubular necrosis is usually good, even though the diagnosis is frequently retrospective. In one series from Ibadan, Nigeria, the factors which correlated directly with a good outcome were a short duration of the oliguric phase and a large urinary output during the diuretic phase. (Ogunlesi *et al.* 1991*a*).

Chronic renal failure

Management of these patients highlights some of the problems of specialist practice in tropical developing countries. Most patients with established renal failure cannot afford private-sector fees for renal replacement therapy, and public hospitals either lack the facility completely, or have long waiting lists because the chronic dialysis service has to complete for allocation of scarce funds from the health budget. The result is that most patients have to be managed conservatively, even though their clinical status warrants more active treatment. Despite this, some patients have done remarkably well, and it is worthwhile reinforcing morale and patient education. (Akinsola *et al.* 1989).

Table 28.2 Causes of chronic renal failure in the tropics

Chronic glomerulonephritis (including quartan malaria nephropathy and schistosomal glomerulopathy)
Accelerated hypertension
Chronic pyelonephritis
Obstructive uropathy (prostatic enlargement and schistosomiasis of the bladder)
Calculus nephropathy
Diabetic nephropathy
Renal dysplasia
Adult polycystic kidney disease

At presentation, some of these patients have acute-on-chronic renal failure. In these cases, the immediate priority is to remove any identifiable precipitating factors. Clinical experience suggests that infection of the renal tract may be responsible for the sudden loss of renal function (Akinsola *et al.* 1989; Akinkugbe O.O., personal communication). Urine microscopy and culture should always be carried out to establish the presence of infection and to guide therapy. Most organisms recovered are Gram-negative, such as *Eschericia coli*, and treatment with a combination of an aminoglycoside and a cephalosporin is very effective (Akinsola *et al.* 1989).

Most patients referred to the renal physician have a history of nocturia, malaise, nausea and vomiting; all non-specific for any diagnosis. A few cases of chronic renal failure are asymptomatic, their referral is prompted by the incidental discovery of an elevated blood pressure. Unfortunately, when most patients are first seen, they are in established renal failure or even end stage renal disease. The most pragmatic approach will be to preserve renal function. Therefore, renal reserve should be estimated: a crude estimate can be made from the creatinine clearance, renal ultrasound and the degree of anaemia.

The control of blood pressure, to prevent development of accelerated hypertension, which rapidly compromises remaining renal function, is imperative (Ojogwu and Anah 1983). In cases with mild to moderate hypertension and early renal failure, diuretics and vasodilators (including ACE inhibitors) would be the preferred antihypertensives (Falase and Salako 1979). Beta-adrenoceptor blockers are not as effective in reducing blood pressure in Nigerian hypertensives as in Caucasians (Salako *et al.* 1979*a*, *b*), presumably because of the more prevalent low-renin state (Osotimehin *et al.* 1984). Plasma renin activity is likely to be high in patients with end stage renal disease and shrunken kidneys, where beta blockers and potent vasodilators (minoxidil) are frequently required. Methyldopa is still widely used in West Africa because it is relatively inexpensive; it is said to increase renal blood flow and therefore may be of benefit to patients with early renal failure.

Dietary restriction of protein, phosphate, sodium, and potassium is the other important factor in the management of patients with chronic renal failure in the tropics. The staple diet of most indigenous peoples is rich in carbohydrates, and low in animal protein and fat, and therefore suited for renal failure. Further reducing the protein content of the diet to conform with the standard renal failure diet of 30–50 g protein and 2000–3000 calories per day rapidly leads to a negative nitrogen balance (Atinmo *et al.* 1985), in addition the unpalatability of the diet only serves to worsen the misery of these patients. There are no convincing controlled trials to assess the impact of locally-derived diets, in the tropics, on the progression of chronic renal failure. Proponents claim that the marginal benefit occurs because most patients recruited actually needed renal replacement therapy, rather than dietary management. Recently, a controlled trial from Italy has questioned the benefit of the low-protein diet in early chronic renal failure (Locatelli *et al.* 1991). At present, it remains uncertain whether dietary management should be one of the main aspects of therapy of chronic renal failure in the tropics.

Fluid restriction may be necessary to relieve oedema, but, more commonly, a reasonably high urinary output is required to optimize excretion of uraemic waste. In early cases (glomerular filtration rate > 40 ml/min), frusemide in large doses (up to 3 g/d) may be very useful. Sodium balance should be monitored through urinary and plasma estimations to give an overall direction, and to assist blood pressure control and prevention of heart failure.

The anaemia of chronic renal failure in these patients is difficult to manage. Recombinant erythropoetin is widely available in most developed countries for the management of this anaemia, but, because of cost, it has not been used to any appreciable degree in West Africa. Transfusion of packed red blood cells is the only alternative for severe, symptomatic anaemia. Folate iron supplements should be given, as most patients are malnourished (anorexia, nausea, and vomiting of uraemia), and will probably be folate deficient.

Renal replacement therapy is the best option, if the patient can afford it privately, or if it is available in government hospitals. Peritoneal dialysis is the logical choice because of its simplicity and cost-effectiveness. Initial experience with continuous ambulatory peritoneal dialysis in Ibadan Nigeria was very disappointing, with almost all patients developing fatal peritonitis (Akinkugbe, O.O., personal communication). The warm humid environment of West Africa may have been an adverse factor, as more success has been recorded in North Africa, where the climate is drier (Moustafa *et al.* 1989).

More haemodialysis centres are now available in West Africa (Fig. 28.3). Services offered include acute haemodialysis, haemofiltration and a few beds for maintenance haemodialysis. Peritoneal dialysis is often available for acute tubular necrosis cases. Reports on long-term experience with haemodialysis for renal replacement therapy in tropical Africa (Odutola *et al.* 1989; McLigeyo *et al.* 1989) are not encouraging. Therefore, unless a renal transplant programme is planned or is in existence, it is unwise to build up a population of dialysis-dependent patients.

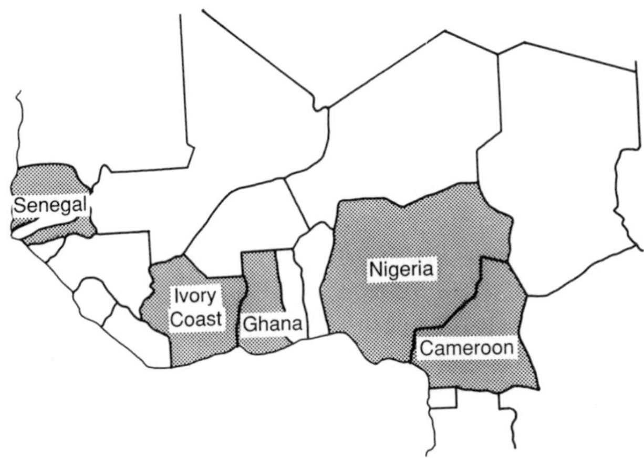

Fig. 28.3 West African countries with haemodialysis centres.

Renal biopsy is mandatory for accurate diagnosis in patients who will receive a transplant (Sobh *et al.* 1988). They are, however, a minority and therefore biopsy is of limited value in most patient with chronic renal failure in the tropics.

Tropical nephropathies

Quartan malaria nephrotic syndrome

More than two decades ago, an association between the nephrotic syndrome in children and *Plasmodium malariae* infection was described in Nigeria and Uganda. (Gilles and Hendrickse 1963; Kibukamusoke 1984). The evidence was mainly clinical, but epidemiological support came from British Guiana, where the previously high incidence of childhood nephrotic syndrome dropped with the eradication of malaria (Giglioli 1962). It has been difficult to prove that quartan malaria causes a nephrosis. A histological picture is said to be diagnostic (Hendrickse *et al.* 1972), but the glomerulus-limited responses to injury and similar lesions have been described in Senegalese children, with no evidence for a malaria aetiology (Morel-Maroger *et al.* 1975). Experimental reproduction of proteinuria and immune deposition has been successful in rats (Cameron 1984), but more convincing chronic renal disease has only been shown occasionally in primates (Hutt *et al.* 1975).

Schistosomal glomerulopathy

Evidence for this as a causative agent of the nephrotic syndrome comes from studies in Brazil, where *Schistosoma mansoni*, particularly when it causes hepatosplenomegaly, may lead to an immune-complex nephritis (Andrade and Rocha 1979), and from Egypt where *Schistosoma haematobium* has been implicated (Ezzat *et al.* 1974). Experimental evidence for causation is strong, but in parts of Nigeria where *S. haematobium* is endemic, no association with glomerular disease has been described. Rather, bladder infection can lead to obstructive uropathy and possibly to bladder carcinoma (Oyediran 1979).

Russell's viper envenoming with acute renal failure

This has been described as one of the most dramatic syndromes in nephrology; following a Russell's viper bite, patients in Burma develop severe defibrination and rapid-onset acute tubular necrosis with oliguria. Renal failure can develop within a few hours, and is heralded by albuminuria (Thein-Than *et al.* 1991). Renal failure accounts for most deaths from Russell's viper, but pituitary failure has also been described in some patients (Tun-Pe *et al.* 1987).

References

Abboud, O.L., Osman, E.M., and Musa, A.R. (1989). The aetiology of chronic renal failure in adult Sudanese patients. *Ann. Trop. Med. Parasitol.* **83**, 411–4.

Adetuyibi, A., Akisanya, J.B., and Onadeko, B.O. (1977). Analysis of the causes of death on the medical wards of the University College Hospital, Ibadan over a 14-year period (1960–1973). *Trans. Roy. Soc. Trop. Med. Hyg.* **70**, 466–73.

Adu, D., Anim-Addo, Y., Foli, A.K. *et al.* (1976). Acute renal failure in tropical Africa. *Br. Med. J.* **1**, 890–2.

Akinkugbe, O.O. and Abiose, P. (1967). Peritoneal dialysis in acute renal failure. *W. Afr. Med. J.* **16**, 165–8.

Akinkugbe, O.O. and Ojo, O.A. (1969). Arterial blood pressures in rural and urban populations in Nigeria. *Br. Med. J.* **2**, 222–4.

Akinsola, W., Odesanmi, W.O., Ogunniyi, J.O., and Ladipo, G.O. (1989). Diseases causing chronic renal failure in Nigerians—a prospective study of 100 cases. *Afr. J. Med. Med. Sci.* **18**, 131–7.

Andrade, Z.A. and Rocha, H. (1979). Schistosomal glomerulopathy. *Kidney International*, **16**, 23–9.

Atinmo, T., Mbofung, C.M., Hussain, M.A., and Osotimehin, B.O. (1985). Human protein requirements: obligatory urinary and faecal nitrogen losses and the factorial estimation of protein needs of Nigerian male adults. *Br. J. Nutr.* **54**, 605–11.

Barnett, A.H. (1991). Diabetes mellitus in the tropics. *Trans. Roy. Soc. Trop. Med. Hyg.* **85**, 327–31.

Beevers, D.G. and Price, J.S. (1991). Hypertension: an emerging problem in tropical countries. *Trans. Roy. Soc. Trop. Med. Hyg.* **85**, 324–6.

Carson, P.E. and Frischer, H. (1966). Glucose-6-phosphate dehydrogenase deficiency and related disorders of the pentose phosphate pathway. *Am. J. Med.* **45**, 401–4.

Cameron, J.S. (1984). Glomerulonephritis in roden malaria. In (ed. J.W. Kibukamasoke), *Tropical nephrology*, Cit. Forge, Camberra.

Edington, G.M. and Mainwaring, A.R. (1966). Nephropathies in West Africa. In *The kidney*, Monograph No. 6, pp. 488–501. Int. Acad. of Path.

Ezzat, E., Osman, R.A., Ahmet, K.Y. *et al.* (1974). The association between

Schistosoma haematobium and infection and heavy proteinuria. *Trans. Roy. Soc. Trop. Med. Hyg.* **68**, 315–18.

Falase, A.O. and Salako, L.A. (1979). Clinical experience with a new vasodilator, binazine in Nigerian hypertensives. *Niger. Med. J.* **9**, 577–84.

Giglioli, G. (1962). Malaria and renal disease with special reference to British Guiana. II: the effect of malarial eradication on the incidence of renal disease in British Guiana. *Ann. Trop. Med. Parasitol.* **56**, 225–41.

Gilles, H.M. and Hendrickse, R.G. (1963). Nephrosis in Nigerian children: role of *Plasmodium malariae* and effect of antimalarial treatment. *Br. Med. J.* **2**, 27–31.

Greenwood, B.M. (1986). Epidemiology of malaria in The Gambia. In (ed. A.A. Buck), *Proceedings of the conference on malaria in Africa: practical considerations on malaria vaccines and clinical trials*, pp. 176–84.

Hill, A.V.S., Allsopp, C.E.M., Kwiatkowski, D. *et al.* (1991). Common West African HLA antigens are associated with protection from severe malaria. *Nature*, **352**, 595–600.

Hutt, M.S.R. and Wing, A.J. (1971). Renal failure in the tropics. *Br. Med. Bull.* **27**, 122–7.

Kibukamasoke, J.W., Hutt, M.S.R., and Wilks, N.E. (1967). The nephrotic syndrome in Uganda and its association with quartan malaria. *Quart. J. Med.* **36**, 393–408.

Kibukamasoke, J.W. (1984). *Tropical nephrology*. Cit. Forge, Camberra.

Kramer, S.L., Campbell, C.C., and Moncrieff, R.E. (1983). Fulminant *Plasmodium falciparum* infection treated with exchange blood transfusion. *J.A.M.A.* **249**, 244–5.

Lagundoye, S.B. (1975). Abdominal X-ray changes in kwashiorkor before and after barium. *J. Trop. Pediatr.* **21**, 55–8.

Looareesuwan, S., Phillips, R.E., Karbwang, J. *et al.* (1990). *Plasmodium falciparum* hyperparasitaemia: use of exchange blood transfusion in seven patients and a review of the literature. *Quart. J. Med.* **75**, 471–81.

Lumlertgul, D., Hutdagoon, P., Sirivanichai, C. *et al.* (1988). Furosemide and dopamine in malarial acute renal failure. *Nephron*, **52**, 40–4.

McLigeyo, S.O., Otieno, L.S., Kinuthia, D.M. *et al.* (1988). *Postgrad. Med. J.* **64**, 783–6.

Morel-Maroger, L.J., Saimot, A.G., Sloper, J.C. *et al.* (1975). Tropical nephropathy and tropical extramembranous glomerulonephritis of unknown aetiology in Senegal. *Br. Med. J.* **i**, 541–6.

Odutola, T.A., Ositelu, S.B., D'Almeida, E.A. *et al.* (1989). *Afr. J. Med. Med. Sci.* **18**, 193–201.

Ogunlesi, A.O., Akanji, A.O., Kadiri, S. *et al.* (1990). Uraemia and adrenocortical function in Nigerians. *Afr. J. Med. Med. Sci.* **19**, 43–8.

Ogunlesi, A.O., Kadiri, S., Akinkugbe, O.O. (1991*a*). Acute tubular necrosis in Ibadan. *Afr. J. Med. Med. Sci.* (In press).

Ogunlesi, A.O., Osotimehin, B., Abbiyessuku, F. *et al.* (1991*b*). Blood pressure and educational level among factory workers in Ibadan, Nigeria. *J. Hum. Hyperten.* **5**, 375–80.

Ohwovoriole, A.E., Kuti, J.A., and Kabiawu, S.I. (1988). Casual blood glucose levels and prevalence of undiscovered diabetes in Lagos metropolis Nigerians. *Diabetes Res. Clin. Prac.* **4**, 153–8.

Ojogwu, L.I. (1983). Peritoneal dialysis in the management of hypertensive acute oliguric renal failure. *Trop. Geog. Med.* **35**, 385–8.

Ojogwu, L.I. and Anah, C.O. (1983). Renal failure and hypertension in tropical Africa—a pre-dialysis experience from Nigeria. *East Afr. Med. J.* **60**, 478–84.

Onwubalili, J.K. (1989). Successful peritoneal dialysis using 0.9% sodium chloride with modified M/6 sodium lactate solution and recycled catheters. *Nephron*, **53**, 24–6.

Osotimehin, B., Erasmus, R.T., Iyun, A.O. *et al.* (1984). Plasma renin activity and plasma aldosterone concentrations in untreated Nigerians with essential hypertension. *Afr. J. Med. Med. Sci.* **13**, 139–43.

Oyediran, A.B.O. and Akinkugbe, O.O. (1970). Chronic renal failure in Nigeria. *Trop. Geogr. Med.* **21**, 41–4.

Oyediran, A.B.O.O. (1979). Renal disease due to schistosomiasis of the lower urinary tract. *Kidney Int.* **16**, 15–22.

Paton, A.M. (1974). The management of chronic renal failure. *East. Afr. Med. J.* **51**, 786–8.

Poulter, N.R., Khaw, K.T., Hopwood, B.E.C. *et al.* (1990). The Kenya Luo migration study: observation of the initiation of a rise in blood pressure. *Br. Med. J.* **309**, 967–72.

Salako, L.A., Falase, A.O., and Aderounmu, A.F. (1979*a*). Comparative beta-adrenoceptor-blocking effects and pharmacokinetics of propanolol and pindolol in hypertensive Africans. *Clin. Sci.* **57**, 393s–6s.

Salako, L.A., Falase, A.O., and Aderounmu, A.F. (1979*b*). Placebo-controlled, double-blind clinical trial of alprenolol in African hypertensive patients. *Curr. Med. Res. Opin.* **6**, 358–63.

Sobh, M., Moustafa, F., and Ghoniem, M. (1988). Value of renal biopsy in chronic renal failure. *Int. Urol. Nephrol.* **20**, 77–83.

Thein-Than, Tin-Tun, Hla-Pe *et al.* (1991). Development of renal fuction abnormalities following bites by Russel's vipers (*Daboia russelii siamenses*) in Myanmar. *Trans. Roy. Soc. Trop. Med. Hyg.* **85**, 404–9.

Tun-Pe, Phillips, R.E., Warrell, D.A. *et al.* (1987). Acute and chronic pituitary failure resembling Sheehan's syndrome following bites by Russell's viper in Burma. *Lancet*, **ii**, 763–6.

Warrell, D.A., Phillips, R.E., and Garrard, C.S. (1991). Intensive care unit management of severe malaria. *Clin. Intens. Care*, **2**, 86–95.

Part 8 Nephrotic Syndrome

29 Pathogenesis of the nephrotic syndrome

J.D. Firth

The nephrotic syndrome consists of the combination of heavy proteinuria, hypoalbuminaemia and oedema, which results from avid renal retention of sodium. This chapter will address two aspects of the pathophysiological process; why the nephrotic kidney retains sodium and leaks protein.

Why does the nephrotic kidney retain sodium?

The traditional hypothesis

Epstein[1] was the first to recognize the close association between proteinuria, hypoproteinaemia and the development of oedema, invoking disturbances in the balance of Starling's forces as the mechanism by which a fall in serum protein concentration led to translocation of fluid from the capillaries into the interstitium. Why did this happen? Three pieces of evidence drew his attention away from the possibility of a renal abnormality: reports of decreased basal metabolism in subjects with chronic nephrosis, 'gratifying results' obtained using thyroid gland therapy, and the recognition that in many cases of nephrosis the glomeruli appeared normal by light microscopy. He subsequently looked for alterations in the structure of plasma proteins, perhaps caused by a forme fruste of hypothyroidism, as the basis for proteinuria and hence oedema. This approach did not prove fruitful.

In 1954, the now traditional view was articulated by Eder *et al.*[2] amongst others. Proteinuria, attributed to changes in the glomerular filter, rather than to abnormalities of the filtered protein, led to hypoproteinaemia. Transudation of salt and water into the interstitial space followed, resulting in oedema and a fall in plasma volume. Neural and humoral responses designed to preserve plasma volume were activated, inducing the kidney—an 'innocent bystander' in this regard—to retain sodium. This hypothesis is now reiterated, with or without any attempt at critical analysis, in the standard textbooks of medicine in answer to the question as to why the nephrotic kidney retains sodium. Is this maintenance of the traditional hypothesis warranted?

Studies relating to the traditional hypothesis

The hypothesis invokes a number of stages in the genesis of renal sodium retention. Amongst other things, whilst the kidney retains sodium, the f ing might be predicted: first, that plasma volume would be reduced; secondly, that the renin-angiotensin-aldosterone system would be activated, and that inhibition of the system would be natriuretic. Do the data support these predictions?

Plasma volume in the nephrotic syndrome

Many studies have described findings in small numbers of poorly-characterized subjects: four satisfy rigorous criteria by providing evidence that at the time of investigation patients were both nephrotic and retaining sodium, and a clear statement that diuretics or other medication had been omitted for at least one week prior to study. All used the distribution volume of radioiodinated albumin to estimate plasma volume. Two of these studies suggest that plasma volume is reduced in nephrotic subjects; two, including the largest, report that it is increased. In 12 patients Brown *et al.* found plasma volume to be elevated by 0.16 l above predicted values.[3] However, in a subsequent smaller study the same authors reported an apparent reduction of 0.14 l in nephrotic individuals. Geers *et al.* estimated plasma volume to be 65 ml/kg lean body mass in 52 nephrotic subjects, compared with 56 ml/kg lean body mass in 51 healthy controls.[5] Kumagai *et al* stated plasma volume to be 79 per cent of 'normal' in 11 nephrotic patients,[6] but it is noteworthy that when the patients were later studied in remission (when, presumably, they should have returned to normal) none were considered to have a plasma volume of greater than 95 per cent of the so-called normal value.

Rather than comparing estimates of plasma volume in nephrotic patients with estimates in controls, a second strategy has involved the comparison of plasma volume measured during the oedematous phase with that determined during subsequent remission in the same individual. This type of study has the obvious advantage of using each subject as his own control. However, the method is only strictly applicable to minimal change disease, in which complete remission can be anticipated. When applied to subjects with nephrotic syndrome of other histological type the 'remission' values seem to have been obtained after periods of diuretic-induced weight loss, perhaps with a brief cessation of diuretic therapy prior to study, but in which a true remission cannot be said to have occurred. Figure 29.1 shows values for plasma volume during the oedematous phase of the nephrotic syndrome in 55 patients, expressed as a percentage of the volume estimated for each individual during subsequent remission. In 16 patients, plasma volume was reduced to below 90 per cent of the remission estimate; in the same number it was increased to more than 110 per cent of that value.

The failure to observe consistent reductions in plasma volume in sodium-retaining nephrotic subjects does not necessarily refute the traditional hypothesis. If reduction in plasma volume is indeed the stimulus for sodium retention, then homoeostatic mechanisms would operate to preserve that volume: in

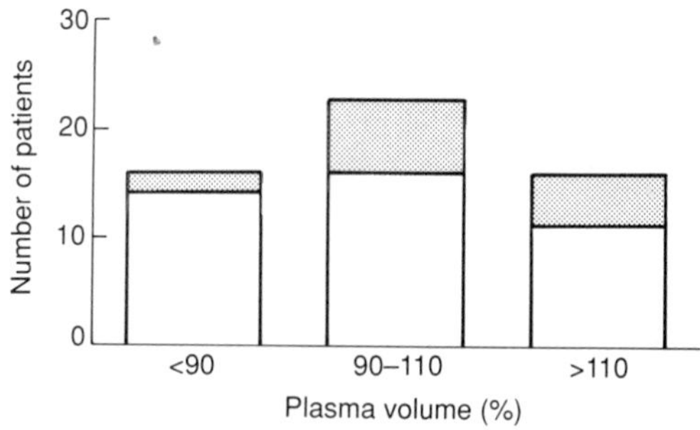

Fig. 29.1 Plasma volume during the oedematous phase of the nephrotic syndrome. Values are expressed as a percentage of the volume estimated during subsequent remission. Open bars, minimal change nephropathy; shaded bars, other forms of the nephrotic syndrome. Data from 55 patients in six studies. (Metcoff *et al.* (1952), *Paediatrics* **10**, 543; Lauson *et al.* (1954). *J. Clin. Invest.* **33**, 657; Garnett *et al.* (1967). *Lancet*, **ii**, 798; Kelsch *et al.* (1972). *J. Lab. Clin. Med.* **79**, 516; Gur *et al.* (1976). *Paediatr. Res.* **10**, 197; Kumagai *et al.* (1985). *Clin. Nephrol.* **23**, 229.

which case, unless the homoeostatic mechanisms were to be overwhelmed, it would be naïve to imagine that reductions in plasma volume could readily be detected. More difficult for the hypothesis, however, are the frequent observations that plasma volume is actually increased.

The renin-angiotensin-aldosterone system in the nephrotic syndrome

In 1955, Luetscher *et al.* isolated crystalline aldosterone from the urine of a nephrotic child.[7] This was clearly a brilliant piece of work, but it perhaps attracted undue emphasis to the proposition that aldosterone was crucially involved in the genesis of sodium retention in all cases of the nephrotic syndrome.

The most quoted study is that of Meltzer *et al.* who related initial ambulatory plasma renin activity to urinary sodium excretion in 17 episodes of nephrosis in 16 subjects, none of whom had received treatment prior to investigation.[8] Some were still accumulating fluid, some had stabilized and some were in a phase of spontaneous diuresis. In 8 instances plasma renin activity fell within the normal range of the 'renin-sodium profile'. In 6 cases, 3 of which were accumulating fluid, plasma renin was depressed. In 3 cases, each associated with fluid retention, plasma renin activity was higher than that found in salt-restricted normal controls. On the basis of these findings they proposed two different pathophysiological forms of nephrotic syndrome. These were a 'classic' form of 'vasoconstriction or hypovolaemic nephrosis' with high renin and aldosterone levels, and a second form of 'hypervolaemic or overfilling nephrosis' with depression of plasma renin and aldosterone. The largest single group of patients, those with normal plasma renin activity, received less consideration. A similarly inconsistent picture of renin-aldosterone activation was observed by Brown *et al.*[3] Furthermore, sequential studies of individual patients with the nephrotic syndrome have failed to reveal any consistent relationships between the propensity to retain sodium and plasma levels of renin or aldosterone.[9–11]

At best, observational studies of the type described above could only ever demonstrate correlation between renin-angiotensin aldosterone activity and renal sodium handling—they could never prove causation. Such proof might come from interventional studies. Dusing *et al.* infused the angiotensin II antagonist saralasin into six nephrotic subjects: they found that sodium excretion fell, rather than rising as might have been predicted.[12] Brown *et al.* administered captopril (50 mg thrice daily) to nephrotic patients who were actively retaining sodium.[4] Sodium retention continued despite the fact that plasma aldosterone levels fell substantially.

Clear evidence that activity of the renin-angiotensin-aldosterone system is crucially involved in the pathogenesis of sodium retention in more than a small proportion of cases of the nephrotic syndrome seems unconvincing.

Evidence for an intrinsic renal defect

The suggestion that nephrotic oedema might arise from a primary impairment of renal ability to excrete solute and water is not new.[13] In 1917, Epstein[1] described Widal's hypothesis of a defect in chloride excretion[13] as having 'received the soundest clinical confirmation' amongst the alternatives to his own on offer. However, Epstein's ideas subsequently dominated thinking on the pathogenesis of nephrotic oedema until the early 1980s, when the accumulation of observations inconsistent with his hypothesis led authors to resurrect the notion that there might be an intrinsic renal abnormality of sodium excretion.[3]

Strong, direct evidence implicating an intrarenal abnormality has come from the study of an animal model. Puromycin aminonucleoside (PAN) is a chemical which induces heavy proteinuria, hypoproteinaemia and oedema when administered systemically to rats. The effect results from direct action on the kidney. Chandra *et al.*[14] and Ichikawa *et al.*[15] used this property of PAN to create a unilateral model of the nephrotic syndrome. They perfused a single kidney with the chemical, leaving the contralateral kidney as an unperfused control. Despite the fact that both organs shared the same systemic milieu, heavy proteinuria and reduced sodium excretion subsequently developed only in the kidney that had been exposed to PAN. Micropuncture suggested that the depression of urinary sodium excretion resulted either from altered behaviour of tubule segments beyond the distal convolution or from altered behaviour of deep nephrons inaccessible to surface micropuncture.

A second piece of evidence strongly implicating an intrarenal abnormality is the behaviour during isolated perfusion of the kidney taken from a rat rendered

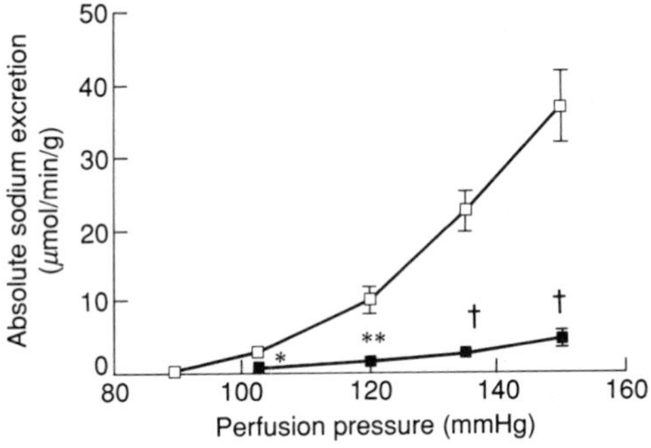

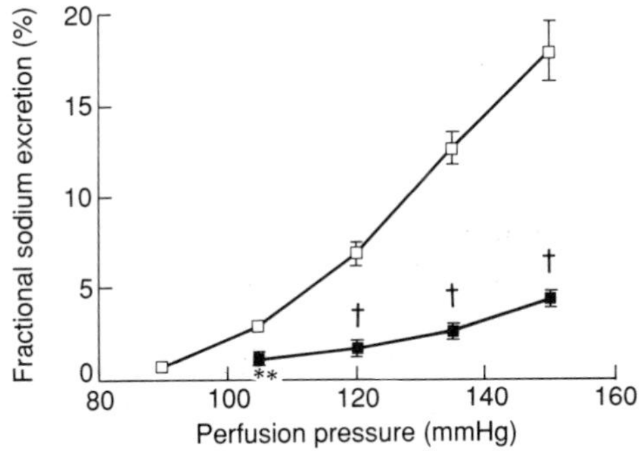

Fig. 29.2 Perfusion of the nephrotic kidney in isolation: effect on absolute and on fractional sodium excretion of increasing perfusion pressure in 6 kidneys from nephrotic animals (solid symbols) and 6 from controls (open symbols). Values are means with bars representing ± SEM. $*p < 0.05$, $**p < 0.01$, $^{\dagger}p < 0.001$ compared with control.

nephrotic by previous exposure to PAN.[16,17] In this preparation the immediate influences of renal nerves and circulating substances are clearly eliminated. Despite this, the kidney of the nephrotic animal retains sodium: at any given arterial perfusion pressure it excretes considerably less sodium than a normal control (Fig. 29.2). The reason for this is not clear: it appears to result from an abnormality of tubular, rather than glomerular, function and is associated with abnormally increased oxygen consumption. The tendency to sodium retention is not overcome by ouabain, acetazolamide, frusemide, hydroflumethiazide, benzamil or atrial natriuretic peptide; but cooling to 8–10°C—used to discern the effect of substantial, if not complete, inhibition of metabolism—renders handling of sodium by the isolated nephrotic organ virtually identical to that of control.

Summary

Recent interest in the possibility of an intrarenal defect of sodium handling in the nephrotic syndrome was born out of frustration that the traditional hypothesis could not adequately explain clinical observations. Clinical study cannot discern whether an intrinsic renal abnormality exists, since the kidney cannot be isolated from extrinsic neural and humoral influences. In one of the best animal models of the nephrotic syndrome (PAN nephrosis) there is substantial evidence of a primary intrarenal abnormality of sodium handling, but the mechanism of this abnormality remains elusive. Clearly, this is relevant to clinical nephrotic syndrome only insofar as the animal preparation is a valid model of the clinical condition.

Why does the nephrotic kidney leak protein?

There is overwhelming evidence that the glomerulus, in addition to functioning as a size-selective filter, also functions in a charge-selective manner.[18] The passage of a negatively charged molecule across the glomerular basement membrane is normally retarded to a greater degree than that of either a neutral or a positively charged molecule of the same size. Fixed anionic groups within the glomerular basement membrane and on the surface of glomerular epithelial cells are thought to be responsible for this phenomenon.[19] Despite dissenting voices,[20] the weight of evidence suggests that in minimal change nephrotic syndrome the occurrence of massive proteinuria is best explained by postulating a reduction in the charge selective properties of the filtration barrier.[21] Why does this occur? Evidence is emerging to suggest that loss of, or masking of, surface negative charge (such as would be expected to cause a diminution of glomerular charge selectivity) might be a generalized phenomenon in the nephrotic syndrome, and that this might be caused by the presence of a circulating polycationic substance(s).

Evidence for a generalized reduction in negative charge in minimal change nephrotic syndrome

Levin *et al.* used binding of a cationic dye, alcian blue, to estimate negative charge on the surface membranes of red blood cells and platelets in children with heavy proteinuria due to steroid-responsive nephrotic syndrome: binding was reduced compared with normal controls.[22] This study has been challenged on methodological grounds; in particular it has been argued that preparations of alcian blue are fickle, often precipitating spontaneously from solution, and that measurements of surface negative charge by alcian blue binding are thereby prone to considerable error.[23] However, others have produced similar findings to the original report.[24]

Ghiggeri *et al.* studied alterations in the electrical charge of serum and urinary albumin in children with minimal change nephropathy.[25,26] Albumin present in the serum of normal individuals has an isoelectric point of 4.7. The presence of less anionic forms was determined by isoelectric focusing. During the proteinuric phase of the illness the serum contained isoalbumins with a less anionic isoelectric point than normal. These anomalously charged species, which were deficient in bound fatty acids, gained preferential access to the urine. It was suggested that the reduced binding of fatty acids could result either from impairment of their access to binding sites, perhaps as a result of conformational change, or as a result of competition with other uncharacterized substances.

Evidence for a circulating polycationic substance in minimal change nephrotic syndrome

Vermylen *et al.* pursued the clinical impression that children with the nephrotic syndrome seemed to require larger doses of heparin than usual to achieve anticoagulation.[27] They added increasing amounts of heparin to plasma and measured the effect that this had on the kaolin partial thromboplastin time. The slope of the relationship between heparin concentration and kaolin partial thromboplastin time was substantially reduced. Variation of factors known to influence heparin sensitivity did not appear to provide an adequate explanation, leading the authors to conclude that there might be an underlying disturbance of charged macromolecules. In further experiments, the Great Ormond Street group produced the first direct evidence of the existence of such a substance.[28] Pooled plasma or urine from children with steroid-responsive nephrotic syndrome (but not from controls), after removal of glycosaminoglycans by ion-exchange chromatography, was capable of inhibiting *in vitro* the binding of alcian blue to heparin. This could be most easily explained by the presence of a highly charged polycation capable of binding to the polyanion heparin sufficiently tightly to resist subsequent displacement by the blue dye. The factor responsible appeared to be a protein, since its activity was lost following digestion with pronase.

Further evidence corroborating the presence of a circulating cationic substance has been obtained by Wilkinson *et al.*[29] Plasma from nephrotic patients with heavy proteinuria was infused into rabbits through an aortic catheter: urinary protein excretion increased and a reduction in the number of anionic sites in the glomerular basement membrane was revealed by polyethyleneamine binding. Plasma from controls was without effect.

Summary

A loss of charge selectivity of the glomerular basement membrane provides the most likely explanation for the occurrence of heavy proteinuria in minimal change nephrotic syndrome. Although the evidence is far from complete, a case can be made to suggest that this may be due to the presence of a circulating cationic substance, producing a generalized reduction in surface negative charge and causing proteinuria by the effect of such reduction on glomerular function.

References

1. Epstein, A.A. (1917). Concerning the causation of edema in chronic parenchymatous nephritis: method for its alteration. *Am. J. Med. Sci.* **154**, 638–47.
2. Eder, H.A., Lauson, H.D., Chinard, F.P., Greif, R.L., Cotzias, G.C., and Van Slyke, D.D. (1954). A study of the mechanism of edema formation in patients with the nephrotic syndrome. *J. Clin. Invest.* **33**, 636–56.
3. Brown, E.A., Markandu, N.D., Roulston, J.E., Jones, B.E., Squires, M., and MacGregor, G.A. (1982). Is the renin-angiotensin-aldosterone system involved in the sodium retention in the nephrotic syndrome? *Nephron*, **32**, 102–7.
4. Brown, E.A., Markandu, N.D., Sagnella, G.A., Jones, B.E., and MacGregor, G.A. (1984). Lack of effect of captopril on the sodium retention of the nephrotic syndrome. *Nephron*, **37**, 43–8.
5. Geers, A.B., Koomans, H.A., Roos, J.C., Boer, P., and Dorhout Mees, E.J. (1984). Functional relationships in the nephrotic syndrome. *Kidney Int.* **26**, 324–30.
6. Kumagai, H., Onoyama, K., Iseki, K., and Omae, T. (1985). Role of renin angiotensin aldosterone on minimal change nephrotic syndrome. *Clin. Nephrol.* **23**, 229–35.
7. Luetscher, J.A. Jr, Dowdy, A., Harvey, J., Neher, R., and Wettstein, A. (1955). Isolation of crystalline aldosterone from the urine of a child with the nephrotic syndrome. *J. Biol. Chem.* **217**, 505–12.
8. Meltzer, J.I., Keim, H.J., Laragh, J.H., Sealey, J.E., Jan, K.M., and Chien, S. (1979). Nephrotic syndrome: vasoconstriction and hypervolemic types indicated by renin-sodium profiling. *Ann. Int. Med.* **91**, 688–96.
9. Chonko, A.M., Bay, W.H., Stein, J.H., and Ferris, T.F. (1977). The role of renin and aldosterone in the salt retention of edema. *Am. J. Med.* **63**, 881–9.
10. Dorhout Mees, E.J., Roos, J.C., Boer, P., Yoe, O.H., and Simatupang, T.A. (1979). Observations on edema formation in the nephrotic syndrome in adults with minimal lesions. *Am. J. Med.* **67**, 378–84.
11. Brown, E.A., Markandu, N., Sagnella, G.A., Jones, B.E., and MacGregor, G.A. (1985). Sodium retention in nephrotic syndrome is due to an intrarenal defect: evidence from steroid-induced remission. *Nephron*, **39**, 290–5.
12. Dusing, R., Vetter, H., and Kramer, H.J. (1980). The renin-angiotensin-aldosterone system in patients with nephrotic syndrome: effects of 1-sar-8-ala-angiotensin II. *Nephron*, **25**, 187–92.
13. Widal, F. and Javal, A. (1903). La chloruremie et la cure de dechloruration dans le mal de Bright. Etude sur l'action dechlorurante de quelques diuretiques. *Presse Méd. Par.* **ii**, 701–5.
14. Chandra, M., Hoyer, J.R., and Lewy, J.E. (1981). Renal function in rats with unilateral proteinuria produced by renal perfusion with aminonucleoside. *Pediatr. Res.* **15**, 340–4.
15. Ichikawa, I., Rennke, H.G., Hoyer, J.R., Badr, K.F., Schor, N., Troy, J.L., Lechene, C.P., and Brenner, B.M. (1983). Role for intrarenal mechanisms in the impaired salt excretion of experimental nephrotic syndrome. *J. Clin. Invest.* **71**, 91–103.

16. Firth, J.D., Raine, A.E.G., and Ledingham, J.G.G. (1989). Abnormal sodium handling occurs in the isolated perfused kidney of the nephrotic rat. *Clin. Sci.* **76**, 335–41.
17. Firth, J.D. and Ledingham, J.G.G. (1990). Effect of natriuretic agents, vasoactive agents and of the inhibition of metabolism on sodium handling in the isolated perfused kidney of the nephrotic rat. *Clin. Sci.* **79**, 559–74.
18. Chang, R.L.S., Deen, W.M., Robertson, C.R., and Brenner, B.M. (1975). Permselectivity of the glomerular capillary wall: III Restricted transport of polyanions. *Kidney Int.* **8**, 212–18.
19. Rosenzweig, L.J. and Kanwar, Y.S. (1982). Removal of sulfated (heparan sulfate) or nonsulfated (hyaluronic acid) glycosaminoglycans results in increased permeability of the glomerular basement membrane to ^{125}I-bovine serum albumin. *Lab. Invest.* **47**, 177–84.
20. Golbetz, H., Black, V., Shemesh, O., and Myers, B.D. (1989). Mechanism of the antiproteinuric effect of indomethacin in nephrotic humans. *Am. J. Physiol.* **256**, F44–51.
21. Bridges, C.R., Myers, B.D., Brenner, B.M., and Deen, W.M. (1982). Glomerular charge alterations in human minimal change nephropathy. *Kidney. Int.* **22**, 677–84.
22. Levin, M., Smith, C., Walters, M.D.S., Gascoine, P. and Barratt, T.M. (1985). Steroid-responsive nephrotic syndrome: a generalized disorder of membrane negative charge. *Lancet*, **ii**, 239–42.
23. Sewell, R.F. and Brenchley, P.E.C. (1986). Red-cell surface charge in glomerular disease. *Lancet*, **ii**, 635–6, (letter).
24. Boulton-Jones, J.M., McWilliams, G., and Chandrachud, L. (1986). Variation in charge on red cells of patients with different glomerulopathies. *Lancet*, **ii**, 186–9.
25. Ghiggeri, G.M., Candiano, G., Ginevri, F., Gusmano, R., Ciardi, M.R., Perfumo, F., Delfino, G., Cuniberti, C., and Queirolo, C. (1987). Renal selectivity properties towards endogenous albumin in minimal change nephropathy. *Kidney Int.* **32**, 69–77.
26. Ghiggeri, G.M., Ginevri, F., Candiano, G., Oleggini, R., Perfumo, F., Queirolo, C., and Gusmano, R. (1987). Characterisation of cationic albumin in minimal change nephropathy. *Kidney Int.* **32**, 547–53.
27. Vermylen, C.G., Levin, M., Lanham, J.G., Hardisty, R.M., and Barratt, T.M. (1987). Decreased sensitivity to heparin *in vitro* in steroid-responsive nephrotic syndrome. *Kidney Int.* **31**, 1396–401.
28. Levin, M., Gascoine, P., Turner, M.W., and Barratt, T.M. (1989). A highly cationic protein in plasma and urine of children with steroid-responsive nephrotic syndrome. *Kidney Int.* **36**, 867–77.
29. Wilkinson, A.H., Gillespie, C., Hartley, B., and Williams, D.G. (1989). Increase in proteinuria and reduction in number of anionic sites on the glomerular basement membrane in rabbits by infusion of human nephrotic plasma *in vivo*. *Clin. Sci.* **77**, 43–8.

30 Management of the nephrotic syndrome

D. Adu

Treatment strategies for the nephrotic syndrome due to idiopathic glomerulonephritis fall into three broad categories. The first is the use of corticosteroids and immunosuppressants, and this is well established in the treatment of minimal change nephrotic syndrome (MCNS). The second is the manipulation of dietary protein and salt intake and the use of diuretics. Thirdly, pharmacological agents such as angiotensin converting enzyme inhibitors and prostaglandin synthetase inhibitors have been used in an attempt to reduce proteinuria by reducing intraglomerular pressures and glomerular filtration rate respectively. In this chapter the use of corticosteroids and immunosuppressive drugs in patients with the nephrotic syndrome and idiopathic glomerulonephritis will be considered. This discussion is restricted to the nephrotic syndrome in temperate countries. The causes and treatment of the nephrotic syndrome in the tropics differ markedly.

Minimal change nephrotic syndrome (MCNS) in children

MCNS accounts for 77 per cent of cases of the nephrotic syndrome in childhood (ISKDC 1970, 1978) (Table 30.1). With treatment with corticosteroids 93 per cent of these children respond with complete loss of proteinuria within eight weeks (ISKDC 1981). The major problem is that 39–55 per cent of these children who initially respond to corticosteroids develop multiple relapses when treatment is discontinued, or become corticosteroid dependent, and relapse when corticosteroid dosage is reduced (ISKDC 1974; Schwartz *et al.*, 1974; ISKDC 1979; Barrett *et al.* 1978). Where corticosteroid toxicity poses problems there is good evidence that short-term treatment with cyclophosphamide can induce sustained or even permanent remission (ISKDC 1974*a, b*; Cameron *et al.* 1974; Barratt and Soothill 1970). Cyclophosphamide is given in a dose of 3 mg/kg/d (ideal height for weight) for 6–8 weeks. Approximately 60 per cent of treated children are in remission at 2 years and 40 per cent at 5 years. Chlorambucil has also been used to treat these patients, but there is no evidence that it is better than cyclophosphamide (Arbeitsgemeinschaft für pädiatrische Nephrologie 1982). Cyclophosphamide has been carefully evaluated in these children and is the drug of choice. The gonadal toxicity of cyclophosphamide,

Table 30.1 Minimal change nephrotic syndrome

	Children	Adults
Proportion of nephrotic syndrome	77%	20–25%
Total remission prednisolone	93%	80%
Partial remission prednisolone	5–10%	12%
Resistant		8%
Time to remission	8 weeks	16 weeks
Frequent relapsers/cortico-steroid dependent	39–55%	21%

and its oncogenic potential are well recognized. At the doses and duration of treatment outlined above it is relatively safe.

MCNS in adults

Some 25–30 per cent of adults with nephrotic syndrome have MCNS (Sharpstone *et al.* 1969). The nephrotic syndrome in these patients responds to corticosteroids slightly less often than in children, and also more slowly. Eighty per cent of adults with MCNS respond to corticosteroids, but remission can take up to 16 weeks to occur (Nolasco *et al.* 1986). The number of relapses in adults is less frequent than in children, at 1.7 per patient, and only 21 per cent of adults develop multiple relapses or are corticosteroid dependent (Nolasco *et al.* 1986). In these patients, cyclophosphamide was effective in inducing a long-lasting remission. In the study of Nolasco *et al.* (1986) 62.5 per cent of patients treated with cyclophosphamide were in remission at 10 years.

Long-term outcome of MCNS

In children 5.5 per cent of patients, all of whom presented before the age of 6 years, continued to relapse into adult life (Trompeter *et al.* 1985). The long-term mortality rate in these children ranges from 2.6 per cent (ISKDC 1984) to 3.2 per cent (Trompeter *et al.* 1985). In one study, 6 per cent of adults were nephrotic after a mean follow-up of 7.5 years (Nolasco *et al.* 1986). The survival in patients aged over 60 years was 50 per cent at 10 years, and in those aged 15–50 years survival was 90 per cent (Nolasco *et al.* 1986). The progression of corticosteroid-responsive biopsy-proven MCNS to end stage renal failure is rare, and is usually due to the subsequent development of focal segmental glomerulosclerosis.

Cyclosporin A in MCNS

There is now good evidence that cyclosporin A is effective in the treatment of MCNS in both adults and children (Meyrier *et al.* 1986; Niaudet *et al.* 1987;

Tejani *et al.* 1988; Maher *et al.* 1988; Meyrier *et al.* 1989). The literature on the use of cyclosporin A in MCNS has been reviewed by Meyrier *et al.* (1989). Approximately 85 per cent of adults and 74 per cent of children with MCNS went into remission on cyclosporin A. The response was better in patients who were corticosteroid responsive (90 per cent in adults and 82 per cent in children) than in patients resistant to corticosteroids (68 per cent in adults and 42 per cent in children).

Several guidelines can be drawn from the studies of cyclosporin A in MCNS. The drug is effective in corticosteroid-responsive MCNS and its use could be considered in patients who run into risks of corticosteroid toxicity because they have multiple relapses or are corticosteroid dependent. Cyclosporin A appears to be effective at blood levels of 100–200 ng/ml, and at these levels significant toxicity is uncommon. Relapses appear to recur with the same frequency after cyclosporin A has been discontinued as before (Brodel and Hoyer 1989). For that reason it would still seem advisable to use cyclophosphamide as the treatment of first choice in patients with a multiple relapsing or corticosteroid dependent MCNS, in the hope of inducing a sustained remission. Cyclosporin can best be viewed as a corticosteroid-sparing agent in patients with MCNS.

Focal segmental glomerulosclerosis

Fewer terms have generated more disagreement amongst pathologists and nephrologists than focal segmental glomerulosclerosis (FSGS). It is important to reiterate that FSGS is not a disease entity, but a histological lesion that is often of unknown aetiology. It has been shown in experimental models that segmental sclerosis can develop from different pathogenic mechanisms (Howie *et al.* 1989). These include toxic injury (puromycin nephropathy), immunological injury (anti-GBM nephritis), the nephritis of NZB/NZW F1 (lupus) and hyperfiltration injury (5/6ths nephrectomy). Some of these models have clinical counterparts, and the diversity of pathogenic mechanisms explains the variability in the clinical presentation of FSGS as well as its responses to therapy.

Classical FSGS

The original description by Rich was of focal lesions in the juxtamedullary glomeruli of children who had a nephrotic syndrome and died within five years (Rich 1957). Approximately 9 per cent of children with the nephrotic syndrome in temperate countries have FSGS at the onset of their illness (Churg *et al.* 1970). The clinical course in these patients has been reviewed (Cameron 1979). Only 30 per cent of these children respond to corticosteroids (Trompeter and Barratt 1988) and these do better than patients who do not respond. Of the remainder of patients resistant to corticosteroids, 30 per cent are in end stage renal failure within five years of onset of the disease (Habib 1973; Cameron

1979). The data from Guys Hospital shows no difference in prognosis between adults and children and emphasizes the point that patients with a nephrotic syndrome fared less well than those with proteinuria only (Cameron 1979).

It is also recognized that some patients whose initial renal biopsies show a minimal change nephropathy, and who are steroid responsive subsequently, are found to have on repeat renal biopsies the changes of FSGS (Hayslett *et al.* 1969). These patients in the main do not have haematuria, they remain responsive to corticosteroids, and only rarely develop end stage renal failure (Hayslett *et al.* 1969; Cameron 1979).

Recently the features in patients with FSGS that might predict their responsiveness to corticosteroids and progression to renal failure have been examined. Initial studies showed that the site of the segmental sclerosing lesions predicted steroid responsiveness. Adult patients with a segmental sclerosing lesion only at the tubular origin, the glomerular tip lesion, (Howie and Brewer 1984) had a corticosteroid or immunosuppressant responsive nephrotic syndrome and did not progress to end stage renal failure (Beaman *et al.* 1987). Similar observations were reported in children by Ito *et al.* (1984), although in both children (Morita *et al.* 1990) and adults (Cameron 1989) these observations have not been confirmed. Nevertheless, the initial observations have recently been confirmed in a further 10 patients with the glomerular tip lesion, all of whom responded to corticosteroids or immunosuppressants (Howie *et al.* unpublished). These studies have emphasized the importance of serial sections in the study of FSGS.

Cyclosporin A in FSGS

Several studies have looked at the effects of cyclosporin A in patients with FSGS and the nephrotic syndrome (Meyrier *et al.* 1986; Niaudet *et al.* 1987; Tehani *et al.* 1988; Maher *et al.* 1988; Meyrier *et al.* 1989). In general, adults with FSGS did not respond to cyclosporin A (Meyrier *et al.* 1986; Maher *et al.* 1988; Meyrier *et al.* 1989). In children, responsiveness to cyclosporin A paralleled corticosteroid responsiveness, and those patients who were corticosteroid resistant achieved little or no benefit from cyclosporin A (Niaudet *et al.* 1987; Tejani *et al.* 1988; Brodehl and Hoyer 1989).

Membranous nephropathy

The twin aims of treating membranous nephropathy are firstly to induce a remission of the nephrotic syndrome, and secondly to prevent the development of end stage renal failure. Despite several careful studies using corticosteroids and immunosuppressants, there is still no agreement that these aims can be achieved (Ponticelli 1986; Donadio *et al.* 1988).

A starting point in the study of treatment for membranous nephropathy is the natural history of this disorder. This has been reviewed by Cameron

(1989). After 5–15 years of follow-up, 23 per cent of patients are in remission, 55 per cent of patients have normal renal function, and 30 per cent are either uraemic or dead. A second point of importance is that idiopathic membranous nephropathy may have a different aetiology in different parts of the world, and this could well affect its varying responsiveness to corticosteroids and immunosuppressants.

Corticosteroid treatment

In the Collaborative Study in the USA (1979), 72 adults with membranous nephropathy were randomized to treatment with either prednisolone (125 mg on alternate days), or placebo for eight weeks (Table 30.2). Deterioration of renal function, as measured by the glomerular filtration rate, was significantly more rapid in untreated than in treated patients. Furthermore, a significantly lower proportion of treated patients than untreated patients developed renal failure (serum creatinine > 440 μmol/l). In the Medical Research Council study in the UK, 107 adult patients with membranous nephropathy were randomized to treatment with either prednisolone (125 mg on alternate days for 8 weeks), or placebo. At 36 months there were no significant differences in plasma creatinine, creatinine clearance and 24-h urine protein between treated and untreated patients (Cameron *et al.* 1990). In the Canadian study, 158 patients

Table 30.2 Prednisolone in membranous nephropathy

Study	Number of patients	Total remission	Partial remission	Unchanged	Worse function
USA Collaborative					
Treated	34	4	8	22	−2% GFR
Untreated	38	4	3	31	−10% GFR
Canadian					
Treated	81	16			32%
Untreated	77	19			25%
MRC					
Treated	52	10			15
Untreated	51	7			26
Ponticelli					
Treated	32	12	11	9	0
Untreated	20	2	7	13	8

USA: Prednisolone (125 mg/alternate day/8 weeks > 45–80 kg;
150 mg/alternate day/8 weeks > 80 kg tapered)
Canadian: Prednisolone (45 mg/M^2/d/6 months)
MRC: Prednisolone (125 mg/alternate day/8 weeks)
Ponticelli: Methyl prednisolone/chlorambucil/alternate months.

were treated with either prednisolone 45 mg/M^{2}body surface area for 6 months or no specific treatment. No benefits were seen in renal function or proteinuria after a mean follow-up of 48 months (Cattran *et al.* 1989). These data do not suggest that corticosteroids are of benefit in the treatment of membranous nephropathy.

The study of Lagrue *et al.* (1975) showed that chlorambucil was more effective than azathioprine or placebo in the treatment of membranous nephropathy. This provided the rationale for the Italian multicentre study, in which patients were randomized to symptomatic treatment only, or to treatment with prednisolone and chlorambucil. The treatment comprised intravenous methylprednisolone for 3 days, followed by oral methylprednisolone for 27 days, alternating with oral chlorambucil for 1 month (Ponticelli *et al.* 1984; Ponticelli 1986). The treatment regimen was given for 6 months. After a mean follow-up of 31–37 months, significantly more treated patients were in remission (either total or partial), 23/32 (72 per cent), than untreated patients, 9/30 (30 per cent). Furthermore, 8 of 30 controls had a 50 per cent rise in serum creatinine and, by contrast, none of the treated patients had a rise in their serum creatinine. Side effects of treatment were minor, and consisted of epigastric pain (2 patients) and leucopenia (2 patients). This study merits replication.

Whether patients with a membranous nephropathy should be treated with corticosteroids and immunosuppresants remains to be established. Deterioration of renal function is more common in adult males, patients with a nephrotic syndrome, and patients with initial poor renal function or whose renal function deteriorated in the first 2.5 years after diagnosis (Davison *et al.* 1984; Donadio *et al.* 1988; Cameron *et al.* 1990). A case could be made for treating such patients with the Italian regime.

Mesangiocapillary glomerulonephritis (MCGN)

The renal lesions of MCGN are well characterized, as are the complement abnormalities in this disorder. In type I MCGN immunohistology shows subendothelial deposits, and in type II MCGN electron dense deposits are found within the glomerular and, at times, tubular basement membranes (Levy *et al.* 1979). A third type has also been described, in which there are subepithelial and subendothelial deposits (Strife *et al.* 1977). Persistently low levels of C3, and the presence of C3 nephritic factor, are characteristically found in dense deposit MCGN (Williams *et al.* 1974) and may also be found in the subendothelial variety, but the relationship between these complement abnormalities and the pathogenesis of the nephritis are unclear.

The prognosis of MCGN is of a slowly progressive history with the development of end stage renal failure in some 50 per cent of patients, 10 years after onset of disease (Habib *et al.* 1973; Magil *et al.* 1979; Cameron *et al.* 1983; Watson *et al.* 1984). Some of these patients were treated with corticosteroids and/or immunosuppressants. Whilst most observers are unconvinced of the benefits of corticosteroids and immunosuppressants in the treatment of MCGN, West's group have consistently supported the value of long-term alternate-day prednisolone in children with this disorder (McAdams *et al.* 1975; McEnery *et al.* 1980; McEnery *et al.* 1985; West 1986). In their hands this regime was associated with a 74 per cent survival after 11 years, and this improved to 90 per cent in patients starting treatment early (West 1986). Although their results are impressive they were uncontrolled, and do not preclude the necessity of a controlled trial in this disorder. Unfortunately, the ISKDC trial was inconclusive because of the large number of withdrawals due to corticosteroid-induced hypertension and encephalopathy (ISKDC 1982). Likewise, evidence of benefit in this disorder from anticoagulants and anti-platelet drugs are inconclusive (Donadio *et al.* 1984).

References

Arbeitsgemeinchaft Für pädiatatrisch Nephrologie (1982). Effect of cytotoxic drugs in frequently relapsing nephrotic syndrome with and without steroid dependence. *New England Journal of Medicine*, **306**, 451–4.

Barratt, T.M. and Soothill, J.F. (1970). Controlled trial of cyclophosphamide in steroid-sensitive relapsing nephrotic syndrome of childhood. *Lancet*, **2**, 479–82.

Barrett, H., Schoenman, M., Bernstein, J., and Edelmann, C.M. (1978). Minimal change nephrotic syndrome. In *Pediatric nephrology*, (ed. C.M. Edelmann), pp. 695–711. Little Brown, Boston.

Beaman, M., Howie, A.J., Hardwicke, J. *et al.* (1987). The glomerular tip lesion: a steroid responsive nephrotic syndrome. *Clinical Nephrology*, **27**, 217–21.

Brodehl, J. and Hoyer, P.F. (1989). Ciclosporin in idiopathic nephrotic syndrome of children. *American Journal of Nephrology*, **9**, (1), 61–64.

Cameron, J.S. (1979). The problem of focal segmental glomerulosclerosis; In (ed. P. Kincaid-Smith, S.J. D'Apice, R.W. Atkins) pp. 209–28. *Progress in glomerulonephritis*. Wiley, New York.

Cameron, J.S. (1989). Treatment of primary glomerulonephritis using immunosuppresive agents. *American Journal of Nephrology*, **1**, 33–40.

Cameron, J.S., Chantler, C., Ogg, C.S., and White, R.H.R. (1974). Long term stability of remission in nephrotic syndrome after treatment with cyclophosphamide. *British Medical Journal*, **4**, 7–11.

Cameron, J.S., Turner, D.R., Hiteman, J. *et al.* (1983). Idiopathic mesangiocapillary glomerulonephritis: Comparison of types 1 and 11 in children and adults and long-term progression. *American Journal of Medicine*, **74**, 175–92.

Cameron, J.S., Healy, M.J.R., and Adu, D. (1990). The Medical Research Council Trial of short-term high-dose alternate day prednisolone in idiopathic membranous nephropathy with nephrotic syndrome in adults. *Quarterly Journal of Medicine*, **274**, 133–56.

Cattran, D.C., Delmore, T., Roscoe, J., Cole, E., Cardella, C., Charron, R., and Ritchie, S. (1989). A randomized controlled trial of prednisone in patients with idiopathic membranous nephropathy. *New England Journal of Medicine*, **320**, 210–15.

Churg, J., Habib, R., and White, R.H.R. (1970). Pathology of the nephrotic syndrome in children. A report for the International study of Kidney Disease in Children. *Lancet*, **1**, 1299–302.

Collaborative Study of the Adult Idiopathic Nephrotic Syndrome (1979). A controlled study of short-term treatment in adults with membranous nephropathy. *New England Journal of Medicine*, **301**, 1301–6.

Davison, A.M., Cameron, J.S., Kerr, D.N.S., Ogg, C.S., and Wilkinson, R.W. (1984). The natural history of renal function in untreated idiopathic membranous glomerulonephritis in adults. *Clinical Nephrology*, **22**, 61–7.

Donadio, J.G., Anderson, F.G., Mitchell, J.C. *et al.* (1984). Membranoproliferative glomerulonephritis. A prospective trial of platelet inhibitor therapy. *New England Journal of Medicine*, **310**, 1421–6.

Donadio, J.V., Torres, V.E., Velosa, J.A., Wagoner, R.D., Holley, K.E., Okanuera, M., Ilstrup, D.M., and Chu, C.P. (1988). Idiopathic membranous nephropathy: the natural history of untreated patients. *Kidney International*, **33**, 708–15.

Habib, R. (1973). Focal glomerular sclerosis. *Kidney International*, **4**, 355.

Habib, R., Kleinknecht, C., Gubler, M., and Levy, M. (1973). Idiopathic membranoproliferative glomerulonephritis in children. Report of 105 cases. *Clinical Nephrology*, **1**, 194–214.

Hayslett, J.P., Kraser, L.S., Klaus, Cr. *et al.* (1969). Progression of lipoid nephrosis to renal insufficiency. *New England Journal of Medicine*, **281**, 181.

Howie, A.J. and Brewer, D.B. (1984). The glomerular tip lesion: a previously undescribed type of segmental glomerular abnormality. *Journal of Pathology*, **142**, 205–20.

Howie, A.J., Kizaki, T., Beaman, M. *et al.* (1989). Different types of segmental sclerosing glomerular lesions in six experimental models of proteinuria. *Journal of Pathology*, **157**, 141–51.

International Study of Kidney Disease in Children (1970). Pathology of the nephrotic syndrome in children. *Lancet*, **1**, 1299–302.

International Study of Kidney Disease in Children (1974*a*). Prospective controlled trial of cyclophosphamide therapy in children with the nephrotic syndrome. *Lancet*, **2**, 423–7.

International Study of Kidney Disease in Children (1974*b*). Prospective controlled trial of cyclophosphamide treatment in children with nephrotic syndrome. *Lancet*, **ii**, 427–37.3.

International Study of Kidney Disease in Children (1978). Prediction of histopathology from clinical and laboratory characteristics at the time of diagnosis. *Kidney International*, **13**, 159–65.

International Study of Kidney Disease in Children (1979). A randomized trial comparing two prednisone regimens in steroid-responsive patients who relapse early. *Journal of Pediatrics*, **95**, 239–43.

International Study of Kidney Disease in Children (1981). The primary nephrotic syndrome in children. Identification of patients with minimal change nephrotic syndrome from initial response to prednisolone. *Journal of Pediatrics*, **98**, 560–4.

International Study of Kidney Disease in Children (1982). Alternate-day steroid therapy in membranoproliferative glomerulonephritis: a randomized controlled clinical trial *Kidney International*, **21**, 150, (Abstract).

International Study of Kidney Disease in Children (1984). Minimal change nephropathy in children: deaths during the first five to fifteen years observation. *Pediatrics*, **73**, 497–50.

Ito, H., Yoshikawa, N., Azoai, F. *et al.* (1984). Twenty-seven children with focal segmental glomerulosclerosis. Correlation between the segmental location of the glomerular lesion and prognosis. *Clinical Nephrology*, **22**, 9–14.

Lagrue, G., Bernard, J., Bariety, P., Arnet, P., and Covenel, J. (1975). Controlled trial of chlorambucil and azathioprine in idiopathic chronic glomerulonephritis (abstract). *Kidney International*, **8**, 274.

Levy, M., Gubler, M.C., and Habib, R. (1979). New concepts on membranoproliferative glomerulonephritis. In *Progress in Glomerulonephritis*. (ed. P. Kincaid-Smith, A.J.F. d'Apice, R.C. Atkins, pp. 177–205. John Wiley, New York.

Magil, A.B., Price, J.D.E., Bower, G. *et al.* (1979). Membranoproliferative glomerulonephritis type 1: comparison of natural history in children and adults. *Clinical Nephrology*, **11**, 239–44.

Maher, E.R., Sweny, P., Chappel, M., Varghese, Z., and Moorhead, T.F. (1988). Cyclosporin in the treatment of steroid-responsive and steroid-resistant nephrotic syndrome in adults. *Nephrology Dialysis Transplant*, **3**, 728–32.

McAdams, A.J., McEnery, P.T., and West, C.D. (1975). Mesangiocapillary glomerulonephritis: changes in glomerular morphology with long-term alternate-day prednisone therapy. *Journal of Pediatrics*, **86**, 23–31.

McEnery, P.T., McAdams, A.J., and West, C.D. (1980). Membranoproliferative glomerulonephritis: improved survival with alternate-day prednisolone therapy. *Clinical Nephrology*, **13**, 117–24.

McEnery, P.T., McAdams, A.J., and West, C.D. (1985). The effect of prednisolone in a high-dose, alternate-day regime on the natural history of idiopathic membranoproliferative glomerulonephritis. *Medicine*, **64**, 401–23.

Meyrier, A., Simon, P., Perret, G., and Condamin-Meyrier, M-C. (1986). Remission of idiopathic nephrotic syndrome after treatment with cyclosporin A. *British Medical Journal*, **292**, 789–92.

Meyrier, A., and Collaborative Group of the Societe de Nephrologie. (1989). Ciclosporin in the treatment of nephrosis. *American Journal of Nephrology*, **1**, 65–71.

Morita, M., White, R.H.R., Coad, N.A.G., and Raafat, F. (1990). The clinical significance of the glomerular sclerosis. *Clinical Nephrology*, **33**, 211–19.

Niaudet, P., Habib, R., Tete, M-J., Hinglais, N., and Broyer, M. (1987). Cyclosporin in the treatment of idiopathic nephrotic syndrome in children. *Pediatric Nephrology*, **1**, 566–78.

Nolasco, F., Cameron, J.S., Heywood, E.F., Hicks, J., Ogg, C., and Williams, D.G. (1986). Adult-onset minimal change nephrotic syndrome. A long-term follow-up. *Kidney International*, **29**, 1215–33.

Ponticelli, C. (1986). Prognosis and treatment of membranous nephropathy. *Kidney International*, **29**, 927–40.

Ponticelli, C., Zucchelli, P., Imbasciati, E., Cagnoli, L., Pozzi, C., Paserini, P., Grassi, C., Limido, D., Pasquali, S., Volpini, T., Sasdelli, M., and Locatelli, F. (1984). Controlled trial of methylprednisolone and chlorambucil in idiopathic membranous nephropathy. *New England Journal of Medicine*, **310**, 946–50.

Rich, A.R. (1957). A hitherto undescribed vulnerability of the juxtamedullary glomeruli in the lipoid nephrosis. *Bull. Johns Hopkins Hosp.* **100**, 173.

Schwartz, M.W., Schwartz, G.J., and Cornfield, D. (1974). A 16 year follow-up of 163 children with nephrotic syndrome. *Pediatrics*, **54**, 547–52.

Sharpstone, P., Ogg, C.S., and Cameron, J.S. (1969). Nephrotic syndrome due to primary renal disease in Adults. A survey of incidence in South East England. *British Medical Journal*, 533–5.

Short, C.D., Soloman, L.R., Mallick, N.P., and Gokal, R. (1987). High-dose methylprednisolone therapy in patients with membranous nephropathy and declining renal function. *Quarterly Journal of Medicine*, **65**, 929–40.

Strife, C.F., McEnery, P.T., McAdams, A.J. *et al.* (1977). Membranoproliferative glomerulonephritis with disruption of the glomerular basement membrane. *Clinical Nephrology*, **7**, 65.

Tejani. A., Butt, K., Trachtman, H., and Suthanthiran, M. (1988). Cyclosporine A induced remission of relapsing nephrotic syndrome in children. *Kidney International*, **33**, 729–34.

Trompeter, R. and Barratt, T.M. (1988). Treatment and management of the nephrotic syndrome in children. In *The nephrotic syndrome*. (ed. J.S. Cameron and R.J. Glassock) pp. 423–60. Marcel Dekker, New York.

Trompeter, R.S., Lloyd, B.W., Hicks, J., White, R.H.R., and Cameron, J.S. (1985). Long term outcome for children with minimal change nephrotic syndrome. *Lancet*, **1**, 368–70.

Watson, A.R., Porcell, S., Thorner, P., *et al.* (1984). Membranoproliferative glomerulonephritis Type 1 in children: correlation of clinical features with pathologic sub-types. *American Journal Of Kidney Disease*, 141–6.

West, C.D. (1986). Childhood membranoproliferative glomerulonephritis: an approach to management. *Kidney International*, **29**, 1077–83.

Williams, D.G., Peters, D.K., Fallows, J. *et al* (1974). Studies of serum complement in the hypocomplementaemic nephritides. *Clinical and Experimental Immunology*, **18**, 391.

Part 9 Complications of Renal Disease

31 Progression of renal failure; have we advanced?

J. Feehally

It has been recognized for many years that progression of renal failure is an inevitable consequence of renal injury in many patients, even if the initial insult is transitory. This progression may not be prevented even by the very best clinical care.

The concept of structural and functional adaptation as a response to nephron loss was advanced more than 40 years ago, to explain the remarkable capacity of the kidney to maintain the internal environment until the very late stages of renal failure.[1] That such changes could also result in unfavourable effects was not viewed with prominence until a seminal paper by Brenner *et al.*,[2] which drew attention to such maladaption as the eventual cause of progressive renal failure, whatever the injury which initiated nephron drop-out. There has now been nearly a decade of vigorous research into the pathophysiological mechanisms which may underlie this inexorable progression of renal disease. Brenner's hypothesis particularly emphasized that dietary protein limitation might ameliorate this maladaptive process and delay progression of renal failure: this approach has now been widely studied in experimental animals and humans.

These studies require critical review in order to appreciate those findings in experimental animals which are not yet proven to be applicable to humans, and to understand the limitations of interpretation which can be made from clinical studies of imperfect design.

Mechanisms of progression—experimental studies

Many factors have been proposed to contribute to experimental progression, but there are six areas of chief interest.

Hyperfiltration

Brenner's proposition was that reduction in nephron numbers from whatever cause would result in hyperfiltration through remnant nephrons, and glomerular hypertension. This hyperfiltration and hypertension would produce proteinuria and mesangial matrix increase, which in turn would result in glomerular sclerosis, starting a vicious circle of further nephron loss. Experimental evidence for this mechanism is powerful and based mainly on studies in the remnant (5/6ths nephrectomy) rat model using the Munich–Wistar strain, in which micropuncture of superficial nephrons can provide direct functional data from individual nephrons.

Such experiments show that glomerular hypertension is a feature of the remnant rat model, and that progressive glomerular sclerosis and proteinuria may be aggravated by high protein intake and lessened by a range of manoeuvres which reduce glomerular hypertension, including dietary protein restriction and control of systemic blood pressure. The choice of hypotensive agents in this setting has been studied, and evidence has been presented that angiotensin converting enzyme (ACE) inhibitors may have particular advantages, because they reduce glomerular efferent arteriolar tone, and may therefore reduce glomerular hypertension independent of any effect on systemic blood pressure.[3] However, not all experimental data supports the primacy of glomerular hypertension. Some models develop glomerular sclerosis without glomerular hypertension and some strategies (including thromboxane synthesis inhibition and exercise) have reduced sclerosis and proteinuria without reducing glomerular pressure.

Hypertrophy

As well as pressure changes, it has been known for many years that reduction in renal mass also produces hypertrophy of remnant nephrons.[4] Experimental studies, particularly those of Ichikawa *et al.*, have suggested that hypertrophy without hyperfiltration may result in glomerular sclerosis and proteinuria. Elegant experiments using peritoneal ureteric diversion show that the contralateral kidney develops glomerular hypertension, but there is no acceleration of glomerular sclerosis unless hypertrophy is also induced by surgical reduction of renal mass.[5] Furthermore, there are strain differences in susceptibility of rats to glomerular sclerosis: PVG/c rats, which have unusually small glomeruli, do not develop glomerulosclerosis even if hyperfiltration is increased by uninephrectomy.[6] The potential importance of renal compensatory growth has led to intensive study of the mechanisms by which it is stimulated and controlled and the importance of a range of extracellular and intracellular messengers on renal growth have recently been reviewed.[7]

Systemic hypertension

Much experimental evidence confirms that systemic hypertension accelerates renal injury in a range of models,[8] provided that the systemic blood pressure rise is transmitted to the glomeruli. Thus, afferent glomerular arteriolar constriction (for example, from the effects of a low protein diet) protects the glomerulus, whereas afferent vasodilatation facilitates glomerular injury.

Furthermore, a wide range of manipulations which have been suggested in animal studies to slow progression are known to lower blood pressure in the rat. These include thromboxane synthetase inhibitors, anticoagulants and antiplatelet agents, lipid-lowering agents, and a diet rich in polyunsaturated fats.

Their success in preventing progression may partly reflect good systemic blood pressure control.

The variety of insults which induce experimental renal injury produce differing changes in glomerular size and pressure. Available evidence does not allow a single mechanism to be invoked for progression; the interplay of glomerular hypertension, systemic hypertension, and glomerular hypertrophy appear to contribute to various degrees in different circumstances.

Proteinuria

Experimental studies regularly use proteinuria as a marker of glomerular injury, although recent evidence suggests that proteinuria may emanate from intact nephrons.[9] Proteinuria is so closely linked to hyperfiltration that it may be impossible to separate the two effects. It is proposed that presence of excess protein in the urinary space is toxic to glomerular epithelial cells and contributes directly to ongoing renal injury. The major support for this view comes from studies of 'overflow' proteinuria, loading animals parenterally with heterologous protein.[10] It should be noted, however, that similar experiments using homologous protein do not produce epithelial cell injury.[11]

Lipid deposition

The morphological analogies between glomerulosclerosis and atherosclerosis are many,[12] and hyperlipidaemia is a consistent feature of progressive uraemia. Glomerular lipid deposition is regarded by some as epiphenomenal, and by others as directly nephrotoxic. Available experimental evidence has not yet resolved this issue.[13]

Calcium phosphate deposition

Hyperphosphataemia in renal insufficiency may allow the calcium phosphate solubility product to be exceeded, resulting in tissue calcification, which may accelerate progression. Phosphate restriction has been shown to delay progression in animal studies.[14]

Glomerular dysfunction

Whatever the interplay of hypertension and hypertrophy, whatever the damage caused by proteinuria or lipid deposition, the events which lead from that initial insult to sclerosis are incompletely understood at the cellular and subcellular level. Work has focused attention on all the major intrinsic glomerular cell types and also on infiltrating cells.[15]

Mesangial cell damage, due to accumulation of macromolecules from increased trafficking, will result in the release of many inflammatory mediators which may produce sclerosis.[16] Endothelial injury may further modify haemodynamic changes and result in the release of cytokines and other mediators which could attract platelets and stimulate cells. Epithelial podocyte retraction is described and presumed to be associated with epithelial cell dysfunction. There is still virtually no direct information on the relevance of these experimental concepts to humans and they have yet to be assessed in clinical studies.

Progression of renal disease in humans

For practical reasons, detailed pathophysiological studies of progression have mainly used the rat remnant kidney model or other rat models of renal injury. There is limited information as to their direct applicability to humans, for several reasons.

First, the rat appears to be uniquely sensitive to reduced renal mass. Remnant models in other species produce little or no glomerulosclerosis and limited functional changes.[17] Mere uninephrectomy in rats will produce proteinuria and glomerulosclerosis, whereas long-term follow up of human kidney donors shows no progressive renal failure.[18] Evidence in humans of the effect of reduction in renal mass equivalent to the rat remnant model is anecdotal.

Secondly, direct measurements of single nephron function by micropuncture cannot be made in humans, so that the presence of hyperfiltration can only be inferred from systemic blood pressure and whole kidney measurements of glomerular filtration rate and renal plasma flow.

Finally, dietary protein manipulations which modify renal scarring adversely or beneficially in the rat are sometimes extreme, the protein content of the diet ranging from 6 to 60 per cent. They are probably inapplicable in humans.

Rather more information is available about glomerular hypertrophy in humans, which was first noted in renal failure 40 years ago.[4] Although unilateral nephrectomy (which always induces compensatory hypertrophy) does not lead to progressive renal failure, the rare inherited condition of oligomeganephronia is associated with progressive renal failure, and it has recently been reported that the glomeruli of nephrotic patients who develop focal glomerulosclerosis are larger than those with minimal change nephrotic syndrome.[19]

Proteinuria is an established marker of poor outcome in patients with chronic glomerular disease,[21] and its resolution indicates improved prognosis. The close relationship between proteinuria and systemic blood pressure means that their investigation as separate risk factors may be extremely difficult. ACE inhibitors lessen proteinuria in many, but not all, patients.[22] Non-steroidal anti-inflammatory agents lessen proteinuria, but usually as a consequence of falling glomerular filtration rate.

Hyperlipidaemia has commanded little therapeutic attention in renal failure, because lipid lowering agents suitable for uraemic patients have not been available. The recent introduction of HMG-CoA reductase inhibitors may result in significant progress, and already one brief report has suggested that lipid control may lessen proteinuria and delay progression in nephrotic patients with early renal insufficiency.[23]

Importance of interstitial injury

Intensive interest in the glomerulus may obscure the fact that tubulo-interstitial injury, both inflammatory and fibrotic, is a universal feature of the end stage kidney in both experimental animals and humans. Although often viewed as a secondary phenomenon, since tubular atrophy inevitably follows glomerular failure, it should be remembered that cellular interstitial infiltrate is a characteristic early feature of the remnant kidney model. Some models, for example adriamycin nephrosis, proceed to uraemia with tubulo-interstitial fibrosis predominating over modest histological glomerular damage, even though the initial injury results in nephrosis and is, at least in part, glomerular. Twenty years ago, a clinicopathological analysis in chronic renal failure showed a better correlation of interstitial lesions than glomerular lesions with fall in glomerular filtration rate.[24] There is a strong case for intensive investigation of the mechanisms of tubulo-interstitial injury, and its role in the progression of renal failure.

Assessment of progression in humans

In the last 10 years many studies manipulating dietary and other parameters in patients with chronic renal failure have been published. However, difficulties in both the design and the interpretation of progression studies have come to light.[25,26]

(1) Up to 15 per cent of patients with chronic renal failure will not progress, over an observation period of several years.[25]

(2) There is a powerful 'placebo' effect in any randomized study, with many patients showing slower progression despite the absence of intervention, presumably because of the close supervision and attention to detail in blood pressure control and other clinical care which is required in these studies.[26]

(3) Assessment of nutritional status is necessary, since restricted diets may lead to significant deficiencies. Anthropometry is a minimum requirement. Serum albumin and tranferrin may remain normal with negative nitrogen balance, and a more sensitive test of muscle turnover, such as ^{3}H-methylhistidine release, may be needed. Acidaemia modifies muscle turnover and has rarely been addressed in these studies.

(4) Measurement of renal function to assess progression is unsatisfactory. The reciprocal creatinine plot, originally described by Mitch and Walser,[28] is valuable in routine clinical practice, but is not sound enough for comparison of groups of patients in clinical studies. As Walser himself has pointed out, the alterations in creatinine metabolism and tubular creatinine handling known to occur in chronic renal failure may both under- and overestimate the true glomerular filtration rate.[25] A clearance method using

inulin, or an appropriate isotope, and involving urine measurement is the 'gold standard'. Plasma isotope disappearance is an excellent and more realistic practical alternative, although less secure at low glomerular filtration rates.

(5) It may be difficult to achieve adequate and comparable control of the many variables other than the one specifically being tested, in different treatment groups. For example, in a study designed to investigate protein restriction it would be necessary to ensure that there was comparable control of phosphate intake, lipid profile, blood pressure, acidosis and (in diabetic patients) glycaemic control.

Because of the uncertainty that pathophysiological mechanisms of progression identified in animals do indeed apply to human disease, well-designed clinical studies are needed. The aim of a perfectly designed randomized prospective study is regarded by some as unrealistic, particularly since patient numbers may need to be high to give worthwhile information. There has been much debate, particularly among proponents of dietary protein restriction, as to the adequacy of historical controls or cross-over studies in which patients act as their own controls.[25,26,29]

Nevertheless, an ideal study of any variable in progression requires true randomization with an adequate run-in period, to assess progression and allow for placebo effect, and adequate assessments of nutrition and compliance, using an accurate measurement of glomerular filtration rate.[26] Worthwhile controlled clinical studies are still only available for two clinical manoeuvres—low protein (and low phosphate) diet, and blood pressure control.

Low protein diet

This is a time-honoured treatment to lessen uraemia, but as much as 40 years ago it was also suggested it may slow progression.[1] Low protein diet has been used widely, particularly in Southern Europe, for more than 20 years. Unfortunately, the very large experience which favours a low protein diet is mostly based on non-randomized trials, and frequently uses historical controls.[16] Nevertheless, there is support for the benefits of a low protein diet in a number of well-designed cross-over studies,[15] although only two randomized controlled studies of low protein diet have been published. The first showed benefit of a reduction to 0.4–0.6 g protein/kg body weight/d, but unfortunately relied on creatinine clearance to assess changes in renal function.[30] The second, the best available study to date, used isotope glomerular filtration rate measurements, and showed the benefit of 0.4 g protein/kg/d.[31] It proved possible to prevent malnutrition in these studies by the provision of adequate calories. Even lower protein intakes, supplemented by essential amino acids and keto-acids, have been proposed, but these supplements are expensive and do not always prevent malnutrition.[32] Two further large studies to confirm the efficacy of a low protein diet are currently underway in USA and Europe.

Dietary phosphate control

It is difficult to discriminate between dietary protein and phosphate effects, since the dietary content of the two is so closely linked. Nevertheless, two studies have addressed this question, and although one showed some slowing of progression with phosphate restriction,[33] in neither study was creatinine clearance significantly different at the end of the study.[33,34] No randomized studies using adequate measurements of glomerular filtration rate have been reported.

Blood pressure control

Control of blood pressure has long been a hallmark of good clinical practice in the treatment of renal impairment.[20] Particular recent interest has been in diabetic nephropathy, where blood pressure control has been shown to delay progression in controlled studies of ACE inhibitors[35] and other agents.[36] However, the particular advantages of ACE inhibitors, suggested by animal studies, have yet to be confirmed in humans by comparison with other hypotensive agents in prolonged randomized studies. Indeed, in the absence of such evidence, they should be used with great caution in patients with uraemia and diabetic nephropathy who may have accelerated atheroma with occult renal artery stenosis; in these patients ACE inhibitors could cause a rapid decline in renal function.

Conclusions

Forty years ago, long before renal replacement therapy was conceived, Platt wrote, 'I shall continue to recommend low protein diet to my renal failure patients, however competent their adaptive processes. By doing so it would seem that we may increase their survival and even if we do not achieve that object we may reduce their blood urea.'[1]

For clinical nephrologists, the advice is unchanged and the only additional confident recommendation is that proper control of blood pressure be achieved. However, the introduction of new agents and good design of powerful clinical studies should tell us in the next few years what other manoeuvres are worthwhile in the amelioration of progressive renal failure.

References

1. Platt, R. (1952). Structural and functional adaptation in renal failure. *Br. Med. J.* 1372–7.
2. Brenner, B.M., Meyer, T.W., and Hostetter, T.H. (1982). Dietary protein intake and the progressive nature of glomerular disease. *New Engl. J. Med.* **307**, 652–9.
3. Anderson, S., Meyer, T., Remke, H., and Brenner, B. (1985). Control of glomerular hypertension limits glomerular injury in rats with reduced renal mass. *J. Clin. Invest.* **76**, 612–19.

4. Oliver, J. (1950). *J. Urol.* **63**, 373.
5. Yoshida, Y., Fogo, A., and Ichikawa, I. (1989). Glomerular hemodynamic changes vs hypertrophy in experimental glomerular sclerosis. *Kidney Int.* **35**, 654–60.
6. Grond, J., Beukers, J., Schilthuis, M., Weening, J., and Elema, J. (1986). Analysis of renal structural and functional features in two rats strains with different susceptibility to glomerular sclerosis. *Lab. Invest.* **54**, 77–83.
7. Fine, L. and Norman, J. (1989). Cellular events in renal hypertrophy. *Ann. Rev. Physiol.* **51**, 19–32.
8. Baldwin, D.S. and Neugarten, J. (1986). Blood pressure control and progression of renal insufficiency. In *The Progressive nature of renal disease*, (ed. W.E. Mitch, B. Brenner, J. Stein) pp. 81–110, Churchill Livingstone, New York.
9. Yoshiaka, T., Shiraga, H., Yoshida, Y. *et al.* (1988). 'Intact' nephrons as the primary origin of proteinuria in chronic renal disease. *J. Clin. Invest.* **82**, 1614–23.
10. Davies, D.J., Brewer, D.B., and Hardwicke, J. (1978). Urinary proteins and glomerular morphometry in protein overload proteinuria. *Lab. Invest.* **38**, 232–43.
11. Schwartz, M., Bidani, A., and Lewis, E.J. (1986). Glomeular epithelial cell function in chronic proteinuria induced by homologous protein loading. *Lab. Invest.* **56**, 673–9.
12. Diamond, J. and Karnovsky M. (1988). Focal and segmental glomerulosclerosis: analogies to atherosclerosis. *Kidney. Int.* **33**, 917–24.
13. Moorhead, J.F., El Nahas Am Chan, M.K., and Varghese, Z. (1982). Lipid nephrotoxicity in chronic progressive glomerular and tubulo-interstitial disease. *Lancet*, **ii**, 1309–11.
14. Lumlertgul, L., Burke, T.J., Gillum, D. *et al.* (1986). Phosphate depletion arrests progression of chronic renal failure independent of protein intake. *Kidney Int.* **29**, 658–66.
15. Klahr, S., Schreiner, G., Ichikawa, I. (1988). The progression of renal disease. *New Engl. J. Med.* **318**, 1657–66.
16. Remuzzi, G.,and Bertani, T. (1990). Is glomerulosclerosis a consequence of altered glomerular permeability to macromolecules? *Kidney Int.* **38**, 384–94.
17. Fine, L.G. (1988). Preventing the progression of human renal disease: have rational therapeutic principles emerged? *Kidney Int.* **33**, 116–28.
18. Taseth, T., Fauchald, P., Skrede, S. *et al.* (1986). Long-term blood pressure and renal function in kidney donors. *Kidney Int.* **29**, 1072–6.
19. Fogo, A., Hawkins, E., and Berry, P. (1990). Glomerular hypertrophy in minimal change disease predicts subsequent progression to focal glomerulosclerosis. *Kidney Int.* **38**, 115–23.
20. Brazy, P.C., Stead, W.W., and Fitzwilliam, J. (1989). Progression of renal insufficiency: role of blood pressure. *Kidney Int.* **35**, 670–4.
21. Williams, P.S., Fass, G., and Bone, J.M. (1988). Renal pathology and proteinuria determine progression in untreated mild chronic renal failure. *Quart. J. Med.* **67**, 343–54.
22. Traindl, O., Pohanka, E., and Kovarik, J. (1988). Progression of chronic renal failure. *Lancet*, **ii**, 962–3 (letter).
23. Rabelink, A., Hene, R., Grkelens, D., Joles, J., and Koomans, H., (1990). Partial remission of nephrotic syndrome in patients on long-term simvastatin. *Lancet*, 1045–6 (letter).
24. Risdon, R.A., Sloper, J.C., and de Wardener, H.E. (1968). Relationship between renal function and histological changes found in renal biopsy specimens from patients with persistent glomerular nephritis. *Lancet*, **ii**, 362–7.

25. Walser, M. (1990). Progression of chronic renal failure in man. *Kidney Int.* **37**, 1195–210.
26. El Nahas, A.M., Coles, G.A. (1986). Dietary treatment of chronic renal failure: ten unanswered questions. *Lancet*, **i**, 597–600.
27. Bergstrom, J., Alvestrand, A., Bucht, H., and Guterriez, A. (1986). Progression of chronic renal failure in man is retarded with more frequent clinical follow-ups and better blood pressure control. *Clin. Nephrol.* **25**, 1.
28. Mitch, W.E., Walser, M., Buffington, G.A., and Lemann, J. (1976). A simple method for estimating progression of chronic renal failure. *Lancet*, **ii**, 1326–8.
29. Giovanetti, S. (1986). Answers to ten questions on the dietary treatment of chronic renal failure. *Lancet*, **ii**, 1140–2.
30. Rosman, J., Terwee, P., Meijer, S., *et al.* (1984). Prospective randomised trial of early dietary protein restriction in chronic renal failure. *Lancet*, **ii**, 1291–6.
31. Ihle, B., Becker, G., Whitworth, J., Charlwood, R., and Kincaid-Smith, P. (1989). The effect of protein restriction on the progression of renal insufficiency. *New Engl. J. Med.* **321**, 1773–7.
32. Lucas, P.A., Meadows, J.H., Roberts, D.E., Coles, G.A. (1986). The risks and benefits of a low protein-essential amono acid-ketoacid diet. *Kidney Int.* **29**, 995–1003.
33. Barsotti, G., Giannoni, E., Morelli, E. *et al.* (1984). The decline of renal function slowed by very low phosphorus intake in chronic renal failure patient following a low nitrogen diet. *Clin. Nephrol.* **21**, 54–9.
34. Barrientos, A., Arteapa, J., Rodicio, J. *et al.* (1982). Role of the control of phosphate in the progression of chronic renal failure. *Miner. Electrolyte Metab.* **7**, 127–33.
35. Parving, H., Hommel, E., and Smidt, U. (1988). Protection of kidney function and decrease in albuminuria by captopril in insulin dependent diabetics with nephropathy. *Br. Med. J.* **297**, 1086–91.
36. Parving, H., Anderson, A., Smidt, U. *et al.* (1987). Effect of antihypertensive treatment on kidney function in diabetic nephropathy. *Br. Med. J.* **294**, 1443–7.

32 Amyloidosis

D.M. Vigushin and M.B. Pepys

Amyloidosis is a disorder resulting from the extracellular deposition of insoluble protein as amyloid fibrils. These occur in association with amyloid P component, the glycoprotein derived from serum amyloid P component, a member of the pentraxin family of plasma proteins, to which C-reactive protein belongs (Pepys 1988).

The deposition of amyloid fibrils may be localized or systemic, and a classification based on the biochemical nature of the amyloid fibril is in common use. The most common forms of systemic amyloidosis are AA amyloidosis, in which the fibril is derived from serum amyloid A protein, an acute phase reactant circulating in increased concentration in many chronic inflammatory diseases, and AL amyloidosis, in which the fibrils derive from monoclonal immunoglobulin light chains (Hawkins *et al.* 1990). The kidney is a frequent target organ in both of these forms of systemic amyloidosis, accounting for significant morbidity and mortality (Pepys 1988). Senile systemic amyloidosis, in which there is a circulating variant of transthyretin (prealbumin), may also result in renal papillary deposits. In addition, there are a number of hereditary forms of systemic amyloidosis in which renal amyloid deposits may form (Pepys 1987).

In recent years, a new form of amyloidosis has been recognized as an important complication of long-term haemodialysis (Maury 1990), and continuous ambulatory peritoneal dialysis (Cornelis *et al.* 1989). A syndrome consisting of polyarthropathy, median nerve compression, bone cysts with associated pathological fractures, soft tissue masses, and tenosynovitis occurring in these patients is now known to result from the deposition of β_2-microglobulin as amyloid deposits at these sites (Stone and Hakim 1989). Attention has also been focused on the possible systemic deposition of amyloid in such patients, and there have now been several reported cases of β_2-microglobulin amyloid found in sites other than the periarticular and bony tissues described above. These deposits have usually been small and perivascular, but more extensive systemic involvement has also been described (Pepys 1988). In the latter, immunohistochemical stains have confirmed the β_2-microglobulin composition of the amyloid deposits (Maury 1990).

Clinical features

The most common presentation of renal amyloidosis is proteinuria resulting from the glomerular deposition of amyloid (Pepys 1987). This is of sufficient magnitude to constitute the nephrotic syndrome in up to 70 per cent of cases. Other components of the nephron may also be affected by systemic amyloidosis. Tubular deposits can cause renal tubular acidosis, and more rarely, nephrogenic diabetes insipidus (Pepys 1987). The complication of renal vein thrombosis is an important cause of acute deterioration in renal function, as well as resulting in progressive proteinuria in some cases (Tribe and Mackenzie 1982). There have also been reports of both vascular and interstitial deposits of amyloid without accompanying involvement of the glomerulus or proteinuria (Falck *et al.* 1983). Haematuria is a rare presenting feature in all forms of renal amyloidosis (Pepys 1987).

In the case of β_2-microglobulin amyloidosis, the syndrome resulting from the periarticular and bony deposits of amyloid fibrils occurs in patients on long-term dialysis with a frequency related to the duration of dialysis (Maury 1990). There has recently been a report of β_2-microglobulin amyloid developing in a patient with chronic renal failure prior to the institution of dialysis, adding strength to the hypothesis that persistently elevated levels of circulating β_2-microglobulin are the causative factor in the development of this condition (Zingraff *et al.* 1990). The incidence of carpal tunnel syndrome is increased in patients treated with haemodialysis for greater than 5 years, reaching 30 per cent at 7 years, and 50 per cent after 12 years (Pepys 1988; Maury 1990). It is frequently bilateral and β_2-microglobulin amyloid is a consistent finding in most cases in which adequate tissue specimens have been obtained following surgical decompression. However, there are a few cases in which amyloid deposits have not been demonstrated (Maury 1990; Stone and Hakim 1989).

The large joint polyarthropathy is characterized by joint space narrowing, effusions and juxtaarticular bone cysts, which may result in pathological fractures in some cases (Stone and Hakim 1989). A contentious issue has been the association of a spondyloarthropathy with dialysis-associated β_2-microglobulin amyloidosis, in which the finding of amyloid deposits has not been as consistent as in the carpal tunnel syndrome. Hyperparathyroidism has been postulated as an aetiological factor, both in the genesis of destructive spondyloarthropathy and in β_2-microglobulin amyloidosis itself (Jadoul and Van Ypersele de Strihou 1990). This remains to be elucidated.

Pathogenesis

The pathogenesis of amyloidosis has not been fully established, but a qualitatively or quantitatively abnormal precursor protein appears to be an essential feature. Some precursors undergo proteolysis prior to polymerization as amyloid fibrils, as in the case of AA amyloidosis (Pepys 1988). In other forms of amyloidosis a subtle genetic variant of the precursor may be necessary for the production of amyloid deposits, as occurs with the substitution of methionine for valine at position 30 in the transthyretin amino acid sequence in most cases of familial amyloid polyneuropathy type I (Pepys 1987).

In the case of β_2-microglobulin amyloid, there is evidence that high levels of circulating normal β_2-microglobulin are responsible (Pepys 1988), although a protein of lower molecular weight than 11.8 kDa has been identified, and proposed as amyloidogenic (Jadoul and Van Ypersele de Strihou 1990). *In vitro* studies however, have demonstrated the formation of amyloid from intact β_2-microglobulin polymers (Maury 1990). Other factors may also play a role in the pathogenesis of the syndrome associated with β_2-microglobulin amyloid. It has been recently suggested that β_2-microglobulin may inhibit bone formation and induce the synthesis of collagenases (Jadoul and Van Ypersele de Strihou 1990). Structural similarities to bone-derived growth factor have also been demonstrated, but its function remains uncertain (Stone and Hakim 1989).

As the synthesis of β_2-microglobulin occurs in all cells which express MHC class I antigens on their surface (Pepys 1988), studies have been performed on the effect of cytokines, including tumour necrosis factor, interferons and interleukins 1 and 2, on the stimulation of these cells to synthesize and release β_2-microglobulin, as well as on various proteinases which may play a role in the enhancement of amyloid fibril deposition (Jadoul and Van Ypersele de Strihou 1990). There is also evidence that the cytokines interleukin 1 and tumour necrosis factor alpha may have even more powerful effects on bone resorption than parathyroid hormone (Stone and Hakim 1989).

The β_2-microglobulin is normally filtered at the glomerulus, and thereafter undergoes catabolism in the proximal tubule (Pepys 1988). There is no mechanism for catabolism of this protein outside the kidney, and therefore with declining glomerular filtration rates, serum levels of β_2-microglobulin rise correspondingly and in parallel with creatinine (Stone and Hakim 1989).

That there is a correlation between the serum β_2-microglobulin level and the type of dialysis membrane used has been well documented. When high flux dialysis membranes made of polysulphone or polyacrylonitrile are used, serum β_2-microglobulin values may be 30 per cent lower than during haemodialysis with cuprophane membranes (Maury 1990; Jadoul and Van Ypersele de Strihou 1990). Haemodiafiltration may achieve a high rate of elimination of β_2-microglobulin, but the volumes required to be exchanged to reduce the serum concentration to normal values are impractical (Pepys 1988). The rise in β_2-microglobulin concentration seen with cuprophane membranes may be secondary to haemoconcentration, but the possible role of cytokines activated during haemodialysis has yet to be fully evaluated (Jadoul and Van Ypersele de Strihou 1990). However, since β_2-microglobulin amyloid deposits are found in patients on continuous ambulatory peritoneal dialysis (Cornelis *et al.* 1989), and have been demonstrated in a patient with chronic renal failure who had not received renal replacement therapy (Zingraff *et al.* 1990), these may prove to be relatively unimportant.

The role of glycosaminoglycans and amyloid P component remains uncertain. All forms of amyloid have been shown to contain amyloid P component

(Hawkins *et al.* 1988), and it may function in retarding the degradation of amyloid fibrils, either by physical means, or possibly by enzyme inhibition (Pepys 1988). The inhibition of elastase has been suggested (Argiles *et al.* 1990), but this property has not been confirmed. Sulphated glycosaminoglycans in the form of dermatan and heparan sulphate are consistently found in association with amyloid fibrils, and amyloid P component has been shown to bind to these in a calcium-dependent manner (Pepys 1988). Furthermore, the half life of amyloid P component is greatly extended once deposited in amyloid, when compared with that of circulating serum amyloid P component (Pepys 1988; Hawkins *et al.* 1988).

Antiproteinases have also been implicated in the pathogenesis of amyloidosis. Alpha-2-macroglobulin has been found in association with β_2-microglobulin amyloid deposits, and may confer resistance to proteolytic degradation (Argiles *et al.* 1990). However, the precise role of proteinase inhibition by alpha-2-macroglobulin and other antiproteinases including cystatin-C found in the Icelandic type of amyloidosis, and the Kunitz type of antiproteinase in Alzheimer's disease, has not been fully elucidated (Argiles *et al.* 1990).

Diagnosis

The characteristic tinctorial properties of amyloid fibrils on staining with Congo red in an alkaline alcoholic medium has made this histological method the gold standard in the diagnosis of amyloidosis. The resulting apple-green positive birefringence seen under polarized light is pathognomonic. Pretreatment with potassium permanganate helps to distinguish AA from AL amyloidosis, and immunohistochemical techniques enable the type of amyloidosis to be determined with certainty (Pepys 1987). However, tissue samples need to be obtained from biopsy or operative resection specimens, and are subject to sampling error (Pepys 1988). There is also an increased risk of haemorrhage resulting from these procedures, which may relate to the perivascular distribution of amyloid and increased vascular fragility, and in a small number of cases, a bleeding diathesis has been ascribed to factor ten deficiency, combined factors nine and ten deficiency, or hyperfibrinolysis (Sane *et al.* 1989). Moreover, histological examination fails to provide any information regarding the extent and distribution of amyloid deposits in the patient. Thus, until recently, little was known about the natural history of amyloid, apart from information obtained from pathological studies. In addition, the effect of any therapeutic intervention on the extent and pattern of deposition of amyloid could hitherto not be accurately determined (Hawkins *et al.* 1988).

A specific and non-invasive method for the diagnosis and localization of amyloidosis has been developed in the form of human serum amyloid P component labelled with ^{123}iodine (Hawkins *et al.* 1988, 1990). This has enabled the scintigraphic imaging of amyloid deposits in β_2-microglobulin amyloidosis

(Hawkins *et al.* 1988), as well as other systemic and localized forms of amyloid (Hawkins *et al.* 1988, 1990). The measurement of whole body retention of labelled serum amyloid P component, and thereby the determination of that present in the extracellular compartment, permits the deposition of amyloid to be quantified, and is useful in monitoring progression and the effect of therapy (Hawkins *et al.* 1988). Asymptomatic at-risk patients may also be identified by this method and appropriate therapy instituted at an early stage.

Therapeutic considerations

To date, the treatment of established amyloidosis is limited to the treatment of the underlying disease process. Encouraging results are emerging with regard to systemic AA amyloidosis, and there is now good evidence that aggressive anti-inflammatory treatment of juvenile chronic arthritis and rheumatoid arthritis using cytotoxic agents, including chlorambucil and cyclophosphamide, can halt the progression, and in some cases promote regression, of established amyloid deposits (Pepys 1988; Hawkins *et al.* 1990).

Various cases of resolution of the nephrotic syndrome resulting from renal amyloidosis have been reported, following treatment of the underlying chronic inflammatory disorders. These have included pulmonary tuberculosis, and chronic suppurative conditions such as those resulting from subcutaneous opiate abuse and osteomyelitis (Pepys 1988; Soler Amigo *et al.* 1990). Cases such as these are unusual, however, and the prognosis in AA amyloidosis correlates closely with the degree of renal involvement (Pepys 1987).

In AL amyloidosis, therapy has been far less successful. Nonetheless, aggressive suppression of the abnormal B-cell clone may retard the otherwise inexorable progression of amyloidosis (Pepys 1987).

The treatment of β_2-microglobulin amyloidosis is best achieved by renal transplantation, and the restoration of normal or near-normal renal function with concomitant reduction in serum β_2-microglobulin levels has resulted in rapid symptomatic improvement (Maury 1990; Jadoul and Van Ypersele de Strihou 1990). This may be due to the effect of corticosteroids administered to suppress rejection, but initial studies have suggested a reduction in the amount of amyloid present in such patients (Jadoul and Van Ypersele de Strihou 1990). Further investigation is currently in progress.

With the membranes currently available for haemodialysis and haemodiafiltration, the routine normalization of serum β_2-microglobulin is hardly feasible given the constraints of time, cost, and exchange volumes necessary (Pepys 1988).

In cases of systemic amyloidosis where the kidney is predominantly affected, renal allografting may be effective, but recurrence may occur in the transplanted organ (Pepys 1987). A recurrence rate of 10 per cent has been cited, most of these cases occurring within a three year period post transplantation (Glassock *et al.* 1986).

Conclusions

A brief review of some aspects of amyloidosis that relate to current practice in nephrology has been presented. The diagnostic targeting of amyloid deposits in β_2-microglobulin amyloidosis should provide further insight into the natural history of this increasingly common condition. Non-invasive monitoring of the amyloid burden in this and other forms of amyloidosis will facilitate the evaluation of therapeutic interventions. Finally, there is the potential for amyloid P component, by virtue of its specific calcium-dependent binding to amyloid fibrils, to be used for the targeting of therapeutic agents (Pepys 1988).

References

Argiles, A., Mourad, G., Atkins, R.C., and Mion, C.M. (1990). New insights into the pathogenesis of haemodialysis associated amyloidosis. *Seminars in dialysis*, **3**, 149–52.

Cornelis, F., Bardin, T., Faller, B., Verger, C., Allouache, M., Raymond, P., Rottembourg, J., Tourliere, D., Benhamou, C., Noel, L.H., and Kuntz, D. (1989). Rheumatic syndromes and β_2-microglobulin amyloidosis in patients receiving long term peritoneal dialysis. *Arthritis Rheum.* **32**, 785–8.

Falck, H.M., Tornroth, T., and Wegelius, O. (1983). Predominantly vascular amyloid deposition in the kidney in patients with minimal or no proteinuria. *Clin. Nephrol.* **19**, 137–42.

Glassock, R.J., Cohen, A.H., Adler, S.G., and Ward, H.J. (1986). Secondary glomerular disease. In *The kidney*, (3rd edn). (ed. B.M. Brenner and F.C. Rector) pp. 1045–50. W.B. Saunders, Philadelphia.

Hawkins, P.N., Lavender, J.P., Myers, M.J., and Pepys, M.B. (1988). Diagnostic radionuclide imaging of amyloid: biological targeting by circulating human serum amyloid P component. *Lancet*, **i**, 1413–18.

Hawkins, P.N., Lavender, J.P., and Pepys, M.B., (1990). Evaluation of systemic amyloidosis by scintigraphy with I-123 labeled serum amyloid P component. *N. Engl. J. Med.* **323**, 508–13.

Jadoul, M. and Van Ypersele de Strihou, C. (1990). New diseases of dialysis patients: amyloidosis. In *International yearbook of nephrology*, pp. 217–33.

Maury, C.P.J. (1990). β_2-microglobulin amyloidosis. *Rheumatol. Int.* **10**, 1–8.

Pepys, M.B. (1987). Amyloidosis. In *Oxford textbook of medicine*, (2nd edn) (D.J. Weatherall, J.G.G. Ledingham, and D.A. Warrell), pp. 9.145–57. Oxford University Press.

Pepys, M.B. (1988). Amyloidosis: some recent developments. *Q.J. Med.* **67**, 283–98.

Sane, D.C., Pizzo, S.V., and Greenberg, C.S. (1989). Elevated urokinase-type plasminogen activator level and bleeding in amyloidosis: case report and literature review. *Am. J. Hematol.* **31**, 53–7.

Soler Amigo, J., Orriols, J., Modol, J., and Garcia, A. (1990). Resolution of nephrotic syndrome secondary heroin-associated renal amyloidosis. *Nephrol. Dial. Transplant.* **5**, 158 (letter).

Stone, W.J. and Hakim, R.M. (1989). Beta-2-microglobulin amyloidosis in long term dialysis patients. *Am. J. Nephol.* **9**, 177–83.

Tribe, C.R. and Mackenzie, J.C. (1982). Amyloidosis. In *Rheumatology I. The kidney and rheumatic disease*, (ed. P.A. Bacon and N.M. Hadler), pp. 297–322. Butterworth Scientific, London.

Zingraff, J.J., Noel, L.H., Bardin, T., Atienza, C., Zins, B., Drueke, T.B., and Kuntz, D. (1990). β_2-microglobulin amyloidosis in chronic renal failure *N. Engl. J. Med.* **323**, 1070–1 (letter).

33 Erythropoietin therapy for renal anaemia

I.C. Macdougall

The advent of recombinant human erythropoietin is without doubt one of the greatest advances in nephrology in recent years. It has transformed the management of the anaemia of end stage renal disease, which previously relied on frequent blood transfusions with considerable disadvantages to the patient. Several large multicentre trials in the USA and Europe have confirmed that erythropoietin is a highly effective therapy with few adverse effects (Eschbach *et al.* 1989; Sundal and Kaeser 1989; Canadian Erythropoietin Study Group 1990).

The rationale for its use came with the recognition that, although the aetiology of renal anaemia is multifactorial, by far the major factor is a relative deficiency of erythropoietin produced by the diseased kidneys. Thus, circulating levels of this hormone are almost always inappropriately low for the degree of anaemia in chronic renal failure (Caro *et al.* 1979). A major breakthrough came in 1977, when human erythropoietin was isolated and purified from the urine of patients with aplastic anaemia (Miyake *et al.* 1977); this allowed the cloning of the gene for human erythropoietin (Lin *et al.* 1985), which was then expressed in a suitable mammalian cell line, making possible the large-scale synthesis of genetically engineered hormone. Animal studies confirmed its efficacy and relative safety, and clinical trials began in Seattle and London/Oxford towards the end of 1985.

Biochemistry and physiology

Erythropoietin is a glycoprotein hormone which is generated mainly by the kidneys (Jacobson *et al.* 1957), although up to 10 per cent may be produced in the liver. The site of synthesis in the kidney remains controversial, with *in situ* hybridization localizing its mRNA to peritubular interstitial cells (Lacombe *et al.* 1988), or to tubular epithelial cells of the renal cortex (Maxwell *et al.* 1989). In its physiologically active form, erythropoietin is a 165 amino acid monomeric protein, with a molecular weight of 30.4 kDa, of which approximately 40 per cent is carbohydrate (Davis *et al.* 1987). The carbohydrate residues of erythropoietin are not required for its biological activity or target cell specificity when measured *in vitro* but, in common with other plasma glycoproteins, prevent its rapid removal from the circulation.

The stimulus for the secretion of erythropoietin is insufficient delivery of oxygen to the tissues to meet metabolic demands (Kurtz *et al.* 1988). Erythropoietin, in turn, enables increased erythropoietic activity in the bone marrow by stimulating the proliferation and differentiation of erythroid precursors and the release of larger numbers of mature red blood cells into the circulation. This process is under negative feedback control (Fig. 33.1).

At a cellular level, erythropoietin exerts its action by binding specifically to receptors on the erythroid progenitor cells in the bone marrow. It then sets in motion a series of events which culminate in the production of the erythrocyte. The first measurable processes are an increase in intracellular calcium concentration and in glucose uptake, followed by α and β globin gene transcription at 6 h, and an increase in transferrin receptor expression. By 12 h, haemoglobin is being produced (Krantz *et al.* 1988).

Erythropoietin, synthesized by recombinant DNA technology has been found to be virtually identical to endogenous human urinary erythropoietin, possessing the same physicochemical, immunological, and physiological/pharmacological properties (Davis *et al.* 1987). It has an activity of greater than 200 000 units/mg protein based on the 2nd International Reference preparation (Recny *et al.* 1987). In view of its extreme hydrophobicity, current

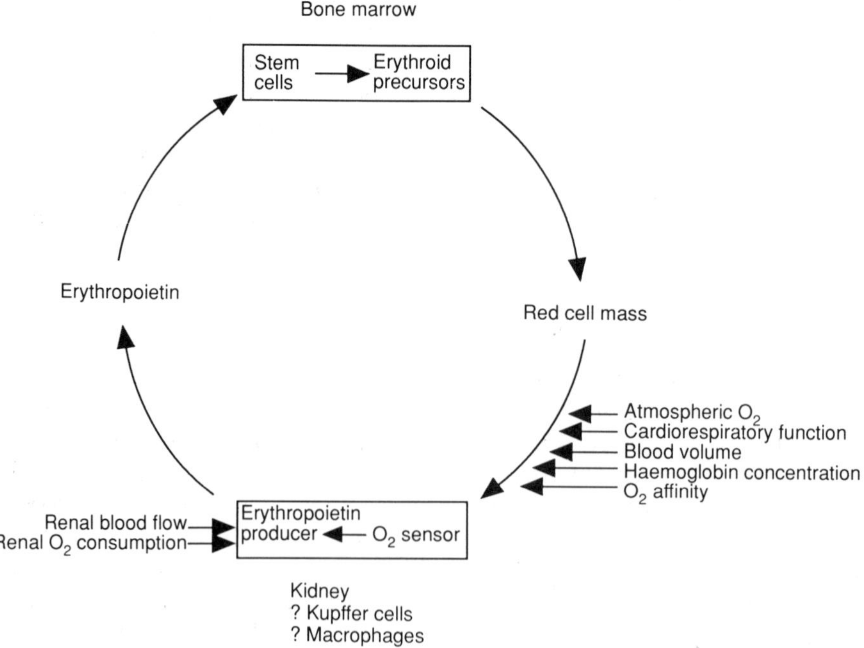

Fig. 33.1 Schematic representation of the feedback system that regulates red cell production.

formulations of erythropoietin in solution contain a carrier protein (human serum albumin) or a mixture of amino acids to reduce adsorptive losses.

Pharmacokinetics and pharmacodynamics

In common with insulin and other therapeutic protein hormones, recombinant erythropoietin is inactivated in the stomach and therefore needs to be given parenterally. The early clinical trials in haemodialysis patients used intravenously-administered erythropoietin given thrice weekly; the subcutaneous and intraperitoneal (in patients on continuous ambulatory peritoneal dialysis (CAPD)) routes were also subsequently investigated. Following intravenous administration, serum erythropoietin levels decay mono-exponentially, with an elimination half-life of around 8 h (Macdougall *et al.* 1989; Neumayer *et al.* 1989); this appears to shorten with repeated administration. The volume of distribution is approximately 1–2 times the plasma volume.

After intraperitoneal administration to CAPD patients, erythropoietin levels in serum begin to rise by 1–2 h and reach a peak at around 18 h (Macdougall *et al.* 1989). The peak levels however are only 2–5 per cent of those obtained after an equivalent intravenous dose, and the bioavailability of intraperitoneal erythropoietin is only 3 per cent. Thus, despite an early report of its efficacy in treating five anaemic CAPD patients (Frenken *et al.* 1988), this route is impracticable and uneconomical, and has largely fallen into disuse.

In contrast, the subcutaneous route has increased in popularity. Peak serum levels following subcutaneous administration of erythropoietin are obtained at around 12 h and decay slowly thereafter, such that levels above baseline are still present 96 h after injection. The bioavailability of subcutaneous erythropoietin is approximately seven times that of intraperitoneal administration, at 20–25 per cent (Macdougall *et al.* 1989), although it is still low compared with insulin, heparin, or growth hormone, probably due to the large molecular size of erythropoietin. There is increasing evidence that lower doses of erythropoietin (about 30 per cent less) are required when given subcutaneously rather than intravenously (Bommer *et al.* 1988), and this route is clearly more practical for CAPD patients who have no vascular access. The starting dose is usually around 40–60 U/kg twice or thrice weekly.

Haematological effects

The multicentre studies have indicated that about 95–98 per cent of dialysis patients treated with erythropoietin will respond with an improvement in their anaemia (Eschbach *et al.* 1989). Following commencement of regular therapy, a significant increase in the reticulocyte count of around 2–3 times baseline can be expected at 1 week, and the rise in haemoglobin concentration is usually evident by 2–3 weeks. The increase in haemoglobin is dose-dependent, and a rise of no more than 1 g/dl/month is sensible in order to minimize the risk

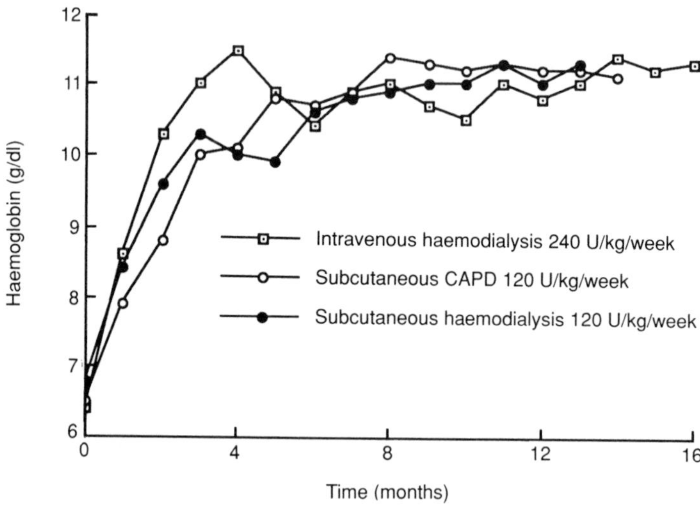

Fig. 33.2 Mean haemoglobin response to erythropoietin in 3 groups of dialysis patients: 9 haemodialysis patients treated intravenously with 240 U/kg/week of erythropoietin, 10 CAPD patients treated subcutaneously with 120 U/kg/week, and 10 haemodialysis patients treated subcutaneously with 120 U/kg/week.

of adverse effects. The target haemoglobin concentration of 10–12 g/dl is usually attained after 4–6 months of therapy (Fig. 33.2) and dose reductions may then be necessary to maintain this level thereafter.

Many factors can influence the response to erythropoietin. Patients with particularly severe anaemia (haemoglobin < 6 g/dl) at the onset of treatment generally require larger doses than those with mild anaemia (haemoglobin 6–8 g/dl) (Sundal and Kaeser 1989). Other conditions which may inhibit the response to erythropoietin are summarized in Table 33.1. All these factors should be considered in any patient failing to respond to treatment, or requiring excessive doses of erythropoietin, or losing a previous haemoglobin response. Functional iron deficiency, in particular, has become increasingly apparent in patients on erythropoietin therapy; many individuals who are iron replete at the start of treatment become deficient under the influence of erythropoietin and require intensive iron supplementation in order to maintain a haemoglobin response (Van Wyck *et al.* 1989).

The rise in haemoglobin concentration is associated with an increase in red cell count; there are no significant changes in the white cell or platelet counts in the individual patient, although small, clinically-insignificant increases in the mean values of these parameters were noted in one of the large multicentre studies (Sundal and Kaeser 1989). Occasionally, an early increase in the mean corpuscular volume is seen in patients receiving erythropoietin, particularly

Table 33.1 Potential causes of resistance to erythropoietin therapy

Decreased red cell production
Iron deficiency
B_{12}/folate deficiency
Aluminium toxicity
Hyperparathyroidism/marrow fibrosis
Infection
acute
chronic
occult
Malignancy
? occult
Poor absorption of erythropoietin (if given subcutaneously)
Marrow dysfunction
Red cell enzyme abnormalities, for example pyruvate kinase deficiency
Increased red cell loss
Blood loss
dialysis
other ? gastrointestinal tract, ? occult
Haemolysis

in those who are iron overloaded at the start of treatment; otherwise the red cell indices do not change during erythropoietin therapy unless iron deficiency supervenes. In this case, in the later stages, there may be a fall in the mean corpuscular volume or mean corpuscular haemoglobin concentration. There is usually a dramatic decline in the serum ferritin concentration and/or the transferrin saturation following commencement of erythropoietin therapy, as large quantities of iron are used up in the manufacture of new red cells.

Blood volume studies have shown that there is an increase in red cell mass after erythropoietin, which is associated with a compensatory reduction in plasma volume such that the whole blood volume remains unchanged. Erythropoietin therapy induces a two-fold increase in marrow erythropoietic activity, as evidenced by a doubling of marrow and red cell iron turnover measured during ferrokinetic studies (Cotes *et al.* 1989; Macdougall *et al.* 1990*a*). There is little or no change in mean red cell life-span after erythropoietin; thus the increased red cell mass is largely accounted for by the production of greater numbers of red cells, rather than by any change in their survival.

Secondary effects of erythropoietin

Cardiovascular and haemodynamic effects

Longstanding severe anaemia, as occurs in chronic renal failure, has profound effects on the cardiovascular system, resulting in an increase in cardiac output, decrease in peripheral resistance due to compensatory hypoxic vasodilatation, and reduced blood viscosity. As a result, patients on regular dialysis have a markedly reduced exercise capacity, impaired cardiorespiratory function, and a high cardiovascular mortality. Not surprisingly, therefore, correction of renal anaemia with erythropoietin has been shown to influence a number of changes in cardiac function.

Firstly, the elevated cardiac output associated with chronic renal anaemia is partially or completely reversed following erythropoietin therapy; this is produced by a reduction in both stroke volume and heart rate (Buckner *et al.* 1989). The compensatory hypoxic vasodilatation that occurs in chronic anaemia is also reversed as the haematocrit increases, thereby producing a rise in total peripheral resistance (Nonnast-Daniel *et al.* 1988). There is an increase in mean arterial blood pressure in approximately one-third of dialysis patients receiving erythropoietin (p. 387), particularly when the cardiac output remains proportionately high for the increased systemic vascular resistance.

Reversal of exercise-induced myocardial ischaemia (as assessed by S-T changes on the exercise electrocardiogram) has also been shown to occur following erythropoietin therapy (Macdougall *et al.* 1990*b*). This suggests that the improved oxygen delivery to the myocardium outweighs the potentially deleterious effect of the increased blood viscosity on coronary blood flow. There is a progressive reduction in left ventricular mass (assessed by echocardiography) in dialysis patients receiving erythropoietin, particularly when this is grossly elevated prior to treatment (Macdougall *et al.* 1990*b*). This latter finding may have long-term implications for cardiovascular mortality, since left ventricular hypertrophy has been shown to be an important, independent determinant of survival in dialysis patients. The internal dimensions of the left ventricle in both systole and diastole also decrease following erythropoietin, and cardiac size therefore progressively diminishes.

Several studies have documented improvements in exercise physiology after erythropoietin therapy (Mayer *et al.* 1988; Macdougall *et al.* 1990*b*). Thus, following correction of the anaemia, patients have a greater exercise capacity, along with increased maximum oxygen consumption and anaerobic threshold, as assessed by analysis of expired gases during exercise.

Coagulation, haemostasis, and blood rheology

As a result of the development of arteriovenous fistula thrombosis in a few of the early clinical trials, interest arose concerning the effect of erythropoietin on blood viscosity and the coagulation system. An early study reported a striking reduction in bleeding time after erythropoietin therapy, along with an increase in platelet adhesion to the subendothelium of human umbilical arteries (Moia *et al.* 1987). Several other studies have confirmed improvements in platelet function (Van Geet *et al.* 1989). The standard coagulation tests are unaffected by erythropoietin therapy, as are measurements of the coagulation factors. One study, however, reported a decrease in plasma levels of the natural anticoagulants, protein C and protein S during the first four months of erythropoietin; this effect was reversed by 8 and 12 months of treatment (Macdougall *et al.* 1991*a*). A further study showed an increase in thrombin–anti-thrombin III complex levels associated with erythropoietin therapy, which again might exacerbate a pro-thrombotic state (Taylor *et al.* 1991).

The haematocrit is the major determinant of whole blood viscosity, and thus an erythropoietin-induced increase in red cell mass inevitably causes a rise in blood viscosity (Schaefer *et al.* 1988). Furthermore, the relationship between haematocrit and blood viscosity is exponential, such that a linear increase in the former results in a disproportionate increase in the latter. Detailed rheological studies have shown a 2.6-, 1.8-, and 1.5-fold increase in whole blood viscosity (measured at $3\,s^{-1}$, $30\,s^{-1}$, and $300\,s^{-1}$ shear rates, respectively) following erythropoietin therapy, and have confirmed that the rise in blood viscosity occurs solely as a result of a larger quantity of circulating red cells without any change in the plasma viscosity or the rheology of the red cells themselves in terms of their deformability or aggregability (Macdougall *et al.* 1991*b*). Nevertheless, the values of blood viscosity occurring at the target haemoglobin concentration are still considerably lower than those present in non-anaemic healthy individuals.

Brain and cognitive function

It has been known for over 150 years that uraemic patients exhibit certain features of brain dysfunction which have been characterized objectively using electrophysiological and neuropsychological testing. These abnormalities are incompletely reversed by chronic dialysis therapy, suggesting that factors other than uraemic toxins may be implicated in their pathogenesis. Elevated parathyroid hormone and aluminium levels have been suggested as playing a role in this context, along with various abnormalities of brain metabolism.

Patients treated with erythropoietin have reported subjective improvements in memory, concentration, and other cerebral functions. Detailed electrophysiological studies, including brain event-related potentials, along with tests of cognitive function, have been monitored in anaemic dialysis patients receiving erythropoietin therapy (Marsh *et al.* 1991). The P3 component of the event-related potential increased in amplitude with treatment, and scores on at least two of four neuropsychological tests improved. These findings suggest that correction of anaemia by erythropoietin improves brain and cognitive function by raising levels of sustained attention, thus increasing speed and efficiency of scanning and perceptual-motion functions, and enhancing learning and memory. It also suggests that anaemia may be an important factor in the aetiology of uraemic brain dysfunction.

Sexual function

Impaired sexual function has been reported among patients of both sexes with end stage renal failure. In females this includes anovulation, disturbances of menstruation, and infertility, while in males impotence, reduced libido, oligospermia, and gynaecomastia are common. The pathogenesis of these features is complex and includes psychological factors, vascular and neurological disorders, drug therapy, uraemic intoxication, and hormonal disturbances.

Several studies have reported improvements in sexual function following erythropoietin therapy. In one study, 4 of 7 males reported an improvement in libido and sexual performance, and 5 of 9 females experienced a return of regular menstruation (Schaefer *et al.* 1989). In another study, self-reported sexual function improved in 4 of 7 male patients, including libido and potency (Bommer *et al.* 1990). In some instances, these improvements have been associated with changes in serum prolactin or testosterone levels, but the results are conflicting (see below).

Endocrine effects

Investigation of endocrine function during erythropoietin therapy began with studies of changes in sex hormone levels as a possible mediator of the improvement in sexual function. Schaefer *et al.* (1989) reported a striking reduction in serum prolactin levels in both sexes after 16 weeks of erythropoietin therapy, but conflicting results were found by Bommer *et al.* (1990), with a similar improvement in sexual function. In both these studies, no changes in serum testosterone were noted, in contrast to Haley *et al.* (1989), who observed a significant rise in testosterone levels in association with an improved sexual performance after erythropoietin.

In an exhaustive study, Kokot *et al.* (1989) reported diverse effects of erythropoietin on endocrine function. Suppressive effects were observed on the renin-angiotensin system, the pituitary-adrenal axis, growth hormone levels, glucagon, gastrin, follicle-stimulating hormone, luteinizing hormone, and prolactin while there were increases in plasma insulin, testosterone, parathyroid hormone, and atrial natriuretic peptide levels. Such widespread endocrine effects require substantiation, and the possible mechanisms of action responsible for these changes are as yet unclear.

A recent study from Italy showed that acute administration of erythropoietin to a group of haemodialysis patients potentiated the growth hormone response to growth hormone-releasing hormone (Cremagnani *et al.* 1990). This potentiation after erythropoietin was not seen in control subjects with normal renal function. Similarly, no potentiation in the thyroid stimulating hormone or prolactin responses to thyrotrophin releasing hormone were seen in the haemodialysis patients, and erythropoietin on its own had no acute effect on growth hormone levels.

Adverse effects

Hypertension is the commonest and potentially most worrying complication of erythropoietin therapy, particularly when associated with encephalopathy or seizures (Eschbach *et al.* 1989; Sundal and Kaeser 1989). Approximately 30–35 per cent of patients treated will manifest a significant rise in blood pressure requiring therapeutic intervention. The risk of developing a significant increase in blood pressure following erythropoietin appears to be independent of whether or not there is a previous history of hypertension (Eschbach *et al.* 1989). Likewise, the rate of rise in the haematocrit does not seem to influence the likelihood of developing hypertensive problems, although most exacerbations occur during the acute correction of anaemia rather than during the maintenance phase.

The mechanism of erythropoietin-induced hypertension remains obscure. Although one study demonstrated a direct pressor effect of erythropoietin, this has not been confirmed in several similar studies. It is generally believed that the hypertension is mediated via a number of pathophysiological changes occurring secondary to the increase in haematocrit (Raine 1990). These include an inadequate reversal of the elevated cardiac output of anaemia, a relative increase in peripheral resistance as the compensatory hypoxic peripheral vasodilatation of anaemia is reversed, and a rise in blood viscosity. In most instances, blood pressure can be easily controlled by attention to fluid status and the use of standard hypotensive agents; it is very rare to have to discontinue erythropoietin therapy for severe uncontrollable hypertension.

In quite a number of the early studies, there were anecdotal reports of seizures or hypertensive encephalopathy occurring in patients receiving erythropoietin. In the USA multicentre trial 18 of 333 patients (Eschbach *et al.* 1989), and in the European multicentre trial 5 of 150 patients (Sundal and Kaeser 1989) developed hypertensive encephalopathy and/or seizures, most frequently within the first 3 months of treatment when the haematocrit was increasing, and often related to a sudden increase in blood pressure. The pathogenesis of these adverse effects remains poorly understood, although loss of autoregulation of cerebral blood flow and/or reduced cerebral perfusion may be contributory (Raine 1990).

Approximately 10 per cent of haemodialysis patients treated with erythropoietin develop thrombosis of their vascular access site, usually during the first few months of therapy (Sundal and Kaeser 1989; Canadian Erythropoietin Study Group 1990). Factors which may be relevant in the pathogenesis of this complication include the rise in blood viscosity (Schaefer *et al.* 1988), shortening of the bleeding time (Moia *et al.* 1987), increase in platelet aggregation and adhesion (Van Geet *et al.* 1989), reduction in protein C and protein S levels, increase in thrombin–anti-thrombin III levels, and a marginal increase in platelet count in some patients.

Occasionally, patients on erythropoietin exhibit an increase in serum potassium, phosphate, and creatinine levels (Eschbach *et al.* 1989; Sundal and Kaeser 1989) which may be due to enhanced dietary intake or alternatively a reduced dialyser clearance of these molecules secondary to the increased haematocrit. Other adverse effects of erythropoietin therapy include transient myalgia or influenza-like symptoms after the first few injections only, conjunctival injection, headache, and loin pain. Genuine intolerance to erythropoietin sufficient to warrant stopping treatment is rare, and there have been no reports of antibody formation to the recombinant hormone.

Conclusions

Erythropoietin therapy has proved to be a highly effective treatment for the anaemia of end stage renal disease. The benefits are considerable and include a subjective improvement in well-being and physical activity, along with objective improvements in quality-of-life parameters, exercise capacity, cardiorespiratory function, bleeding dysfunction, brain and cognitive function, endocrine profiles, and sexual function. Adverse effects are few and are rarely serious. Trials are underway to examine a possible role for erythropoietin in the management of other anaemias, such as those associated with rheumatoid arthritis, malignant disease, cancer chemotherapy, sickle cell disease, myelodysplastic syndrome, acquired immunodeficiency syndrome, and zidovudine therapy, prematurity, and multiple myeloma.

References

Bommer, J., Ritz, E., Weinreich, T., Bommer, G., and Ziegler, T. (1988). Subcutaneous erythropoietin. *Lancet*, **ii**, 406.

Bommer, J., Kugel, M., Schwöbel, B., Ritz, E., Barth, H.P., and Seelig, R. (1990). Improved sexual function during recombinant human erythropoietin therapy. *Nephrol. Dial. Transplant.* **5**, 204–7.

Buckner, F.S., Eschbach, J.W., Haley, N.R., Davidson, R.R., and Adamson, J.W. (1989). Correction of the anemia in hemodialysis patients with recombinant human erythropoietin: hemodynamic changes and risks for hypertension. *Kidney Int.* **35**, 190.

Canadian Erythropoietin Study Group (1990). Association between recombinant human erythropoietin and quality of life and exercise capacity of patients receiving haemodialysis. *Br. Med. J.* **300**, 573–8.

Caro, J., Brown, S., Miller, O., Murray, T., and Erslev, A.J. (1979). Erythropoietin levels in uremic nephric and anephric patients. *J. Lab. Clin. Med.* **93**, 449–58.

Cotes, P.M., Pippard, M.J., Reid, C.D.L., Winearls, C.G., Oliver, D.O., and Royston, J.P. (1989). Characterization of the anaemia of chronic renal failure and mode of its correction by a preparation of human erythropoietin (r-Hu EPO). An investigation of the pharmacokinetics of intravenous erythropoietin and its effects on erythrokinetics. *Q. J. Med.* **262**, 113–37.

Cremagnani, L., Cantalamessa, L., Orsatti, A., Vigna, L., Bianchi, M.L., and Buccianti, G. (1990). Growth hormone response to growth hormone releasing hormone in haemodialysed patients. Effect of acutely administered recombinant human erythropoietin. *Nephrol. Dial. Transplant.* **5**, 738.

Davis, J.M., Arakawa, T., Strickland, T.W., and Yphantis, D.A. (1987). Characterization of recombinant human erythropoietin produced in Chinese hamster ovary cells. *Biochemistry*, **26**, 2633–8.

Eschbach, J.W., Downing, M.R., Egrie, J.C., Browne, J.K., and Adamson, J.W. (1989). USA multicenter clinical trial with recombinant human erythropoietin. *Contrib. Nephrol.* **76**, 160–5.

Frenken, L.A.M., Coppens, P.J.W., Tiggeler, R.G.W.L., and Koene, R.A.P. (1988). Intraperitoneal erythropoietin. *Lancet*, **ii**, 1495.

Haley, N.R., Matsumoto, A.M., Eschbach, J.W., and Adamson, J.W. (1989). Low testosterone levels increase in male hemodialysis patients treated with recombinant human erythropoietin. *Kidney Int.* **35**, 193.

Jacobson, L.O., Goldwasser, E., Fried, W., and Pizak, L. (1957). Role of the kidney in erythropoiesis. *Nature*, **179**, 633–4.

Kokot, F., Wiecek, A., Grzeszczak, W., Klepacka, J., Klin, M., and Lao, M. (1989). Influence of erythropoietin treatment on endocrine abnormalities in haemodialyzed patients. *Contrib. Nephrol.* **76**, 257–72.

Krantz, S.B., Sawyer, S.T., and Sawada, K.I. (1988). The role of erythropoietin in erythroid cell differentiation. *Contrib. Nephrol.* **66**, 25–37.

Kurtz, A., Eckardt, K.-U., Tannahill, L., and Bauer, C. (1988). Regulation of erythropoietin production. *Contrib. Nephrol.* **66**, 1–16.

Lacombe, C., DaSilva, J.L., Bruneval, P., Fournier, J.G., Wendling, F., Casadevall, N., Camilleri, J.P., Bariety, J., and Varet, B. (1988). Peritubular cells are the site of erythropoietin synthesis in the murine hypoxic kidney. *J. Clin. Invest.* **81**, 620–3.

Lin, F.K., Suggs, S., Lin, C.H., Browne, J.K., Smalling, R., Egrie, J.C., Chen, K.K., Fox, G.M., Martin, F., Stabinsky, Z., Badrawi, S.M., Lai, P.H., and Goldwasser, E. (1985). Cloning and expression of the human erythropoietin gene. *Proc. Natl. Acad. Sci. USA*, **82**, 7580–5.

Macdougall, I.C., Roberts, D.E., Neubert, P., Dharmasena, A.D., Coles, G.A., and Williams, J.D. (1989). Pharmacokinetics of recombinant human erythropoietin in patients on continuous ambulatory peritoneal dialysis. *Lancet*, **i**, 425–7.

Macdougall, I.C., Davies, M.E., Hutton, R.D., Cavill, I., Lewis, N.P., Coles, G.A., and Williams, J.D. (1990*a*). The treatment of renal anaemia in CAPD patients with recombinant human erythropoietin. *Nephrol. Dial. Transplant.* **5**, 950–5.

Macdougall, I.C., Lewis, N.P., Saunders, M.J., Cochlin, D.L., Davies, M.E., Hutton, R.D., Fox, K.A.A., Coles, G.A., and Williams, J.D. (1990*b*). Long-term cardiorespiratory effects of amelioration of renal anaemia by erythropoietin. *Lancet*, **335**, 489–93.

Macdougall, I.C., Davies, M.E., Hallett, I., Cochlin, D.L., Hutton, R.D., Coles, G.A., and Williams, J.D. (1991*a*). Coagulation studies and fistula blood flow during erythropoietin therapy in haemodialysis patients. *Nephrol. Dial. Transplant.* **6**, 862–7.

Macdougall, I.C., Davies, M.E., Hutton, R.D., Coles, G.A., and Williams, J.D. (1991*b*). Rheological studies during treatment of renal anaemia with recombinant human erythropoietin. *Br. J. Haem.* **77**, 550–8.

Marsh, J.T., Brown, W.S., Wolcott, D., Carr, C.R., Harper, R., Schweitzer, S.V., and Nissenson, A.R. (1991). rHuEPO treatment improves brain and cognitive function of anemic dialysis patients. *Kidney Int.* **39**, 155–63.

Maxwell, A.P., Lappin, T.R.J., Bridges, J.M., Johnston, C.F., and McGeown, M.G. (1989). Renal tubular cell production of erythropoietin co-localised by immunohistochemistry and in situ hybridisation. *Nephrol. Dial. Transplant.* **4**, 420.

Mayer, G., Thum, J., Cada, E.M., Stummvoll, H.K., and Graf, H. (1988). Working capacity is increased following recombinant human erythropoietin treatment. *Kidney Int.* **34**, 525–8.

Miyake, T., Kung, C.K.H., and Goldwasser, E. (1977). Purification of human erythropoietin. *J. Biol. Chem.* **252**, 5558–64.

Moia, M., Mannucci, P.M., Vizzotto, L., Casati, S., Cattaneo, M., and Ponticelli, C. (1987). Improvement in the haemostatic defect of uraemia after treatment with recombinant human erythropoietin. *Lancet*, **ii**, 1227–9.

Neumayer, H.-H., Brockmoller, J., Fritschka, E., Roots, I., Scigalla, P., and Wattenberg, M. (1989). Pharmacokinetics of recombinant human erythropoietin after SC administration and in long-term IV treatment in patients on maintenance hemodialysis. *Contrib. Nephrol.* **76**, 131–42.

Nonnast-Daniel, B., Creutzig, A., Kuhn, K., Bahlmann, J., Reimers, E., Brunkhorst, R., Caspary, L., and Koch, K.M. (1988). Effects of treatment with recombinant human erythropoietin on peripheral hemodynamics and oxygenation. *Contrib. Nephrol.* **66**, 185–94.

Raine, A.E.G. (1990). Seizures and hypertension events. *Semin. Nephrol.* **10**, (1), 40–50.

Recny, M.A., Scoble, H.A., and Kim, Y. (1987). Structural characterisation of natural human urinary and recombinant DNA-derived erythropoietin (identification of des-Arginine 166 erythropoietin). *J. Biol. Chem.* **262**, 17156–63.

Schaefer, R.M., Leschke, M., Strauer, B.E., and Heidland, A. (1988). Blood rheology and hypertension in hemodialysis patients treated with erythropoietin. *Am. J. Nephrol.* **8**, 449–53.

Schaefer, R.M., Kokot, F., Wernze, H., Geiger, H., and Heidland, A. (1989). Improved sexual function in hemodialysis patients on recombinant erythropoietin: a possible role for prolactin. *Clin. Nephrol.* **31**, 1–5.

Sundal, E. and Kaeser, U. (1989). Correction of anaemia of chronic renal failure with recombinant human erythropoietin: safety and efficacy of one year's treatment in a European multicentre study of 150 haemodialysis-dependent patients. *Nephrol. Dial. Transplant.* **4**, 979–87.

Taylor, J.E., McLaren, M., Henderson, I.S., Belch, J.J.F., and Stewart, W.K. (1991). Prothrombotic effect of erythropoietin in dialysis patients. *Nephrol. Dial. Transplant.* (In press).

Van Geet, C., Hauglustaine, D., Verresen, L., Vanrusselt, M., and Vermylen, J. (1989). Haemostatic effects of recombinant human erythropoietin in chronic haemodialysis patients. *Thromb. Haemost.* **61**, 117–21.

Van Wyck, D.B., Stivelman, J.C., Ruiz, J., Kirlin, L.F., Katz, M.A., and Ogden, D.A. (1989). Iron status in patients receiving erythropoietin for dialysis-associated anemia. *Kidney Int.* **35**, 712–16.

34 Modern management of urinary tract stones

H.N. Whitfield

During the 1960s and 1970s renal surgery advanced as the result of the development of techniques for preserving renal function during periods of ischaemia. Ischaemia was necessary if complex intra-renal surgery was to be performed, such as the complete removal of staghorn stones and the correction of arterial and arterio-venous malformations. Renal preservation with hypothermia allowed surgeons 3 h of operating time with the renal artery clamped, without jeopardizing renal function. Both *in vivo* and *in vitro* intra-renal surgery was developed to a very great extent. Towards the end of the 1970s Fernström performed the first percutaneous nephrolithotomy, a technique which was rapidly developed and adopted by urologists all over the world (Whitfield 1983). In the 1980s lithotripsy, first developed in Munich (Hausler and Kiefer 1973; Chaussy *et al.* 1980), opened a new dimension in the surgical management of renal and ureteric stones. The integration of these two techniques (Eisenberger *et al.* 1985), together with uretero-renoscopic instruments has revolutionized urinary tract stone surgery.

Percutaneous renal surgery

This technique depends on establishing a track between the skin surface and the collecting system. The usual site of access into the kidney is through a lower pole calix. It is important that the puncture should be transparenchymal, and as peripheral as possible, so that major vessels are avoided.

With the patient under general anaesthesia, the first step is to pass a retrograde ureteric catheter. This enables the collecting system to be filled with contrast media. The patient is then turned prone-oblique and the puncture is guided either radiologically, ultrasonically, or with a combination of both. A guidewire is passed down the shaft of the puncture needle, and over this guidewire a series of graded dilators, which may be either Teflon or metal, are passed sequentially. The track is dilated to between 26 F and 30 F (a diameter of 8.5 mm and 10.0 mm). A plastic Amplatz sheath is passed over the largest dilator, to provide secure access between the skin surface and the collecting system. Operating instruments of various kinds can be introduced into the collecting system through this sheath.

Stones which are small enough to be removed intact through the sheath are grasped with an instrument passed through the instrument channel of a nephroscope. Stones which are too large to be removed in this way can first be disintegrated, using either an electrohydraulic high tension spark discharge method, or by utilizing ultrasound energy delivered to the end of a hollow probe. Stone clearance is checked visually and radiologically. At the end of the procedure the track is tamponaded with a large size of nephrostomy tube. It is usually necessary for this tube to remain in place for 24–48 h. A nephrostogram may be performed to check that the ureter is patent. Alternatively, the nephrostomy tube may be clamped and, providing that no leakage occurs around the tube, that the patient remains apyrexial, and without any pain, it is safe to remove the nephrostomy tube. The transparenchymal puncture track usually dries within a few hours.

The majority of renal stones can be removed using this technique. It may not always be possible to visualize all parts of the collecting system through one puncture site. If stones lie in an inaccessible calix a second or even a third track may be necessary. However, since the advent of lithotripsy, multiple tracks are used less often.

Results

Before the introduction of extracorporeal shock wave lithotripsy, centres practising percutaneous techniques were able to remove more than 90 per cent of all renal stones without resorting to open surgery (Whitfield 1987). Complex stones could be dealt with, either by means of multiple tracks or, on occasions, by staging the removal of stones over treatment sessions.

Complications

Bleeding

Bleeding can occur during track dilatation, but this is usually insufficient to necessitate either conclusion of the procedure or open surgery. During stone extraction it is possible to injure the collecting system and cause bleeding which is sufficiently severe to impair vision. In this case it is best to introduce a nephrostomy tube and complete the procedure on a subsequent occasion.

Secondary haemorrhaging can occur, particularly after the removal of a nephrostomy tube in a patient whose stone was infective in origin. Secondary haemorrhaging may settle spontaneously, but occasionally will provoke the need for selective renal arterial embolization.

Septicaemia

Many patients have stones which are caused by infection, and, in this case, the administration of antibiotic prophylaxis prior to and after percutaneous surgery is mandatory. Even metabolic stones may have an infective component and antibiotic prophylaxis is also indicated. The length of time for which antibiotics are continued post-operatively will depend on the results of culturing the stones. If a proteus organism is found, an appropriate antibiotic should be given for between 6 weeks and 3 months.

Perforation

It is possible to perforate the collecting system during percutaneous surgery. Large volumes of irrigating fluid can then be extravasated. Such extravasation can occur extraperitoneally or intraperitoneally. It is therefore important to monitor the amount of fluid being used during the operation, so that any fluid loss is noticed quickly. The irrigating fluid should always be normal saline, to reduce the chances of a TUR reaction from fluid absorption.

Ureteric blockage

If fragments of stone become lodged in the ureter, obstruction may occur. It is possible to prevent this by leaving a ureteric catherter *in situ* during percutaneous stone removal. Such catheters may be of the balloon variety, to obstruct the ureter completely, but any large size ureteric catheter will have the same beneficial effect. If fragments do pass down the ureter, they need to be treated in the same way as any other ureteric stone. Clearly, it is unsafe to remove a nephrostomy tube while a ureteric stone remains.

Residual stones

Stone fragments can pass into parts of the collecting system which are inaccessible to rigid instruments. Flexible nephroscopes have a limited place, but most often symptomatic residual stones can be treated subsequently with lithotripsy. Stones which are smaller than 3 mm in diameter may be treated conservatively.

Disadvantages of percutaneous surgery

The learning curve

Percutaneous techniques for the removal of renal stones are difficult to learn. Only those departments in which there is a sufficiently high enough stone referral rate can ever hope to acquire the necessary expertise. The insertion of an accurately placed track under radiological control can result in prolonged exposure times, which can be hazardous to both the patient and the surgeon. The time taken for the removal of a renal stone by percutaneous means can be prolonged during the learning phase.

Radiation exposure

Since percutaneous techniques require radiological screening, the potential hazards from excessive radiation must be recognized (Bowsher and Whitfield 1990). All possible ways of minimizing exposure must be used routinely, such as lead aprons, lead glasses and lead neck protection. Lead protective gloves can also be used with advantage, although their thickness may cause some difficulties. Received doses must be monitored and recorded. Methods of reducing the radiation include the use of coning, monitor screens with a memory, and under-couch rather than over-couch screening units. It is also possible to reduce screening times when the operator's foot is on the pedal.

Theatre staff involvement

The operator works without an assistant and it is therefore easy for theatre staff to become bored. The provision of video monitoring equipment helps to overcome this problem.

Auditory trauma

Ultrasound generators for stone disintegration can cause tinnitus, particularly to the operator, but also to nearby theatre staff (Samuels 1988). Ear protection is therefore advisable.

Advantages

From the patient's point of view there are significant advantages to percutaneous surgery compared with open renal surgery. The post-operative pain is significantly less, the hospital stay is short, usually only 72 h for an uncomplicated case, and a return to full activities is possible after one week (Whitfield and Mills 1985; Das *et al.* 1988).

In terms of renal function, less damage is caused by the percutaneous approach than by conventional open operations which require incisions into the parenchyma. The tracks from a percutaneous operation heal as fine scars.

Lithotripsy

Lithotripsy was developed in the 1960s and 1970s, and introduced into clinical practice in the 1980s (Chaussy and Fuchs 1989). However, it was not until 1985 that centres outside Germany had the opportunity to benefit from the introduction of the Dornier HM3 lithotripter. This was the first commercially available lithotripter and remains a gold standard with which all other lithotripters are compared. A variety of manufacturers have produced a bewildering range of hardware. Nevertheless, the principles underlying lithotripsy have remained the same.

Dornier HM3

In this original lithotripter the patient required either a general or an epidural anaesthetic. An underwater high tension spark discharge was used to create an energy focus. The patient was secured on a frame, which was then lowered into a water bath containing de-ionized and de-gassed water at body temperature. The stone was imaged using two-dimensional X-ray screening units, and the frame could be positioned with an accuracy of 1 mm in any direction (Fig. 34.1). The patient was manoeuvred by means of the frame until the

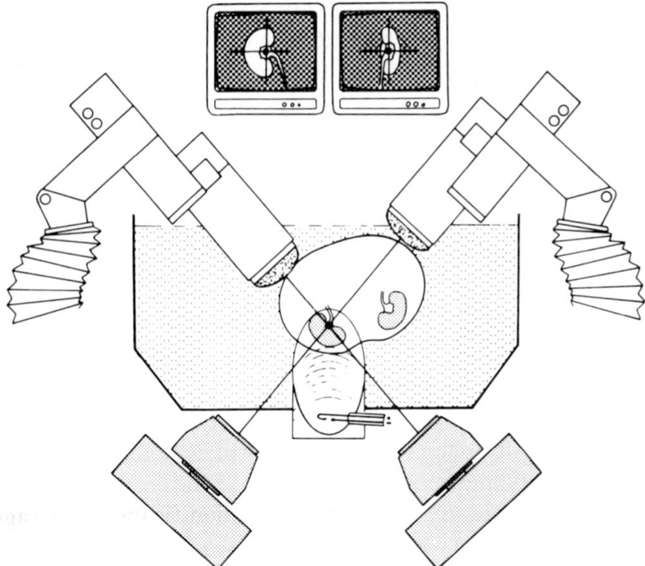

Fig. 34.1 Diagram of the essential features of the Dornier HM3 extracorporeal shock wave lithotripter, showing bi-planar fluoroscopy, spark keep generator in a hemi-ellipsoid, control panel and water bath in which the patient lies.

stone was brought into the zone of the shock wave energy focus. Shock waves were then triggered by the R wave of the electrocardiogram, to minimize the risk of inducing cardiac dysrhythmia. The number of shock waves required varied depending on the size and composition of the stone, but it was unusual for more than one treatment session to be required for anything other than the very largest or hardest stones. Following fragmentation, the disintegrated particles were passed spontaneously, though it was recognized that complete stone clearance would often take up to 3 months.

Results

Most centres reported a stone clearance rate at 3 months of 85 per cent (Chaussy and Fuchs 1989). Some 10 per cent of patients required an interventional endoscopic procedure either preoperatively or postoperatively. Such procedures included the insertion of JJ stents, percutaneous de-bulking of a large stone, a needle nephrostomy, or an endoscopic ureteric procedure.

Complications of lithotripsy

Ureteric obstruction

If fragments of stone too large to pass spontaneously are produced by lithotripsy, ureteric obstruction occurs. One large fragment may then cause more fragments to build up proximally, the so-called ***steine strasse***. These may pass spontaneously but lithotripsy, endoscopic ureteroscopy, or percutaneous needle nephrostomy may be required to resolve the problem.

Septicaemia

If ureteric obstruction occurs in the presence of infection the patient may become very ill and renal function severely jeopardized. Prophylactic antibiotics are therefore given to all patients whose stones are infective in origin. The use of prophylactic antibiotics for patients whose stones are metabolic remains controversial, but in the author's opinion is mandatory.

Ureteric colic

The amount of pain the patient experiences during the spontaneous passage of stone fragments varies. The size of the stone particles will obviously influence the incidence of ureteric colic. The aim of treatment is to reduce the stone into fragments of 2 mm diameter or less, which will pass without any significant discomfort. However, about 30 per cent of patients develop pain sufficiently severe to require analgesia. A non-steroidal anti-inflammatory analgesic is very effective. Anti-muscarinic agents play no useful part in the management of ureteric colic.

Hypertension

There was a report suggesting that the incidence of hypertension following lithotripsy was increased (Lingeman and Kulb 1987). However, subsequent reports (Montgomery and Shuttleworth 1989) have provided evidence that hypertension occurs no more commonly in patients who have had lithotripsy for their renal stones than in a matched control population.

Miscellaneous problems

Skin bruising occurs in some patients, but this rarely causes symptoms. Subcapsular haematomas have been reported, but they are rare (Kaude *et al*. 1985). Intestinal ileus may arise, but it usually settles spontaneously within 36 h.

Second generation lithotripters

There are now a large number of lithotripters available from different manufacturers (Table 34.1). Every lithotripter that is available clinically is effective, but the different features incorporated have clinical implications.

The basis of any lithotripter is to have an imaging device, an energy shock wave focus and a positioning potential, so that the shock wave focus and the patient's stone can be brought into alignment.

Imaging

Ultrasound technology has now improved to the extent where it is possible to image most renal stones with ultrasound. That is not to say that screening for

Table 34.1 Lithotripters currently available

Diasonics Tetrasonic
Direx - Tripter X-1
Dornier HM3
HM4
MPL 9000
Compact
EDAP LT-01
Medstone
Northgate
Phillips MFL 5000
Siemens Lithostar
Siemens Lithostar Plus
Storz Modulith
Technomed Sonolith
Wolf Piezolith 2300
2500
Yachiyoda SZ-1

renal stones by ultrasound is to be recommended. However, when a plain abdominal X-ray reveals the site of the stone it is a very different proposition to locate the stone using ultrasound and to differentiate the echo patterns which result from similar echo patterns that can cause confusion without the additional information available from radiology. Ultrasound imaging of ureteric stones, can be very difficult, however. In the upper third of the ureter, when there is significant dilatation above the stone, it may be possible to image a stone with ultrasound. In the intramural or juxta-vesical ureter, when the full bladder may be used as an acoustic window, stones may also be visible on ultrasound. Elsewhere in the ureter it is not possible to image stones reliably with ultrasound.

Radiological screening is incorporated into some lithotripters as the imaging modality of choice. Since over 90 per cent of renal and ureteric stones are radio-opaque this poses no problem. Even radiolucent stones can be imaged as negative shadows after contrast. All ureteric stones can thus be located and treated. However, the radiation risk factor cannot altogether be ignored.

Many lithotripters now incorporate both radiological and ultrasound imaging facilities. This widens the scope of patients that can be treated, and few, if any, renal, ureteric, gall bladder, or common bile duct stones cannot be imaged in one or other way.

Shock wave energy

Different methods of producing focused shock waves have been utilized (Table 34.2). Laser energy has proved unsatisfactory, since sufficiently high

Table 34.2 Methods of shock wave production by different lithotripters

Method	Lithotripter
Electrohydraulic	
	Direx - Tripter X-1
	Dornier - HM3
	HM4
	MPL 9000
	Medstone
	Northgate
	Phillips - MFL 5000
	Technomed - Sonolith
Piezoelectric	
	Diasonics - Tetrasonic
	EDP - LT-01
	Wolf - 2300
	2500
Electromagnetic	
	Dornier - Compact
	Siemens - Lithostar
	Storz - Modulith

peak focal pressures cannot be realised. Micro-explosion techniques have not gained popularity, although they are incorporated into one Japanese lithotripter. The original Dornier HM3, and many subsequent lithotripters, have made use of an electrohydraulic high tension spark discharge under water. Piezo-electric crystals can be aligned so that small individual energy sources are focused to an area of sufficient peak focal pressure to cause stone disintegration. An electromagnetic energy source has also been utilized. There is no one method of shock wave production which has inherent advantages over the others. However, the characteristics of a shock wave vary and the clinical implications of these varying shock wave forms must be appreciated (Coleman and Saunders 1987). Piezo-electric lithotripters have a small focal area of high pressure, though the peak pressure at that focus is higher than in the larger focus produced by electrohydraulic and electromagnetic methods. Piezo-electric lithotripsy therefore requires very accurate targetting, and several treatment sessions are often required to completely disintegrate a stone which would be fully disintegrated in one session by other energy sources (Bowsher *et al.* 1989). However, the size of the fragmented particles after piezo-electric lithotripsy tends to be very small and the associated risk of ureteric obstruction by stone fragments is therefore reduced.

Shock wave coupling

The original Dornier HM3 required the patient to be immersed in a water bath. Second and third generation lithotripters have incorporated either a partial water bath or a water cushion, making treatment more convenient.

Analgesia/anaesthesia requirement

Pain from lithotripsy is dependent on a number of factors. The larger the size of the surface area over which the shock wave enters, the less pain is produced. Lithotripter manufacturers have accordingly increased the size of the hemi-ellipsoid in the electrohydraulically generated shock wave lithotripters, thereby increasing the surface area of the skin entry site. The shape of the shock wave itself, the rise in time to peak pressure, is also a variable which influences pain production. The negative component of the pressure wave varies and may be significant.

Treatment using the Dornier HM3 lithotripter was always performed under general or epidural anaesthesia. Manufacturers now claim that lithotripsy can be performed without more than local anaesthesia and/or intravenous sedation and analgesia with all subsequent lithotripters (Rassweiler *et al.* 1988). The amount of pain produced varies depending on the site of the stone, the voltage used to create the shock wave (which is variable for all lithotripters), and on the patient's reaction to pain. For example, treatment of stones situated near a vertebral transverse process or in an upper pole calix adjacent to a rib may cause more pain than a stone in the renal pelvis or in a lower pole calix. However, most patients with most lithotripters can tolerate treatment very satisfactorily with local anaesthesia, analgesia and/or sedation. The small percentage of patients who require general anaesthesia can still be treated as day cases, which makes lithotripsy on an out-patient basis feasible.

Indications for stone treatment

The classical indications for treating renal and ureteric stones are pain, infection, obstruction, and renal impairment (Whitfield 1990). The combination of obstruction and infection is particularly damaging and has to be treated as a matter of great urgency. However, the advent of lithotripsy has made it possible to consider the potential for any of these indications to exist as sound reasons for prophylactic treatment. Before lithotripsy was available, caliceal stones which were of a size to be passed spontaneously, i.e. less than 5 mm in diameter, would not have been treated. Nowadays, it is logical to offer such patients lithotripsy before any trouble has arisen.

The contra-indications to treating stones are very few. Patients who are being treated with anti-coagulants are generally considered unsuitable, since the risk of subcapsular haematoma formation is increased. Lithotripsy alone cannot be used if there is obstruction in association with a stone, for example pelvi-ureteric junction obstruction. Such a combination is an indication for either a percutaneous approach, when the pelvi-ureteric junction obstruction can be treated by an endopyelotomy, or even for an open operation.

Integration of stone management

The advances in technology have produced a bewildering list of hardware that is necessary if patients are to be offered a comprehensive stone treatment service. This has lead to the development of stone centres with a catchment area of a size to justify such investment.

Early calculations suggested that a population of 5 million would generate enough urinary tract stones to justify the purchase of a lithotripter. However, this was certainly an underestimate of the number of lithotripters which could with benefit be provided, since the calculation was based on the number of patients treated for stones before the advantages of lithotripsy were fully recognized. The indications for treatment have broadened widely and, since the cost of lithotripters has begun to fall, it is reasonable to aim to provide one lithotripter per 2 million population. However, a lithotripter on its own is of no value; it is essential that the necessary expertise in percutaneous renal surgery and endoscopic ureteroscopic techniques is available.

It is recognized that any stone in the kidney which is larger than 2.5 cm diameter cannot be treated by lithotripsy alone. There are advocates of treatment even of staghorn stones by lithotripsy combined with the insertion of a JJ stent. Others recommend that as much as possible of the stone should be removed percutaneously, and lithotripsy reserved for inaccessible caliceal fragments. This is the author's preferred choice. There is still a place for open surgery; this is largely confined to those patients who have staghorn calculi which have not previously been operated upon, and in whom the bulk of the stone lies in calices rather than in the renal pelvis. The term staghorn calculus is often used loosely, and the term covers a wide range of differing stones. Generalizations in the management of stone disease are always difficult, and best avoided.

Ureteroscopic techniques

Instrument manufacturers have designed rigid and flexible endoscopes with which to view the upper urinary tract from below. Such endoscopes, when rigid, incorporate a solid rod lens system of the same variety as is found in a cystoscope. However, the instrument channel, which also serves as an irrigating channel, is inevitably significantly smaller, since the calibre of these instruments is in the region of 11 French (3.5 mm diameter). Smaller flexible cystoscopes exist, some of which have manoeuvrable tips, but again the instrument/irrigating channel is small and this can provide difficulties. Nevertheless, small flexible laser fibres are available which can be used to disintegrate ureteric stones. Small electrohydraulic fibres can also be used. Since an ultrasound probe is necessarily straight, a ureteroscope with an offset eyepiece has to be used if an ultrasound probe is necessary. The amount of energy available from laser, electrohydraulic or ultrasound energy sources within the ureter is usually sufficient to fragment most stones, though slowly (Miller 1987).

There are methods of dilating the ureteric orifice and the ureter itself if the calibre of the ureter is too small to accommodate the ureteroscope. There is some discussion about the size to which it is safe to dilate a normal ureter. Ischaemic damage can occur from over-dilation (Boddy *et al.* 1988). The blood supply can also be damaged when the ureter is raised off the posterior abdominal wall by a rigid endoscope. Nevertheless, in practice, ureteric strictures arising as a result of ischaemic injury are uncommon. Perforation of the ureteric wall can occur either with the instrument itself or with an instrument passed down the ureteroscope. When recognized, such injuries are best dealt with by the insertion of a JJ stent.

Conclusions

The advent of new techniques has revolutionized the management of urinary tract stones. Success depends on the correct choice of procedure and the effective integration of different treatment methods. Ninety-five per cent of urinary tract stones, both in the kidney and the ureter, can be managed without resorting to open surgery. This has significant implications in terms of patient morbidity, hospital stay, and therefore health economics.

References

Boddy, S.A., Nimmon, C.C., Jones, S. *et al.* (1988). Acute ureteric dilatation for ureteroscopy. An experimental study. *Brit. J. Urol.* **61**, (1), 27–31.

Bowsher, W., Carter, S., Philip, T. *et al.* (1989). Clinical experience using the Wolf Piezolith device at two British stone centres, *J. Urol.* **142**, 679–82.

Bowsher, W. and Whitfield, H.N. (1990). Radiation exposure during percutaneous renal surgery. *Brit. J. Urol.*, (In press).

Chaussy, C.G., Brendel, W., and Schmiedt, E. (1980). Extracorporeally induced destruction of kidney stones by shock waves. *Lancet*, **2**, 1265.

Chaussy, C.G. and Fuchs, G. (1989). Current state and future developments of non-invasive treatment of human urinary stones with extracorporeal shock wave lithotripsy. *J. Urol.* **141**, 782–9.

Coleman, A.J. and Saunders, J.E. (1987). Comparison of extracorporeal shock wave lithotripters. In *Lithotripsy II* (ed. Coptcoat, Miller, and Wickham), pp. 121–31. BDI Publishing.

Das, G., Dick, J., Bailey, M.J., Fletcher, M.S. *et al.* (1988). 1500 cases of renal and ureteric calculi treated in an integrated stone centre. *Brit. J. Urol.* **62**, 301–5.

Eisenberger, F., Fuchs, G., Miller, K. *et al.* (1985). Extracorporeal shock wave lithotripsy and endourology: an ideal combination for the treament of kidney stones. *World J. Urol.* **3**, 41–3.

Fernstrom, I. and Johannson, B. (1976). Percutaneous pyelolithotomy: a new extraction technique. *Scandinavian Journal of Urology and Nephrology*, **4**, 257–9.

Hausler, E. and Kiefer, W. (1973). Beruhrungsfreie Zerstorung Fester Einschlusse in flussiger Umgebung. *Verh. Deutsch Physikal. Gesellscaft.* **6**, 692.

Kaude, J.V., Williams, C.M., Scott, K.N. *et al.* (1985). Renal morphology and function immediately after extracorporeal shock wave lithotripsy. *Amer. J. Roentgen.* **145**, 305.

Lingeman, J.E. and Kulb, T.B. (1987). Hypertension following extracorporeal shock wave lithotripsy. *J. Urol.* (part 2), **137**, 142a, (Abstract).

Miller, R.A. (1987). Instrumentation for upper tract endoscopy. In *Recent advances in urology/andrology 4*. (ed. W.F. Hendry), pp. 23–40. Churchill Livingstone, London.

Montgomery, B.S.I. and Shuttleworth, K.E.D. (1989). Does extracorporeal shockwave lithotripsy cause hypertension. *Brit. J. Urol.* **64**, 567–71.

Rassweiler, J., Westhauser, A., Bub, P., and Eisenberger, F. (1988). Second generation lithotripters—a personal study. *J. Endourology*, **2**, 193–204.

Samuels, M.A. and Frost, G.P. (1989). Auditory risks in percutaneous lithotripsy. *Lancet*, **2**, 447–8.

Whitfield, H.N. (1983). Percutaneous nephrolithotomy. *Brit. J. Urol.* **55**, 609–12.

Whitfield, H.N. (1987). Stone destruction and removal. In *Recent advances in urology/andrology 4*. (ed. W.F. Hendry), pp. 41–60. Churchill Livingstone, London.

Whitfield, H.N. (1990). Stone disease—progress with lithotripsy. In *Recent advances in urology/andrology 5*. (ed. W.F. Hendry), Churchill Livingstone, London.

Whitfield, H.N. and Mills, V.A. (1985). Percutaneous nephrolithotomy: a report of 150 cases. *Brit. J. Urol.* **57**, 603–4.

Part 10 Dialysis and Transplantation

35 High-flux short-duration haemodialysis; a new approach

R.N. Greenwood

Maintenance haemodialysis has been an established form of therapy for end stage renal failure for more than 25 years. At present, 250 000 patients are treated world-wide, with survival rates reaching 75 per cent at 10 years in experienced centres (Laurent *et al.* 1983). This success is surprising, in the sense that renal function is incompletely replaced and the process of dialysis has many bio-incompatible features. While survival is one measure of the success of any treatment, long-term morbidity is another. Patients on haemodialysis are at risk of developing disabling skeletal complications after a number of years, and at risk of premature death from cardiovascular disease. The optimal duration of haemodialysis has long been a subject of debate. A balance has to be struck between the safety and well-being of patients, both in the short and long term, and allowing them to live a life as close to normal as possible.

The average weekly dialysis time of 24–40 h 25 years ago was reduced to approximately 12–15 h in the 1970s and 1980s. Technological advances, in particular the advent of biocompatible synthetic high-flux membranes and inexpensive delivery of bicarbonate dialysis fluid, have allowed a further reduction of treatment time, without worsening intradialytic symptoms. However, whether long-term health will be threatened by such a reduction is not known. This chapter examines some of the arguments for and against high-flux dialysis with reduced treatment time and describes the introduction of the technique in a new renal unit.

The factors which determine how and for what length of time patients are dialysed include patient comfort and convenience, as well as long-term survival and long-term morbidity. The economics of the health care system and methods of remuneration to physicians in charge of patients are also important. The reduction of treatment session time to an average of 3.25 h in the USA almost certainly relates to restrictions on reimbursement rates. This dialysis time contrasts with the 5 h sessions commonplace in Japan, where payment to physicians is greater if dialysis time exceeds 4 h. While nephrologists who care for patients undergoing chronic dialysis are often asked by patients to reduce their dialysis time or frequency, it is likely that the factors reducing dialysis time, particularly in the USA, have their origin in health economics. Few countries will be able to escape such economic pressures in the future, and the UK is no exception.

In May 1989, a new renal service was introduced at the Lister Hospital in Stevenage, UK, to serve the northern part of the North West Thames health region. The facilities included a 10 station haemodialysis unit. Commissioning of this new National Health Service unit provided an ideal opportunity to introduce the most appropriate treatment strategy for the 1990s. The main factors which determined the final direction taken were the demographic trends perceived to be underway, not only in the UK, but also in Europe and the USA. Also important were the financial and quality implications for renal services embodied in the forthcoming reorganization of the National Health Service. It was decided to introduce high-flux dialysis, with reduced treatment times, with a commitment to ongoing audit of outcomes.

Historical contrasts in dialysis demography

At the end of the 1970s it was clear that only relatively young and relatively fit patients were being accepted for dialysis in the UK. The UK had pioneered, and then commited itself to, self-supervised home haemodialysis. The home programme was run from less than 60 renal units, which functioned essentially as training centres for patients destined for the home. Nursing staff were heavily commited to training patients to dialyse themselves, rather than simply to treating them. By contrast, there were several hundred renal units in West Germany, France, and Italy, which were established as haemodialysis treatment centres. Selection in the UK against the elderly and those with co-morbid conditions, such as diabetes, was becoming a national scandal and a reappraisal of strategy was urgent.

At about this time, continuous ambulatory peritoneal dialysis (CAPD) was successfully developed in Canada. This seemed to offer a solution to the problem in the UK. Its simplicity allowed patients who were unsuitable for home haemodialysis to go home, self-supervised, after only a short training period. This meant that the UK could expand its dialysis programme without increasing the number of renal units or haemodialysis stations. Adoption of CAPD was very rapid and enthusiastic in the UK, contrasting with the indifference shown towards this treatment in the rest of Europe (Fig. 35.1).

In the 1980s the acceptance rate of new patients for dialysis increased dramatically, outstripping the transplantation rate. There is now a large waiting list for transplant in the UK (approximately 4000 patients), with the result that many new patients are destined to spend the rest of their lives on dialysis treatment.

The failure of the UK to expand facilities for in-centre haemodialysis during the 1980s may prove to be a costly mistake. As home haemodialysis fades as a realistic option in all countries, other European countries have expanded their hospital dialysis facilities, while the UK continues to put all its resources into CAPD. The true costs of CAPD have probably been underestimated, as has the dependence of CAPD programmes on haemodialysis back-up. These questions and others had to be addressed as the new renal service at the Lister

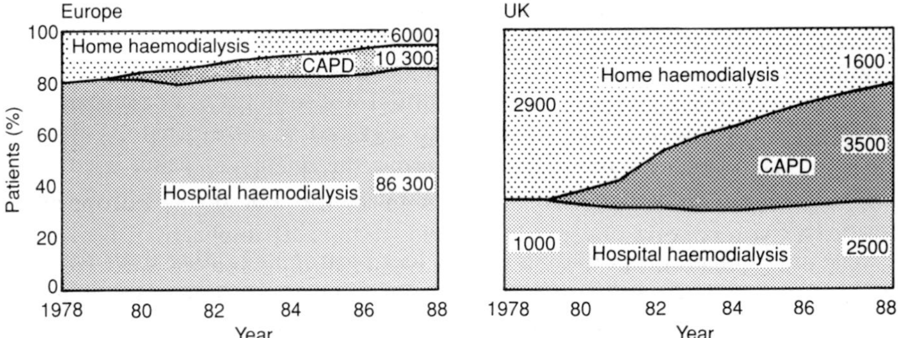

Fig. 35.1 Contrasts in dialysis demography between the UK and Europe.

Hospital was begun in 1989. The catchment population, of approximately 1.5 million, is one of the most rapidly ageing in the UK. The purchaser–provider split was underway, and most of the new haemodialysis units in the UK were being built, managed, staffed and supplied by commercial companies.

Strategic and economic considerations

Home haemodialysis

This has long been argued to be the most cost-effective treatment. However, estimates tend to ignore the high costs of staffing renal units with nurses dedicated to training, rather than treatment, and large technical back-up teams. Home haemodialysis is becoming less common in all European countries (only 6 per cent of European patients are on home haemodialysis), and it is declining particularly rapidly in Germany and the UK. The patients currently on home haemodialysis in the UK tend to have been on this treatment for many years, many with no upgrade in equipment, thus they are receiving treatment using increasingly obsolescent and ageing technology. It will be difficult to pass on the benefits of new developments in haemodialysis, which it is hoped will occur, to patients now treating themselves at home. Almost equally worrying is the placing of patients in satellite units, staffed by lay personnel, where patients are essentially self-supervised. It will be similarly difficult to re-train these people to be self-sufficient using more advanced therapies.

Dialysis costs and quality

With increasing pressure on resources renal units tend to offer what they are able to, rather than what they should, provide. Haemodialysis as currently practised (standard dialysis), although compatible with long survival, leads to disabling arthropathy. The aetiology of dialysis amyloid is poorly understood, and there is no immediately obvious solution. This is a biocompatibility issue in the broadest sense. It is known that acetate dialysis fluids, cellulosic membranes, bacteria and pyrogens in the water are particularly bio-incompatible features of standard dialysis, and they may play a role in this and other complications. There is a danger that these features will continue to be accepted as standard, with superior alternatives being regarded as extra. There is also a danger that procedures necessary to improve the quality of water used for dialysis will not be addressed in costing exercises for dialysis services (Smith *et al.* 1989).

The use of bicarbonate dialysis (which uses bicarbonate rather than acetate as the base) and non-cellulosic, synthetic membranes, such as polysulphone and polyacrylonitrile, have been adopted much less enthusiastically in the UK than in the rest of Europe. There are still no official UK standards for dialysis water, and consequently regular monitoring of water and dialysis fluid for bacterial contamination and pyrogen levels is practised in very few centres.

The costs of dialysis in the UK

The pattern of health regions funding out to districts, with comprehensive combined renal services, has resulted in complex budgets. Some of the costs of dialysis, for example those of CAPD, are currently borne by general practitioners. Costs in some of the heavily centralized renal units, with staffing little changed since the heyday of home dialysis, are likely to be high. Transport costs for dialysis will also be very significant in the future.

CAPD and haemodialysis as interchangeable options

The preparation of patients for CAPD and the treatment of CAPD failures (usually due to peritonitis) occupies a proportion of any haemodialysis facility. Some UK CAPD programmes are so large that the haemodialysis unit essentially backs up CAPD. Many units in the UK can no longer offer maintenance haemodialysis as a separate treatment modality. It is interesting that CAPD, which often looks competitive, is costed on a marginal basis in some units. The danger of having too few haemodialysis stations is that patients with recurrent peritonitis will have to remain on CAPD, thereby increasing the costs.

Lessons from the USA

The mortality of all patient groups in the USA is increasing (Hull and Parker 1990), probably due to widespread under-dialysis, as treatment times have been reduced without appropriate quality control. It is generally accepted that urea kinetic modelling, in which the dialysis 'dose' is prescribed and monitored using computer techniques, whilst not yet ideal, provides the best available method of quality assurance if treatment times are reduced to less than 5 h. However, this is still not employed in most centres adopting short hours in the USA.

Cost containment has been a major stimulus to innovation in dialysis technology. Hospital haemodialysis in the USA now costs a third of what it did in 1973. The re-use of dialysers, which is practised for 80 per cent of patients in the USA has been automated. Re-use of biocompatible non-cellulosic membranes is becoming more common, and implementation of water standards from the Association for the Advancement of Medical Instrumentation (AAMI) is mandatory.

The registered nurse/non-licensed personnel ratio in the USA has reduced dramatically. Support workers will have to take up much of the practical burden of haemodialysis in the UK in the future.

Dialysis and staffing strategies at the Lister Hospital

The following are the key features of the programme at the Lister Hospital in Stevenage.

(1) No home haemodialysis.

(2) Attempt to maintain a free movement of patients between CAPD and hospital haemodialysis, favouring CAPD for the relatively fit, and well-supported haemodialysis for the elderly and infirm.

(3) Disconnect CAPD systems in an attempt to keep peritonitis rates low.

(4) Bicarbonate dialysis fluid and synthetic high flux membrane (polysulphone) standard in all patients.

(5) A quality assurance programme for dialysis water, using AAMI standards as guidelines, and upgrading local standards in the light of experience.

(6) Re-use of all dialysers using automated equipment employing peracetic acid rinsing and disinfection. High quality water used for re-use.

(7) Minimize treatment time on dialysis by employing high blood and dialysis fluid flow rates using urea kinetic modelling for prescription and quality assurance in all patients.

(8) Three 12.5 h nursing shifts per week. Support workers also on a 12.5 h shift, but offset from the nursing shift, so that support workers set up equipment for dialysis at the start of the day and clear up and disinfect equipment at the end of the day. Aim to treat 3 or 4 patients per machine per 12.5 h shift.

(9) Strict appointment system, with emphasis on service and none on patient training.

(10) To ensure that financial resources available are used to maximum effect.

Clinical experience and outcomes

Practical aspects

The Lister Renal Unit became operational in May 1989. There are (June 1992) 63 patients on haemodialysis (mean age 61 years) and 58 patients on CAPD

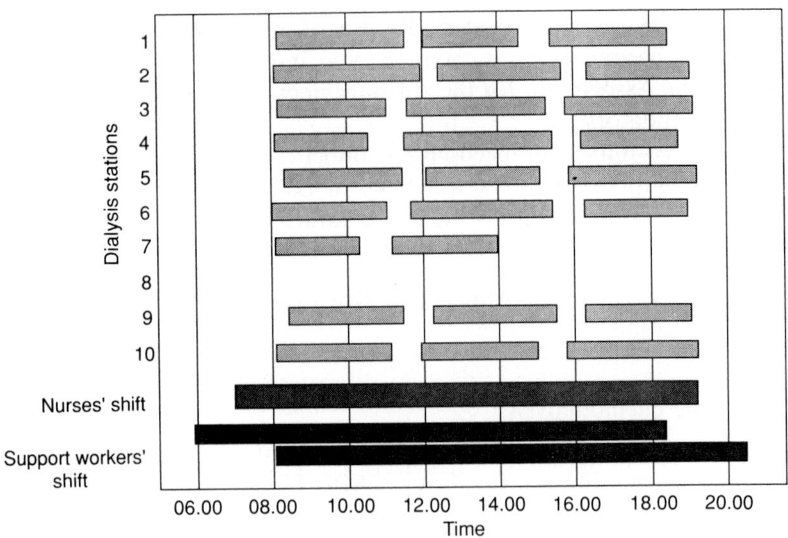

Fig. 35.2 Shift patterns of nurses and support workers related to the usage of dialysis stations.

(mean age 59 years). The CAPD peritonitis rate is quite low, being 1 episode every 40 patient months.

The proposed shift patterns have proved to be popular with staff and the planned pattern of treatment in the haemodialysis unit has been achieved

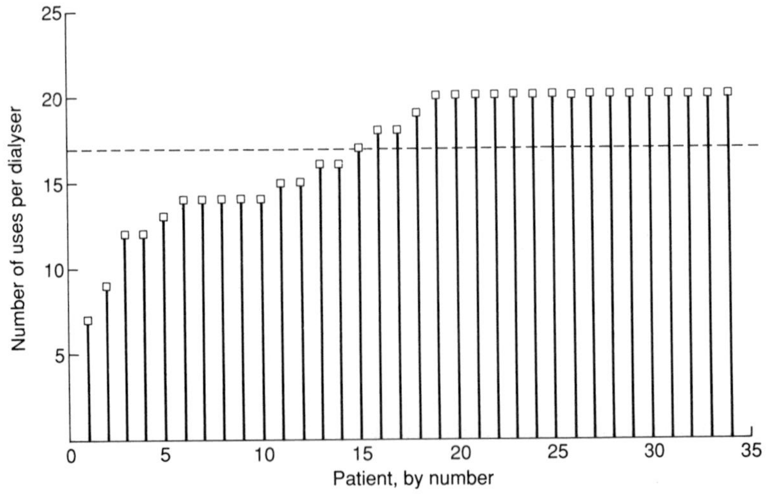

Fig. 35.3 Re-use of dialysers in the Lister Renal Unit.

(Fig. 35.2). Three, or sometimes 4, patients are treated at each dialysis station per shift.

Dialysers are re-used on average 17 times. Sixteen patients re-use the dialyser 20 times, then the dialyser is routinely discarded. Only two patients re-use the dialyser less than 10 times (Fig. 35.3). In dialysers re-used up to 20 times 'normalized' urea clearance is well maintained.

The mean time on dialysis is 2 h 43 min (range 71–241 min). The 'dialysis day' is impressively short for patients at the Lister Renal Unit. The mean time between leaving home and returning is 4.6 h (Fig. 35.4).

Clinical outcomes

Tolerance of high-flux bicarbonate dialysis, at least anecdotally, is impressive. This is particularly so in the elderly, in patients with severe heart failure, and in patients who experience angina off dialysis. There is very little nausea and vomiting during dialysis, although headache has caused dialysis time to be extended in three patients with resolution of symptoms.

Six per cent of haemodialysis treatments are complicated by hypotension. This contrasts favourably with the overall incidence of 11 per cent estimated in Europe in 1983 (Shaldon *et al.* 1983).

There are currently 15 admissions for non-access reasons per 1000 dialyses. Data are available on admission rates from Giessen, a unit in Germany where there was early experience with maintenance haemodiafiltration (Wizemann and Kramer 1987). In 1981, the mean time on haemodiafiltration was 7.5 h/week, and there were 13 admissions per 1000 treatments. In 1985 there had been a return to dialysis, the mean treatment time having been increased to 12.3 h. Admissions were reduced to 9 per 1000 dialyses. Although not quoted, it is likely that the mean age of these patients was lower than that of the

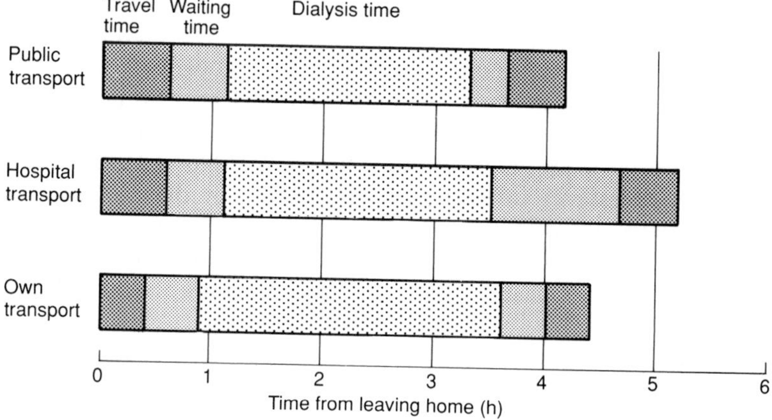

Fig. 35.4 The dialysis day for patients attending for haemodialysis.

patients currently treated at the Lister Hospital.

The urea kinetic modelling targets are being achieved in all patients (KT/V 1.06 and 1.80 in thrice and twice weekly dialysis respectively).

Control of hypertension

This is a cause for concern. Sixty-five per cent of patients take hypotensive drugs. This problem is not restricted to the elderly.

The reservation most frequently put forward regarding short dialysis schedules concerns their possible negative effect on the control of blood pressure. The excellent control of blood pressure without the use of anti-hypertensive drugs obtained in patients treated with long dialysis sessions (over 20 h) provides good evidence that the dialytic process itself can be a powerful hypotensive tool (Laurent *et al.* 1983). A decrease in the number of patients taking hypotensive drugs was reported by Wizemann after the short haemo-diafiltration times of less than 10 h/week were replaced by dialysis sessions of over 12 h/week. Survival in all patient groups on dialysis in the USA, where there has been a gradual but definite reduction of treatment time to less than 4 h, is decreasing (Hull and Parker 1990).

One of the main features of dialysis practice in the 1970s was the normalization of high blood pressure, which was present in the majority of patients starting on dialysis, by slow salt and water removal. This process of 'drying out' usually took several months and was often painful and laborious, but the end result was a patient who was normotensive without the need for drug therapy. As treatment times have been reduced, this goal seems to have been pursued much less aggressively. One of the factors which helped in the 'drying out' of patients was that removal of more than 1.5–2 kg of fluid weight using acetate/cuprophane dialysis often resulted in unpleasant intradialytic symptoms. As a result, the weight gains of patients tended to be fairly strictly controlled.

With the advent of modern membranes, bicarbonate dialysis fluid and purified water, symptoms are undoubtedly fewer, and patients tend to drink more between dialysis sessions, so making the normalization of blood pressure more difficult. It is also probable that physicians feel more comfortable with modern hypotensive drugs, arguing that correction of blood pressure by drug therapy is not necessarily inferior to normalization by dialysis. It certainly appears that physicians now put less energy into achieving true dry weight in dialysis patients. In two large centres in London employing standard treatment times, the percentages of patients on hypotensive drugs are 55 and 48 (personal communication).

It is the author's view that reduction of treatment time to less than 4 will exacerbate the widespread problem of hypertension in dialysis patients. It may be that dialysis for 4, 5 or more hours would be necessary to reduce dependence on hypotensive therapy. Patients are unlikely to agree to return to these long treatment times, and they are also probably not economically viable in some countries. Although it cannot be directly presumed, it is possible that the presence of hypertension, albeit controlled with antihypertensive therapy, will have an adverse effect on cardiovascular mortality. A significantly higher mortality resulting from myocardial infarction was documented in patients in Europe who were dialysed for 12 h/week or less, compared with those dialysed for 14 h/week or more (Broyer *et al.* 1982). While there is limited concern about patients over the age of 65 years requiring hypotensive drugs, the causes of hypertension in younger patients require further study. It may be that better understanding of sodium shifts, and tailoring of dialysis fluid sodium to the individual patient, may provide a way forward.

The costs of high-flux dialysis

The development of a new National Health Service renal service provided the opportunity of controlling the renal budget from the beginning. The Renal Unit was set up as a Directorate with a Renal Unit Management Group. The amount of money flowing into the District for renal services for 1990/91 was £973 000. A financial plan was set out, as illustrated in Table 35.1.

The division of the renal budget into compartments allowed a cost analysis to be made of the hospital haemodialysis programme (Greenwood *et al.* 1990). The cost of a haemodialysis session was derived as follows:

Total revenue budget (haemodialysis)	£513 000
DHQ apportionment resource management/nurse training	£20 000
Depreciation charge	£41 000
Capital charge	£87 000
Total haemodialysis running costs	£661 000
Projected haemodialysis sessions 1990/91	4200
Cost per session (bicarbonate/polysulphone/quality controlled water)	£157.38

This figure includes the costs of doctors, business manager, dietitian and technicians. It also includes the costs of pathology, theatres, finance, advertising,

Table 35.1 Money (pounds, sterling) allocated to the four service compartments, and training and education

Unit	Pay	Non-pay	Budget
Haemodialysis	280328	232849	513177
Nephrology ward	241213	52742	293955
CAPD	23418	95240	118658
Outpatients	13179	6000	19179
Training and education	0	28000	28000
Total	**558138**	**414831**	**972969**

telephones, publications, fuel, light, heat, personnel, administration, works and medical records. Also included in the cost of nurse training (ENB renal course and continuing education courses).

Comparison with commercial sector dialysis

Most of the dialysis centres built in the UK in the last 5 years are commercial sector units. In this situation a commercial company builds, supplies and staffs the dialysis unit, cross-charging the District Health Authority for its services. The costs of these units have been examined in a report from the Welsh Office (Smith *et al.* 1989).

The commercial sector price usually includes nursing and disposable costs, CSSD, cleaning, and machine maintenance. However, it often excludes medical, operating theatre, and administrative costs, and, to a large extent, nurse education costs.

A rough estimate was made of the cost per session at the Lister Hospital including only these items. On this basis, the cost per session was approximately £76, which compares favourably current prices for standard dialysis in commercial units (Editorial, Lancet 1990).

Conclusions

High-flux dialysis with reduced treatment times has been introduced successfully at the Lister Hospital. The treatment is well tolerated. Intradialytic morbidity is low and short-term health of the patients is satisfactory, as judged by the numbers of admissions to hospital. Objective criteria for dialysis adequacy, determined by urea kinetic modelling, are being met in all patients, in line with currently accepted norms. The staff shift patterns are successful, and the dialysis day is short for most patients.

The main concern is the prevalence of hypertension in the haemodialysis population and this requires careful audit and merits further research.

The costings suggest that high-flux dialysis can be delivered at a competitive price, but more experience is necessary before firm conclusions can be drawn. This new dialysis strategy may eventually prove to be the most effective way of combining quality and costs, while maintaining the flexibility to pass on the benefits of technical advances to large numbers of patients.

References

Broyer, M., Brunner, F.P., Brynger, H. *et al.* (1982). Combined report on regular dialysis and transplantation in Europe, XII, 1981. *Proceedings of the European Dialysis and Transplant Association*, **19**, 2–59.

Editorial. (1990). Quality assurance and dialysis services. *Lancet*, **8724**, 1160–1.

Greenwood, R.N., Sanders, B., and Powell, D. (1990). Costing a new approach to dialysis. *THS Health Summary*, Oct, 7–11.

Hull, A.R., and Parker, T.F. (1990). Proceedings from the Mortality and Prescription of Dialysis Symposium, Dallas, September 15–17, 1989. *American Journal of Kidney Diseases*, **15**, 5.

Laurent, G., Calemard, E., and Charra, B. (1983). Long dialysis. A review of fifteen years experience in one centre 1968–1983. *Proceedings of the European Dialysis and Transplant Association*, **20**, 122.

Shaldon, S., Baldamus, C.A., Koch, K.M., and Lysaght, M.J. (1983). Of sodium, symptomatology and syllogism, *Blood Purification*, **1**, 16–24.

Smith, W.G.J., Cohen, D.R., and Asscher, A.W. (1989). *Evaluation of renal services in Wales with particular reference to the role of subsidiary renal units. Report to the Welsh Office 1989*. Cardiff, KRUF Institute of Renal Disease, Royal Infirmary.

Wizemann, V. and Kramer, W. Morbidity during short and longer dialysis treatment. *Blood Purification*, **5**, 193–201, 1987.

36 Continuous ambulatory peritoneal dialysis; current role and future prospects

J. Michael

Peritoneal dialysis was first described as a treatment for renal failure in 1923, when Ganter detailed a single clinical case and reported studies on experimental acute renal failure in guinea-pigs. During the 1940s sporadic case reports appeared in the literature. In 1950, O'Dell reviewed the 101 patients treated by peritoneal dialysis that had been reported in the world literature between 1923 and 1948 (O'Dell *et al.* 1950). Of these patients, 63 had acute renal failure, 32 surviving the episode, 32 patients had irreversible chronic renal failure, and 2 remained unclassified. The use of intermittent peritoneal dialysis for chronic renal failure started in the early 1960s, but was handicapped by the need repeatedly to pierce the peritoneum with hard catheters. The development of a soft permanent catheter by Tenckhoff in 1973 allowed permanent access to the peritoneum. In 1976, Popovich and Moncrief described the process of slow turn-over 'equilibrium peritoneal dialysis' (Popovich *et al.* 1976). In 1978, Oreopoulos and colleagues made use of developments in plastic manufacture to describe a disposable delivery system for dialysis fluid, which is now an integral part of a technique known as continuous ambulatory peritoneal dialysis (CAPD) (Oreopoulos *et al.* 1978). Since then there has been much debate about the relative merits of CAPD and haemodialysis, and the viability of CAPD as a technique for long-term dialysis treatment.

Relative merits of CAPD and haemodialysis

CAPD has a number of theoretical advantages compared with conventional haemodialysis. Equilibrium dialysis allows a steady state, rather than fluctuating levels of biochemistry. Middle molecule clearance is more efficient than with haemodialysis using conventional dialysis membranes. Salt and water control is generally easier, and dietary restriction frequently less severe than on haemodialysis. Average haemoglobin levels on CAPD are higher. There is more flexibility over the location of dialysis, and treatment is more user-friendly, with few, if any, mechanical aids. Self-treatment is much easier for the majority of patients, and independent self-dialysis can be established relatively quickly.

There are also clear financial advantages in initial or capital costs compared with establishing independent community-based haemodialysis, although the revenue expenditure on CAPD is comparable to non-hospital based haemodialysis. With all these advantages one is then forced to ask why CAPD has not become the universal mode of dialysis for chronic renal failure. The percentage of dialysis patients being treated by CAPD in different countries is shown in Table 36.1. The reasons for the great range in percentage of patients receiving CAPD in national and local dialysis populations are varied. Those countries with a long-established tradition for extensive haemodialysis provision have seen a smaller role for CAPD than those countries whose haemodialysis facilities were more limited. CAPD allowed a number of countries and individual programmes to expand total dialysis numbers relatively quickly and cheaply because of the limited requirement for capital and fixed resources. In the UK, the Birmingham programme expanded from 20 dialysis patients, all on haemodialysis, to 330 dialysis patients, of whom 70 per cent were on CAPD, in less than 10 years. Programmes heavily committed to CAPD are also very dependent on the long-term survival of CAPD as a technique.

One of the difficulties in the management of end stage renal failure is the assessment of the adequacy of dialysis. The simple assessment of biochemical variables is inadequate, and methods of assessing adequate dialysis, although increasingly defined on the basis of urea kinetic modelling for haemodialysis, are as yet less well established in routine practice for CAPD. Many nephrologists assess 'adequacy of dialysis' from factors other than direct biochemical analyses. How does CAPD compare with haemodialysis by these non-biochemical measures of adequate treatment?

Table 36.1 CAPD as a percentage of treatment for end stage renal failure

Country	CAPD (%)
Germany	4
Japan	4
Italy	7
Brazil	16
USA	17
Switzerland	20
Israel	24
Australia	30
Denmark	33
South Africa	38
Canada	39
UK	42
Finland	44
Venezuela	53
Mexico	76

Anaemia has been shown to be improved in patients who have been converted from haemodialysis to CAPD. There are no data to support the idea that the change in dialysis technique *per se*, rather than the reduction in blood losses, is associated with the difference in haemoglobin or the occurrence of severe anaemia.

The National Institutes of Health Registry in the USA showed that 35 per cent of patients starting CAPD had been taken off all hypotensive drugs within one year of starting treatment (NIH 1987). However, control of hypertension was shown to be as effective, but no more so than that obtainable on haemodialysis, as long as effective ultrafiltration was maintained. There have been a number of studies of cardiac function in patients on CAPD, but few make direct comparisons with haemodialysis. There are a number of reasons why CAPD could be associated with fewer cardiac stresses, as it is a treatment which avoids the problems of an extra-corporeal circulation. In one study, Alpert demonstrated abnormal cardiac indices in 54 per cent of haemodialysis patients and 39 per cent of CAPD patients. The absence of any randomized controlled studies of haemodialysis and CAPD introduces an inevitable selection bias into any comparisons of the two treatment modalities.

Only four studies of any size have examined the instance of renal osteodystrophy in patients on CAPD compared with haemodialysis. There were no clear advantages of one treatment over the other. Some series have demonstrated lower levels of β_2-microglobulin in patients on CAPD compared with haemodialysis, but no difference was shown in β_2-microglobulin associated with amyloidosis, and in some cases the lower levels of β_2-microglobulin were associated with better preservation of residual urine volumes. Uraemic neuropathy has been studied in a limited number of series, and no differences between the treatment modalities shown. In summary, there is no evidence that CAPD is better or worse than haemodialysis in controlling the clinical manifestations of uraemia.

It might be expected that the differences in patient survival would give some indication of the advantage of one form of dialysis over another. Unfortunately, the problems of patient selection and the lack of a randomized controlled trial of CAPD versus haemodialysis means that there are no clear data on this point. A number of studies have used Cox's proportional hazard regression model to attempt to allow for differing risk factors between groups on different treatments. No significant differences have been shown in survival between groups of diabetic and non-diabetic patients treated by either haemodialysis or CAPD.

Technique survival in CAPD

One of the problems with CAPD highlighted by early EDTA Registry data was the very high failure rate, even after allowing for drop-out due to death or transplantation. This high technical failure rate may have been associated with early negative selection policies, or by early learning curves with this technique. What are the factors that influence the technique survival of CAPD? Peritonitis has always been the major limiting factor in CAPD programmes. World-wide, peritonitis rates are very variable. Large multi-centre series usually report higher rates of peritonitis than do smaller single centres. The seven centre study in the UK reported 2.4 episodes of peritonitis per patient year (Gokal *et al.* 1987). A large study from the USA reported an instance of 1.5 episodes per patient year on CAPD. By contrast, some large series, notably from Italy, have reported an incidence of peritonitis as low as 0.33 episodes per patient year (Maiorca *et al.* 1988, 1989). This may be related to the greater use in Italy of the Y or disconnect CAPD systems. Peritonitis may lead to an immediate or delayed, temporary or permanent failure of the peritoneum to allow adequate dialysis.

Peritoneal membrane failure can be divided into two types. Type I membrane failure is associated with loss of ultrafiltration, but preservation of solute removals. There is hyperpermeability of the membrane to glucose, with absorption of glucose and loss of the osmotic gradient between blood and dialysate. This type of membrane failure is, in some cases, reversible if the peritoneum is rested. A very transient disruption of peritoneal membrane function of this type is frequently seen and associated with an acute episode of peritonitis.

Type II membrane failure features loss of both ultrafiltration and solute removal. Macrophage stimulation triggers fibroblast proliferation. The peritoneal membrane is thickened due to an increase in fibrous connective tissue. There has been much consideration of the mechanism involved. Amongst possible causative factors are osmotic damage and the effect of peritonitis, in particular Gram-negative and fungal infections, although Gram-positive infections, especially those due to *Staphylococcus aureus*, can have the same effect. Chemical effects from differing buffer agents (lactate versus acetate), contamination of dialysate with plasticizer from the bags or tubing sets, or with antiseptics, such as chlorhexidine or iodine, have all been suggested causes. An extreme form of this progressive peritoneal thickening results in an encapsulating sclerosing peritonitis, which causes significant morbidity and mortality. This has been reported in association with particular buffering agents, and as a result of contamination with antiseptics. The incidence of irreversible failure of the peritoneal membrane has been reported variously as between 1.7 and 10 per cent of CAPD patients per year.

Ultrafiltration is positively correlated with the concentration of glucose in the dialysate, and there is a negative correlation with the degree of glucose absorption from the dialysate, and with the overall time spent on CAPD. The associations are, however, far from straightforward. Lactate-buffered dialysate from one manufacturer seems to be associated with a greater glucose absorption, and hence reduced ultrafiltration, than is seen with the apparently identical lactate buffered dialysate from other manufacturers.

Nutritional factors can play a major role in the viability of CAPD as a long-term technique for dialysis, in contrast with the situation with haemodialysis. Protein and amino acid losses in the peritoneal dialysate can amount to between 5 and 15 g protein/d. Between 1.3 and 3.4 g free amino acids/d can be lost into the dialysate. In order to prevent a progressive fall in serum proteins and amino acids, a dietary protein intake of at least 1.2 g protein/kg body weight/d is required. The achievement of this dietary target is inhibited or prevented quite frequently by the suppression of appetite caused by absorption of glucose from the dialysate. Similarly, the sensation of abdominal fullness and distension associated with the volumes of dialysate within the peritoneal cavity, and the effect of this fluid on delaying gastric emptying, may both discourage adequate dietary intake.

Obesity can become a factor which limits the continuation of CAPD. The use of hypertonic glucose as an osmotic agent results in the absorption of significant quantities of carbohydrate, thus causing obesity. This is particularly so in those patients in whom a reduction in ultrafiltration rate has already been identified with isotonic exchanges, resulting in the increased frequency of hypertonic exchanges. The risk of hyperlipidaemia is a significant additional factor when frequent hypertonic exchanges are required. Associated with the increased prevalence of hyperlipidaemia, a frequent occurrence with chronic renal failure, the effects of chronic glucose loading on lipid metabolism may become an important factor in development of cardiovascular morbidity.

Patient failure

Patient failure is a term used to describe the inability of a patient to continue with a mode of dialysis which is technically successful. The incidence of patient failure is related to the selection of patients for CAPD treatment. Some of the earlier reports of patient and technique survival came from units practising negative selection for CAPD. Only patients who were deemed unsuitable for haemodialysis were placed on CAPD, as an alternative to being not treated.

The converse can produce better results. Positive selection for CAPD can lead to a group of highly motivated, technically competent patients, who are less likely to fail in the technique, and more likely to be willing and capable of continuing long-term. The ability to select the correct patient for the technique and the correct technique for the patient is fundamental to the running of a proper dialysis programme, yet it is not possible for many units because of constraints on resources or facilities (Burton and Walls 1987).

There are a number of particular problems associated with CAPD as far as the patient is concerned. It is a continuous process, day in day out with little respite although some units have allowed patients a day off to try to relieve this problem. Some patients have difficulty with their body image when asked to accept a permanent catheter protruding from the abdominal wall, however discretely. There is a psychosexual component to this problem. The very nature of CAPD is that of an independent form of dialysis treatment which, while attractive to many patients, is in itself associated with the burden of independence and of self-reliance, with which not all patients can cope.

A number of abdominal or intra-abdominal problems not directly related to the cause of renal failure or to the technique of CAPD catheter insertion may also influence technique survival. Previous surgery may limit the available peritoneal cavity, by producing adhesions, or more directly by partitioning the peritoneal cavity, for instance in patients in whom ileal conduit urinary diversions have been fashioned. Diverticular disease may be a relative contra-indication to CAPD and is undoubtedly a risk factor for Gram-negative peritonitis. Polycystic disease, with gross enlargement of the liver and kidneys, may result in technical difficulties, even in the face of bilateral nephrectomy. Vascular disease, especially when associated with aortic aneurysm formation and surgery, can result in difficulties in the short-term, or indeed permanently, in CAPD treatment.

How does CAPD technique survival today compare with early experience? Pooled data from the European Dialysis and Transplant Registry are shown in Table 36.2. It can be seen that the best technique survival is to be found in older patients who opt for CAPD as a first choice; however, even in this group 62 per cent had been withdrawn from CAPD within 5 years. The reasons for abandoning CAPD are listed in Table 36.3. Peritonitis is by far the largest cause

Table 36.2 EDTA Registry (1983–84): CAPD technique survival

	Survival %			
Option (age group, years)	1 year	2 years	3 years	5 years
First choice (15–64)	75	50	35	25
First choice (>65)	75	58	50	38
Enforced (15–64)	70	50	38	28

Table 36.3 EDTA registry (1983–84): reasons for abandoning CAPD

Reason	Number (%)
Peritonitis	52
Peritoneal failure	14
Other abdominal complications	12
Other CAPD complications	8
Failure to cope	6
Patient request	6
Family request	1
Other	1

ıl failure in this group. The seven centre study from the UK, albeit ıonstrated almost double the rate of change of dialysis mode in arting on CAPD, compared with those starting on haemodialysis (Table 36.4). The importance of peritonitis in causing technical failure can be shown from the data of the Italian multi-centre study, which demonstrated over 52 per cent of patients remaining on CAPD at 3.5 years (Table 36.5). Peritonitis was again the most common cause of technique failure, but not to such an extent as in the overall European experience (Table 36.6).

Table 36.4 UK seven centre trial

Outcome (end point or 12 months)	Haemodialysis (n = 47)	CAPD (n = 92)
On original therapy	12 (26%)	27 (29%)
Transplanted	23 (49%)	39 (42%)
Deaths	5 (11%)	9 (10%)
Changed dialysis therapy	4 (9%) (CAPD)	15 (16%) 5 home Haemodialysis, 9 hospital haemodialysis, 1 intermittent peritoneal dialysis

Table 36.5 Italian multicentre study: patients remaining on CAPD

Time (months)	Patients remaining on CAPD (%)	
	Excluding deaths transplantation	All causes
12	83	66
18	75	56
24	70	45
42	52	22

Table 36.6 Italian multicentre study of CAPD: reasons for drop-out

Reason	Number	(percent)
Peritonitis	67	(42)
Medical reasons	29	(18)
Loss of ultra filtration or solute removal	20	(13)
Catheter failure	11	(7)
Patient choice	12	(8)
Failure to cope	12	(8)
Other	6	(4)

Conclusions

What then is the present position of CAPD in the treatment of end stage renal failure? What is quite clear is that the extremes of views expanded by enthusiastic proponents and opponents are probably misplaced. It should also be remembered that the technique of CAPD is, after 10 years development, being compared with haemodialysis, a technique which has been refined over 3–4 times that period. Technical advances are occurring in the delivery and understanding of both techniques, and it is to be hoped that some of the problems with CAPD, such as the high peritonitis rate, will be addressed by these technical advances. The wider use of some of the technical aids and disconnect systems initially developed in Italy are already showing encouraging signs of improving the situation, but at the expense of increasing the cost of treatment.

Technique survival in haemodialysis is excellent, yet patient acceptance is variable. Patient acceptance of CAPD is often excellent, but there is a full range of proponents and opponents of the technique amongst patients, as there is amongst their medical advisers. There would appear to be inherent limitations to technique survival in CAPD centred around the vulnerability of the peritoneal membrane to irreversible damage. Even if CAPD is not to be a technique for long-term treatment in an individual patient, it offers a very important option for selected patients. Amongst these are patients who are likely to be transplanted within a relatively short time, those with cardiac instability and other vascular problems especially diabetes, and many older patients and those wishing for enhanced flexibility and mobility while on dialysis.

CAPD should be regarded neither as second class nor as superior to haemodialysis. It is different, and should be seen as one therapeutic option available to all units offering dialysis treatment for end stage renal failure. Given the relatively short technique survival time on CAPD, over-reliance on this technique to the detriment of the provision of adequate haemodialysis facilities results in uncomfortable pressures on the units or countries concerned.

References

Burton, P.R. and Walls, J. (1987). Selection adjusted comparison of life expectancy of patients on continuous ambulatory peritoneal dialysis, haemodialysis and renal transplantation. *Lancet*, **1**, 115–19.

Gokal, R., King, J., Boyle, S. *et al.* (1987). Outcome in patients on continuous ambulatory peritoneal dialysis and haemodialysis: four year analysis of a prospective multi-centre study. *Lancet*, **2**, 1105–9.

Maiorca, R., Vonesh, E., Cancarini, G.C. *et al.* (1988). A six year comparison of patient and technique survival in continuous ambulatory peritoneal dialysis. *Kidney International*, **34**, 518–24.

Maiorca, R., Cancarini, G.C., Camerini, C. *et al.* (1989). CAPD competitive with haemodialysis for long term treatment of uraemic patients. *Nephrology Dialysis Transplantation*, **4**, 244–54.

O'Dell, H.M., Ferris, D.O., Power, M.K. (1950). Peritoneal lavage as an effective means of extrarenal excretion. *Am. J. Med.* **9**, 63–77.

Oreopoulos, D.G., Robson, M., Izatt, G., Clayton, S., De Vebbe, G.A. (1978). A simple and safe technique for continuous ambulatory peritoneal dialysis (CAPD). *Trans. Am. Soc. Artif. Intern. Organs*, **24**, 481–9.

Popovich, R.P., Moncrief, J.W., Deckerd, J.F., Bomar, J.J.B., Pyle, W.K. (1976). The definition of a novel portable—wearable equilibrium peritoneal technique. *Abst. Am. Soc. Artif. Intern. Organs*, **64**.

Report of the National CAPD Registry of the National Institutes of Health, (1987).

37 HLA matching; its relevance in the 1990s

M.C. Jones and G.R.D. Catto

The major histocompatibility complex (MHC) evolved as a concept from studies into the effects of blood transfusions in various animal species towards the end of the last century. Landsteiner not only discovered the A, B and O blood groups in 1901,[1] but speculated in his Nobel lecture 30 years later that a similar system for tissue cells might govern the acceptance or rejection of a transplant.[2] While considerable progress was made during the next few years in defining the mouse MHC, its human equivalent was not discovered until the 1950s.

This delay was due to the fact that human erythrocytes express only low levels of class I histocompatibility antigens[3] and, thus, the haemagglutination assay, which was of such value to investigators of the mouse MHC, could not be used to study the human MHC. Leucoagglutination, however, was demonstrated after incubation with serum from patients who had received multiple blood transfusions, and was shown by Dausset to be caused by alloantibodies.[4] Further studies demonstrated that leucocyte antigens were inherited, and the first specific antigen was discovered by the reactions of six sera (from patients who had received multiple transfusion) with white cells from normal donors.[5] The recognition that sera from parous women frequently contained leucocyte antibodies enabled sera of greater specificity to be identified, and eventually led to the discovery of the various antigens and genetic loci.

In 1967, the human MHC was designated HLA (human leucocyte locus A—the first system described in humans—although later the A was changed to mean antigen). By 1971, HLA-A, B and C antigens had been identified using the complement-dependent microlymphocytotoxicity assay.[6]

HLA-D antigens were detected by the mixed lymphocyte reaction, using cells from individuals who were apparently matched for the known HLA types.[7] Subsequently, similar antigens termed HLA-DR were detected serologically.[8]

It is now known that the genes coding for the human MHC occupy a two centimorgan segment on the short arm of chromosome 6 (Fig. 37.1). By 1988, a total of 148 HLA antigens had been identified (Table 37.1), making the task of identifying HLA identical donors and recipients more, rather than less, difficult.[9]

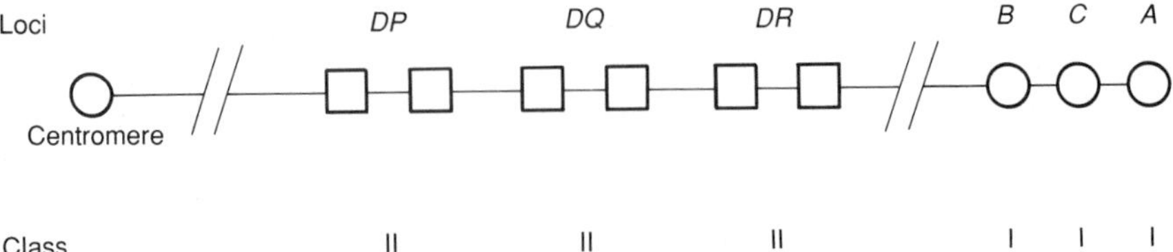

Fig. 37.1 Genetic map of the HLA gene complex on chromosome 6.

Table 37.1 HLA antigens

HLA-A	24
-B	52
-C	11
-D	26
-DR	20
-DQ	9
-DP	6
Total	148

Patient survival rates

During the last decade, patient survival rates have steadily improved, and they now exceed 94 per cent at one year after renal transplantation.[10] These good results may, however, not be maintained as increasing numbers of elderly and debilitated patients, often with secondary forms of renal disease (e.g. diabetic nephropathy, polyarteritis nodosa, etc.), are accepted on to transplant programmes.

Renal allograft survival rates

One-year renal allograft survival rates declined at approximately 2 per cent per year from 1966 to 1974, and were less than 50 per cent for cadaver grafts. This decrease occurred at a time when pre-transplant blood transfusions were restricted, both to limit the spread of hepatitis B and to prevent the potential recipient developing HLA antibodies, which might subsequently cause a positive cross-match test to donor cells. Since 1978, by which time most units had instituted elective transfusion programmes, graft survival rates at one year after transplantation have gradually increased. Currently many centres report rates in excess of 85 per cent for the numerically important group of first cadaver allografts (92 per cent for living related grafts).[10] As has been known for many years, renal allografts are lost in three principal ways after transplantation. Hyperacute rejection, now less common with improvements in cross-match techniques, and acute rejection together account for the 10–15 per cent of kidneys lost within the first three months of surgery; chronic rejection is held responsible for the slower but inexorable decline in graft survival thereafter. Within these broad categories are concealed a variety of less common, but still important, causes of graft loss, including surgical problems, infection, and vascular thromboses. Nevertheless, all data continue to demonstrate a relatively high rate of graft loss within the first few weeks or months after transplantation, and slower but continuing loss thereafter (Fig. 37.2).

The causes of the recent welcome improvement in graft survival are complex. A recent analysis from the UK Transplant Service[10] indicated clearly that the improvement occurred not in the first 15 days after surgery, but predominantly in the periods 16–100 days and 101–365 days after transplantation; there was little change noted in graft survival rates after the first year.

One factor contributing to this increase in allograft survival has been patient survival. It seems likely that changes in immunosuppressive therapy, an overall decrease in steroid dosages, standardization of anti-lymphocytic globulin, and the introduction of cyclosporin A have decreased mortality rates, and thus caused an increase in graft survival. During the 1970s the beneficial effect of pre-transplant blood transfusion was widely appreciated, and by 1978 a majority of transplant centres had instituted elective transfusion programmes; by 1983 only 4 per cent of patients awaiting transplantation were untransfused. Despite the many clinical and experimental studies undertaken during the last 20 years, the mechanisms by which blood transfusions improve

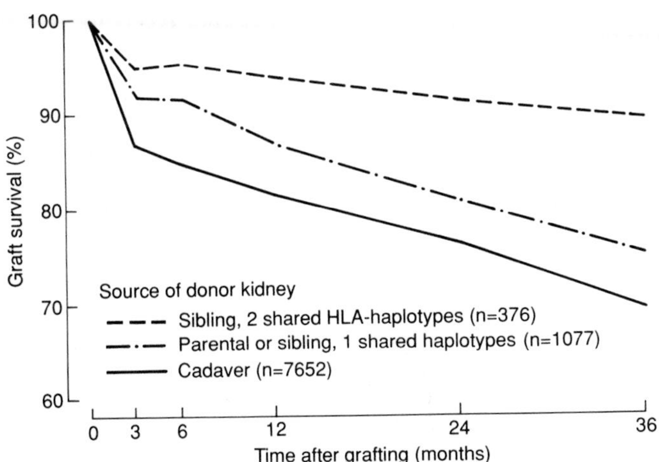

Fig. 37.2 Survival of first kidney transplants from 1986 to 1988, according to the type of donor. Data from patients aged 15–44 years with standard primary renal disease.[16]

allograft survival remain controversial.

More recently, it has been increasingly difficult to detect this beneficial transfusion effect;[11] several investigators have reported excellent graft survival results without pre-transplant transfusion and in the presence of blood group compatibility only between donor and recipient. Whether or not the transfusion effect has disappeared remains a matter for debate, but it would appear now to be clinically less relevant with recent improvements in graft survival. The potential benefits are no longer so obvious as they were a decade ago. Concerns over sensitization and transmission of viral disease will inevitably result in more transplant units abandoning elective transfusion programmes. Indeed the proportion of units with a policy of elective blood transfusions fell from 72 per cent in 1982 to 40 per cent in 1988; those advocating no transfusion rose from 5 per cent to 35 per cent during that time, despite evidence that blood transfusions and HLA matching might produce an additive effect.[12]

Effect of HLA matching

Living related renal transplantation

In transplantation from a living related donor there is a good correlation between the number of antigens shared and graft survival. One-year graft survival for HLA identical grafts is approximately 90–95 per cent—similar to that observed for grafts from an identical twin; completely mismatched grafts have a survival rate of 60–70 per cent,[14] and one haplotype mismatched grafts (commonly parent to child) are between these two figures. While haplotype matching is of principal importance in living related transplantation, the value of DR matching has been emphasized in patients receiving one haplotype matched kidneys; when donor and recipient were matched for both DR antigens, graft survival was significantly better than if one DR antigen was mismatched.[15] The recent decision to introduce sophisticated genetic techniques using appropriate restriction fragment-length polymorphisms into living related transplantation is based more upon a need for an objective genetic link between recipient and donor than any primary wish to improve graft survival rates.

Both patient and graft survival rates one year after surgery have been better in living related than in cadaver donor transplantation. These results have also improved during the last 10 years, and are now more than 96 per cent and 92 per cent, respectively. Graft losses, however, follow the pattern outlined above, and it is relevant to note (Fig. 37.2) that the rate of attenuation more than one year after transplantation is reduced only in the group comprising transplants between siblings sharing two haplotypes.[16]

Cadaver donor renal transplantation

The early attempts at renal transplantation were all unsuccessful because of immunologically-mediated graft rejection. To avoid this problem, the first successful graft was performed between monozygotic twins.[17] Attempts to decrease the immunological dissimilarity by HLA matching between donor and recipient have been more successful in living related than in cadaver donor transplantation.

Matching for class I antigens

As early as 1967, matching donor and recipient for the then known HLA-A and HLA-B antigens was claimed to result in improved graft survival. Although not all reports supported these claims, most large studies indicated that matching for class I antigens improved not only graft, but also patient survival—presumably because less immunosuppression was required with fewer and less severe episodes of rejection.[14] When the effects of matching for the different class I antigens were investigated, matching for B antigens conferred greater benefit than matching for A antigens;[18] matching for C antigens did not produce a significant beneficial effect on graft survival.[19] The whole question of the relative importance of matching for class I antigens has been overtaken both by the logistical problems caused by the large number of private specificities now recognized (Table 37.1), and by the recognition of the serologically-recognized class II antigens.

Matching for class II antigens

The beneficial effect of matching donor and recipient for class II antigens on both patient and graft survival rates has been appreciated since 1978 and confirmed on many occasions—an effect that remains detectable even when the patient has received blood transfusions.[11]

Matching for class I and class II antigens

In 1984 D'Apice *et al.* reported that survival rates for grafts fully matched for class I (HLA-A and HLA-B) or for class II (HLA-DR) antigens were similar, but matching for HLA-B and HLA-DR antigens together enhanced survival further.[20] These data have been confirmed by other investigators, although all studies of the significance of HLA matching have been impeded by the small proportion of patients who received fully matched grafts.

The introduction of cyclosporin into clinical practice during the 1980s further confused the debate. With the overall improvement in graft survival rates noted recently it has become increasingly difficult to determine the effect of HLA matching in single-centre studies. Multicentre co-operative studies, with all their disadvantages for accurate date collection are, perhaps unfortunately, necessary to provide the requisite numbers of patients in the various categories to allow statistically significant analyses. Reports from such large studies continue to confirm the importance of matching donors and recipients for HLA class I and class II antigens—with the greatest benefit being observed for matching at both HLA-DR and HLA-B loci.[13]

Current policy

With the recent improvement in graft survival rates it is generally impossible for a single centre to confirm the beneficial effect of HLA matching. Because of this, there has been a conflict between those who wish to harvest and transplant donor organs locally (thus decreasing cold ischaemic times and potentially improving the likelihood of both immediate graft function and survival), and those wishing to transplant better matched kidneys (thus accepting the need for an organ sharing service with the attendant delays and bureaucracy).

The benefits of HLA matching, and hence organ sharing, have been emphasized by the UK Transplant Service, which has calculated that on average 24 per cent of patients receive a graft matched at the DR loci with only one A or B mismatch, whereas 60 per cent of the current pool of over 3000 potential transplant recipients could receive a well-matched graft through improved organ sharing.[21] It is now apparent that transfused patients receiving a graft matched at the B and DR loci experience significantly better graft survival rates than all other groups.[22]

The UK Transplant Service[10] has defined a 'beneficial' match as one in which there are either no mismatches at the HLA-A, HLA-B and HLA-DR loci, or only one HLA-A or HLA-B mismatch. Defined in that way, graft survival is indeed better for recipient of a 'beneficial' match (Fig. 37.3). The beneficial matches have been subdivided into two groups, one with DR homozygous donors, and the other with DR heterozygous donors. The advantage of beneficial matching is stronger in the DR heterozygous group because of incom-

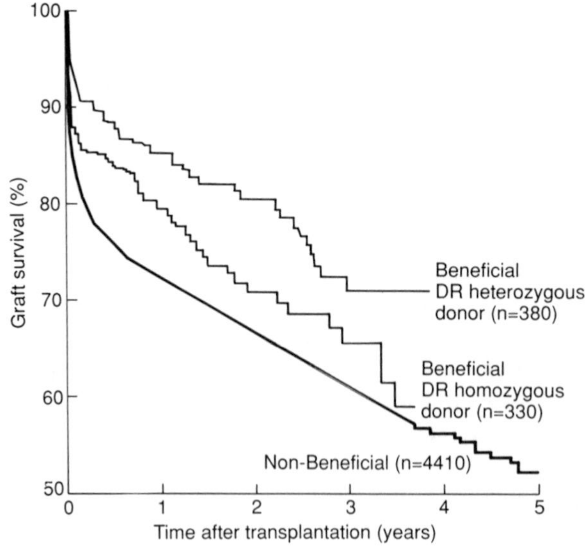

Fig. 37.3 Graft survival after first cadaveric transplantation according to match.[10]

plete or inaccurate donor DR typing; many donors typed as 'DR homozygous' are in reality DR heterozygous, with one DR antigen undetected. Thus a considerable number of grafts apparently in the 'beneficially matched DR-homozygous' groups are really not beneficially matched.

It is apparent that the benefit of 'beneficial' HLA matching remains statistically detectable and clinically relevant. It is disappointing, however, in view of the postulated importance of HLA matching to long-term graft survival, that the slopes of the lines do not diverge (Fig. 37.3). One year after transplantation it would appear that the rate of graft loss is not influenced by beneficial matching, except perhaps in the special category of grafts between HLA identical siblings[16] (Fig. 37.2). Nevertheless, despite the logistical difficulties involved, most but by no means all of the transplant centres in the UK have agreed to expand organ sharing arrangements by contributing at least one of a harvested pair of kidneys to the national pool in an attempt to improve the chances of a successful transplant for the specific recipient and the overall renal transplant success rate.

HLA matching in transplantation of other organs

Although there are numerous studies of the effect of HLA matching in renal transplantation, there have been relatively few in heart, liver or pancreas grafting. Recent studies from the USA have suggested a role for HLA matching in cardiac transplantation, but these have small patient populations and need to be confirmed by larger, multi-centre trials.[23,24] In hepatic transplantation, patients who received grafts poorly matched at the class I HLA loci were more likely to suffer 'vanishing bile duct' syndrome, a form of chronic liver graft rejection, than the patients who received well-matched grafts.[25] The International Pancreas Registry has data from 524 of 855 cadaveric grafts performed since 1982, which suggest that results could be improved by minimizing HLA mismatches between donor and recipient.[26]

Immune response and HLA

Regions of the MHC govern the strength of the immune response; in humans this is believed to lie within the DR region. It has been suggested that HLA-DRw6 is a marker of high immune responsiveness, because patients expressing that antigen had poorer graft survival than those who were DRw6 negative.[27] This finding has not been universally confirmed, and at present it remains unclear whether or not HLA-DRw6 is a true marker of high immune responsiveness.

A more recent report has suggested that HLA-DR2 is also a marker of immune responsiveness to class I antigens. In an analysis of 423 dialysis patients and 461 transplant recipients, those who were DR2 positive were more likely to develop cytotoxic antibodies to class I antigens after blood transfusion, and were also found to have poorer graft survival rates.[28] Conversely, it has been proposed that HLA-DR1 is a marker for weak immune responders, as patients with that antigen have better graft survival rates.[29] These observations await confirmation. It is, however, important to remember that such studies become increasingly difficult with the recent improvement in overall graft survival rates and the expanding diversity of HLA alleles distinguished by molecular biological techniques.

What happened to monoclonal antibodies?

Despite the recent major advances in biotechnology, HLA tissue typing still depends upon testing and exchanging serum samples between the various laboratories. The identification of HLA antigens remains dependent upon suitable antisera from parous women. Such sera almost invariably contain several antibodies, requiring computer analysis to determine the specificities involved. When the technology became available it was widely predicted that monoclonal antibodies to the relevant HLA antigens would resolve many of the outstanding problems and establish tissue typing on a more objective basis. Unfortunately these predictions have not been realized.

When either whole human cells expressing the specific HLA molecule on their surface or, less commonly, purified HLA proteins were used for fusions, both monomorphic and polymorphic antibodies were produced.[30] Unfortunately, certain specificities, particularly HLA-A2, DR monomorphic and some DQ specificities, were preferentially produced. It is now apparent that monoclonal antibodies are directed to dominant epitopes resulting, for example in at least 9 sub-types of HLA-A2 and 6 sub-types of HLA-B27.[31] The scientific knowledge engendered by such studies has been considerable, but has not as yet resulted in monoclonal antibodies which can be used as reliable reagents to explore HLA polymorphisms and improve the objectivity of clinical tissue typing.

Conclusions

Current data indicate that HLA matching is essential for living related transplantation. For cadaver donor transplantation the policy of beneficial matches proposed by the UK Transplant Service should be accepted. The proportion of transplant centres, however, attaching strong clinical importance to HLA matching has increased between 1981 and 1988 only from 33–37 per cent.[12] Perhaps the potential value of the HLA antigen system has yet to be fully realized.

References

1. Landsteiner, K. (1901). Uber agglutination—serscheginungen normalen menschlichen Blutes. *Weiner Klinische Wochenschrift*, **14**, 1132–4.

2. Landsteiner, K. (1931). Individual differences in human blood. *Science*, **73**, 403–9.
3. Nordhagen, R. and Orjsaeter, H. (1974). Association between HL-A and red cell antigens: an autoanalyser study. *Vox Sanguis*, **27**, 124–33.
4. Dausset, J. (1954). Leucoagglutination. IV: Leucoagglutinins and blood transfusions. *Vox Sanguis*, **4**, 190–8.
5. Dausset, J. (1958). Iso-leuco anticorps. *Acta Haematologica*, **20**, 156–66.
6. Kissmeyer-Neilsen, F., Svejgaard, A., and Thorsby, E. (1971). Human transplantation antigens. The HLA system. *Bibliotheca Haematologica*, **38**, 276–81.
7. Thorsby, E. and Piazza, A. (1975). Joint report from the sixth international histocompatibility workshop conference II typing for HLA-D (LD-1 and MLC) determinants. In *Histocompatibility testing 1975* (ed. F. Kissmeyer-Neilsen), pp. 414–22.
8. Bodmer, W.F. and Bodmer, J. (1978). Evolution and function of the HLA system. *British Medical Bulletin*, **34**, 309–16.
9. Nomenclature Committee. (1988). Nomenclature for factors of the HLA system 1987. *Immunogenetics*, **28**, 391–8.
10. UK Transplant Service. (1989). *Annual report, 1989*. Bristol, UK.
11. Opelz, G. (1989). The role of HLA matching and blood transfusion in the cyclosporine era. *Transplantation Proceedings*, **21**, 609–12.
12. Tufveson, G., Geerlings, W., Brunner, E.P., Brynger, H., Dykes, S.R., Ehrich, J.H.H., Fassbinder, W., Rizzoni, G., Selwood, N., and Wing, A.J. (1989). Combined report on regular dialysis and transplantation in Europe, XIX, 1988. *Nephrology Dialysis and Transplantation*, **4**, (4), 5–29
13. Opelz, G. (1987). Improved kidney graft survival in non transfused recipients. *Transplantation Proceedings*, **19**, 149–52.
14. Ting, A. (1988). In *Kidney transplantation: principles and practice* (ed. P.J. Morris), pp. 123–214.
15. Sutherland, D.E.R., Morrow, C.E., Kaufman, D. *et al* (1983). The effect of matching for HLA-DR in recipients of renal allografts sharing one HLA ABC haplotype with related donors. *Transplantation*, **36**, 643–7.
16. Fassbinder, W., Brunner, E.P., Brynger, H., Ehrich, J.H., Geerlings, W., Raine, A.E.G., Rizzoni, G., Selwood, N.H., Tufveson, G., Wing, A.J. (1991). Combined report on regular dialysis and transplantation in Europe, XX, 1989. *Nephrology, Dialysis and Transplantation*, **6**, Suppl 1, 5–35.
17. Merrill, J.P., Murray, J.R., Harrison, J.H., and Guild, W.R. (1956). Successful homotransplantation of human kidney between identical twins. *Journal of the American Medical Association*, **160**, 277.
18. Dewar, P.J., Wilkinson, R., Elliott, R.W., Ward, M.K., Kerr, D.N.S., Kenward, D.H., Proud, G., and Taylor, R.M.R. (1982). Superiority of B locus matching over other HLA matching in renal graft survival. *British Medical Journal*, **284**, 779–82.
19. Solheim, B.G., Flatmark, A., Enger, E. *et al.* (1977). Influence of HLA-A, -B, -C and -D matching on the outcome of clinical kidney transplantation. *Transplantation Proceedings*, **9**, 475–8.
20. D'Apice, A.J.F., Sheil, A.G.R., Tait, B.D., and Bashir, H. (1984). A prospective randomised trial of matching for HLA-A and -B versus HLA-DR in renal transplantation. *Transplantation*, **38**, 37–41.
21. Gilks, W.R., Bradley, B.A., Gore, S.M., and Klouda, P.T. (1987). Substantial benefits of tissue matching in renal transplantation. *Transplantation*, **43**, 669–74.
22. Ciccarelli, J., Terasaki, P.I., and Mickey, M.R. (1987). The effect of zero HLA class I and II mismatching in cyclosporine-treated transplant recipients. *Transplantation*, **43**, 636–40.

23. DiSesa, V.J., Kuo, P.C., Horvath, K.A., Mudge, G.H., Collins, J.J. Jr, and Cohn, L.H. (1990). HLA histocompatibility affects cardiac transplant rejection and may provide one basis for organ allocation. *Annals of Thoracic Surgery*, **49**, 220–3.
24. Kerman, R.H., Van Buren, C.T., Lewis, R.M., Frazier, O.H., Cooley, D., and Kahan, B.D. (1988). The impact of HLA-A, B and DR, blood transfusions and immune responder status on cardiac allograft recipients treated with cyclosporine. *Transplantation*, **45**, 333–7.
25. Donaldson, P.T., Alexander, G.J., O'Grady, J. *et al.* (1987). Evidence for an immune response to HLA class I antigens in the vanishing-bile duct syndrome after liver transplantation. *Lancet*, **i**, 945–51.
26. Squifflet, J.P., Moudry, K., Sutherland, D.E.R. (1988). Is HLA matching relevant in pancreas transplantation? A registry analysis. *Transplantation International*, **1**, 26–9.
27. Hendriks, G.F., Schreuder, G.M., d'Amaro, J., and van Rood, J.J. (1986). The regulatory role of HLA-DRw6 in renal transplantation. *Tissue Antigens*, **27**, 121–30.
28. Kreisler, J.M., Ramentaria, M.C., de Pablo, R., and Moreno, M.F. (1988). HLA-DR2 a marker for class I antigen sensitisation. *Transplantation*, **45**, 1071–4.
29. Cook, D.J., Cecka, J.M., and Terasaki, P.I. (1987). HLA-DR1 recipients have the highest kidney transplant survival. *Transplantation Proceedings*, **9**, 675–7.
30. Marsh, S.G.E, and Bodmer, J.G. (1989). HLA-DR and -DQ epitopes and monoclonal antibody production. *Immunology Today*, **10**, 305–12.
31. Lopex de Castro, J.A. (1989) HLA-B27 and HLA-A2 subtyes: structure, evolution and function. *Immunology Today*, **10**, 239–46.

38 Recurrent primary disease following renal transplantation

J.S. Cameron

The principal, indeed the overwhelming, cause of renal graft loss is either an acute immunological attack on the allograft during the first few months, or the slower process usually called chronic rejection, although its natrue is in fact far from clear. However, there is a small proportion of cases in which graft failure is the direct result of recrudescence of the original disease in the allograft. This is likely to be more common in children than adults, because of the higher proportion of inherited metabolic diseases as a cause of renal failure in children, and the greater proportion of types of nephritis known to recur. Even so, in a recent report from the American paediatric transplant registry, only 7 of 152 (4.6 per cent) grafts which were lost failed because of recurrent disease (Alexander *et al.* 1990). The topic has, however, an importance above its frequency as a cause of graft loss, because of what these observations on experiments of nature indicate about the nature of the diseases.

The topic of recurrent disease has been reviewed in the past (Cameron 1982, 1983), and this chapter will concentrate on more recent data. General reviews of recurrent glomerulonephritis or primary disease have also been published in the past few years by Mathew (1988a,b), Habib *et al.* (1987), and Leumann and Briner (1984), whilst further series of cases from individual units have been published by Honkanen *et al.* (1984), O'Meara *et al.* (1989), and Vangelista *et al.* (1990).

For a disease to recur in the allograft following renal failure implies that a milieu persists in that individual which leads to renal involvement. Table 38.1 lists most of the circumstances in which recurrence might be expected. Oxalosis is the only common metabolic disorder in which recurrence of the disease is the result of accumulation of an abnormal metabolite, although theoretically this would also be seen in renal failure resulting from tyrosinaemia, galactosaemia, Fabry's disease (Popli *et al.* 1987), or any other inherited metabolic defect which is not cured by transplantation of a normal kidney bearing the deficient enzyme. In cystinosis, of course, the kidney is not affected with recurrent disease, since it has normal transport of cystine out of lysosomes; the problem here is one of progressive failure in other organs. However it has been known for 30 years that cystine crystals may be seen occasionally throughout the graft, in host cells invading the organ (Spear *et al.* 1989).

Other diseases in which the milieu is not changed by renal transplantation

Table 38.1 Reappearance of disease in transplanted kidney depends upon persistence of a milieu damaging to the kidney

Cause	Disease
Metabolic product toxic to kidney	Oxalosis etc.
Organ-specific antibody	Anti-GBM disease ? Membranous nephropathy
Amyloidogenic proteins	Amyloid
Immune aggregates	Other forms of glomerulonephritis
Sickling red cells	Sickle cell disease
Unknown	FSGS
	Type II MCGN
	Diabetes
	Haemolytic-uraemic syndrome

per se are homozygous sickle cell disease and amyloidosis. In glomerulonephritis, a putative mechanism of damage in the form of organ-specific antibody is available in anti-GBM disease, and possibly some forms of recurrent membranous nephropathy, although the autoantigen in humans has not yet been identified. In focal segmental glomerulosclerosis (FSGS), in dense 'deposit' disease, in the haemeolytic uraemic syndrome, and in diabetes mellitus, the route by which the transplanted kidney is affected is not yet clear.

Oxalosis

Type I primary hyperoxaluria results from a deficiency of the hepatic peroxisomal enzyme alanine-glyoxalate aminotransferase (Watts *et al.* 1987). Thus, overproduction of oxalate persists in patients given renal transplants, and although the plasma oxalate levels are lower than the extreme levels found in patients on dialysis, the concentration is still elevated to the point at which accumulation of oxalate in the vascular tree and elsewhere in the tissues persists. Combined liver and kidney transplantation is apparently curative, judged from the 4-year survival to date of the first successful combined operation (Watts *et al.* 1987; Watts and Mansell 1990), and by 1990 some 20 patients had been so treated world-wide. The value of renal transplantation alone remains controversial. Clearly, results are somewhat better than at first supposed (Broyer *et al.* 1990), but whether it is useful to do renal transplantation alone, with the option of later liver transplantation, is unclear (McDonald *et al.* 1990; Watts and Mansell 1990). Even with good transplant function, plasma oxalate levels may be ten times normal, and the long-term survival of these patients must remain in doubt.

Broyer *et al.* (1990) presented a large pan-European series of 98 patients transplanted with first renal allografts from 1965 to 1986: 79 received a cadaver donor graft and 15 living donor grafts (in 4 the source was unknown), at age 6–55 years (mode 24 years). After 3 years, only 23 per cent of living donor graft and 17 per cent of cadaver donor grafts survived, and 39 recipients (40 per cent) had died. Neither the time on dialysis nor the age at end stage renal failure influenced results.

If renal transplantation alone is performed, then attempts should be made to minimize subsequent damage by oxalate to the allograft. These include administration of large doses of pyridoxine (which is a cofactor for the defective enzyme) and phosphate (Scheinman *et al.* 1984), together with magnesium chloride and the maintenance of a high urinary output. With these measures, graft survival of up to 10 years can be achieved, but the really long-term outcome for these patients remains in doubt.

Glomerulonephritis

Although recurrence of virtually all types of nephritis has been reported on occasion, the risk of recurrence varies greatly according to the type of glomerular disease, as summarized in Table 38.2. This table also emphasizes that, in several forms of glomerulonephritis, there may be histological recurrence with few or no clinical consequences.

In some types of nephritis there is today little more information than was available 10 years ago. This is true of anti-GBM disease. Recurrence of disease is usual in patients with elevated anti-GBM antibody titres, although antibody can deposit without damage. Conversely, in those with normal anti-GBM antibody titres, either the result of spontaneous disappearance (Flores *et al.* 1986) or accelerated by treatment, recrudescence of antibody and recurrent disease have not been recorded. Thus, management centres on continuing dialysis until antibody levels are normal. How long one should then wait is not clear—most clinicians wait 6–12 months.

Table 38.2 Recurrent glomerulonephritis after renal transplantation

Glomerulonephritis	Histological	Clinical
Anti GBM disease (with antibody)	High	High
FSGS	High	High
MCGN type I	Moderate	Moderate
MCGN type II	High	Low
IgA nephropathy	High	Low
HSP nephritis	High	Low
Membranous (NB de novo)	Low	Low
Crescentic	Low	Low
Vasculitis	Low	Low
Lupus glomerulonephritis	Almost never	

The main concern today is over FSGS, in which waiting appears not to reduce the possibility of recurrence, and the transmitting agent is unknown. Recent reviews (Senggutuvan *et al.* 1990; Cameron 1990) provide a good database on 250 patients; these papers may be consulted for a review of the rather extensive literature on the subject. Overall recurrence is about 25 per cent, and the graft loss rate is about half this. However, the risk varies greatly from patient to patient, the main risk factors for recurrence being age, especially if less than 15 years, a rapid evolution into renal failure (less than 3 years), and mesangial expansion in the renal biopsy. Thus, in those aged less than 5 years the recurrence rate is about 50 per cent, and in those aged less than 15 at onset who go into renal failure within 3 years and also have mesangial expansion, the recurrence rate is 80–100 per cent. It has been suggested that haploidentity or living donor transplantation makes recurrence more likely, but our data do not support this suggestion.

The severity of recurrence varies greatly from patient to patient, from immediate massive proteinuria to milder forms of later onset. In the precocious onset with massive proteinuria, immediate transplant nephrectomy is necessary or the recipient's life may be put at risk. Even in these patients, it has proved impossible in many attempts to induce proteinuria by injecting serum into experimental animals, with the possible exception of the report by Zimmerman and Mann (1984).

Plasma exchange, although theoretically attractive, does not appear to give any benefit, although the report of Laufer *et al.* (1988) is at variance with this conclusion. Likewise, treatment with NSAIDs has been recommended (Torres *et al.* 1984) but many have not noted any benefit from these drugs. The recurrence rate appears to be identical in recipients treated with cyclosporin compared with those receiving azathioprine, despite some limited success in the primary disease. Ingulli *et al.* (1989) suggest that high-dose cyclosporin may reduce proteinuria and may induce remission in recurrent disease.

Second grafts in those who have had recurrence in their first graft is usually, but not invariably (Hosenpud *et al.* 1985), accompanied by a further recurrence (Cameron 1990). It is our feeling, shared by many, that where possible, living donor grafts should be avoided in those patients with FSGS who are judged to be at high risk, unless a first cadaver graft is lost by rejection without evidence of recurrence.

The situation is less clear with regard to mesangiocapillary glomerulonephritis (MCGN), type I or II (dense 'deposit' disease), because fewer recurrences have been reported and there is no extensive database to judge the frequency of recurrence accurately. Since 1982 (Cameron 1982) only anecdotes have been reported (including descriptions of recurrence in two successive grafts (Glicklich *et al.* 1987; Mathew *et al.* 1988*a*)). The previously reported data suggested a recurrence rate in type I MCGN of about 20–25 per cent, with about 10–12.5 per cent graft loss. Again, it seems prudent to avoid living donor grafts in these patients where possible. There are no published data on recurrence in recipients treated with cyclosporin, but we have seen this in one patient, who lost two successive grafts.

In type II MCGN (dense 'deposit' disease) it seems clear that histological recurrence of electron-dense material in the basement membranes of the transplant occurs in about 85 per cent of cases, whilst clinical recurrence is rare (Cameron 1982), except in patients showing extensive crescent formation, in whom there is a suggestion of more frequent clinical recurrence (Eddy *et al.* 1984; De Castro *et al.* 1988). There are no data on recipients treated with cyclosporin. In neither type of MCGN does the concentration of C3 or C4 in the serum correlate with recurrence, suggesting that this is an epiphenomenon. Mathew (1988*a*) reports successful regrafting in one patient, albeit with histological recurrence of 'deposits'.

In IgA nephropathy and Henoch–Schönlein purpura (HSP) nephritis recurrence of the mesangial IgA is seen in 25–45 per cent of recipients, but clinical recurrence is rare (Bachman *et al.* 1986), probably in only 1 per cent (2 of 200) recipients (Mathew 1988*a*). Again, the rather rare subjects with extensive crescent formation may be an exception (Brensilver *et al.* 1988; Diaz-Tejeiro *et al.* 1990) with a higher risk of recurrence. In HSP nephritis also, transplantation into those still having recurrent attacks of purpura also seems to be relatively hazardous (Nast *et al.* 1987; Hasegawa *et al.* 1989). Recurrence seems to occur as readily under cyclosporin treatment as with azathioprine.

Recurrence of membranous nephropathy has been reported in only about 25 cases (see Cameron 1982; Berger *et al.* 1983; First *et al.* 1984; Obermiller *et al.* 1984), but it is not a common cause of renal failure in most series. The proportion of recurrence is difficult to determine, especially as *de novo* membranous nephritis is much more common (see below), but we have seen 3 cases in 14 patents transplanted (22 per cent). Again, recurrence is not apparently inhibited by cyclosporin (Montanigno *et al.* 1989). The appearance of the lesion is earlier in those with recurrent (mean 10 months, range 1 week to 2 years) than in those with *de novo* disease (mean 22 months, range 4 months to 6 years). About half the grafts affected by recurrent membranous nephropathy have failed, but it is difficult to assess what role the glomerulonephritis may have played in these failures. Renal venous thrombosis may be seen in allografted kidneys in both recurrent and *de novo* membranous nephropathy. Although it has been suggested that more aggressive disease is more likely to recur (Obermiller *et al.* 1984) we have not seen recurrence in one patient who went from normal renal function (glomerular filtration rate (110 ml/min) to end stage renal disease in only 15 months and received a transplant only a few months later.

No further data on the recurrence of crescentic glomerulonephritis, other than that depending upon anti-GBM antibody, have appeared in recent years (Cameron 1982). The risks, surprisingly in view of the rapid evolution of this type of nephritis, appear to be small in this group. Recurrence is not avoided by cyclosporin (Turney *et al.* 1985). A few papers describing recurrence of crescentic vasculitis in allografts have appeared (Curtis *et al.* 1983).

Two further cases of probable recurrence of lupus nephritis have been published recently (Kumano *et al.* 1987; Nyberg *et al.* 1987), but the striking feature of lupus nephritis is the almost complete absence of recurrence. This surely indicates that the kidney itself has some determinant which allows the appearance of nephritis, and which is missing from the allografted kidney, since some patients (including those in the author's unit) have been transplanted at times of activity, with circulating immune complexes, antiDNA antibody, and requirement for steroid treatment.

Amyloidosis

Recently, a large single-centre experience (Pasternack *et al.* 1986) has been published, in which 4 of 45 patients (8 per cent) showed recurrence, 3 of whom had rheumatic disorders (Light and Hall-Craggs 1979) and one 'primary' amyloid. Recurrence has been recorded also in patients with familial Mediterranean fever (Benson *et al.* 1977; Metaxas 1981), but does not seem to be a major deterrent to transplantation, although no systematic survey of the frequency has been carried out; unfortunately, the recent large survey conducted by the EDTA-European Renal Association did not contain an estimate of the frequency of amyloid in the allografts.

Diabetes mellitus

Patients with diabetes face major problems after transplantation compared with non-diabetic controls. However, recurrent hyaline vasculopathy—although almost invariable after 5 years or so (Mauer *et al.* 1976; Najarian *et al.* 1979)—seems not to be a major problem, at least during the first decade following transplantation. Bohman *et al.* (1987) showed thickened capillary basement membranes in all grafts *in situ* for longer than 4 years. Presumably, the disease in the allograft has the same slow evolution with a long latent period before the clinical appearance of clinical disease. As in the native kidney, hyalinosis may affect both afferent and efferent arterioles. Surprisingly, only two cases of frank nodular glomerulosclerosis have been reported in allografted kidneys (Mauer *et al.* 1976; Maryniak *et al.* 1985).

Haemolytic-uraemic syndromes

It is somewhat difficult to assess the true significance of recurrent haemolytic-uraemic syndromes in patients bearing renal allografts, since both the histological and haematological features of the syndrome may be seen in severe allograft rejection, and cyclosporin itself may give rise to a similar picture in individuals with normal kidneys, such as after bone marrow transplantation. At one extreme, 5 of 11 children whose renal failure arose from haemolytic-uraemic syndrome lost their grafts with a similar syndrome (Hebert *et al.* 1986);

at the other extreme, none of 12 adults and 4 children experienced recurrence (Pirson *et al.* 1986). Eijenraam *et al.* (1990), in probably the largest single-centre experience reported, found probable recurrence of haemolytic-uraemic syndrome in only 2 first grafts out of 24 placed into 20 children. In adults, recurrence seems to be rare in patients who remained on dialysis for several months at least (Arias-Rodriguez *et al.* 1977). The Australian–New Zealand data showed only 3 graft losses in 27 cases (Mathew 1988*a*).

Possible factors favouring recurrence are early placement of grafts in acute disease, although this is not at all certain, and use of living donors rather than cadaver donors (as in the series of Hebert *et al.* 1986). Whether cyclosporin should be used in these patients has been debated (Leithner *et al.* 1982; Hamilton *et al.* 1982), and one patient who developed haemolytic-uraemic syndrome after transplantation on cyclosporin has been transferred to FK 506 (McCauley *et al.* 1989). Certainly, cyclosporin has been used without problems on a number of occasions (including in the author's unit) without recurrent haemolytic-uraemic syndrome, but whether the incidence of recurrence is higher in those so treated has not been established.

Miscellaneous conditions

A number of other conditions have been recorded to recur in renal allografts from time to time. These include sickle cell disease (Miner *et al.* 1987; Barber *et al.* 1987), both interstitial disease and glomerulonephritis in sarcoidosis (Beaufils *et al.* 1983; Shen *et al.* 1986), and a group of plasma cell dyscrasias, including Waldenstrom's macroglobulinaemia (Bradley *et al.* 1987), light chain nephropathy (Gerlag *et al.* 1986; Alpers *et al.* 1989), together with 'fibrillary' nephritis (Korbet *et al.* 1990), and mixed essential cryoglobulinaemia (Hiesse *et al.* 1989). The almost complete absence of myeloma itself from this list (De Lima *et al.* 1981) probably results from the fact that very few patients with myeloma receive allografts.

De novo glomerulonephritis

Although not strictly recurrent diseases, two *de novo* conditions are worth considering in parallel with recurrent disease: membranous nephropathy, and the appearance of anti-GBM nephritis in patients with Alport's syndrome.

De novo *membranous nephropathy*

In about 1–2 per cent of most series of renal transplants, after about 1–2 years, profuse proteinuria or a frank nephrotic syndrome heralds the appearance of a membranous nephropathy in patients whose original disease was different; in a recent analysis of 95 cases in the literature (Truong *et al.* 1989) there was no predilection for any particular type of chronic renal failure. About half of the cases are nephrotic, and both living and cadaver donors appear to be equally affected. Evolution into graft failure is slow, with half the grafts surviving a further 3 years.

In the single series in which immunofluorescence was done on routine biopsies of all allografts, that of Antignac *et al.* (1987), a much higher prevalence of over 9 per cent was noted, symptomless in about one quarter and with only minor proteinuria in another quarter of patients. Evolution was correspondingly favourable. This series was in children, and it is not known if a similar investigation in adults would reveal a similarly high incidence.

The pathogenesis of the condition is obscure: suggestions that it might arise from alloantibody immune complex formation are made less likely by its appearance in a isograft between conjoined twins separated after birth (Bansal *et al.* 1986). As yet, no autoantibodies have been identified in patients with *de novo* membranous nephropathy.

Anti-GBM antibody nephritis in Alport's syndrome

Alport's syndrome arises from mutations in a gene located at Xq 22 coding for an alpha 5 (IV) collagen of the glomerular and other basement membranes (Barker *et al.* 1990). Thus, male Alport hemizygotes lack a portion of the normal collagen structure, whilst allografts into such patients possess this epitope. In some patients this is sufficient to evoke an immune response with production of antibody directed against it, which is capable of fixing and inducing a severe crescentic glomerulonephritis: the 12 cases reported since its first description in 1982 are summarized in the recent papers of Van der Heuvel *et al.* (1989), Kashtan *et al.* (1990), and Rassoul *et al.* (1990).

All of the patients developed renal failure early in life and all were deaf. There was a latent period of some months before the nephritis appeared, but all grafts except one were lost to severe nephritis, despite plasma exchange in three patients. One unexpected feature is that one of the patients reported by Van der Heuvel *et al.* (1989) was female. Heterozygotes would normally be equal mosaics of normal and abnormal genes, resulting from random inactivation of one of each the pairs of X chromosomes, and thus should have tolerance of the antigen in the allograft. It may be that in this individual the great majority of cells expressed the abnormal chromosome.

References

Alexander, S.R. *et al.* (1990). The 1989 report of the North American pediatric renal transplant cooperative study. *Ped. Nephrol.* **4**, 542–53.

Alpers, C.E., Marchioro, T.L., and Johnson, R.J. (1989). Monoclonal immunoglobulin deposition disease in a renal allograft: probable recurrent disease in a patient without myeloma. *Am. J. Kidney Dis.* **13**, 418–23.

Antignac, C., Hinglais, N., Gubler, M.-C., Gagnadoux, M.-F., Broyer, M., and Habib, R. (1988). De novo membranous glomerulonephritis in renal allografts in children. *Clin. Nephrol.* **30**, 1–7.

Arias Rodriguez, M. *et al.* (1977). Renal transplantation and immunological abnormalities in thrombotic microangiopathy of adults. *Transplantation*, **23**, 360–5.

Axelsen, R.A., Seymour, A.E., Mathew, T.H., Fisher, G., Canny, A., Pascoe, V. (1984). Recurrent focal glomerulosclerosis in renal transplants. *Clin. Nephrol.*, **21**, 110–14.

Bachman, U. *et al.* (1986). The clinical course of IgA-nephropathy and Henoch-Schönlein purpura following renal transplantation. *Transplantation*, **42**, 511–15.

Bansal, V.K., Koseny, G.A., Fresco, R., Vertuno, L.L., and Hano, J.E. (1986). De novo membranous nephropathy following transplantation between conjoint twins. *Transplantation*, **41**, 404–5.

Barber, W.H. *et al.* (1987). Renal transplantation in sickle anemia and sickle cell disease. *Clin. Transpl.* **1**, 169–75.

Barker, D.F. *et al.* (1990). Identification of mutations in the COL4A5 collagen gene in Alport syndrome. *Science*, **248**, 1224–7.

Beaufils, H., Gompel, A., Gubler, M.-C., Lucsko, M., and Guedon, J. (1983). Pre- and posttransplant in a case of sarcoidosis. *Nephron*, **35**, 124–9.

Benson, M.D., Skinner, M., Cohen, A.S. (1977). Amyloid deposition in a renal transplant in familial Mediterranean fever. *Ann. Intern. Med.* **87**, 31–4.

Berger, B.E., Vincenti, F., Biava, B., Amend, W.J., Feduska, N., and Salvatierra, O. (1983). De novo and recurrent membranous glomerulopathy following kidney transplantation. *Transplantation*, **35**, 315–19.

Bohman, S.-O. *et al.* (1987). Recurrent diabetic nephropathy in renal allografts placed in diabetic patients and protective effect of simultaneous pancreatic transplantation. *Transplant. Proc.* **19**, 2290–3.

Bradley, J.R., Thiru, S., Bajallan, N., and Evans, D.B. (1988). Renal transplantation in Waldenstrom's macroglobulinaemia. *Nephrol. Dial. Transpl.* **2**, 214–16.

Brensilver, J.M., Mallat, S., Scholes, J., and McCabe, R. (1988). Recurrent IgA nephropathy in living-related donor transplantation: recurrence or transmission of familial disease? *Am. J. Kidney. Dis.* **12**, 147–51.

Broyer, M. *et al.* (1990). Kidney transplantation in primary oxalosis: data from the EDTA registry. *Nephrol. Dial. Transpl.* **5**, 332–6.

Cameron, J.S. (1982). Glomerulonephritis in renal transplants. *Transplantation*, **34**, 237–45.

Cameron, J.S. (1983). Effect of the recipient's disease on the results of kidney transplantation (other than diabetes mellitus). *Kidney Int.* **23**, (14), S-24–33.

Cameron, J.S. (1990). Ask the expert. *Ped. Nephrol.* **3**, 300.

Curtis, J.J., Diethelm, A.G., Herrera, G.A., Crowell, W.T., and Whelchel, J.D. (1983). Recurrence of Wegener's granulomatosis in a cadaver renal allograft. *Transplantation*, **36**, 452–4.

De Castro, S.S. *et al.* (1988). Recidiva de glomerulonefritis mebranoproliferativa tipo II en riñon trasplantado con evolución rapdiamente progresiva. *Nefrologia*, **8**, 70–3.

De Lima, J.J.G., Kourilsky, O., Meyrier, A., Morel-Maroger, L., and Sraer, J.-D. (1981). Kidney transplant in multiple myeloma. Early recurrence in the graft with sustained normal renal function. *Transplantation*, **31**, 223–4.

Diaz-Tejeiro, R. *et al.* (1990). Loss of graft due to recurrent IgA nephropathy with rapidly progressive course: an unusual clinical evolution. *Nephron*, **54**, 431–3.

Eddy, A., Sibley, R., Mauer, S.M., and Kim, Y. (1984). Renal allograft failure due to recurrent dense intramembranous deposit disease. *Clin. Nephrol.* **21**, 305–13.

Eijenraam, F.J., Donckerwolcke, R.A., Monnens, L.A.H., Proesmans, W., Wolff,

E.D., and van Damme, B. (1990). Renal transplantation in 20 children with hemolytic uremic syndrome. *Clin. Nephrol.* **33**, 87–93.

First, M.R., Mendoza, N., Maryniak, R.K., and Weiss, M.A. (1984). Membranous glomerulopathy following kidney transplantation: association with renal vein thrombosis in two of nine cases. *Transplantation*, **38**, 603–7.

Flores, J.C. *et al.* (1986). Clinical and immunological evolution of oligoanuric anti-GBM nephritis treated by haemodialysis. *Lancet*, **i**, 5–8.

Gerlag, P.G.G., Koene, R.A.P., Berden, J.M.H. (1986). Renal transplantation in light chain nephropathy: case report and review of the literature. *Clin. Nephrol.* **25**, 101–4.

Glicklich, D. *et al.* (1987). Recurrent membrano-proliferative glomerulonephritis type I in successive renal transplants. *Am. J. Nephrol.* **7**, 143–99.

Habib, R., Gagnadoux, M.F., and Broyer, M. (1987). Recurrent glomerulonephritis in transplanted children. *Contr. Nephrol.* **55**, 123–35.

Hamilton, D., Calne, R.Y., and Evans, D.B. (1982). Haemolytic-uraemic syndrome and cyclosporin A. *Lancet*, **ii**, 151–2.

Hasegawa, A. *et al.* (1989). Fate of renal grafts with recurrent Henoch-Schönlein pupurpura nephritis. *Transpl. Proc.* **21**, 2130–3.

Hebert, D., Sibley, R.K., and Mauer, S.M. (1986). Recurrence of hemolytic uremic syndrome in renal transplant recipients. *Kidney Int.* **30**, S51–8.

Hiesse, C. *et al.* (1989). Recurrent essential mixed cryoglobulinemia in renal allografts. Report of two cases and review of the literature. *Am. J. Nephrol.* **9**, 150–4.

Honkanen, E., Tornroth, T., Pettersson, E., and Kuhlback, B. (1984). Glomerulonephritis in renal allografts: results of 18 years of transplantation. *Clin. Nephrol.* **21**, 210–19.

Hosenpud, J., Piering, W.F., Grancis, J.C., and Kauffman, H.M. (1985). Successful second kidney transplantation in a patient with focal glomerulosclerosis. *Am. J. Nephrol.* **5**, 299–304.

Ingulli, E. *et al.* (1990). High-dose cyclosporine therapy in recurrent nephrotic syndrome following renal transplantation. *Transplantation*, **49**, 219–21.

Kashtan, C., Kippel, M.M., Butkowski, R.J., Michael, A.F., and Fish, A.J. (1990). Alport syndrome, basement membranes and collagen. *Ped. Nephrol.* **4**, 523–32.

Korbet, S.H., Rosenberg, B.F., Schwartz, M.M., and Lewis, E.J. (1990). The course of renal transplantation in immunotactoid glomerulopathy. *Am. J. Med.* **89**, 91–5.

Kumano, K., Sakai, T., Mashimo, S., Endo, T., Koshiba, K., Elises, J.S., and Iitaka, K. (1987). A case of recurrent lupus nephritis after renal transplantation. *Clin. Nephrol.* **27**, 94–8.

Laufer, J., Ettenger, R.B., Ho, W.G., Cohen, A.H., Marik, J.L., and Fine, R.N. (1988). Plasma exchange for recurrent nephrotic syndrome following renal transplantation. *Transplantation*, **46**, 540–2.

Leithner, C. *et al.* (1982). Recurrence of haemolytic uraemic syndrome triggered by cyclosporin A after renal transplantation. *Lancet*, **1**, 1470.

Leumann, E.P., Briner, J. (1984). Rezidive der Grundkrankheit im Nierentransplantation *Klin. Wschr.* **62**, 289–98.

Light, P. and Hall-Craggs, M. (1979). Amyloid deposition in a renal allograft in a case of amyloid secondary to rheumatoid arthritis. *Am. J. Med.* **66**, 532–5.

McCauley, J., Bronster, O., Fung, J., Todo, S., and Starzl, T.E. (1989). Treatment of cyclosporin-induced haemolytic-uraemic syndrome with FK 506. *Lancet*, **300**, 1516.

McDonald, J.C., Landeranu, M.D., Rohr, M.S., and Devault, G.A. Jr. (1989). Reversal by liver transplantation of the complications of primary hyperoxaluria as well as the metabolic defect. *N. Engl. J. Med.* **321**, 1100–3.

Maryniak, R.K., Mendoza, N., Clyne, D., Balakrishnan, K., and Weiss, M.A. (1985). Recurrence of diabetic nodular glomerulosclerosis in a renal transplant. *Transplantation*, **39**, 35–8.

Mathew, T.H. (1988*a*). Recurrence of disease following renal transplantation. *Am. J. Kidney Dis.* **12**, 85–96.

Mathew, T.H. (1988*b*). Nephrotic syndrome following transplantation. In *The nephrotic syndrome* (ed. J.S. Cameron, R.J. Glassock), Marcel Dekker, New York.

Mauer, S.M. *et al.* (1976). Development of diabetic vascular lesions in normal kidneys transplanted into patients with diabetes mellitus. *N. Engl. J. Med.* **295**, 916–20.

Metaxas, P. (1981). Familial Mediterranean fever and amyloidosis. *Kidney Int.* **20**, 676–85.

Miner, D.J., Jorkasky, D.K., Perloff, L.J., Grossman, R.A., and Tomaszewski, J.E. (1987). Recurrent sickle cell nephropathy in a transplanted kidney. *Am. J. Kidney Dis.* **10**, 306–13.

Montagnino, G., Colturi, C., Banfi, G., Arnoldi, A., Tarantino, A., and Ponticelli, C. (1989). Membranous nephropathy in cyclosporine-treated renal transplant recipients. *Transplantation*, **47**, 725–7.

Moritz, M.J., Burke, J.F., and Carabasi, R.A. (1987). The incidence of membranoproliferative glomerulonephritis in renal allografts. *Transplantation Proceedings*, **19**, 2206–7.

Najarian, J.S. *et al.* (1979). Ten years' experience with renal transplantation in juvenile onset diabetics. *Ann. Surg.* **190**, 487–500.

Nast, C.C., Ward, H.J., Koyle, M.A., and Cohen, A.H. (1987). Recurrent Henoch-Schonlein purpura following renal transplantation. *Am. J. Kidney Dis.*, **9**, 39–43.

Nyberg, G., Blohme, I., Svalander, C., Persson, H., and Brynger, H. (1987). Rejection and recurrence of SLE nephritis in cyclosporine treated kidney transplant recipients. *Transpl. Proc.*, **19**, 1637–8.

Obermiller, L.E., Hoy, W.E., Eversole, M., and Sterling, W.A. (1985). Recurrent membranous glomerulonephritis in two renal transplants. *Transplantation*, **40**, 100–3.

O'Meara, Y. *et al.* (1989). Recurrent glomerulonephritis in renal transplants: fourteen years' experience. *Nephrol. Dial. Transpl.* **4**, 730–4.

Pasternack, A., Ahonen, J., and Kuhlback, B. (1986). Renal transplantation in 45 patients with amyloidosis. *Transplantation*, **42**, 598–601.

Pirson, Y. *et al.* (1986). Good prognosis of the hemolytic uremic syndrome after renal transplantation. *XXIIIrd Congress of the EDTA-ERA*, Budapest, (Abstract).

Popli, S., Molnar, Z.V., and Leehy, D.J. (1987). Involvement of renal allograft by Fabry's disease. *Am. J. Nephrol.* **7**, 316–18.

Rassoul, Z., Alkhader, A.A., Al-Sulaiman, M., Dhar, J.M., and Coode, P. (1990). Recurrent allograft antiglomerular basement membrane glomerulonephritis in a patient with Alport's syndrome. *Am. J. Nephrol.* **10**, 73–6.

Senggutuvan, P. *et al.* (1990). Recurrence of focal segmental glomerulosclerosis in transplanted kidneys: analysis of incidence and risk factors in 59 allografts. *Ped. Nephrol.* **4**, 21–8.

Scheinman, J.I., Najarian, J.S., and Mauer, S.M. (1984). Successful strategies for renal transplantation in oxalosis. *Kidney International*, **25**, 804–11.

Shen, S.Y., Hall-Craggs, M., Posner, J.N., and Shabazz, B. (1986). Recurrent sarcoid granulomatous nephritis and ractive tuberculin skin test in a renal transplant recipient. *Am. J. Med.* **80**, 699–702.

Spear, G.S., Gubler, M.C., Habib, R., and Broyer, M. (1989). Renal allografts in cystinosis and mesangial demography. *Clinical Nephrology*, **32**, 256–61.

Tornroth, T. (1987). Recurrent and de novo glomerulonephritis in allografted kidneys. Aspects of ultrastructural diagnosis. Ultrastructural pathology of the kidney. *Appl. Pathol.* **5**, 88–94.

Torres, V.E., Velosa, J.A., Holley, K.E., and Frohnert, P.P. (1984). Meclofenamate treatment of recurrent idiopathic nephrotic syndrome with focal segmental glomerulosclerosis after renal transplantation. *Mayo Clin. Proc.* **59**, 146–52.

Truong, L. *et al.* (1989). De novo membranous glomerulopathy in renal allografts: a report of ten cases and review of the literature. *Am. J. Kidney Dis.* **14**, 131–44.

Turney, J.H. *et al.* (1985). Recurrent crescentic glomerulonephritis in renal transplant recipient treated with cyclosporin. *Lancet*, **1**, 1104.

Van der Heuvel, L.P.W.J. *et al.* (1989). The development of anti-glomerular basement membrane nephritis in two children with Alport's syndrome after renal transplantation: characterization of the antibody target. *Ped. Nephrol.* **3**, 406–13.

Vangelista, A., Frasca, G.M., Martella, D., and Bonomini, V. (1990). Glomerulonephritis in renal transplants. *Nephrol. Dial. Transpl.* **i**, 42–6.

Watts, R.W.E. and Mansell, M.A. (1990). Oxalate, livers and kidneys. Combined renal and hepatic transplants transform the outlook in primary hyperoxaluria type I. *Br. med. J.* **301**, 772–3.

Watts, R.W.E. *et al.* (1987). Successful treatment of primary hyperoxaluria type I by combined hepatic and renal transplantation. *Lancet*, **ii**, 474–5.

Zimmerman, S.W. and Mann, S. (1984). Increased urinary protein excretion in the rat produced by serum from a patient with recurrent focal glomerulosclerosis after renal transplantation. *Clin. Nephrol.* **22**, 32–8.

39 Dialysis and transplantation in diabetic nephropathy

D.W.R. Gray

The introduction of dialysis as a treatment for renal failure was initially restricted to patients who were expected to regain renal function within a few weeks. The restriction was mainly due to the short life of vascular access techniques, as well as to limitations on funding, equipment and staff. As vascular access techniques for long-term dialysis were developed, chronic dialysis was introduced, initially with restriction of its use to those patients in whom benefit was most easily demonstrable, namely young patients with non-progressive disease as a cause of renal failure. The extension of chronic dialysis to less favourable groups of patients with renal failure was carried out at varying rates in different countries, depending on the availability of resources, and on complex political pressures.

Introduction of dialysis for end stage diabetic nephropathy

Renal failure due to diabetes mellitus was seen as one of the least favourable indications for long-term dialysis. The reasons are not difficult to understand, since almost all patients with end stage diabetic nephropathy have evidence of retinopathy, and a considerable proportion have neuropathy and vascular disease (Friechman and Esperance 1982). The advance of these complications was not slowed by introduction of dialysis, indeed there was some evidence that haemodialysis accelerated complications such as vitreous haemorrhage, leading to early blindness. Maintainance of vascular access was a particular problem in the presence of extensive vascular disease, and was made worse by the increased susceptibility of these patients to infection of access sites. For these reasons, countries with limited resources where chronic dialysis was restricted dialysed few diabetic patients, in contrast to countries, such as the USA, where less financial restriction and more patient pressure resulted in an increasing number of diabetic patients on chronic dialysis, rising to between 10 and 20 per cent of the total dialysis population in the early 1970s. Most European countries were dialysing considerably fewer diabetic patients at this time, the figure being 1.2 per cent in the UK. The incidence of diabetes varies considerably between different countries, and the exception to the rule in Europe was Finland, where 13 per cent of patients on the long-term dialysis program were diabetic. The generally nihilistic attitude to commencement of renal replacement therapy in diabetic patients with end stage renal failure in the early 1970s was reinforced by reports from the USA of cumulative survival figures of approximately 40 per cent at 2 years after commencement of dialysis for diabetes.

Increasing dialysis for diabetic patients

The decade from the mid-1970s saw a rapid change in attitude to diabetic patients, probably initiated by the introduction of continuous ambulatory peritoneal dialysis (CAPD), which allowed an alternative technique for chronic dialysis that did not rely on vascular access. Increasing affluence, in addition to moral, ethical and political pressures, has resulted in a steady increase in the percentage of new patients referred to dialysis with a diagnosis of diabetes, from an average of 2.5 per cent in Europe in 1976 to over 10 per cent in 1985. Over the same period, the number in the UK has risen from 1.2–11.4 per cent.

Coincident with the rise in diabetic patients taken on for dialysis, has been a steady improvement in overall cumulative survival, although analysis has shown that this improvement is greater for insulin dependent diabetes (IDDM) than for non-insulin dependent diabetes (NIDDM). Particularly poor prognostic signs are age at commencement of diabetes (a variable linked to NIDDM), the presence of vascular disease and current smoking. Deaths are mainly from cardiac and vascular disease.

Renal transplantation

The hesitancy with which dialysis was introduced for diabetic nephropathy was reflected in a similar reluctance to take diabetic patients onto renal transplant programmes. There were good theoretical reasons for supposing that diabetic patients would do less well after transplantation, since the immunosuppression required would be likely to enhance greatly the susceptibility to infection already present, and corticosteroid therapy would be likely to make the control of the diabetes even more difficult. To some extent, these fears have been justified, but by careful selection of recipients and judicious management it is possible to obtain good results after renal transplantation, as was first demonstrated conclusively by the Minneapolis group in the early 1980s (Najarian *et al.* 1989). The Minneapolis results encouraged European transplant centres to take on an increasing number of diabetic patients for renal transplantation. The remarkable results of the Minneapolis group have not been repeated in the European centres, and there is still an excess mortality over non-diabetic patients, again mainly from vascular disease, with age and smoking as strong risk factors. The overall 1 and 2 year graft survival in diabetic patients is above 70 per cent in most centres, not significantly different from that of non-diabetics; however the long-term mortality becomes evident, with patient survival dropping to around 60 per cent at 5 years, in comparison with 90 per cent in non-diabetics.

Recurrent diabetic nephropathy

A particularly interesting observation has been the incidence of recurrent diabetic nephropathy. Biopsy studies have shown that virtually all kidney grafts into diabetic patients develop glomerular changes indicative of diabetic nephropathy, the most consistent of which is glomerular basement membrane thickening (Bilous *et al.* 1989). However, the rapidity with which these changes develop varies considerably, and relatively few patients develop overt signs of diabetic nephropathy with proteinuria, nephrotic syndrome and eventual renal failure before 10 years after transplantation. Since this does not differ significantly from the expected survival of non-diabetic cadaveric renal allografts (the half life of which varies from 8 to 12 years in most studies), the risk of recurrent disease is not seen as a contraindication to transplantation in most centres. Occasional patients do develop rapid recurrence and at the author's department in Oxford, biopsy proven recurrent nephropathy has been seen to progress to nephrotic syndrome and renal failure within 3 years.

Pancreas transplantation

The success of transplantation as a treatment for the failure of a number of organs is now beyond dispute. Since diabetes is a disease caused by 'organ' failure, namely failure of the B cells within the islets of Langerhans of the pancreas, the proposal that this might be treated by transplantation is entirely logical. The approach that is most immediately attractive to the transplant surgeon is to employ the techniques that are already well established for other organs, and transplant the pancreas as a vascularized organ, based on the coeliac artery and splenic artery for its arterial supply and the splenic vein and portal vein for venous drainage.

The first pancreas transplant, performed by Lillehei in Minneapolis in 1966, was followed by a small number of pancreas transplants performed each year until the late 1970s. The results were very poor, with few patients surviving beyond 1 year after transplantation. The major problem was how to deal with the exocrine enzymes draining from the pancreatic duct. A number of techniques were tried, but fistulae, abscesses and sepsis remained major causes of morbidity and mortality, until a technique of injecting a variety of latex glue into the exocrine duct, to suppress exocrine secretion, was developed, as described by Dubernard from Lyon in 1977. Although the technique did not completely solve the problems associated with exocrine leakage, the improved safety and simplicity of the technique justified the application of pancreas transplantation to fitter patients, with consequent improvement in mortality and graft survival. The annual number of pancreas transplants performed has increased each year since that time, with the world total exceeding 2000 in 1989.

Results of pancreas transplantation

The results of pancreas transplantation have been more carefully documented than virtually any other surgical activity, co-ordinated by the Pancreas Transplantation Registry in Minneapolis, USA. Many changes to the technique have been introduced (Sutherland *et al.* 1981), but the most significant modifications have been the introduction of bladder drainage of exocrine secretions, first described by Sollinger from Madison, USA, and the improved immunosuppression provided by the addition of cyclosporine, particularly when used as triple therapy in addition to azathioprine and steroids. Bladder drainage of exocrine secretions has become the technique of choice, particularly in the USA, although duct injection and drainage to a loop of gut is still practised in Europe. The overall graft survival has now improved such that some units are describing 1 year graft survival (defined as patient off insulin) of better than 80 per cent, with patient survival of 95 per cent. These results (Sutherland *et al.* 1981) are now comparable with those of other organs, and are a testament to the perseverance and skill of those who continued to develop the technique through the learning curve. However, the morbidity of pancreas transplantation still remains more than that for transplantation of the kidney alone, with the average hospital stay more than twice as long, and complications such as vascular thrombosis and exocrine leakage remain a problem.

Effect of pancreas transplantation on complications

The benefit of most organ transplants is obvious, but in the case of a non-lifesaving procedure, such as pancreas transplantation, the benefit has been more difficult to demonstrate. The procedure has been carried out mainly in patients with end stage diabetic nephropathy, and virtually all patients at this stage also have diabetic retinopathy with long-standing hypertension, and more than half the patients have neuropathy and vascular disease.

Although numerous studies have reported the effect of vascularized pancreas transplantation on the progression or stabilization of diabetic complications, most are simply anecdotes and cannot be used to discern any true effect. Only four centres (Minneapolis, Lyon, Munich, and Stockholm) have reported studies with sufficient numbers to make any comment meaningful. Diabetic retinopathy is a major cause of blindness and disability, and it was hoped that pancreas transplantation would have a major impact in this population. The Munich group have reported visual acuity improvement in more than half of their patients, but diabetic retinopathy has a highly variable progression, and vision may also be influenced by successful kidney transplantation alone. A study from Minneapolis, USA, which has a group of patients with failed pancreas transplants as controls has not shown any significant different at 4 years in retinopathy grade or visual acuity between those patients that have a functioning pancreas graft and those that do not (Ramsay *et al.* 1988). There is a trend to advantage of pancreas transplantation after 4 years, which may become significant as more patients are added to the study, but this remains to be seen.

Diabetic neuropathy has been shown to improve dramatically after combined pancreas and kidney transplantation, but most of this improvement is due to the clearing of uraemia associated with kidney transplantation. As for retinopathy, there is a suggestion from the Minneapolis group that further deterioration may be prevented after 2 years, but the advantage is not striking (Van der Vlict 1988). Some patients in Minneapolis have also been given pancreas transplants at a relatively early stage of diabetic nephropathy in order to try to preserve renal function (Sutherland *et al.* 1988). Most patients have shown an immediate decline in creatinine clearance, usually related to the use of cyclosporin for immunosuppression, and a few patients have required dialysis as a consequence. Again, there is some evidence of stabilization of progression two years after transplantation, but these preliminary findings await expansion and confirmation.

The major cause of mortality from diabetes is vascular disease, most often expressed as coronary occlusion, and there is no evidence that pancreas transplantation affects the progression of vascular disease and subsequent mortality. Myocardial infarction remains the commonest cause of death with a functioning pancreas transplant.

Current indications for pancreas transplantation

To summarize, the effect of pancreas transplantation on the complications of diabetes can only be described as disappointing. Despite this, most units with a large pancreas transplant programme have pointed to less well-defined benefits, such as a sense of freedom and psychological well-being experienced by many of their patients, and use the argument that if these benefits can be obtained with relatively little additional morbidity to the kidney transplant or alteration to the immunosuppression that they already require, then why not perform a pancreas transplant? These arguments appear to be sufficient to maintain the activity in societies with adequate resources, such as the USA, but insufficient to justify continuation in countries such as the UK, where vascularized pancreas transplantation has virtually ceased.

One interesting positive benefit of pancreas transplantation has been the observation that the recurrence of diabetic nephropathy in kidneys transplanted into diabetic patients can be prevented by a successful pancreas transplant. As described above, the recurrence of nephropathy is seldom rapid enough to justify pancreas transplantation for this indication alone, but the observation has implications for the likely aetiology of diabetic complications and the probability that complications could be prevented were it possible to perform transplantation at a much earlier stage. Support for this concept comes also from transplantation experiments in animals, but these studies suggest in addition that the most benefit would be likely to come from transplantation soon after development of diabetes. The idea of performing such a large operation, with the added requirement of immunosuppression, in otherwise healthy young diabetic patients is not attractive.

Isolated islet transplantation

The search for a technique of transplantation with low morbidity that might be applied to young diabetics has led to the suggestion that the insulin secreting tissue be separated from the pancreas and transplanted alone. It is certainly illogical to transplant the whole pancreas when only approximately 2 per cent of the tissue mass is required. Although a number of potential sources of tissue are available to fill this need, the most practical approach is to separate the islets of Langerhans from the pancreas of adult human cadaveric donors. Although this idea was first suggested even before the discovery of insulin, the practical possibility was first raised by the demonstration of successful islet isolation and transplantation in the rat by Lacy and colleagues in St Louis in 1972 (Lacy and Kostianovsky 1967). Unfortunately, these early reports were soon followed by premature attempts at islet transplantation in humans, which failed. Since then, considerable effort has been expended in experimental laboratories around the world to define techniques for islet isolation, the correct implantation site, and how to prevent rejection with or without immunosuppression (Gray and Morris 1988).

The lack of an efficient method for retrieval of islets of the human pancreas was the major problem for many years, but since 1984 considerable advances have been made (Gray *et al.* 1984) and human islet isolation sufficient for clinical transplantation is now possible (Scharp *et al.* 1988, Warnock *et al.* 1989, and Tzakis *et al.* 1990). The technique is currently at an exciting phase of development, with the recent description of successful clinical islet transplantation from three separate groups, all three having obtained insulin independence in patients documented as being previously insulin-dependent (Scharp *et al.* 1990, Warnock *et al.* 1989, and Tzakis *et al.* 1990). The longevity and metabolic efficiency of these islet transplants remains to be seen and the application of the technique to the young diabetic patient may still be a long way off. However, there is reason to hope that islet transplantation will eventually be the treatment of choice for young diabetic patients, who will then not require the subsequent attentions of the nephrologist.

Bibliography

Bilous R.W., Mauer, S.M., Sutherland, D.E., Najarian, J.S., Goetz, F.C., and Steffes, M.W. (1989). The effects of pancreas transplantation on the glomerular structure of renal allografts in patients with insulin-dependent diabetes. *N. Engl. J. Med.* **321**, 80–5.

Friedman, E.A. and L'Esperance, F.A. (ed.). (1982). *Diabetic renal syndrome. Volume 2. Prevention and management*. Grune and Stratton, New York.

Gray, D.W. and Morris, P.J. (1988). Transplantation of isolated pancreatic islets. In *Pancreatic transplantation*. (ed. C.G. Groth), pp. 363–90, W.B. Saunders, Philadelphia.

Gray, D.W.R., McShane, P., Grant, A., and Morris, P.J. (1984). A method for isolation of islets of Langerhans from the human pancreas. *Diabetes*, **33**, 1055–61.

Heidland, A., Koch, K.M., and Heidbreder, E. (eds). (1989). Diabetes and the kidney. In *Contributions to nephrology 1989 No 73*. (ed. G.M. Berlyne, S. Giovannetti) Karger, Basel.

Lacy, P.E. and Kostianovsky, M. (1967). Method for the isolation of intact islets of Langerhans from the rat pancreas. *Diabtes*, **16**, 35–9.

Najarian, J.S., Kaufman, D.B., Fryd, D.S. *et al.* (1989). Long-term survival following kidney transplantation in 100 type I diabetic patients. *Transplantation*, **47**, 106–13.

Ramsay, R.C., Goetz, F.C., Sutherland, D.E. *et al.* (1988). Progression of diabetic retinopathy after pancreas transplantation for insulin-dependent diabetes mellitus. *N. Engl. J. Med.* **318**, 208–14.

Scharp, D.W., Lacy, P.E., Santiago, J.V. *et al.* (1990). Insulin independence after islet transplantation into type I diabetic patient. *Diabetes*, **39**, 515–18.

Sutherland, D.E. (1981*a*). Pancreas and islet transplantation.II. Clinical trials. *Diabetologia*, **20**, 435–50.

Sutherland, D.E. (1981*b*). Pancreas and islet transplantation.I. Experimental studies. *Diabetologia*, **20**, 161–85.

Sutherland, D.E., Najarian, J.S., Greenberg, B.Z., Senske, B.J., Anderson, G.E., Francis, R.S., and Goetz, F.C. (1981). Hormonal and metabolic effects of a pancreatic endocrine graft. Vascularized segmental transplantation in insulin dependent diabetic patients. *Ann. Intern. Med.* **95**, 537–41.

Sutherland, D.E., Kendall, D.M., Moudry, K.C. *et al.* (1988). Pancreas transplantation in nonuremic, type I diabetic recipients. *Surgery*, **104**, 453–64.

Sutherland, D.E., Dunn, D.L., Goetz, F.C. *et al.* (1989). A 10-year experience with 290 pancreas transplants at a single institution. *Ann. Surg.* **210**, 274–85.

Tzakis, A.G., Ricordi, C., Alejandro, R. *et al.* (1990). Pancreatic islet transplantation after upper abdominal exenteration and liver replacement. *Lancet*, **336**, 402–5.

Van der Vliet, J.A., Navarro, X., Kennedy, W.R., Goetz, F.C., Barbosa, J.J., Sutherland, D.E., and Najarian, J.S. (1988). Long-term follow-up of polyneuropathy in diabetic kidney transplant recipients. *Diabetes*, **37**, 1247–52.

Warnock, G.L., Kneteman, N.M., Ryan, E.A. *et al.* (1989). Continued function of pancreatic islets after transplantation in type I diabetes. *Lancet*, **2**, 570–2.

40 Cyclosporin nephrotoxicity

G.H. Neild

When cyclosporin was first discovered to be nephrotoxic it appeared to cause acute tubular necrosis, and it was assumed that it was a tubular toxin in the manner of gentamicin or amphotericin. It is now clear, however, that there are essentially two separate components to cyclosporin nephrotoxicity: an acute, reversible phase, which is haemodynamically mediated, and a chronic, irreversible component, associated with vascular pathology and ischaemic injury. Cyclosporin increases renal vascular resistance in all patients, and this reverses when cyclosporin is withdrawn. There is no good evidence for direct tubular toxicity.

Clinical aspects

There are no major differences between nephrotoxicity affecting the native kidneys, as in heart allograft recipients, and renal allografts, although it is possible that the renal allograft, being denervated, is relatively protected.

Acute nephrotoxicity

Acute renal dysfunction is dose-dependent and invariable but reversible.[1,2] The glomerular filtration rate falls, there is retention of salt and water, and hyperuricaemia may occur.[1,3,4] With more severe toxicity, systemic acidosis and hyperkalaemia occur, which are disproportionate to the degree of renal failure.[5,6]

In the kidney, morphological lesions of acute tubular necrosis are seen. These include vacuolization and foamy changes to the tubular epithelial cytoplasm, and necrosis and exfoliation of tubular cells.[7-10] A vasculopathy affecting arterioles may also occur, with myocyte necrosis leading to hyalinosis of the wall.[7,8,10]

Chronic nephrotoxicity

Chronic nephrotoxicity may be defined as the induction of irreversible renal dysfunction associated with morphological injury. There is patchy (focal) tubular atrophy and interstitial fibrosis. In severe examples, and particularly when they are a consequence of severe prolonged acute tubular necrosis, these changes may be diffuse. Vascular hyalinosis and sclerosis are also prominent.[7,10,11] In native kidneys, but not in renal allografts, glomerular sclerosis is also a major feature. This has been seen especially in heart transplant recipients with severe cyclosporin toxicity.[11]

Induction of cyclosporin nephrotoxicity

In humans, it is unusual for oral cyclosporin to cause nephrotoxicity in less than five days,[1,3] although intravenous therapy may cause profound oliguria within three days.[12] Intravenous infusion of cyclosporin into animals produces an immediate fall in glomerular filtration rate and renal blood flow.[13]

Cyclosporin is very lipophilic, and it appears that it takes certain time to saturate body stores.[14] Thus, nephrotoxicity may take several weeks to develop when patients are started on low doses (for example 5 mg/kg/d).

Monitoring cyclosporin

The majority of the toxicity resides with the parent compound. The concentration of parent compound in whole blood can be measured either by high pressure liquid chromatography (HPLC) or with radioimmunoassay using an appropriate monoclonal antibody. Two or three of the 17 metabolites of cyclosporin may have minimal nephrotoxic activity, but there is no evidence in humans that renal failure and retention of metabolites further enhances renal dysfunction.

Cyclosporin metabolism

Cyclosporin is metabolized in the liver by the P450 cytochrome system.[15] This enzyme is also present in parts of the renal tubules, although renal metabolism is not known to play a role in nephrotoxicity. A variety of drugs may induce or inhibit this enzyme system. Thus, phenytoin will greatly enhance enzyme activity, and the cyclosporin dose may have to be increased three-fold to maintain the same blood concentration. Conversely, drugs such as ketoconazole or erythromycin may necessitate a major reduction in dosage. If a patient is on a stable dose and the blood levels start, for example, to rise, a reason must be sought; it is usually the addition of a further drug, or mild hepatic dysfunction.

Drug modification of nephrotoxicity

Drugs such as gentamicin, which by themselves are known to be nephrotoxic, act synergistically with cyclosporin. For this reason, a course of gentamicin will almost inevitably cause a rise in creatinine, even if the cyclosporin trough levels remain in the therapeutic range, where they would not normally cause renal dysfunction. Plasma creatinine concentrations invariably rise when patients are given trimethoprim, but this is due to the inhibition of tubular secretion of creatinine, and not to a fall in glomerular filtration rate.

Non-steroidal anti-inflammatory drugs, by their inhibition of vasodilatory prostaglandins, will always reduce glomerular filtration rate in patients with renal dysfunction. Patients with cyclosporin nephrotoxicity may be particularly susceptible, and severe hyperkalaemia may also occur.[16]

Does cyclosporin therapy lead invariably to progressive interstitial fibrosis?

High doses and levels of cyclosporin lead to renal fibrosis.[11] An early study in renal allografts showed that at one year there could be an alarming degree of fibrosis, but subsequent biopsies from the same patients showed little or no progression of the fibrosis.[17] A second study of renal allografts biopsied at 1 and 12 months showed very slight progression of interstitial fibrosis,[18] and these patients were receiving doses in excess of those generally used today. It seems unlikely that the lower maintenance doses of cyclosporin now being used will cause progressive renal damage in renal allografts.

Native kidneys seem to be more susceptible to progressive injury, and patients have reached end stage renal failure after heart transplantation, despite cyclosporin being reduced or withdrawn.[11]

Pathogenesis

Mechanism of acute nephrotoxicity

Several initial clues suggested that cyclosporin was not a tubular toxin, but that it might act in a haemodynamic manner. First, when cyclosporin was withdrawn after a prolonged delay in function of a renal allograft, renal function improved within 48 h, unlike the longer period of non-function occurring after the withdrawal of a toxin such as gentamicin.[19] Secondly, reduction of the cyclosporin dose in cases of nephrotoxicity would result similarly in an improvement of glomerular filtration rate in 24–48 h.

Histological features of tubular injury similar to those described in humans were seen when rats were given toxic doses of cyclosporin,[20] but none of the features was unique to cyclosporin and all could be reproduced in experimental models of acute tubular necrosis induced by ischaemia.[21] Despite the morphological evidence of injury, tubular function remained intact. A series of experiments had subsequently clarified these mechanisms.

Toxicology

Humes *et al.* showed *in vitro* that although cyclosporin was able to inhibit mitochondrial metabolism in tubular segments in a dose-dependent manner, it was not toxic to tubular segments cultured in hypoxic conditions, whereas nephrotoxins such as gentamicin were invariably toxic in such conditions.[22]

Tubular physiology

Dieperink *et al.*, in a classic series of experiments measuring lithium and inulin clearance in rats, showed that when rats were given doses of cyclosporin from 0–50 mg/kg there was a dose-dependent reduction in glomerular filtration rate, but an increase in fractional proximal tubular reabsorption of lithium, which is handled by the proximal tubule, as for sodium.[23] They also found that proximal tubular hydrostatic pressure was low, and concluded that the fall in glomerular filtration rate was due to a decreased net untrafiltration pressure secondary to reversible spasm of the afferent glomerular arteriole.[23,24]

Renal blood flow and glomerular haemodynamics

Murray and Paller showed that the infusion of 20 mg/kg cyclosporin into rats caused a 48 per cent reduction in renal blood flow with an 80 per cent increase in renal vascular resistance.[13] Half this dose did not reduce renal blood flow unless indomethacin was also given. The inference that cyclosporin caused both afferent and efferent arteriolar constriction was later confirmed in micropuncture studies in the rat,[25] which also showed that there was a marked fall in the glomerular ultrafiltration coefficient (Kf).[25] Cyclosporin appears to increase renal allograft vascular resistance in all patients, and this reverses when cyclosporin is withdrawn.[26]

Mechanism of vasospasm

Renin angiotensin system

In acute experiments there is prompt activation of this system in association with activation of the sympathetic nervous system.[27,28] Cyclosporin will also release renin when incubated with slices of renal tissue, and this can be inhibited by beta-blockers.[28] However, it is unlikely that this system is important in the maintenance of reduced glomerular filtration rate (or hypertension), since captopril will not modify acute nephrotoxicity, and after several weeks of cyclosporin therapy in humans, when hypertension and hyperkalaemia may occur, plasma renin activity is depressed.[6,29]

Sympathetic nervous system

Vasoconstriction in acute experiments is mediated in part by activation of the sympathetic nervous system. In the rat, the fall in renal blood flow following infusion of cyclosporin can be reduced by alpha-adrenergic blockade and by denervation;[13] and there is an increase in sympathetic nerve activity.[30] The latter has now also been demonstrated in humans[31] and may be a major factor contributing to the development of hypertension, and the more severe nephrotoxicity in native (un-denervated) kidneys.

Vasodilatory prostaglandins

It was proposed that cyclosporin might act by inhibiting vasodilatory prostaglandins, particularly prostacyclin.[32] Although there has been conflicting experimental evidence, it now appears that in cyclosporin nephrotoxicity there is an increase in urinary prostaglandin excretion but less than predicted when compared with appropriate controls,[33] i.e. if the kidney is ischaemic a marked increase in the release of vasodilatory prostaglandins is expected. For example, when rats were given cyclosporin for 45 d the renin-angiotensin system was stimulated with an increase in plasma renin activity, but these changes were not accompanied by the expected, parallel increase in the renal synthesis of the vasodilatory prostaglandins PGE2 and PG12.[34]

Cyclosporin inhibits prostacyclin production by endothelial cells in a dose-dependent manner,[35] and there is evidence for a direct action on cyclo-oxygenase.[35,36] Thus, vasodilatory prostaglandins play, at least a secondary role in offsetting the vasospasm induced by cyclosporin, and for this reason the addition of non-steroidal anti-inflammatory drugs may seriously enhance toxicity.[16]

Thromboxane synthesis

The role of vasoconstrictor thromboxanes appears more convincing. Several studies in animals have shown a direct correlation between thromboxane excretion in urine and the fall in glomerular filtration rate.[37,38] Not only does thromboxane rise, but the ratio of thromboxane to vasodilatory prostaglandins rises in favour of vasoconstriction. Thromboxane synthetase inhibitors slightly modify the fall in glomerular filtration rate;[38] specific thromboxane antagonists are being evaluated at present.

Endothelin

This is a very potent constrictor peptide, which is synthesized by the endothelium. Infusion of endothelin causes sustained hypertension and a fall in renal blood flow and glomerular filtration rate. It is an obvious candidate for the primary mediator in cyclosporin nephrotoxicity. Two groups have recently produced evidence that would argue strongly in favour of this mechanism.[39,40] Both showed, in rat models, that cyclosporin nephrotoxicity was associated with raised levels of circulating endothelin and that the renal haemodynamic changes could be reversed or prevented by infusion of anti-endothelin antibody.

Other mediators

Platelet activating factor is a renal vasoconstrictor and its role is being evaluated.[41] The idea that cyclosporin might inhibit endothelium-derived relaxing factor was an attractive one, but cyclosporin appears to blunt the vasodilator reponse to both endothelium-dependent and endothelium-independent vasodilators.[42] These experiments suggest that cyclosporin is acting beyond the endothelium, i.e. at the level of the vascular smooth muscle cell, and this has been confirmed in other studies.[43,44] Cyclosporin will cause strips of isolated rat aorta to contract, possibly by inducing a release of noradrenaline from adrenergic terminals.[45]

Mechanism of chronic nephrotoxicity

Chronic nephrotoxicity would appear to be due to a combination of vasospasm and vascular pathology, which together result in ischaemia from the narrowing and obliteration of arterioles. Toxic doses of cyclosporin enhance all forms of vascular injury, although there is no evidence that it initiates vascular injury *per se*. In addition, there is increasing evidence that cyclosporin may stimulate fibrosis, possibly by enhancing or stimulating the release of cytokines, such as transforming growth factor beta (TGF-beta).[46]

Experimental vascular injury

When rabbits with serum sickness were treated with cyclosporin, a severe vascular injury was produced, and a glomerular injury similar to the haemolytic uraemic syndrome.[47] It was concluded that cyclosporin profoundly altered the inflammatory component of normal vascular injury, and it was proposed that this was due to an effect on platelet-endothelial interactions, possibly mediated by failure of prostacyclin production.[32]

Cyclosporin will enhance the vascular injury that occurs in renal vessels in spontaneously hypertensive rats.[48] However, cyclosporin given alone to animals has never produced a vascular lesion.

Mechanism of vacular injury

When cyclosporin nephrotoxicity occurs in humans, the earliest, and most important, morphological change seen in arterioles is one of myocyte necrosis.[7] In severe cases, endothelial cell necrosis may also occur.[7,8] It is assumed that the myocyte necrosis is principally a response to vascular spasm, and that this spasm, with or without endothelial injury, leads to local activation of platelets and coagulation.[8,49] These lesions are now seen much less often as lower cyclosporin regimens are used. Platelet-fibrin thrombi in glomerular capillaries were also seen frequently in the first few weeks after transplantation,[50] but were invariably associated with high cyclosporin levels and are now rarely seen.

A haemolytic-uraemic syndrome was first reported following bone marrow transplantation[51] and subsequently has been reported following both liver and renal transplantation. Haemolytic-uraemic syndrome has usually occurred within the first few weeks after transplantation, while blood levels are in the toxic range.

These instances of vascular injury suggest that endothelial damage is usual in cyclosporin nephrotoxicity. Evidence of endothelial injury was obtained in renal allograft recipients by demonstration of very high circulating levels of factor VIII-related antigen (Von Willebrand factor) during periods of cyclosporin nephrotoxicity. Levels of this endothelially-derived protein fell as toxicity resolved and doses and blood concentrations of cyclosporin fell.[49] However, cyclosporin has not been shown to be toxic to cultured endothelial cells in doses which inhibit prostacyclin synthesis.[35] Recent longitudinal studies of renal allograft recipients have shown that after one year they still have evidence of platelet hyperaggregability, platelet activation and decreased fibrinolysis—all features which would contribute to a prothrombotic state.[52,53]

Cyclosporin does enhance procoagulant synthesis by macrophages,[54,55] which may be the most important factor in the prothrombotic state.

Management

It should be noted that all patients receiving cyclosporin have a degree of functional nephrotoxicity, as demonstrated by increased renal vascular resistance,[26] and lower glomerular filtration rates than patients with similar plasma creatinine concentrations who are not receiving cyclosporin.[56] This accounts for the sensitivity of these patients' kidneys to blood volume contraction (hypovolaemic) and to nephrotoxins.[57]

Prevention of toxicity

Clearly, enough cyclosporin has to be given to ensure adequate immunosuppression. The lowest effective doses have still not been defined, but it is now known that the toxic doses that were originally given are not necessary. Triple and quadruple therapy regimens have also allowed the use of lower doses of cyclosporin. Nevertheless, the regular measurement of trough levels of cyclosporin in whole blood are very helpful. At any particular time after transplantation and for any particular assay, a range of blood levels can be defined, above which toxicity is very likely, and below which immunosuppression is likely to be inadequate.[58] Nephrotoxicity may occur within the therapeutic range, in which case the cyclosporin dose can be reduced and the level monitored. It is self-evident that the level has to be reduced as well as the dose, and similarly if the level starts to rise and the plasma creatinine increases in a patient who was established on a stable dose of cyclosporin, then some explanation for the change must be sought.

Diagnosis of toxicity

A cause for renal dysfunction must always be established. When the plasma creatinine is in the normal range, more sensitive indices of renal function, such as creatinine clearance and isotopic glomerular filtration rate may be necessary to detect genuine changes in function. A diagnosis of cyclosporin nephrotoxicity is more likely, (1) if no other explanation can be found, (2) if the blood levels are high, (3) if there is other evidence of toxicity, such as a tremor in the early stages, or hypertrichosis or gingival hypertrophy in later stages, or (4) if the degree of hyperkalaemia, acidosis, or hyperuricaemia is disproportionate to the relatively mild degree of renal insufficiency.

Summary

Nephrotoxicity must always be excluded as a cause of renal dysfunction in allograft recipients receiving cyclosporin. Prolonged nephrotoxicity will lead to impairment of renal function. This is a consequence of the renal vasospasm and vascular pathology that leads to ischaemia and atrophy of the renal cortex.

References

1. Klintmalm, G.B.G., Iwatsuki, S., and Starzl, T.E. (1981). Nephrotoxicity of cyclosporin A in liver and kidney transplant patients. *Lancet*, **1**, 470–1.
2. Feutren, G., Assan, R., Karsenty, G. *et al.* (1986). Cyclosporin increases the rate and length of remissions in insulin-dependent diabetes of recent onset. *Lancet*, **ii**, 119–24.
3. Hows, J.M., Chipping, P.M., Fairhead, S., Smith, J., Baugham, A., and Gordon-Smith, E.C. (1983). Nephrotoxicity in bone-marrow transplant recipients treated with cyclosporin A. *Brit. J. Haematol.* **54**, 69–78.
4. Chapman, J.R., Griffiths, D., Harding, N.G.L., and Morris, P.J. (1985). Reversibility of cyclosporin nephrotoxicity after three months treatment. *Lancet*, **i**, 128–30.
5. Hamilton, D.V., Evans, D.B., Henderson, R.G., Thiru, S. *et al.* (1981). Nephrotoxicity and metabolic acidosis in transplant patients on cyclosporin A. *Proc. EDTA*, **18**, 400–9.
6. Adu, D., Turney, J., Michael, J., and McMaster, P. (1983). Hyperkalaemia in cyclosporin-treated renal allograft recipients. *Lancet*, **2**, 370–2.
7. Mihatsch, M.J., Thiel, G., Spichtin, H.P. *et al.* (1983). Morphological findings in kidney transplants after treatment with Cyclosporin. *Trans. Proc.* **15**, 2821–35.
8. Mihatsch, M.J., Thiel, G., and Ryffel, B. (1988). Histopathology of cyclosporin nephrotoxicity. *Trans. Proc.* **20**, (3), 759–71.
9. Sibley, R.K., Rynasiewicz, J., Ferguson, R.M. *et al.* (1983). Morphology of cyclosporin nephrotoxicity and acute rejection in patients immunosuppressed with cyclosporin and prednisolone. *Surgery*, **94**, 225–34.
10. Neild, G.H., Taube, D.H., Hartley, R.B. *et al.* (1986). Morphological differentiation between rejection and cyclosporin nephrotoxicity in renal allografts. *J. Clin. Pathol.* **39**, 152–9.
11. Myers, B.D., Sibley, R., Newton, L. *et al.* (1988). The long term course of cyclosporine-associated chronic nephropathy. *Kidney Int.* **33**, 590–600.
12. Powell-Jackson, P.R., Young, B., Calne, R.Y., and Williams, R. (1983). Nephrotoxicity of parenterally administered cyclosporin after orthotopic liver transplantation. *Transplantation*, **36**, 505–8.
13. Murray, B.M., Paller, M.S., and Ferris, T.F. (1985). Effect of acute and chronic cyclosporine administration on renal hemodynamics in conscious rats. *Kidney Int.* **28**, 767–74.
14. Wood, A.J., and Lemaire, M. (1985). Pharmacologic aspects of cyclosporin therapy: pharmacokinetics. *Trans. Proc.* **17**, 27–32.
15. Kahan, B.D. (1985). Individualization of cyclosporine therapy using pharmacokinetic and pharmacodynamic parameters. *Transplantation*, **40**, 457–76.
16. Berg, K.J., Forre, O., Djoseland, O., Mikkelsen, M., Narverud, J., and Rugstad, H.E. (1989). Renal side effects of high and low cyclosporin A doses in patients with rheumatoid arthritis. *Clin. Nephrol.* **31**, 232–8.
17. Klintmalm, G., Bohman, S.-O., Sundelin, B., and Wilczek, H. (1984). Interstitial fibrosis in renal allografts after 12 to 46 months of cyclosporin treatment:

beneficial effect of low doses in early post-transplantation period. *Lancet*, **2**, 950–4.

18. Bignardi, L., Neild, G.H., Hartley, R.B. *et al.* (1987). Histopathological changes in cyclosporin treated renal allografts biopsied at one and twelve months. *Nephrol. Dial. Transplant.* **2**, 366–70.
19. Calne, R.Y., White, D.J.G., Evans, D.B., and Wight, C. (1982). Three years' experience with cyclosporin A in clinical cadaveric kidney transplantation. In *Proceedings of an International Conference on Cyclosporin A.* (ed. D.J.G. White), pp. 347–53. Elserier Biomedical Press, Cambridge.
20. Thomson, A.W., Whiting, P.H., Blair, J.T., Davidson, R.J.L., and Simpson, J.G. (1981). Pathological changes developing in the rat during a 3-week course of high dosage Cyclosporin A and their reversal following drug withdrawal. *Transplantation*, **32**, 271–7.
21. Thiel, G. (1986). Experimental cyclosporine A nephrotoxicity: a summary of the international workshop (Basle, April 24–26, 1985). *Clin. Nephrol.* **25**, 205–10.
22. Jackson, N.M., O'Connor, R.P., and HUmes, H.D. (1988). Interactions of cyclosporine with renal proximal tubule cells and cellular membranes. *Transplantation*, **46**, 109–14.
23. Dieperink, H., Leyssac, P.P., Kemp, E., and Steinbrueckl, D. (1986). Glomerulotubular function in cyclosporin treated rats. *Clin. Nephrol.* **25**, S70–4.
24. Dieperink, H., Leyssac, P.P., Starklint, H., and Kemp, E. (1986). Nephrotoxicity of cyclosporin A. A lithium clearance and micropuncture study in rats. *Eur. J. Clin. Invest.* **16**, 69–75.
25. Barros, E.J.G., Boim, M.A., Ajzen, H., Ramos, O.L., and Schor, N. (1987). Glomerular hemodynamics and hormonal participation on cylosporine nephrotoxicity. *Kidney Int.* **32**, 19–25.
26. Curtis, J.J., Luke, R.G., Dubovsky, E., Diethelm, A.G., Whelchel, J.D., and Jones, P. (1986). Cyclosporin in therapeutic doses increases renal allograft vascular resistance. *Lancet*, **ii**, 477–9.
27. Siegl, H., Ryffel, B., Petric, R. *et al.* (1983). Cyclosporin, the renin-angiotensin-aldosterone system, and renal adverse reactions. *Trans. Proc.* **15**, 2719–25.
28. Duggin, G.G., Baxter, C., Hall, B.M., Horvath, J.S., and Tilter, D.J. (1986). Influence of cyclosporin A on intrarenal control of GFR. *Clin. Nephrol.* **25**, (1), S43–5.
29. Thompson, M.E., Shapiro, A.P., Johnsen, A.M., Reeves, R. *et al.* (1983). New onset of hypertension following cardiac transplantation. *Trans. Proc.* **15**, 2573–7.
30. Moss, N.G., Powell, S.L., and Falk, R.J. (1985). *Proc. Natl. Acad. Sci. USA*, **82**, 8222–6.
31. Scherrer, U., Vissing, S.F., Morgan, B.J. *et al.* (1990). Cyclosporine-induced sympathetic activation and hypertension after heart transplantation. *N. Engl. J. Med.* **323**, 693–9.
32. Neild, G.H., Rocchi, G., Imberti, L. *et al.* (1983). Effect of cyclosporin A on prostacyclin synthesis by vascular tissue. *Thrombosis Research*, **32**, 373–9.
33. Neild, G.H. (1988). Vasodilatory prostaglandins and cyclosporin nephrotoxicity. *Prostaglandins, Leukotrienes and Essential Fatty Acids—Reviews 1988*, **33**, 207–12.
34. Perico, N., Zoja, C., Benigni, A., Ghilardi, F., Gualandris, L., and Remuzzi, G. (1986). Effect of short-term cyclosorin administration in rats on renin-angiotensin and thromboxane A2: possible relevance to the reduction in glomerular filtration rate. *J. Pharmacol. Exp. Ther.* **239**, 229–35.

35. Brown, Z., and Neild, G.H. (1987). Cyclosporine inhibits prostacyclin production by cultured human endothelial cells. *Trans. Proc.* **19**, 1178–80.
36. Brown, Z., Neild, G.H., and Lewis, G.P. (1988). Mechanism of cyclosporine inhibition of prostacyclin synthesis by cultured human umbilical vein endothelial cells. *Trans. Proc.* **20**, (3), 654–7.
37. Kawaguchi, A., Goldman, M.H., Shapiro, R., Foegh, M.L., Ramwell, P.W., and Lower, R.R. (1985). Increase in urinary thromboxane B2 in rats caused by Cyclosporin. *Transplantation*, **40**, 214–16.
38. Perico, N., Benigni, A., Zoja, C., Delaini, F., and Remuzzi, G. (1986). Functional significance of exaggerated renal thromboxane A2 synthesis induced by cyclosporin A. *Am. J. Physiol.* **251**, F581–7.
39. Kon, V., Sugiura, M., Inagami, T., Hoover, R.L. *et al.* (1990). Cyclosporine causes endothelin-dependent acute renal failure. *Kidney Int.* **37**, 486.
40. Dadan, J., Perico, N., and Remuzzi, G. (1990). Role of endothelin in cyclosporine-induced renal vasoconstriction. *Kidney Int.* **37**, 479.
41. Pavao dos Santos, O.F., Boim, M.A., Bregman, R. *et al.* (1989). Effect of platelet-activating factor antagonist on cyclosporin nephrotoxicity. *Transplantation* **47**, 592–5.
42. Cairns, H.S., Fairbanks, L.D., Westwick, J., and Neild, G.H. (1989). Cyclosporin therapy in vivo attenuates the response to vasodilators in the isolated perfused rabbit kidney. *Br. J. Pharmacol.* **98**, 463–8.
43. Pfeilschifter, J., and Ruegg, U.T. (1987). Cyclosporin A augments angiotensin II-stimulated rise in intracellular free calcium in vascular smooth muscle cells. *Biochem. J.* **248**, 883–7.
44. Meyer-Lehnert, H., and Schrier, R.W. (1988). Cyclosporine A enhances vasopressin-induced Ca^{2+} mobilization and contraction in mesangial cells. *Kidney Int.* **34**, 89–97.
45. Xue, H., Bukoski, R.D., McCarron, D.A., and Bennett, W.M. (1987). Induction of contraction in isolated rat aorta by cyclosporin. *Transplantation*, **43**, 715–18.
46. Kopp, J.B., and Klofman, P.E. (1990). Cellular and molecular mechanisms of cyclosporin nephrotoxicity. *J. Am. Soc. Nephrol.* **1**, (2), 162–79.
47. Neild, G.H., Ivory, K., and Williams, D.G. (1984). Severe systemic vascular necrosis in cyclosporin-treated rabbits with acute serum sickness. *Br. J. Exp. Path.* **65**, 731–43.
48. Ryffel, B., Siegl, H., Mueller, A.M., Hauser, R., and Mihatsch, M.J. (1985). Nephrotoxicity of cyclosporine in spontaneously hypertensive rats. *Transplant. Proc.* **17**, 1430–31.
49. Brown, Z., Neild, G.H., Willoughby, J.J., Somia, N.V., and Cameron, J.S. (1986). Increased factor VIII as an index of vascular injury in cyclosporin nephrotoxicity. *Transplantation*, **42**, 150–3.
50. Neild, G.H., Reuben, R., Hartley, R.B., and Cameron, J.S. (1985). Glomerular thrombi in renal allografts associated with Cyclosporin therapy. *J. Clin. Pathol.* **38**, 253–8.
51. Shulman, H., Striker, G., Deeg, H.J., Kennedy, M., Storb, R., and Thomas, E.D. (1981). Nephrotoxicity of Cyclosporin A after allogeneic marrow transplantation: glomerular thromboses and tubular injury. *N. Engl. J. Med.* **305**, 1392–5.
52. Cohen, H., Neild, G.H., Patel, R., Mackie, I.J., and Machin, S.J. (1988). Evidence for chronic platelet hyperaggregability and in vivo activation in cyclosporin-treated renal allograft recipients. *Thrombosis Research*, **49**, 91–101.
53. Cohen, H., Neild, G.H., Mackie, I.J., and Machin, S.J. (1988). Persistent

decreased fibrinolytic activity in cyclosporin-treated renal allograft recipients. *Fibrinolysis*, **2**, 197–201.

54. Carlsen, E., Mallet, A.C., and Prydz, H. (1985). Effect of cyclosporin A on pro-coagulant activity in mononuclear blood cells and monocytes in vitro. *Clin. Exp. Immunol.* **60**, 407–17.
55. Carlsen, E., Gaudernack, G., Filion-Myklebust, C., Pettersen, K.S., and Prydz, H. (1989). Allogenic induction of thromboplastin synthesis in monocytes and endothelial cells. Biphasic effect of cyclosporin A. *Clin. Exp. Immunol.* 428–33.
56. Cairns, H.S., Raval, U., and Neild, G.H. (1988). Failure of cyclosporin-treated renal allograft recipients to increase glomerular filtration rate in response to an infusion of amino-acids. *Transplantation*, **46**, 79–82.
57. Laskow, D.A., Curtis, J., Luke, R. *et al.* (1988). Cyclosporin impairs the renal response to volume depletion. *Trans. Proc.* **20**, (30), 568–71.
58. Holt, D.W., Marsden, J.T., Johnston, A., Bewick, M., and Taube, D.H. (1986). Blood cyclosporin concentrations and renal allograft dysfunction. *Br. Med. J.* **293**, 1057–9.

41 Renal replacement therapy—too little or too much?

A.J. Wing

It is because of their success that dialysis and transplantation pose ethical problems for nephrologists. Renal replacement therapy not only prevents death from end stage renal failure, but provides a reasonable quality of life, and patients have an expectation that it will be offered to them if they need it.

In the UK, sparse resources have effectively rationed treatment, and nephrologists have been united in campaigning for more facilities (Wing 1990). It seems clear that insufficient resources have been available.

The plight of patients with end stage renal failure in developing countries is even more pitiable (Brunner *et al.* 1989). However, experience in countries which have been able to afford the unrestrained use of dialysis services is beginning to raise questions concerning the appropriateness of treatment (Rennie *et al.* 1985), especially as the average age of patients accepted for treatment increases (Brunner and Selwood 1990). Is the technological imperative driving us to apply these treatments inappropriately? Are commercial motives unbalancing professional judgement? Do we add to patients' suffering? Is there too much renal replacement therapy?

There has been too little renal replacement therapy in the UK

The UK has lagged behind other Western European countries in the number of new patients per million population accepted each year for treatment. In 1982, most countries accepted more than 40 patients per million population, while the UK accepted half this number (Table 41.1). On December 20th 1984 the Minister for Health in the UK proposed a target of 40 new patients per million population for each of the health regions (Patten 1984). By 1988, this target had been passed but other countries were treating between 70 and 80 new patients per million population. The Renal Association continued to be concerned by the poor performance in the UK, and sponsored studies of the rate of preventable death due to renal failure in Northern Ireland (McGeown 1990), and in Devon and Blackburn (Feest *et al.* 1990). These studies suggested that a need for 75–80 new patients per million population per year should be anticipated.

The low rates of treatment appear to have been determined by small numbers of renal units (Wing 1990). Market forces operating through insurance or

Table 41.1 Acceptance of new dialysis patients in Europe (Numbers of patients per million population per year.)

	1982	1988
Austria	38.1	95.7
Belgium	39.6	85.2
Israel	58.2	80.0
West Germany	44.0	77.0
Holland	27.4	65.3
Sweden	41.3	64.3
Switzerland	44.5	61.8
Greece	18.8	59.2
Spain	32.7	57.1
France	30.9	56.3
Great Britain	19.9	55.1
Italy	33.8	54.7
Norway	37.3	52.7
Denmark	26.9	52.5
Ireland	19.4	33.8

government reimbursed services have resulted in France having 4.4 centres per million population, Federal Republic of Germany 6.3 and Italy 7.1, whereas bureaucratic constraints in the UK have held the number to 1.3 centres per million population (Tufveson *et al.* 1989). This means that four out of every five district general hospitals in the UK have no dialysis unit. They are also very unlikely to have the services of a nephrologist. Periodic surveys of the workforce in adult renal medicine in the UK have been carried out by Jones on behalf of the renal disease committee of the Royal College of Physicians and the Renal Association (Jones *et al.*, in press). In 1975 there were 57 full-time equivalent senior staff, each treating around 60 patients; by 1989 there were 87, each responsible for an average of nearly 200 endstage renal failure patients (Fig. 41.1). The ratio between patients and senior staff is between five and ten times higher in the UK than in other large Western European countries.

The pattern of development of renal replacement therapy in the UK (Fig. 41.2) is different from that in other countries. The uniqueness of the British pattern results from rationing of resources and from the mechanism through which the rationing has been implemented, i.e. restricted centres and nephrologists. In the 1970s the only way of increasing the numbers of dialysis patients was by training the patients to dialyse themselves outside the centres, and over that decade more than twice as many British patients performed home haemodialysis as were treated in centres. By the early 1980s continuous ambulatory peritoneal dialysis (CAPD) provided an alternative approach to out-patient treatment, and this grew rapidly to become, by 1985, the most

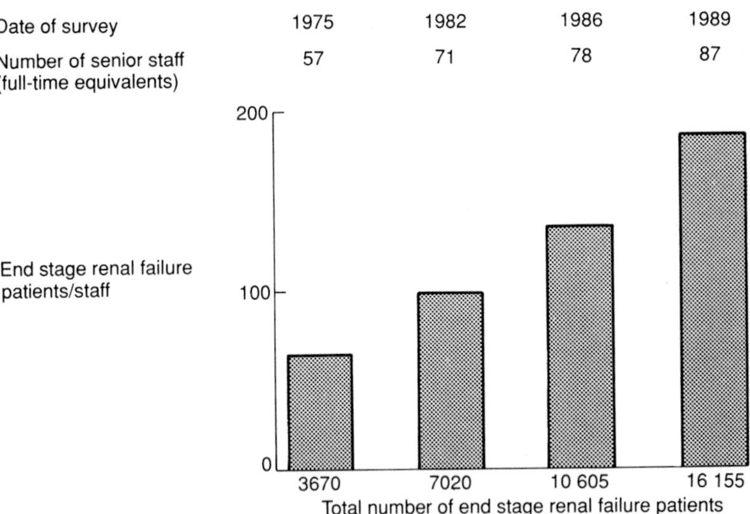

Fig. 41.1 Workforce and workload in adult renal medicine in the UK, 1975–89. The relationship between the numbers of senior staff in full-time equivalents and the total number of end stage renal failure patients alive on treatment given in four surveys carried out in 1975, 1982, 1986, and 1989. Data provided by N.F. Jones (Jones *et al.*, in press).

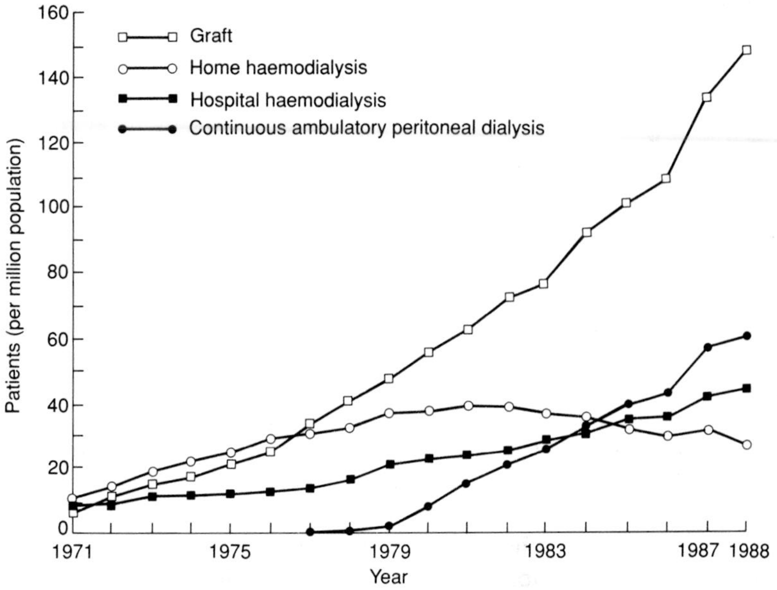

Fig. 41.2 Number of patients alive in the UK on 31st December according to method of treatment, for each of the years 1971–88. Data from EDTA Registry.

important type of dialysis in the UK, while home haemodialysis declined. Figure 41.2 makes clear the overall reliance on renal transplantation, now sustaining more than 150 patients per million population and illustrates how the net number on this form of therapy has increased in recent years, due as much to longer graft survival as to an increased availability of cadaver organs.

British nephrologists have always pointed out that the shortfall in patient numbers was not due to their turning patients away, and it appears that they have been protected by 'negative selection' practised by general practitioners and consultant physicians (Challah *et al.* 1984). That this practice still denies patients the referrals necessary to reach specialist care, under a system in which the generalist is the gatekeeper, is exemplified by the fates of some of the patients in the recently published Renal Association surveys (McGeown 1990; Feest *et al.* 1990). However, there is an interesting difference between the age-specific rates of detection of advanced chronic renal failure in Northern Ireland, where notification depended on a doctor's response, and in Devon and Blackburn, where the patients were identified through laboratory results (Fig. 41.3).

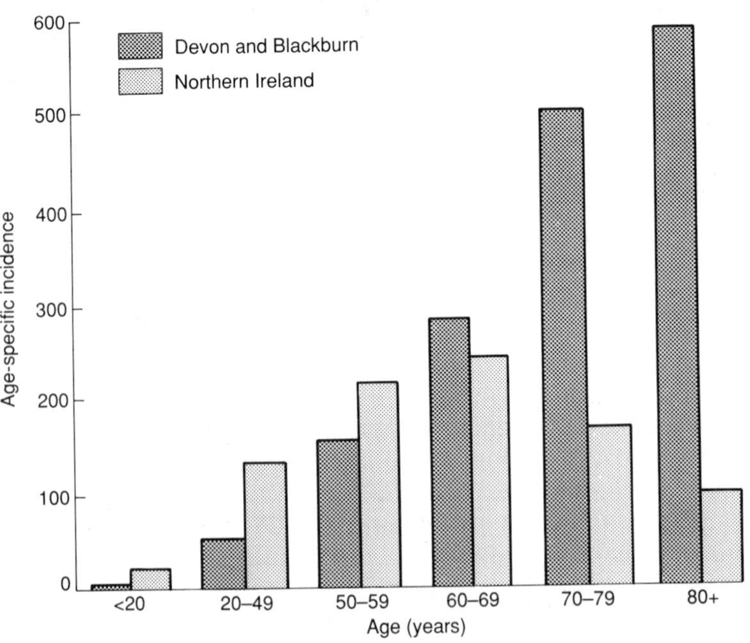

Fig. 41.3 Comparison of age-specific incidence of chronic renal failure in Devon and Blackburn, and in Northern Ireland, 1985–86. Data of McGeown (1990) and of Feest *et al.* (1990) have been compared. The differences could be accounted for by the different methods used to collect cases (see text).

Ministers have accepted the validity of the studies from Northern Ireland and Devon and Blackburn. How will renal replacement therapy increase in the UK?

There could be more renal replacement therapy in the UK

The mechanisms for rationing renal replacement therapy through the planned economy of the National Health Service are evident. The remedy is now proven. At about the same time that English Regions were given a target of 40 new patients per million population per year, the Secretary of State for Wales announced a target of 50 patients per million population for the principality (Edwards 1984). This target was to be met by opening new renal units. In 1985, the existing pair of centres in Cardiff and Rhyl in Wales were joined by another main renal unit at Swansea and subsidiary renal units were opened in Carmarthen and Bangor. Rapid commissioning of these units was achieved by contracting the service to commercial operators. National Health Service consultants in each of the sites remained in clinical control, but decisions about buildings and equipment were strongly influenced by commercial considerations. The new patient acceptance rate in Wales rose from 34 per million population in 1983 to 61 per million population in 1987 (Smith *et al.* 1989).

The inevitable consequence of increasing the rate of patient acceptance has been a liberalization of previous selection criteria—many would say this was long overdue—with more elderly patients and diabetic patients included. Patients aged over 55 years, or those with diabetes, have been termed high risk and Fig. 41.4 shows that in Wales and three English health regions, the increase in numbers of new patients between 1979 and 1988 was achieved largely by increasing the acceptance of patients in this category.

High risk patients have a higher incidence of co-morbid conditions, are more prone to complications, are likely to have a shorter survival, are less easy to train for home haemodialysis, are less 'transplantable', and more dependent. They therefore need more hospital centre haemodialysis.

Need has expanded to fill the new facilities available for it. However, other benefits have accrued. Towns which were unaware that their citizens were dying of a treatable disease are now proud of their own facilities; local practitioners make good use of them and the quality of nephrological care and allied pathological services has been raised in an exemplary fashion.

Accepting that an elderly patient population is appropriately treated by centre haemodialysis, a recently opened unit in Stevenage has deliberately forsaken any attempt to train patients to perform self-dialysis and made its objective to provide high quality haemodialysis (Greenwood 1990). The philosophy of this approach is in contrast to the tradition in the UK of training patients to self-reliance and independence. It will be interesting to observe whether passivity and dependence prove more acceptable over the long-term, or bring the patient more quickly to the state of having had too much of it.

Whichever model is followed, it seems likely that more units will be opened

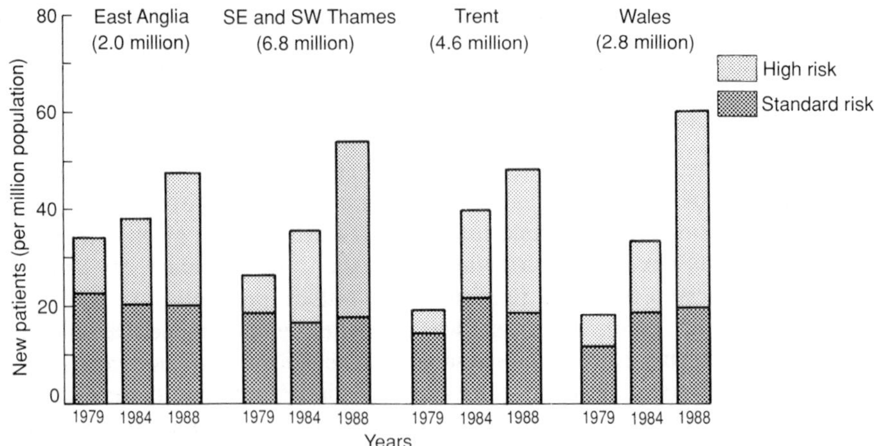

Fig. 41.4 Classification of patients in four UK regions in 1979, 1984 and 1988 into standard risk (age less than 55 years and not diabetic) and high risk (over 55 years, or diabetic). Note that in each of the three English regions and in Wales the increase in acceptance of new patients was due largely to increased numbers of high risk patients.

during the 1990s, bringing dialysis nearer to centres of population which are at present poorly served and must bear with all the disadvantages of long travel to and from treatment centres (Dalziel and Garrett 1987). The UK will have more renal replacement therapy. Will it ever approach 'too much'?

There might be too much renal replacement therapy in the UK

People who work in all branches of medicine and those who are concerned with the distribution of resources available to the National Health Service share a concern that too much money could be spent on the small number of patients with renal failure. Health enconomists are concerned that provision of renal replacement therapy might distort our health services, by drawing excessive funds from less glamorous areas (Williams 1985). These fears are not unfounded. Wood *et al.* (1987), using a steady state model to predict patient numbers resulting from a modest acceptance rate of 40 patients per million population, estimated that the number of hospital beds required to support a regional programme would be equivalent to almost half the acute medical facilities of a small district general hospital. However, as de Wardener (1966) pointed out long ago, it is not without precedent for our society to spend heavily on a single disease, as it did when tuberculosis was managed in sanatoria. Furthermore, it is a harsh decision to deny a life-saving treatment which costs less than society is prepared to pay to deny criminals their liberty.

Inevitably, macro-allocative options depend on the wealth of the society in which patients live. It does not appear to be political persuasion or medical repute of peoples, but rather their gross domestic product which correlates with the numbers of patients who receive renal replacement therapy (Brunner *et al.* 1989). If choice is to be forced upon us, we shall either subconsciously, or in a calculating fashion, take some account of the quality of life that is bought for the high cost paid. The quality adjusted life year (QALY) is a measurement advocated by some health economists (Williams 1985), but the scale of values on which it has been based (Rosser and Kind 1978) was gleaned from a sample of healthy people who might not have had the same priorities as those at imminent risk of death.

Nevertheless, it must be conceded that there is 'an economic threshold that will inherently limit the development and diffusion of new methods of treatment' (Evans *et al.* 1986). The words of this debate will never erase 'the indelible realisation that there, but for the grace of God, go us all. The easiest misfortunes to bear are surely the misfortunes of others' (Halper 1989).

The clinical nephrologist is concerned when the renal replacement therapy given to the individual patient is too much for the patient to bear. In this instance dialysis is no longer appropriate. The decision to stop therapy is taken more frequently in the USA than in Europe, and it is taken particularly in those who are old and have complicating debilitating diseases. Neu and Kjellstrand (1986) reported that dialysis was discontinued in 155 (9 per cent) of 1766 patients. Half of the patients were competent at the time the decision to withdraw treatment was taken; amongst incompetent patients it was more common in the 1980s for the patient's family to initiate the decision than it was in the 1970s. They conclude that, because of the increasing age of patients on dialysis, withdrawal of treatment will become more common in the future. Families and physicians may be relieved of this decision by the advent of a fatal catastrophe, and the high death rate due to cardiovascular complications (Tufveson *et al.* 1989) may sometimes offer an exit before harder tragedy strikes. Physicians in the UK sometimes turn to renal transplantation as providing a definitive answer to their quandary. If it succeeds it brings a better quality of life, if it fails a further period on dialysis need not be offered. Surgical colleagues resent this approach. At the present time there are insufficient cadaver organs to support it.

Returns from British centres to the EDTA Registry showed that of 975 deaths in 1986, 9 occurred because the 'patient refused further treatment', 5 were due to 'suicide' and 34 because 'therapy ceased for any other reason'. Thus, 4.9 per cent of deaths were classified under social causes. Coded entry of cause of death permitted only one cause to be recorded.

It was felt that social factors might contribute, albeit indirectly to other deaths and therefore a study was carried out of deaths in 1987 (Tasker and Wing, unpublished study). The doctors' opinion about the degree of disability/handicap, of distress (pain), of poor compliance, of lack of social support, of depression and of suicidal intent which was thought to have contributed to

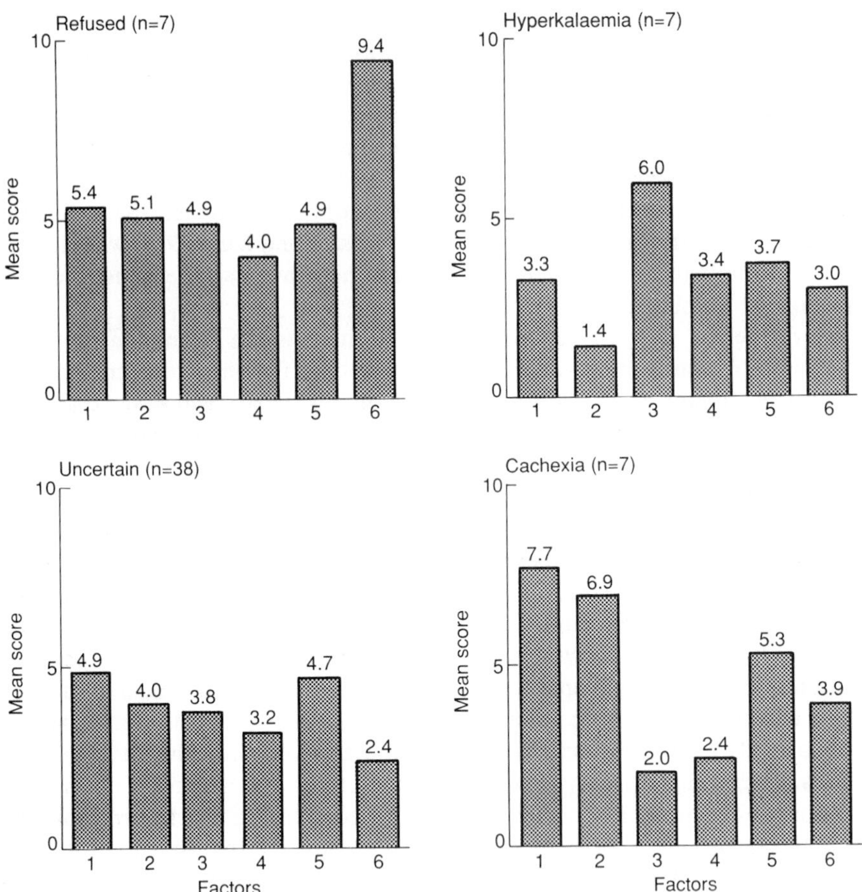

Fig. 41.5

or been mitigating factors in the deaths was investigated. Each factor was envisaged as a continuous variable scaled 0–10. Examples of the profiles of replies for patients whose death was recorded as 'patient refused further treatment', 'hyperkalaemia', 'uncertain' and 'cachexia' are shown in Fig. 41.5. Although suicidal intent scored high amongst the 7 patients who refused further treatment, high scores were also given for disability/handicap, distress (pain), poor compliance and depression. Amongst another 7 patients whose death was attributed to hyperkalaemia, poor compliance scored highly, with a notably low score for distress (pain). A large number of deaths (38) had 'uncertain' cause, but the patients were recognized as suffering disability/handicap and depression. The 7 patients who died cachexic deaths had high scores for disability/handicap, distress (pain) and depression, but low scores for poor compliance and lack of social support.

Clearly, physicians have considerable sympathy for adverse factors detracting from their patients' quality of life and which may contribute to, or be, mitigating circumstances in their deaths. None of us wish to drive our patients into a corner from which the only escape is overt suicide. That would be too much renal replacement therapy.

There has been too little, there might be too much: how can we arrive at a balance and provide the right amount of renal replacement therapy?

Neither too little nor too much: the right amount of renal replacement therapy in the UK?

The balance between too little and too much renal replacement therapy is a dynamic equilibrium acted upon by considerations of what constitutes appropriate medical intervention and what is affordable. The balance has changed as technical advances have occurred, expectations have grown and wealth has increased.

It was the ebullience of technology and burgeoning economies in the 1960s which provided the springboard for renal replacement therapy. In the 1970s it was generally considered appropriate to treat young patients, whose circumstances supported them and motivated them to endure demanding therapies. In the 1980s it became acknowledged that such selection of patients was no longer appropriate.

What is affordable has usually been outpaced by medical advances. Resource levels repeatedly frustrate the provision of appropriate rates of treatment. Different countries have found different answers. End stage renal disease programmes in the USA could not be supported by individual medical insurance, and the cruelty of this fact could not be endured by a society which was capable of putting a man on the moon; thus the umbrella of Medicare was exceptionally invoked to cover this expensive therapy. In the UK, central funding launched an imaginative national programme, only to embarrass the regional authorities to which responsibility for further development was devolved. It is hoped that the new National Health Service, in which money follows patients will find a way to overcome bureaucratic obstacles and provide the capital and revenue funding to increase the numbers of renal units in the UK (Wing 1990).

Whenever there is an increment in what is affordable there is a danger that the care of the individual could at times become inappropriate. Wise physicians will need to preserve an ethical balance (Wing 1979) when technological imperatives, personal ambition and commercial incentives combine to influence their decisions in directions which are not appropriate for their patients.

References

Brunner, F.P. and Selwood, N.H. (1990). Results of renal replacement therapy in Europe, 1980 to 1987. *Amer. J. Kidney diseases*, **15**, 384–96.

Brunner, F.P., Wing, A.J., Dykes S.R. *et al.* (1989). International review of renal replacement therapy: strategies and results. In, *Replacement of renal function by dialysis*, (ed. J.F. Maher). pp. 697–719, Kluwer Academic, Dordrecht.

Challah, S., Wing, A.J., Bauer, R., Morris, R.W., and Schroeder, S.A. (1984). Negative selection for dialysis and transplantation in the United Kingdom. *Br. Med. J.* **288**, 1119–22.

Dalziel, M. and Garrett, C. (1987). Intraregional variation in treating end stage renal failure. *Br. Med. J.* **294**, 1382–3.

Edwards, N. (1984). Kidney patients. House of Commons Official Report (Hansard) Dec 5, **710**, col. 23.

Evans, R.W., Manninen, D.L., Garrison, L.P., and Maier, A.M. (1986). Donor availability as the primal determinant of the future of heart transplantation. *JAMA*, **255**, 1892–8.

Feest, T.G., Mistry, C.D., Grimes, D.S., and Malik, N.P. (1990). Incidence of advanced chronic renal failure in the United Kingdom and the need for end stage renal replacement treatment. *Br. Med. J.* **301**, 897–900.

Greenwood, R., Sanders, B., and Powell, D. (1990). Costing a new approach to dialysis. *THS Health Summary*, **7**, 7–11.

Halper, T. (1989). *The misfortunes of others; end stage renal disease in the United Kingdom*. Cambridge: Cambridge University Press.

Jones, N.F., Mallick, N.P., Taube, H.D., and Walls, J. (in press). Manpower and workload in adult renal medicine in the United Kingdom 1975–1989.

McGeown, M.G. (1990). The prevalence of advanced renal failure in Northern Ireland. *Br. Med. J.* **301**, 900–3.

Neu, S. and Kjellstrand, C.M. (1986). Stopping long term dialysis; an empirical study of withdrawal of life supporting treatment. *New Eng. J. Med.* **314**, 14–20.

Patten, J. (1984). Kidney patients. *House of Commons Official Report (Hansard)* Dec. 20, **710**, cols. 309–10.

Rennie, D., Rettig, R.A., and Wing, A.J. (1985). Limited resources and the treatment of end stage renal failure in Britain and the United States. *Q.J. Med. N.S.* **56**, 321–36.

Rosser, R. and Kind, P. (1978). A scale of valuation of states of illness: Is there a social concensus? *Int. J. Epidemiology*, **7**, 347–58.

Smith, W.G.J., Cohan, D.R., and Asscher, A.W. (1989). *Evaluation of renal services in Wales with particular reference to the role of subsidiary renal units: report to the Welsh Office 1989*. KRUF Institute of Renal Disease, Cardiff, Royal Infirmary.

Tufveson, G., Geerlings, W., Burnner, F.P., Brynger, H., Dykes, S.R., Ehrich, J.H.H., Fassbinder, W., Rizzoni, G., Selwood, N.H., and Wing, A.J. (1989). Combined report on regular dialysis and transplantation in Europe, XIX, 1988. *Nephrology Dialysis Transplantation*, **4**, (Suppl. 4), 5–29.

De Wardener, H.E. (1966). Some ethical and economic problems associated with intermittent haemodialysis. In *Ethics and medical progress: with special reference to transplantation*. pp. 104–18. Churchill, London.

Williams, A. (1985). Economics of coronary artery bypass grafting. *Br. Med. J.* **291**, 326–9.

Wing, A.J. (1979). The impact of financial constraint. In *Decision making in medicine:*

the practice of its ethics. (ed. G. Scorer and A.J. Wing). pp. 151–64 Edward Arnold, London.

Wing, A.J. (1990). Can we meet the real need for dialysis and transplantation? *Br. Med. J.* **301**, 885–6.

Wood, I.T., Mallick, N.P., and Wing, A.J. Prediction of resources needed to achieve the national target for the treatment of renal failure. *Br. Med. J.* **294**, 1467–70.

Index